Plumbing
Level One

Trainee Guide

 Pearson

NCCER

Chief Executive Officer: Don Whyte
President: Boyd Worsham
Chief Operations Officer: Katrina Kersch
Plumbing Curriculum Project Manager: Chris Wilson
Director, Product Development: Tim Davis
Senior Production Manager: Erin O'Nora
Senior Manager of Projects: Chris Wilson
Testing/Assessment Project Manager: Elizabeth Schlaupitz
Project Assistant: Lauren Corley

Technical Writers: Gary Ferguson, Troy Staton, Veronica Westfall
Managing Editor: Graham Hack
Desktop Publishing Manager: James McKay
Art Manager: Kelly Sadler
Multimedia Project Manager: Alan Youngblood
Digital Content Coordinator: Rachael Downs
Production Specialists: Gene Page, Eric Caraballoso
Editors: Jordan Hutchinson, Karina Kuchta

Writing and development services provided by Active Voice Writing and Editorial Services, Annapolis, Maryland; Integra Software Services Pvt. Ltd
Lead Writer/Project Manager: Paul Lagasse

Pearson

Director of Alliance/Partnership Management: Kelly Trakalo
Content Producer: Alexandrina B. Wolf
Assistant Content Producer: Alma Dabral
Digital Content Producer: Jose Carchi
Senior Marketing Manager: Brian Hoehl

Composition: NCCER
Printer/Binder: LSC Communications
Cover Printer: LSC Communications
Text Fonts: Palatino and Univers

Credits and acknowledgments for content borrowed from other sources and reproduced, with permission, in this textbook appear at the end of each module.

10 9 8 7 6 5 4 3 2 1

Perfect Bound: 978-0-13-663791-2
Case Bound: 978-0-13-663801-8

Preface

To the Trainee

Most people are familiar with plumbers who come to their home to unclog a drain or install an appliance. In addition to these activities, however, plumbers install, maintain, and repair many different types of pipe systems. For example, some systems move water to a municipal water treatment plant and then to residential, commercial, and public buildings. Other systems dispose of waste, provide gas to stoves and furnaces, or supply air conditioning. Pipe systems in generation plants carry the steam that powers turbines. Pipes also are used in manufacturing plants, such as wineries, to move material through production processes.

Plumbers and their associated trades constitute one of the largest construction occupations, holding about 500,000 jobs. The occupation continues to grow, and nearly 66,000 job openings are expected to be available by 2028. Plumbers are also among the highest paid construction occupations.

New with *Plumbing*

This revised fourth edition of *Plumbing Level 1* has been reformatted to provide a better experience to both the Plumbing trainee and the instructor. NCCER is proud to release *Plumbing* with our latest instructional systems design, linking learning objectives to each module's content. *Plumbing Level One* contains design improvements and content updates to several technologies that plumbers use every day. The environmental effects of plumbing are outlined through new Going Green features. Coverage of low flow fixtures, backflow preventers, graywater systems, and brownwater systems has been enhanced. Details of the *Globally Harmonized System of Classification and Labeling of Chemicals* are also included.

Several new joining technologies and their associated equipment have been added, as well as an array of new hand and power tools. Vital information from the *Corrugated Stainless Steel Tubing* module was folded into *Steel Pipe and Fittings*, creating more concise and trade-appropriate learning outcomes.

We wish you success as you embark on your first year of training in the plumbing craft and hope that you'll continue your training beyond this textbook. As many of the craftspeople employed in this trade will tell you, there are many opportunities awaiting those with the skills and the desire to move forward in the plumbing profession.

We invite you to visit the NCCER website at **www.nccer.org** for information on the latest product releases, training information, and much more. You can also reference the Pearson product catalog online at **www.crafttraining.com**.

Your feedback is welcome. You may email your comments to **curriculum@nccer.org** or send general comments and inquiries to **info@nccer.org**.

NCCER Standardized Curricula

NCCER is a not-for-profit 501(c)(3) education foundation established in 1996 by the world's largest and most progressive construction companies and national construction associations. It was founded to address the severe workforce shortage facing the industry and to develop a standardized training process and curricula. Today, NCCER is supported by hundreds of leading construction and maintenance companies, manufacturers, and national associations. The NCCER Standardized Curricula was developed by NCCER in partnership with Pearson, the world's largest educational publisher.

Some features of the NCCER Standardized Curricula are as follows:

- An industry-proven record of success
- Curricula developed by the industry, for the industry
- National standardization providing portability of learned job skills and educational credits
- Compliance with the Office of Apprenticeship requirements for related classroom training (*CFR 29:29*)
- Well-illustrated, up-to-date, and practical information

NCCER also maintains the NCCER Registry, which provides transcripts, certificates, and wallet cards to individuals who have successfully completed a level of training within a craft in NCCER's Curricula. *Training programs must be delivered by an NCCER Accredited Training Sponsor in order to receive these credentials.*

Special Features

In an effort to provide a comprehensive and user-friendly training resource, this curriculum showcases several informative features. Whether you are a visual or hands-on learner, these features are intended to enhance your knowledge of the construction industry as you progress in your training. Some of the features you may find in the curriculum are explained below.

Introduction

This introductory page, found at the beginning of each module, lists the module Objectives, Performance Tasks, and Trade Terms. The Objectives list the knowledge you will acquire after successfully completing the module. The Performance Tasks give you an opportunity to apply your knowledge to real-world tasks. The Trade Terms are industry-specific vocabulary that you will learn as you study this module.

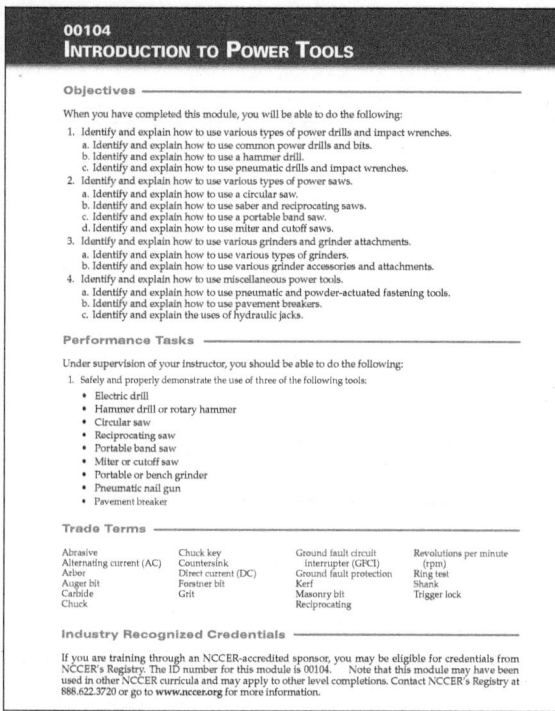

Trade Features

Trade features present technical tips and professional practices based on real-life scenarios similar to those you might encounter on the job site.

Bowline Trivia

Some people use this saying to help them remember how to tie a bowline: "The rabbit comes out of his hole, around a tree, and back into the hole."

Figures and Tables

Photographs, drawings, diagrams, and tables are used throughout each module to illustrate important concepts and provide clarity for complex instructions. Text references to figures and tables are emphasized with *italic* type.

Notes, Cautions, and Warnings

Safety features are set off from the main text in highlighted boxes and categorized according to the potential danger involved. Notes simply provide additional information. Cautions flag a hazardous issue that could cause damage to materials or equipment. Warnings stress a potentially dangerous situation that could result in injury or death to workers.

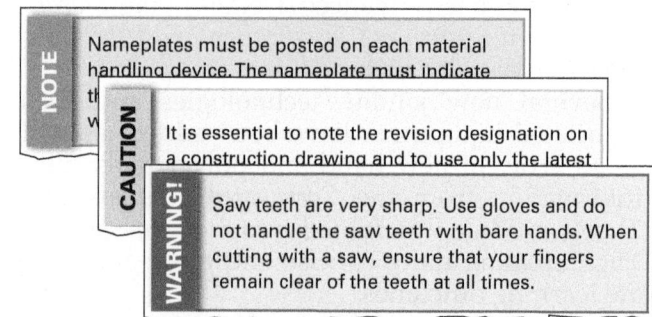

NOTE — Nameplates must be posted on each material handling device. The nameplate must indicate...

CAUTION — It is essential to note the revision designation on a construction drawing and to use only the latest...

WARNING! — Saw teeth are very sharp. Use gloves and do not handle the saw teeth with bare hands. When cutting with a saw, ensure that your fingers remain clear of the teeth at all times.

Case History

Case History features emphasize the importance of safety by citing examples of the costly (and often devastating) consequences of ignoring best practices or OSHA regulations.

Case History

Requesting an Outage

An electrical contractor requested an outage when asked to install two bolt-in, 240V breakers in panels in a data processing room. It was denied due to the 24/7 worldwide information processing hosted by the facility. The contractor agreed to proceed only if the client would sign a letter agreeing not to hold them responsible if an event occurred that damaged computers or resulted in loss of data. No member of upper management would accept liability for this possibility, and the outage was scheduled.

The Bottom Line: If you can communicate the liability associated with an electrical event, you can influence management's decision to work energized.

Going Green

Going Green features present steps being taken within the construction industry to protect the environment and save energy, emphasizing choices that can be made on the job to preserve the health of the planet.

GOING GREEN

Reducing Your Carbon Footprint

Many companies are taking part in the paperless movement. They reduce their environmental impact by reducing the amount of paper they use. Using email helps to reduce the amount of paper used,

Did You Know

Did You Know features introduce historical tidbits or interesting and sometimes surprising facts about the trade.

Did You Know?

Safety First

Safety training is required for all activities. Never operate tools, machinery, or equipment without prior training. Always refer to the manufacturer's instructions.

Step-by-Step Instructions

Step-by-step instructions are used throughout to guide you through technical procedures and tasks from start to finish. These steps show you how to perform a task safely and efficiently.

Perform the following steps to erect this system area scaffold:

Step 1 Gather and inspect all scaffold equipment for the scaffold arrangement.

Step 2 Place appropriate mudsills in their approximate locations.

Step 3 Attach the screw jacks to the mudsills.

Trade Terms

Each module presents a list of Trade Terms that are discussed within the text and defined in the Glossary at the end of the module. These terms are presented in the text with bold, blue type upon their first occurrence. To make searches for key information easier, a comprehensive Glossary of Trade Terms from all modules is located at the back of this book.

During a rigging operation, the load being lifted or moved must be connected to the apparatus, such as a crane, that will provide the power for movement. The connector—the link between the load and the apparatus—is often a sling made of synthetic, chain, or wire rope materials. This section focuses on three types of slings:

Section Review

Each section of the module wraps up with a list of Additional Resources for further study and Section Review questions designed to test your knowledge of the Objectives for that section.

Additional Resources

Materials Handling Handbook, The American Society of Mechanical Engineers (ASME) and The International Material Management Society (IMMS), Raymond A. Kulwiec, Editor-in-Chief. 1985. New York, NY: Wiley-Interscience.

Manufacturing Facilities Design & Material Handling, Matthew P. Stevens, Fred E. Meyers. 2013. West Lafayette, IN: Purdue University Press.

1.0.0 Section Review

1. For material handling tasks, it is just as important to be mentally fit as it is to be _____.
   a. physically fit
   b. physically aggressive
   c. closely supervised
   d. over 200 pounds

2. Which of the following is a type of knot that is often used to join the ends of two ropes in non-critical, low-strain applications?
   a. Bowline
   b. Clove hitch
   c. Half hitch
   d. Square knot

Review Questions

The end-of-module Review Questions can be used to measure and reinforce your knowledge of the module's content.

Review Questions

1. Identification tags for slings must include the _____.
   a. type of protective pads to use
   b. type of damage sustained during use
   c. color of the tattle-tail
   d. manufacturer's name or trademark

2. The type of wire rope core that is susceptible to heat damage at relatively low temperatures is the _____.
   a. fiber core
   b. strand core
   c. independent wire rope core
   d. metallic link supporting core

3. Synthetic slings must be inspected _____.
   a. once every month
   b. visually at the start of each work week
   c. before every use
   d. once wear or damage becomes apparent

4. An alloy steel chain sling must be removed from service if there is evidence that _____.
   a. the sling has been used in different hitch configurations
   b. replacement links have been used to repair the chain
   c. the sling has been used for more than one year
   d. strands in the supporting core have weakened

5. A piece of rigging hardware used to couple the end of a wire rope to eye fittings, hooks, or other connections is a(n) _____.
   a. eyebolt
   b. hitch
   c. shackle
   d. U-bolt

6. A lifting clamp is most likely to be used to move loads such as _____.
   a. steel plates
   b. piping bundles
   c. concrete blocks
   d. plastic tubing

7. Chain hoists are able to lift heavy loads by utilizing a _____.
   a. rope and pulley system
   b. rigger's strength
   c. stationary counterweight
   d. gear system

8. Before attempting to lift a load with a chain hoist, make sure that the _____.
   a. hoist is secured to a come-along
   b. load is properly balanced
   c. tag lines are properly anchored
   d. tackle is connected to its power source

9. A hitch configuration that allows slings to be connected to the same load without using a spreader beam is a _____.
   a. double-wrap hitch
   b. choker hitch
   c. bridle hitch
   d. basket hitch

10. To make the emergency stop signal that is used by riggers, extend both arms _____.
    a. horizontally with palms down and quickly move both arms back and forth
    b. directly in front and then move both arms up and down repeatedly
    c. vertically above the head and wave both arms back and forth
    d. horizontally with clenched fists and move both arms up and down

NCCER Standardized Curricula

NCCER's training programs comprise more than 80 construction, maintenance, pipeline, and utility areas and include skills assessments, safety training, and management education.

Boilermaking
Cabinetmaking
Carpentry
Concrete Finishing
Construction Craft Laborer
Construction Technology
Core Curriculum: Introductory
 Craft Skills
Drywall
Electrical
Electronic Systems Technician
Heating, Ventilating, and Air
 Conditioning
Heavy Equipment Operations
Heavy Highway Construction
Hydroblasting
Industrial Coating and Lining
 Application Specialist
Industrial Maintenance Electrical
 and Instrumentation Technician
Industrial Maintenance Mechanic
Instrumentation
Ironworking
Manufactured Construction
 Technology
Masonry
Mechanical Insulating
Millwright
Mobile Crane Operations
Painting
Painting, Industrial
Pipefitting
Pipelayer
Plumbing
Reinforcing Ironwork
Rigging
Scaffolding
Sheet Metal
Signal Person
Site Layout
Sprinkler Fitting
Tower Crane Operator
Welding

Maritime

Maritime Industry Fundamentals
Maritime Electrical
Maritime Pipefitting
Maritime Structural Fitter
Maritime Welding
Maritime Aluminum Welding

Green/Sustainable Construction

Building Auditor
Fundamentals of Weatherization
Introduction to Weatherization
Sustainable Construction
 Supervisor
Weatherization Crew Chief
Weatherization Technician
Your Role in the Green
 Environment

Energy

Alternative Energy
Introduction to the Power Industry
Introduction to Solar Photovoltaics
Power Generation Maintenance
 Electrician
Power Generation I&C
 Maintenance Technician
Power Generation Maintenance
 Mechanic
Power Line Worker
Power Line Worker: Distribution
Power Line Worker: Substation
Power Line Worker: Transmission
Solar Photovoltaic Systems Installer
Wind Energy
Wind Turbine Maintenance
 Technician

Pipeline

Abnormal Operating Conditions,
 Control Center
Abnormal Operating Conditions,
 Field and Gas
Corrosion Control
Electrical and Instrumentation
Field and Control Center
 Operations
Introduction to the Pipeline
 Industry
Maintenance
Mechanical

Safety

Field Safety
Safety Orientation
Safety Technology

Supplemental Titles

Applied Construction Math
Tools for Success

Management

Construction Workforce
 Development Professional
Fundamentals of Crew Leadership
Mentoring for Craft Professionals
Project Management
Project Supervision

Spanish Titles

Acabado de concreto: nivel uno
 (*Concrete Finishing Level One*)
Aislamiento: nivel uno
 (*Insulating Level One*)
Albañilería: nivel uno
 (*Masonry Level One*)
Andamios (*Scaffolding*)
Carpintería: Formas para
 carpintería, nivel tres
 (*Carpentry: Carpentry Forms, Level
 Three*)
Currículo básico: habilidades
 introductorias del oficio
 (*Core Curriculum: Introductory Craft
 Skills*)
Electricidad: nivel uno
 (*Electrical Level One*)
Herrería: nivel uno
 (*Ironworking Level One*)
Herrería de refuerzo: nivel uno
 (*Reinforcing Ironwork Level One*)
Instalación de rociadores: nivel uno
 (*Sprinkler Fitting Level One*)
Instalación de tuberías: nivel uno
 (*Pipefitting Level One*)
Instrumentación: nivel uno, nivel
 dos, nivel tres, nivel cuatro
 (*Instrumentation Levels One through
 Four*)
Orientación de seguridad
 (*Safety Orientation*)
Paneles de yeso: nivel uno
 (*Drywall Level One*)
Seguridad de campo
 (*Field Safety*)

Acknowledgments

This curriculum was revised as a result of the farsightedness and leadership of the following sponsors:

ABC Southern California Chapter
Calculated Industries
College of Southern Maryland
Corinthian Colleges, Inc.
Hughes Supply
Industrial Management & Training Institute, Inc
Ivey Mechanical Company, LLC
Lake Mechanical Contractors, Inc.

Lee Company
Maryland Correctional Training Center
Mississippi Construction Education Foundation
MultiCraft Construction LLC
Southern Plumbing
Sundt
Vision Quest Academy
Wat-Kem Mechanical, Inc.

This curriculum would not exist were it not for the dedication and unselfish energy of those volunteers who served on the Authoring Team. A sincere thanks is extended to the following:

Doug Allen
Ken Allen
Jonathan Byrd
Jesse Coyle
Frank Guertler
Steve Guy
Terry Lunt
W.B. Noble

Jan Prakke
Bob Redd
Ed Rimbey
Brad Sims
John Stronkowski
Brent Thompson
Ray Thornton
Tony Vazquez

NCCER Partners

American Council for Construction Education
American Fire Sprinkler Association
Associated Builders and Contractors, Inc.
Associated General Contractors of America
Association for Career and Technical Education
Association for Skilled and Technical Sciences
Construction Industry Institute
Construction Users Roundtable
Design Build Institute of America
GSSC – Gulf States Shipbuilders Consortium
ISN
Manufacturing Institute
Mason Contractors Association of America
Merit Contractors Association of Canada
NACE International
National Association of Women in Construction
National Insulation Association
National Technical Honor Society
National Utility Contractors Association
NAWIC Education Foundation
North American Crane Bureau
North American Technician Excellence
Pearson

Prov
SkillsUSA®
Steel Erectors Association of America
U.S. Army Corps of Engineers
University of Florida, M. E. Rinker Sr., School of Construction Management
Women Construction Owners & Executives, USA

NCCER Business Partners

JUDGMENT INDEX
MEASURING, BUILDING
AND STRENGTHENING
GOOD JUDGMENT

Contents

Module Nine

Steel Pipe and Fittings

Discusses threading, labeling, and sizing of steel pipe and reviews the differences between domestic and imported pipe. Covers the proper techniques for measuring, cutting, threading, joining, and hanging steel pipe. Also reviews corrugated stainless steel tubing. (Module ID 02109; 12.5 Hours)

Module Ten

Introduction to Plumbing Fixtures

Discusses the proper applications of code-approved fixtures in plumbing installations. Reviews the different types of fixtures and the materials used in them. Also covers storage, handling, and code requirements. (Module ID 02110; 7.5 Hours)

Module Eleven

Introduction to DWV Systems

Explains how DWV systems remove waste safely and effectively. Discusses how system components, such as pipe, drains, traps, and vents work. Reviews drain and vent sizing, grade, and waste treatment. Also discusses how building sewers and sewer drains connect the DWV system to the public sewer system. (Module ID 02111; 10 Hours)

Module Twelve

Introduction to Water Distribution Systems

Identifies the major components of water distribution systems and describes their functions. Reviews water sources and treatment methods and covers supply and distribution for the different types of systems that trainees will install on the job. (Module ID 02112; 10 Hours)

Glossary

PLUMBING LEVEL ONE

Module Twelve
Introduction to Water Distribution Systems
(02112)

Module Eleven
Introduction to Drain, Waste, and Vent (DWV) Systems (02111)

Module Ten
Introduction to Plumbing Fixtures (02110)

Module Nine
Steel Pipe and Fittings
(02109)

Module Eight
Cast-Iron Pipe and Fittings
(02108)

Module Seven
Copper Tube and Fittings
(02107)

Module Six
Plastic Pipe and Fittings
(02106)

Module Five
Introduction to Plumbing Drawings (02105)

Module Four
Introduction to Plumbing Math (02104)

Module Three
Tools of the Plumbing Trade (02103)

Module Two
Plumbing Safety
(02102)

Module One
Introduction to the Plumbing Profession
(02101)

**Core Curriculum:
Introductory Craft Skills**

This course map shows all of the modules in *Plumbing Level One*. The suggested training order begins at the bottom and proceeds up. Skill levels increase as you advance on the course map. The local Training Program Sponsor may adjust the training order.

This page is intentionally left blank.

Introduction to the Plumbing Profession

OVERVIEW

Plumbers protect the health, safety, and comfort of people. Training and critical-thinking skills are essential to being a good plumber. A professional work ethic and good safety habits go a long way toward adding to the success of a plumber.

Module 02101

Trainees with successful module completions may be eligible for credentialing through NCCER's National Registry. To learn more, go to *www.nccer.org* or contact us at 1.888.622.3720. Our website, *www.nccer.org*, has information on the latest product releases and training.

Your feedback is welcome. You may email your comments to *curriculum@nccer.org*, send general comments and inquiries to *info@nccer.org*, or fill in the User Update form at the back of this module.

02101 V4.5

11 2023

02101
INTRODUCTION TO THE PLUMBING PROFESSION

Objectives

Successful completion of this module prepares trainees to:

1. Describe the plumbing profession.
    a. Describe the history of the plumbing profession.
    b. Describe the plumbing profession today.
2. Identify the responsibilities of a person working in the plumbing industry.
    a. State the personal characteristics of a professional.
    b. Identify career opportunities in plumbing.

Performance Tasks

This is a knowledge-based module; there are no Performance Tasks.

Trade Terms

Aboveground rough-in
Appurtenances
Aqueduct
Aquifer depletion
Backflow
Backflow preventer
Bioswale
Chlorine
Code
Cross-connection
Disinfection
Drain, waste, and vent (DWV)
Ethics
Filtration
Finish
Fixtures
Geothermal
Graywater
Journey plumber

Leadership in Energy and Environmental Design (LEED)
Model codes
On-the-job learning (OJL)
Plumbarius
Plumber
Plumbing
Plumbum
Polyvinyl chloride (PVC)
Potable
Rainwater harvesting
Reclaimed water
Softening
Solar hot water
Thermoplastic
Thermoset
Underground rough-in
United States Green Building Council (USGBC)
Water efficiency

Industry Recognized Credentials

If you're training through an NCCER-accredited sponsor you may be eligible for credentials from NCCER's Registry. The ID number for this module is 02101. Note that this module may have been used in other NCCER curricula and may apply to other level completions. Contact NCCER's Registry at 888.622.3720 or go to *www.nccer.org* for more information.

CODE NOTE

Codes vary among jurisdictions. Because of the variations in code, consult the applicable code whenever regulations are in question. Referring to an incorrect set of codes can cause as much trouble as failing to reference codes altogether. Obtain, review, and familiarize yourself with your local adopted code. Safety codes are developed by the US Occupational Safety and Health Administration (OSHA).

Contents

Figures

This page is intentionally left blank.

1.0.0 THE PLUMBING PROFESSION

Objective

Describe the plumbing profession.
  a. Describe the history of the plumbing profession.
  b. Describe the plumbing profession today.

Trade Terms

Aboveground rough-in: The second phase of a plumbing project. During this phase, holes are cut in walls, ceilings, and floors. Then, supply and waste pipes are attached or hung so they can be connected to fixtures. Also referred to as stack out, top out, or in-wall rough-in.

Appurtenances: Accessories or apparatus that require no demand from the water supply side and add no load to the waste side.

Aqueduct: A man-made channel used to carry water.

Aquifer depletion: The use of underground fresh water at a rate faster than it can be replenished.

Backflow: The flow of contaminated water into the freshwater system resulting from a cross-connection between potable and nonpotable water systems.

Backflow preventer: A device that prevents nonpotable water from entering a potable supply system.

Bioswale: A depression in the ground that filters pollutants from stormwater.

Chlorine: A heavy, greenish-yellow gas used as a disinfectant in water treatment. Chlorine should be handled only when wearing appropriate personal protective equipment.

Code: A requirement published by state and local governments to establish minimum standards for various types of construction. A code carries the force of law.

Cross-connection: An arrangement between a potable water system and a nonpotable water system in which an accidental pressure differential between the two systems causes backflow of contaminated water into the freshwater system.

Disinfection: The process of destroying harmful organisms in potable water.

Drain, waste, and vent (DWV): A piping system that combines sanitary drainage with venting.

Filtration: The process of cleansing water to remove particles and chemicals.

Finish: The third phase of a plumbing project. During the finish phase, plumbers install fixtures, appliances, water purification systems, water heaters, and controls. Also referred to as trim-out or trim finish.

Fixtures: Devices that receive water from a water supply line. Common fixtures include sinks, shower stalls, and toilets.

Geothermal: Heat that is generated below the earth's surface.

Graywater: Water that comes from baths and washing machines.

Journey plumber: A plumber who has successfully completed an apprenticeship-training program.

Leadership in Energy and Environmental Design (LEED): A system for certifying that buildings have been designed and constructed to environmental standards.

Model codes: Construction ordinances that are written by a national construction organization according to suggested national plumbing standards. Model codes that have not been adopted by a jurisdiction do not have the force of law.

On-the-job learning (OJL): Field experience used in conjunction with classroom lessons in an apprenticeship program. OFFICE OF APPRENTICESHIP requires 144 hours of classroom instruction per year and 2,000 hours of OJL per year.

Plumbarius: The Roman term for someone who works with lead. The root of the modern word plumber.

Plumber: One who installs or repairs plumbing systems and fixtures.

Plumbing: According to the National Standard Plumbing Code, plumbing is "the practice, materials, and fixtures within or adjacent to any building structure or conveyance, used in the installation, maintenance, extension, alteration, and removal of all piping, plumbing fixtures, plumbing appliances, and plumbing appurtenances... ."

Plumbum: Latin word for lead.

Polyvinyl chloride (PVC): A thermoplastic material frequently used in tubing for cold water systems and the first type of plastic approved for use in plumbing.

Potable: Water that is safe for cooking and drinking.

Rainwater harvesting: The collection and storage of rainwater for irrigation.

Reclaimed water: Wastewater that has had impurities and solids removed from it so that it can be reused for non-potable purposes.

Softening: The process of removing magnesium and sodium salts that cause scale on the inside of pipes and fittings.

Solar hot water: Water that has been directly or indirectly heated by sunlight.

Thermoplastic: A plastic material used in plumbing and sanitary systems that is soft and pliable when heated and hard and rigid when cooled.

Thermoset: A plastic material used in plumbing and sanitary systems that becomes substantially infusible and insoluble when treated by heat or chemicals.

Underground rough-in: The phase of a plumbing project during which the plumber locates all supply and waste connections from the building systems to public utilities, and establishes where these systems will enter or leave the building.

United States Green Building Council (USGBC): The non-profit construction trade organization responsible for the development of LEED.

Water efficiency: The managed use of drinkable water to reduce waste.

Plumbing has profoundly influenced the development of modern society by improving public health and safety. Plumbing systems allow people to have safe, healthy, fresh water for drinking, washing, cooking, and many other uses. Additionally, plumbing reduces the spread of disease by safely draining away wastewater that contains harmful organisms. Improved sanitation contributes to longer life expectancies for men, women, and children. Plumbing systems also allow people to fight fires, water their lawns, fill their pools, and perform many daily activities.

1.1.0 History of Plumbing

Plumbing is the result of thousands of years of improvements, inventions, and innovations. *Figure 1* illustrates how plumbing systems have evolved since ancient times.

Plumbing systems, including drain, waste, and vent (DWV) systems, have been essential to the protection of public health and safety. These systems have evolved over time to the point where, today, many common water-borne diseases are rare in the United States.

1.1.1 Early Plumbing Systems

Archaeologists have determined that rudimentary plumbing systems were in use as early as 2900 BC. Earthenware pipes, masonry sewers, water closets, and drainage systems have been found in Mesopotamia (modern-day Iraq) to prove this.

In 312 BC, the Romans began bringing water into Rome through an aqueduct system, or a system of channels used to carry water. Most aqueducts were open, stone-lined trenches that used gravity to move water downhill. The more famous arched aqueducts were not nearly as commonly used as the simpler trench systems. By 100 AD, the aqueduct system was so advanced that Rome built and maintained public bathhouses and fountains throughout the city. The aqueducts were also used to drain wastes and discharge them into the river downstream from the city.

Archaeologists and historians believe that the widespread use of lead pipe in ancient Roman cities may have contributed to the downfall of the Roman Empire. Lead is harmful to the brain, the nervous system, and vital organs such as the liver. Water carried through lead pipes may have gradually poisoned many people in the Roman Empire.

The Romans also gave us the word that we use today to describe people who install and maintain water supply and waste systems. The Latin word for lead is plumbum, and a person who worked with lead was called a plumbarius. Over the centuries, this ancient word has come down to us as a word you know very well: plumber.

During the fifth century AD, the Goths invaded Rome and the Roman Empire came to an end. Roman cities and their infrastructure fell into disrepair. Much of the science, math, and technology of the Roman and Greek golden age were gradually forgotten. The period after the fall of Rome (500 AD to 1400 AD) is often referred to as the Dark Ages because of the social and cultural decline that resulted from this loss.

During that time, people neglected sanitation and hygiene. They emptied sewage into the streets and did not store or prepare their food in a healthful way. As a result, more people died from disease than from wars during those years. Some historians estimate that one-third of the population of Europe died from diseases spread by unsanitary living conditions and led to widespread epidemics like the Bubonic plague and smallpox, which killed hundreds of thousands across the continent.

By the mid-1300s, however, sanitation began to return to the cities of Europe. The first water supply pipe was laid in London during that time.

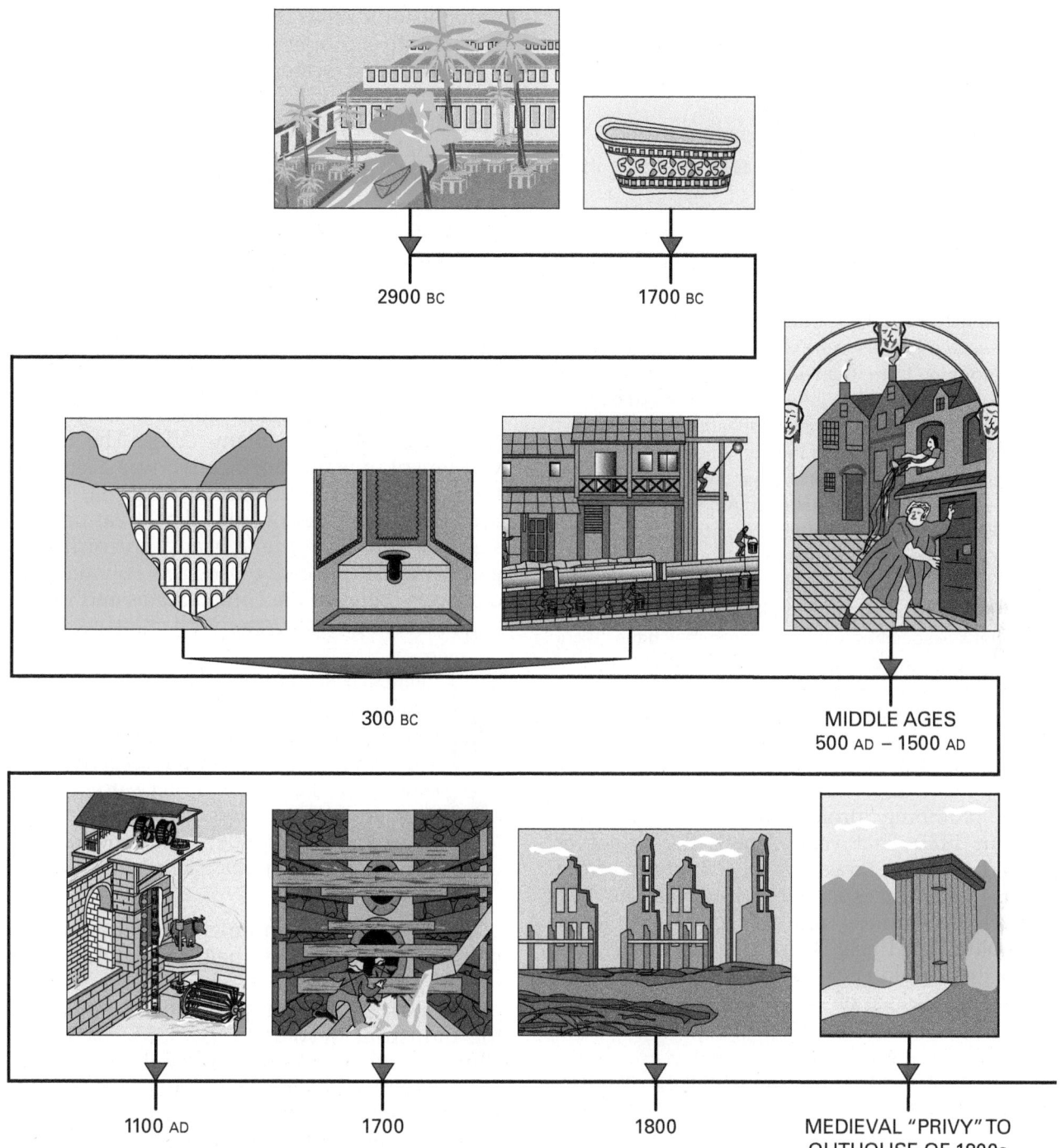

Figure 1 The evolution of plumbing.

2900 BC

1700 BC

300 BC

MIDDLE AGES
500 AD – 1500 AD

1100 AD

1700

1800

MEDIEVAL "PRIVY" TO
OUTHOUSE OF 1800s

The ancient baths and spas built by the Romans in England were put to use again during the sixteenth century. During that time, plumbing re-emerged as a profession. In 1625, the first apprenticeship laws for plumbing were enacted in England. Other countries began to develop water supply systems. A new era of sanitary awareness slowly emerged.

One of the first known water closets was built in England in 1596 by a godson of Queen Elizabeth I, Sir John Harrington. However, it was another 200 years before the use of this invention became widespread. In 1775, Alexander Cumming invented the S-trap, the precursor of the toilet traps used today. Three years later, Joseph Bramah patented an improved version of Cumming's device that incorporated two hinged valves. The original is still in use today in the House of Lords in London's Parliament building. In 1848, England passed the National Public

Safety Act, which served as model legislation for the rest of the world. The Act required every house to have sanitary facilities, whether a flush toilet or an outhouse.

Invention of Flush Toilets

London plumber Thomas Crapper has been credited with the invention of the self-contained flush toilet. He was a plumber who was responsible for many innovations in his profession. During the late nineteenth century, he was granted patents for everything from improvements to drains to manhole covers to pipe joints. Around the same time, the first "modern" toilet appeared: a U-bend siphoning system to flush the pan where waste was deposited. Records show that the patent for this device was actually issued to Albert Giblin. It is likely that Crapper bought the patent rights from Giblin and then marketed the device himself. Crapper also served as the sanitary engineer for many members of the English royal family. Crapper's association with the device may have originated during World War I, when soldiers used his name as a synonym for the toilets they saw as they passed through England on the way to the front lines in Europe.

The plumbing profession received a royal boost in 1871. That year, Albert, the Prince of Wales (later King Edward VII), almost died from typhus. Later investigation found that he had contracted the illness from contact with contaminated plumbing lines. The problem was corrected before they could cause more illness. A relieved and healthy Prince of Wales was later quoted as saying, "If I could not be a prince, I'd rather be a plumber."

Inspiration and innovation played a large role in the development of modern plumbing, but so did disasters. In 1666, after the Great Fire of London nearly destroyed the city, its citizens organized the first firefighting brigades and expanded the city's water supply system. In the United States, deadly cholera epidemics in Philadelphia in 1793 and New York City in 1832 spurred construction of improved water supply systems. Earthquakes in California led to the modernization of municipal water systems there as well.

The United States, in fact, has been responsible for many of the most notable advances in plumbing technology as well as in health and safety standards. Early settlers of North America brought with them the idea that baths and spas had curative properties. In 1652, Boston developed the first waterworks, using wood pipes for both firefighting and private use. In 1804, Philadelphia began using cast-iron piping for its water mains. By the mid-1800s, American homes often had outdoor privies supplied from wells or cisterns.

It was not until the middle of the nineteenth century, however, that the United States began to develop practical water and sewage systems. In 1857, an engineer named John Adams designed a sewer system for Brooklyn, New York. When it proved to be successful, he published his design, and other cities adopted it.

Wastewater treatment, however, still proved to be a problem for many cities. The first advancement in water purification was a slow sand filter installed in Richmond, Virginia, in 1832. Scientists later discovered that treating water supplies with chlorine (a heavy, greenish-yellow gas) would kill deadly bacteria. Following that discovery, cities in the United States and England began treating wastewater with chlorine.

> **CAUTION**
>
> Chlorine is highly toxic and corrosive. It also poses a serious fire risk. Chlorine will react violently with many chemicals, including water. Always wear appropriate personal protective equipment (PPE) when working with liquid or gaseous chlorine. Always consult the material safety data sheet (MSDS) before moving, storing, or working with chlorine.

Despite these advancements, by the mid-1800s plumbing systems were neither complex enough nor large enough to keep up with the rapid growth of cities and populations spurred by the Industrial Revolution. By 1855, for example, Chicago still relied on individual wells and the Chicago River to support a population of 75,000. Densely populated factory cities in the United States threatened to deplete local water supplies and overwhelm rivers and streams with wastewater. The plumbing profession had to change with the times to meet these new demands.

1.1.2 A New Professional Organization

By the late nineteenth century, plumbing technology and practices were recognizably modern. Manufacturers and wholesale dealers of new plumbing devices sold them over the counter as separate components. They ignored the fact that these components would have to be combined and installed into a properly designed plumbing system in order to work. Dealers claimed no responsibility for the proper installation of plumbing systems. The results were predictable. People suffered from shoddy and unsafe plumbing systems that possibly made sanitary conditions worse instead of better. And plumbers were usually blamed—unfairly—when things went wrong. Plumbers had the knowledge and ability to install safe and efficient plumbing systems, but the manufacturers and dealers completely dominated the trade.

Accordingly, at the association's second annual convention in Baltimore in 1884, the attendees adopted a set of professional guidelines that became famous as the Baltimore Resolutions. The resolutions succeeded at last in putting plumbers in charge of the profession. They helped ensure that plumbing installations in the United States would be safe and sanitary, and that plumbers would be held to rigorous professional standards.

1.1.3 Plumbing in the Twentieth Century

Sanitation, along with medical science, continues to be largely responsible for the maintenance of public health. In the United States, great progress in the development of plumbing methods and technologies has been made since 1910. The reliability of traditional piping materials, such as copper and cast iron, has been dramatically improved. A new plastic compound called thermoplastic (which is soft and pliable when heated) and thermoset (which becomes solid when heated) have been developed. The physical properties of these plastics make them ideal for use in sanitary systems. Polyvinyl chloride (PVC), developed in the 1930s, was the first plastic used in plumbing. PVC is a rigid pipe with high-impact strength that is manufactured from a thermoplastic material. The material has an indefinite life span under most conditions. Plumbers frequently use PVC in cold water systems. Manufacturers have also improved plumbing fixtures, which are devices that receive water from a water supply line. Common fixtures include sinks, faucets, shower stalls, and toilets.

In 1911, the Kohler Company developed the first one-piece recessed bath. Before this, baths had been built in two separate sections: the tub itself and the surrounding apron. Kohler's one-piece tub was much more sanitary and attractive. In 1926, Kohler introduced the electric sink, a combination of a conventional sink and electric dishwasher.

Other inventors established companies whose names are familiar to today's plumbers. Al Moen, who held more than 75 patents, invented the single-handle mixing faucet. In 1906, William Sloan

developed a flush valve in which water flows under pressure from the supply pipe directly to the fixture, so that it is always ready to be used. Halsey W. Taylor established the company that bears his name in 1912 to manufacture and sell hygienic drinking fountains.

1.2.0 Plumbing Today

Fresh water accounts for only about 3 percent of the water on Earth. In many places, the overuse of groundwater supplies has led to aquifer depletion, the use of underground fresh water at a rate faster than it can be replenished. Many areas of the United States are dependent on aquifers such as the Ogallala, which stretches from Texas to South Dakota.

Moving into the twenty-first century, societies worldwide began to recognize the need to make wise use of natural resources, especially fresh water. The efforts to do so will continue to impact the plumbing profession for many years to come. Those efforts include practices such as rainwater harvesting (the collection and storage of rainwater for reuse), the reuse of graywater (laundry and bath water), and technologies such as the creation of a bioswale, which is a constructed depression in the ground that filters pollutants from stormwater. Many places are also using reclaimed water for reuse in systems that are not meant for drinking, such as flushing fixtures. Reclaimed water has had impurities and solids removed from it.

Founded in 1993, the non-profit United States Green Building Council (USGBC) developed the Leadership in Energy and Environmental Design (LEED) green building certification system in 2000. LEED provides a way for contractors to identify and implement improved methods for the design, construction, operation, and maintenance of green buildings. Out of 110 possible points on the LEED rating scale, between 10 and 14 fall under the category of water efficiency, or the managed use of drinkable water to reduce waste, with water management addressed in other categories as well. Both federal agencies and state governments have adopted LEED certification as a requirement for new construction, so it is important for plumbers to know and be able to work within these standards.

In some areas of the country, plumbers are also responsible for installing and maintaining geothermal (heated by energy from below the earth's surface) and solar hot water (water that has been heated directly or indirectly by the sun) systems. Refer to your local applicable code to determine whether plumbers in your area install and maintain these systems.

1.2.1 Responsibilities in the Plumbing Profession

Plumbing is practically everywhere: in kitchens and bathrooms; at golf courses; in parks; and at municipal sanitary, sewage, and water systems. Because the manmade environment relies on plumbing, plumbers play a very important role in society. The systems that plumbers install do many things, including the following:

- Make drinking water safe and pure
- Protect homes and businesses from fires
- Help keep basements dry
- Help keep lawns and gardens green

The plumbing profession is made up of talented craftworkers who draw on a set of skills that are well suited to the field. Plumbers are good at working with their hands. They can use precise measuring and testing tools with skill. They get a lot of satisfaction from solving complex problems. Plumbers take pride in their work.

A plumber must have critical thinking skills to be able to read and interpret construction drawings, review manufacturer's specifications, understand and apply legal standards, rough-in entire plumbing systems, and troubleshoot problems. Plumbers must have a firm grasp of basic mathematics. They need to be fairly fit and flexible, because their work can be physically demanding. Plumbers often work in cramped spaces and do a lot of work while standing up.

Plumbers need these skills because they perform a wide range of tasks. Plumbers design, install, repair, and maintain potable (drinkable) water supply lines and piping layouts; drainage systems; drain, waste, and vent (DWV) systems; and gas systems. There are many professions related to plumbing that require the same base knowledge that a plumber has, including the following:

- Sprinkler fitters, who install fire-sprinkler and other fire-protection systems in buildings
- Pipefitters, who install piping for steam, water heating and cooling, lubrication, and other uses
- Steamfitters, who install pipe systems to move liquids or gases under high pressure
- Irrigation system installers, who design, install, and maintain irrigation systems for small areas such as residential gardens and large areas such as golf courses

Plumbers also specialize in commercial and residential design, construction, remodeling, renovation, and maintenance. Plumbers can own and operate their own contracting business. They can also branch out into related fields by becoming plumbing inspectors, instructors, estimators, or safety managers.

Many plumbers specialize in servicing specific types of systems, such as air conditioning, heating, and water distribution systems. They become experts in troubleshooting and repairing malfunctioning equipment. If you specialize in the service and repair of specific systems, you will always be in demand.

Whatever their specialty, plumbers work with the latest technology, tools, and equipment. Being a plumber means keeping up-to-date on the technological developments in plumbing, on better methods to do plumbing work, on the qualities and advantages of new materials, and on changes to the codes and standards that govern plumbing. A code is a legal document adopted by a jurisdiction that establishes the minimum acceptable standards, rules, and regulations for all materials, practices, and installations used in buildings and building systems.

Most states and localities require plumbers to be licensed as a journey plumber, master plumber, and/or plumbing contractor. Refer to your local code for the guidelines that apply where you work.

1.2.2 What is Plumbing?

Plumbing has different meanings to different people. The National Standard Plumbing Code (NSPC) defines plumbing as follows:

- The practice, materials, and fixtures within or adjacent to any building structure or conveyance, used in the installation, maintenance, extension, alteration, and removal of all piping, plumbing fixtures, plumbing appliances, and plumbing appurtenances (accessories that don't require water supply or add a load to the waste side) connected with the following:
    a. sanitary drainage systems and related vent systems;
    b. storm water drainage facilities and venting systems;
    c. public or private potable water supply systems;
    d. the initial connection to a potable water supply upstream of any required backflow prevention devices (which prevent nonpotable water from entering the potable supply system) and the final connection that discharges indirectly into a public or private disposal system;

e. medical gas and medical vacuum systems;

f. indirect waste piping, including refrigeration and air conditioning drainage;

g. liquid waste or sewage, and water supply, of any premises to their connection with the approved water supply system or to an acceptable disposal facility.

Both the *International Plumbing Code (IPC)* and the *Uniform Plumbing Code (UPC)* define plumbing as:

- The practice, materials and fixtures utilized in the installation, maintenance, extension and alteration of all piping, fixtures, plumbing appliances and plumbing appurtenances, within or adjacent to any structure, in connection with sanitary drainage or storm drainage facilities; venting systems; and public or private water supply systems.

You can see that these definitions largely cover the same range of activities, although the NSPC definition is more specific in including medical gas and vacuum systems and refrigeration and air conditioning drainage systems.

1.2.3 Protecting Public Health, Safety, and Comfort

The goal of plumbers is to provide adequate piping systems that protect the health, safety, and comfort of the nation. Plumbers accomplish this goal in two ways:

- By installing safe, reliable plumbing systems
- By properly maintaining and repairing existing systems

Federal agencies such as the Environmental Protection Agency (EPA) enforce rules and regulations related to environmental issues. Laws such as the Clean Water Act and the Clean Drinking Water Act have been passed to develop programs to help ensure the enforcement of these rules and regulations. State and local health agencies also establish their own complementary standards and regulations that apply to their jurisdictions. A plumber's responsibility is to follow all of the rules and safeguards related to the work. Plumbers should also report any accidents or events that may be hazardous to health or to the environment.

One of the biggest health and safety issues facing plumbers today is contamination caused by an improper cross-connection between plumbing systems. A cross-connection is an arrangement between a potable water system and a nonpotable water system in which an accidental pressure differential between the two systems causes backflow of contaminated water into the freshwater system. Backflow can force wastes through a cross-connection. Several things can cause backflow:

- Cuts or breaks in the water main
- Failure of a pump
- Injection of air into the system
- Accidental connection to a high-pressure source

When wastewater or other liquids are siphoned into the fresh water supply, they can cause contamination, sickness, and even death. Escherichia coli, also known as E. coli, and Legionella (the bacteria that causes Legionnaire's disease) can be spread through backflow from contaminated systems. A safety device called a backflow preventer keeps wastewater from entering the water supply system. You will learn more about backflow and backflow prevention later in the Plumbing curriculum.

To prevent the spread of harmful organisms, chemicals, and materials through water supply systems, water is made safe through disinfection, filtration, and softening. To disinfect water means to destroy harmful organisms in the water. Filtration is the process of cleansing water to remove particles and chemicals. Softening removes magnesium and sodium salts that cause scale on the inside of pipes and fittings. You will learn more about these methods of water treatment later in the Plumbing curriculum.

The safety and health of the plumber is just as important as that of the customer. One of the dangers facing plumbers is injury caused by scalding. For many years, plumbing codes have usually required water to be available at temperatures of up to 140°F (60°C). The American Society of Plumbing Engineers (ASPE) has found that this temperature is necessary to prevent the growth of Legionella bacteria. At this temperature, however, water can scald exposed skin almost immediately. In response to this danger, many codes include specific requirements to reduce the threat of scalding.

> **CAUTION**
>
> Incorrectly sized pipe in a plumbing system can cause the system to work incorrectly and could even damage the pipe and fixtures. Ensure that all pipe used in a plumbing system is sized according to the applicable local code. Codes include sizing charts that should be used for reference.

Low-Flow Fixtures

As you have learned, one of the most important goals of the plumbing profession is to protect the health, safety, and comfort of the nation. Another important goal is to act as a steward of the environment. Water conservation is one of the most important environmental issues, and water-conserving fixtures, such as toilets and showerheads, help preserve this vital natural resource.

Toilets consume about 30 percent of the water in a typical household. In 1992, the federal government required that toilets installed in new and renovated plumbing systems use no more than 1.6 gallons per flush, or about half as much as older toilets. These low-flow toilets have since been followed by high-efficiency toilets that use as little as 1.28 or even 1.1 gallons per flush. Advocates of low-flow toilets claim that the water savings from low-flush toilets can save millions of gallons a year and extend the operating life of water treatment facilities. Waterless urinals use a cartridge filter containing a liquid seal rather than water to remove waste and odor. They are installed in locations where access to a water supply is difficult, or when water conservation considerations require their use.

However, these fixtures must be properly sized and installed to work. When low-flow or high-efficiency toilets are connected to DWV systems that are designed for higher flow rates, the systems may clog and overflow, or they may require multiple flushes to empty the bowls, defeating the purpose of the fixture. The installation of low-flow fixtures may even require replacement of the sewer lines to accommodate the lower fluid flows—an expensive proposition. Contractors report that they receive more callbacks regarding low-flow toilets than for any other reason.

Always refer to your local code when installing plumbing fixtures. Read the manufacturer's instructions and follow the installation instructions carefully. Getting the installation right the first time will help save the contractor time and money and will ensure satisfied customers.

1.2.4 The Three Phases of a Plumbing Project

Plumbing can be divided into three broad phases:

- Underground rough-in
- Aboveground rough-in (also called *top-out, stack-out,* or *in-wall rough-in*)
- Finish (also called *trim-out* or *trim finish*)

During the underground rough-in phase, the plumber locates all supply and waste connections from the building systems to public utilities and establishes where these systems enter or leave the building. During the aboveground rough-in phase, the plumber cuts holes in walls, ceilings, or floors to attach or hang pipe for connection to fixtures. Then the plumber installs the pipe for the building's various supply and waste systems. To join pipe runs, a plumber might use welding tools, soldering equipment, or special chemicals for plastic pipe. The plumber may operate power threading machines, propane torches, and other power tools during this phase.

> **NOTE**
> Many plumbing terms are used interchangeably in some areas. Be sure you are using the terms the way they are typically used where you work.

Finally, in the finish phase, the plumber installs fixtures such as sinks, showers, and toilets, and appliances such as dishwashers, water purification systems, and water heaters. The plumber may be called upon to install the automatic controls that regulate pressurized pipe systems as well.

During each of the three stages, you must ensure that the system is tight and safe. Do not let your customers discover leaks behind the walls after you have completed your work. Callbacks are expensive and can hurt your reputation as a professional.

Regardless of how well a plumbing system is installed, it must be maintained and periodically repaired. You could even think of service and maintenance as a fourth phase of a plumbing project. Plumbers perform a variety of tasks as part of service and maintenance, including the following:

- Checking lubrication levels in pumps, test gauges, and meters
- Repairing faulty fixtures and components
- Verifying operating systems
- Regulating flow and usage rates

Additionally, plumbers must ensure that the drain, waste, and vent (DWV) system for each project is installed and working properly. DWV systems, which are discussed in greater depth later, are essential to protect public safety.

1.2.5 Plumbing Codes and Licenses

Plumbing codes have grown increasingly comprehensive to accommodate new developments in technology and materials. Municipalities throughout the United States adopt ordinances based on suggested national plumbing standards, called model codes. They often modify or interpret these model codes to reflect local conditions. For instance, a city that is prone to earthquakes, such as Los Angeles, would incorporate standards to protect installations from earthquake damage. Likewise, a city that is subject to flooding, such as Houston, would add special requirements to guard against contamination from floodwater.

The three primary model codes associated with plumbing are the International Plumbing Code (IPC), the Uniform Plumbing Code (UPC), and the National Standard Plumbing Code (NSPC). You will learn how codes are developed and implemented later in the Plumbing curriculum.

Although model codes act as guidelines, municipal ordinances have the force of law. As an apprentice plumber, you need to know the applicable plumbing code that applies to the area where you work. Your local plumbing code covers the installation, alteration, repair, replacement, additions to, and use of plumbing systems. It establishes standards for many things, including the following:

- Materials commonly used in plumbing systems
- Fixtures, joints, and connections
- Hangers and supports
- Water supply and distribution
- Vents
- Storm water and sanitary drainage systems
- Potable water supply systems

All states require plumbers to be licensed. Many states require completion of an apprenticeship program and/or on-the-job learning (OJL), or field experience, before testing for a license. When you apply for a license (*Figure 2*), you will be tested on your knowledge of both your profession and the local plumbing ordinances.

Figure 2 Examples of plumbing licenses.

Additional Resources

Plumbing a House, 1998. Peter A. Hemp. Newtown, CT: Taunton Press.

The National Standard Plumbing Code, current edition. Falls Church, VA: Plumbing-Heating-Cooling Contractors Association.

1.0.0 Section Review

1. Philadelphia began using cast-iron piping for water mains in the early _____.
   a. 1600s
   b. 1700s
   c. 1800s
   d. 1900s

2. Softening may be added to a water supply in order to _____.
   a. prevent scale
   b. improve flavor
   c. reduce bacteria
   d. prevent tooth decay

2.0.0 RESPONSIBILITIES OF A PLUMBING PROFESSIONAL

Objective

Identify the responsibilities of a person working in the plumbing industry.

a. State the personal characteristics of a professional.
b. Identify career opportunities in plumbing.

Trade Terms

Ethics: A set of principles and values that guide an individual's conduct.

To be a successful plumber, you must be able to produce finished products of high quality in a minimum amount of time. You have probably heard the phrase "on time and on (or under) budget." Take a moment to really think about what this phrase means. It means that the job is completed by the deadline, and that it does not cost more than what was originally expected.

2.1.0 Keys to Professional Success

Successful plumbers must be flexible enough to adapt their methods of working to the particular needs of each job. As you gain experience as a plumber, you will develop the flexibility necessary to meet changing job requirements. You will also improve your knowledge of the different tools and materials needed to accomplish certain jobs, and you will learn more about the technologies and methods that are available. Over time, you will improve your ability to do your job on time, on budget, and to the satisfaction of your customers. Satisfied customers mean repeat business.

Keep current with the latest methods, materials, and equipment in the plumbing profession. This knowledge will help you and the company you work for stay competitive. You should never stop learning. When you complete your apprenticeship, you will simply be moving into a bigger classroom: the world.

Never take chances with your own safety or the safety of others. Always demonstrate a high level of professionalism. Remember, you are part of a proud heritage, and you represent a profession that is vital to modern society. As a plumber, you will want your customers to think well not only of your work as an individual, but also of the profession in general.

Construction professionals adhere to a code of professional ethics. Ethics are principles and values that guide conduct. As a plumber, you are responsible for understanding and following the code of ethics for your chosen profession. Appendix A lists the ethical principles that are followed by members of the construction trades.

Successful plumbers exhibit all of the following qualities:

- Positive attitude
- Honesty
- Loyalty
- Willingness to learn
- Responsibility
- Cooperation
- Attentiveness to rules and regulations
- Promptness and reliability

You were introduced to many of these concepts in the *Core Curriculum*. In this module, you will learn how they apply specifically to the plumbing trade. In the classroom and on the job, you will see that the successful plumbers you meet will demonstrate all of these qualities. Learn from them and become a better plumber.

2.1.1 Positive Attitude

A positive attitude entails being energetic, motivated, attentive, and alert. It means understanding the importance of safety and responding well to change. A positive attitude will help you fulfill your obligations and take pride in your job.

An employee with a positive attitude contributes to others' productivity by setting an example. You can probably recall a situation where you had to work with someone who was complaining and slacking off. Do you remember how difficult and unpleasant it was to work in a situation like that? Try to keep that in mind when you are on the job, and ask yourself: Am I demonstrating a positive attitude?

A positive work environment not only leads to greater productivity, but also makes the job more enjoyable. Employers pay attention to your attitude. Supervisors see your attitude in your approach to the job, in your reactions to instructions, and in the way you handle problems, especially unforeseen ones.

2.1.2 Honesty

Honesty and personal integrity are important traits of the successful professional. Honesty means more than just telling the truth and never cheating or stealing. It also means doing a fair day's work for a fair day's pay. It means holding up your side of a bargain. It means backing up promises with action.

You should take pride in performing a job well and you should never cut corners. Never lie to clients, supervisors, or co-workers. Respect the property of others. Ensure that the tools and materials belonging to your employer (or sponsor), customers, and other trades are not damaged or stolen from the shop or the job site. Treat their property with the same respect that you give to your own.

Your reputation will always precede you. A reputation for honesty and integrity will make employers want to hire you because they know that you can be trusted. On the other hand, if you lie, cheat, or steal, that reputation will precede you wherever you apply for work and your career will suffer accordingly.

If you want to be gainfully employed, earn a good salary, and be well regarded by teammates and supervisors, then you should start by being honest. It's that simple.

2.1.3 Loyalty

As an employee, you expect your employer or sponsor to look out for your interests, keep you steadily employed, and promote you as job openings occur. Your employers and sponsors, in turn, have a right to expect your loyalty in return.

Remember, however, that loyalty cannot be demanded from someone; it can only be earned.

Loyalty means different things to different people, but in a professional setting, loyalty usually means the following:

- Keeping the interests of your employer or sponsor in mind
- Speaking well of your employer or sponsor to others
- Keeping any minor troubles strictly within the job site
- Respecting the confidentiality of all matters that pertain to your employer's or sponsor's business

A loyal employee fulfills commitments and meets obligations. Never go to work for someone else without first ensuring that you have met all of your commitments to your current employer or sponsor. A prospective employer will be impressed with your commitment and be willing to wait for you. If not, you might want to reconsider whether you should work for that company.

2.1.4 Willingness to Learn

A good plumber never stops learning. New machines, tools, materials, and methods are constantly being introduced. Even experienced plumbers need to be willing to learn about these advancements. If you keep your methods and tools up-to-date, your company will be able to use your skills to compete successfully with other companies. The better a company competes, the more profit it makes. The more profit the company makes, the more likely it is to stay in business, provide jobs, and raise salaries, all of which benefit you as an employee.

In addition, each workplace has its own way of doing things. Your employer or sponsor will expect you to learn how things are done where you work. You must be able to adapt to change and be willing to learn new methods and procedures as quickly as possible.

2.1.5 Responsibility

As a trainee, your primary responsibility is to do what is asked of you. As a professional plumber, however, your responsibility will extend beyond assigned tasks. You should be able to see what needs to be done and do it without having to be asked. Your supervisor expects you to do the work well without being asked twice. Assuming responsibility makes a favorable impression on an employer/sponsor.

Always take responsibility for your actions, including any work done under your supervision. Do not make excuses or blame others when you make an error. Doing so will cause people to distrust you and lose respect for you. If that happens, the quality of your work, and that of everyone who works with you, will suffer.

2.1.6 Cooperation

As a plumber, you will probably work with other people as part of a team. Every day, you interact with members of your work crew and with your supervisor. The ability to cooperate with your teammates is an important skill in the construction industry, but simply working and playing well with others is not enough. The success of any team depends on all of its members doing their parts. Everyone must contribute to the team to ensure that it achieves its goals. Such cooperation will help get projects done on time and on budget.

2.1.7 Attention to Rules and Regulations

Employers want workers who respect the rules and regulations that apply to their company, their work, and their profession. Rules and regulations exist to keep people safe and to keep a project on schedule. Companies establish rules pertaining to work times, administrative issues, and professional conduct. Local, state, and federal authorities develop regulations that apply to a wide range of issues, including safety and health, materials, tools and equipment, and legal issues.

People can work well together only if they all understand the work that needs to be done, when it needs to be done, and who will perform the work. Rules and regulations are necessary in any work situation. The rules might vary from employer to employer, so it is important to make sure you understand your employer's or sponsor's regulations each time you start a new job.

2.1.8 Promptness and Reliability

Imagine that you are a member of a three-person team that is laying pipe. According to the schedule, you must lay a certain amount of pipe in two weeks. If a member of your team were absent the whole time, you and the other team member would have to do that person's work in addition to your own, which means that each of you would have to do 50 percent more work than anticipated over the two-week period. However, if your teammate notified your supervisor, then arrangements could be made to provide another worker or to extend the completion date of the pipe-laying task.

Professionals are always prompt and reliable when they report to work. A prompt employee shows up to work on time; a reliable worker shows up every day. Employees who are chronically late or absent have poor work habits. They are being unprofessional and are showing that they lack commitment.

The clock governs your work life. Everyone is required to be at work and ready to start at a definite time. Failure to get to work on time results in confusion, lost time, and the resentment of your teammates. The good news, however, is that bad habits can be changed. If you are chronically late for work, then change your morning routine to leave earlier or get up earlier to take care of your personal responsibilities.

Sometimes you need to take time off from work because you or a family member are sick, you have suffered a death in your family, or you might get stuck in a traffic jam or have car trouble. When something like this happens, let your employer or sponsor know as soon as possible so that your supervisor can make arrangements to replace you while you are out.

Your employer or sponsor has the right to expect you to be on the job unless you have a good reason. You have an obligation to your supervisor and your teammates to be at your job because they are relying on you. Indeed, you should be proud of this; it means that they trust you and respect you. When you fulfill that obligation, you will be trusted by your co-workers and valued by your supervisor.

2.1.9 Other Important Qualities

Demonstrate personal integrity and the courage of your convictions by doing what is right even where there is pressure to do otherwise. Do not sacrifice your principles because it seems easier. Be fair and just in all dealings. Do not take undue advantage of another's mistakes or difficulties. Fair people are open-minded and committed to justice, equal treatment of individuals, and tolerance for and acceptance of diversity. Be courteous and treat all people with equal respect and dignity. Abide by laws, rules, and regulations relating to all personal and business activities.

Pursue excellence in performing your duties, be well informed and prepared, and constantly try to increase your proficiency by gaining new skills and knowledge. By your own conduct, seek to be a positive role model for others.

2.2.0 Career Opportunities in Plumbing

The construction industry employs more people and contributes more to the nation's economy than any other industry. Our society will always need people to build new homes, roads, airports, hospitals, schools, factories, and office buildings, maintain existing structures, and replace old and damaged ones. This means there will always be a source of well-paying jobs and career opportunities for construction trade professionals, including plumbers. Both large and small contracting firms employ plumbers. Plumbers who master the principles of small-business management and develop and hone their entrepreneurial skills can be self-employed.

Plumbers can choose from a number of career paths depending on their skills and interests (see *Figure 3*). Where you go depends on the pride and professionalism you show in your work. How well you succeed depends on your willingness to develop your skills and obtain the right training. With time and experience, plumbers progress from apprentices to one or more different specialties. These are some of the many positions and careers in the plumbing profession. Note that the career paths shown in *Figure 3* are examples; your path may end up being different from those shown.

An apprentice is someone who is learning the plumbing trade. As you have learned, when a plumber successfully completes an apprenticeship, he or she becomes a journey plumber. Journey plumbers can remain at that level or advance in the profession. Journey plumbers might supervise less experienced plumbers or be called on to estimate the materials and labor for a specific job. With larger companies and on larger jobs, journey plumbers often become specialists.

Foremen are leaders who direct a single team of apprentices and/or journey plumbers. Large construction projects with more than one team require supervisors who oversee the work of all teams. They are responsible for assigning, directing, and inspecting the work of each team.

A master craftsman is someone who consistently demonstrates the highest level of skill at a profession. Master plumbers are proven experts in the plumbing profession. They can act as mentors and teachers to less experienced members of the profession. Master plumbers often start their own businesses, no matter what role they play.

The safety manager is another key, on-site person. Safety managers are responsible for the safety and health of the workers on a project. They are responsible for developing a safety plan and procedures, training workers in safety procedures, and ensuring that the project is in compliance with applicable health and safety regulations.

Project managers, often called project administrators, control the scope and direction of a plumbing business. Managers ensure that the right workers are being assigned to each task and are responsible for handling project finances. Large contracting firms may have several project managers.

Plumbing estimators work for contractors and building supply companies. They assess the materials and labor required to complete a project and contractors submit bids for jobs based on these estimates. Plumbing estimators must have a thorough understanding of construction methods as well as knowledge of the availability and prices of materials and supplies.

Plumbing estimators are experienced plumbers with superior mathematical abilities and the patience to prepare detailed, accurate estimates. Today's estimators also need solid computer skills. Estimators have a great deal of responsibility, because errors in estimates can cost the contractor significant amounts of money. Depending on the size and type of the business, the estimating job may also be done by the contractor or the project manager.

Mechanical engineers research, develop, design, manufacture, and test a wide range of mechanical devices, including the following:

- Machine tools
- Internal combustion engines
- Electric generators
- Steam and gas turbines
- Refrigeration and air conditioning equipment
- Material-handling systems
- Elevators and escalators
- Industrial production equipment
- Robots used in manufacturing

Construction contractors and owners establish and run contracting businesses. Typically, they hire apprentices, journey plumbers, and master plumbers. Depending on the size of their business, contractors may work with the crew or may manage the business full-time. Small contractors may have only one or two employees who manage the business, prepare estimates, obtain supplies, and perform the actual contracted work. Subcontractors and contractors/owners may specialize in installing particular systems, such as sprinkler systems, high-pressure pipe systems, or municipal water-treatment facilities.

Plumbing inspectors may work for a contractor or for a municipal jurisdiction such as a local or state government. Inspectors ensure that all

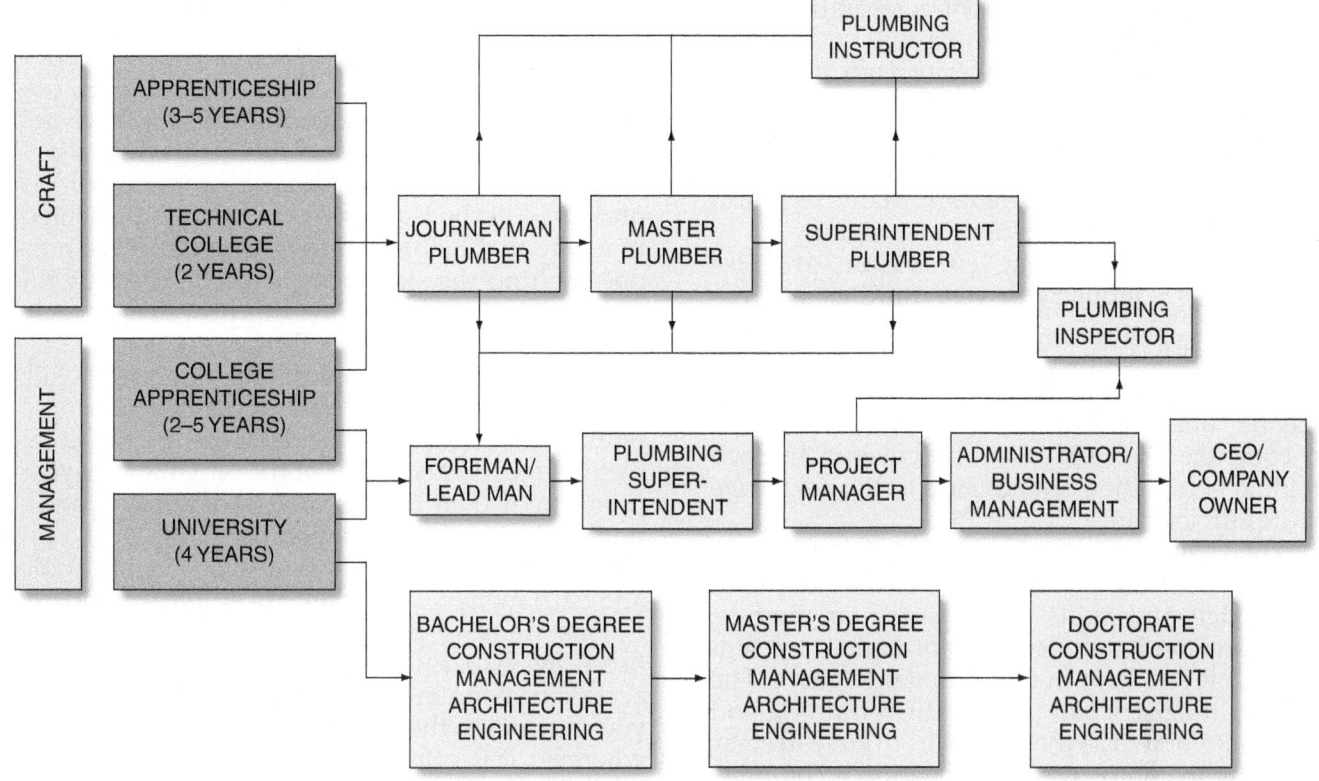

Figure 3 Opportunities in the construction industry.

plumbing installations follow applicable codes and quality requirements. Plumbing inspectors review and approve work at various stages of construction, depending on the installation.

Plumbing instructors train apprentices to become the next generation of journey and master plumbers, managers, inspectors, and engineers. Many dedicated and experienced plumbers find that becoming an instructor is a natural evolution in their career. Master plumbers often become instructors as their careers progress.

No matter what career path you choose, you will be embarking on a lifelong learning process. Effective plumbers along any career path need to keep up-to-date with new tools, materials, and methods. If you choose to become a manager or if you intend to start your own construction business someday, you will need to learn the appropriate skills and continue to hone your skills as a plumber.

Apprenticeship is your first step down an exciting and challenging career path. Remember, every successful manager and business owner started as an apprentice. Successful plumbers have at least one thing in common: a willingness to keep on learning. That process of learning begins with apprentice training.

2.2.1 Formal Construction Training

The Department of Labor's Office of Apprenticeship sets the minimum standards for training programs across the country. Office of Apprenticeship programs rely on mandatory classroom instruction and OJL. Office of Apprenticeship requires 144 hours of classroom instruction per year and 2,000 hours of OJL per year. In a typical Office of Apprenticeship program, trainees spend 576 hours in classroom instruction and 8,000 hours in OJL before receiving journey plumber certificates.

To address the training needs of the professional communities, NCCER developed a four-year plumbing training program. NCCER uses the minimum Office of Apprenticeship standards as a foundation for comprehensive curricula that provide trainees with in-depth classroom and OJL experience.

The primary goal of NCCER is to standardize construction craft training throughout the country so that employers and employees benefit from it, no matter where they are located. As a trainee in an NCCER program, you receive benefits to help you prepare for your career in the plumbing profession. You will become part of the National Registry. You will receive a certificate for each level of training you complete. If you apply for a job with any participating contractor in the country, a transcript of your training will be available for that contractor to help verify your qualifications. If your training is incomplete when you make a job transfer, you can pick up where you left off, because every participating training center is using the same program. Many technical schools and colleges also use this program.

The NCCER Plumbing curriculum provides trainees with industry-driven training and education. It adopts a purely competency-based teaching philosophy. This means that trainees must demonstrate to the instructor that they possess the understanding and the skills necessary to complete the tasks that are covered in each module before they can advance to the next stage of the curriculum.

When the instructor is satisfied that a trainee has successfully demonstrated the required knowledge and skills for a particular module, test scores and completion information are sent to NCCER and kept in the National Registry. The National Registry can then confirm training and skills for workers as they move from state to state, company to company, or even within a company (see Appendix B).

Whether you enroll in an NCCER program or another Office of Apprenticeship -approved program, ensure that you work for an employer or sponsor who supports a nationally standardized training program that includes credentials to confirm your skill development.

A good apprenticeship program effectively combines competency-based, hands-on training with classroom instruction. This combination is the most effective way for a beginning plumber to learn and advance through the plumbing profession. Successful apprentices will develop the skills and knowledge they need to carry out a wide range of tasks.

If you possess these professional and personal skills, you can become a top-notch, highly productive journey plumber. Journey plumbers have successfully completed an apprenticeship-training program. The term comes from the fifteenth-century word journeyman. Journeymen were apprentices who left their masters after learning their craft and worked alone or for someone else.

Successful journey plumbers who practice their craft with pride and skill are valued employees and co-workers. Once you have mastered the professional and personal skills necessary, you will be a good candidate to advance to an exciting and challenging career as a master plumber, superintendent plumber, plumbing supervisor, instructor, inspector, project manager, or contractor and business owner.

2.2.2 Youth Apprenticeship Program

The concept of apprenticeship training goes back thousands of years. Over that time, the basic principles of apprenticeship training have not changed. Apprentices learn their craft from those who have mastered it. Although some theory is presented in the classroom, it is always presented in a way that helps trainees understand the purpose behind the skill to be learned. The NCCER training curriculum follows this proven approach.

Youth Apprenticeship Programs allow students to begin their apprentice training while they are still in high school. Students entering a plumbing program in the 11th grade may complete as much as one year of the NCCER four-year program by their high school graduation. In addition, the program, in cooperation with local craft employers, allows students to work in the trade and earn money while they are still in school.

This training program is similar to the one used by NCCER-accredited training sponsors across the country. With your official transcripts, you can enter the second year of the program wherever it is offered. You may also apply these credits to a two- or four-year college that offers degree or certificate programs in the construction profession.

You may become an employee at age 16, but the US Department of Labor (DOL) limits your work to occupations that are not specifically defined as hazardous. Hazardous jobs include such things as explosives manufacturing, mining, meatpacking, and power-driven sheet metal work. You may work on a project site as an apprentice or journey plumber when you are 16, but only if you are enrolled and in good standing in a plumbing apprenticeship program.

You must be at least 18 before you can legally work in a hazardous occupation. When you have graduated from your apprenticeship program, you can enter the industry at a higher level and with more pay than people who are just starting the apprenticeship program.

2.2.3 Apprenticeship Standards

All apprenticeship standards call for certain work-related training, or OJL. OJL is broken down into specific tasks in which the apprentice receives hands-on training. Each task requires a specified number of hours. The total number of hours in a plumbing apprenticeship is traditionally 8,000, which amounts to four years of training. In a competency-based program, this time may be shortened by testing out of specific tasks through a series of performance exams.

Child Labor Laws

Federal law establishes the minimum standards for workers under age 18. Some municipal jurisdictions may enforce stricter regulations. Employers are required to abide by the laws that apply to them.

The Child Labor Provisions of the Fair Labor Standards Act forbid employers from using illegal child labor, and also forbid companies from doing business with any other business that does. DOL investigates alleged abuses of the law. In such cases, employers have to provide proof of age for their employees.

In addition to the Child Labor Provisions, employers in the construction trades are required to follow DOL's Child Labor Bulletin No. 101, Child Labor Requirements in Nonagricultural Occupations Under the Fair Labor Standards Act. Bulletin No. 101 does the following:

- Explains the coverage of the Child Labor Provisions
- Identifies minimum age standards
- Lists the exemptions from the Child Labor Provisions
- Sets employment standards for 14- and 15-year-old workers
- Defines the work that can be performed in hazardous occupations
- Provides penalties for violations of the Child Labor Provisions
- Recommends the use of age certificates for employees

In a traditional program, you may complete the required OJL at the rate of 2,000 hours per year. If you are laid off or get sick, it may take longer.

Sponsors need to receive, log in, and report an apprentice's time records, and have them reviewed by the federal government and/or the state apprenticeship program. Therefore, apprentices should keep accurate and up-to-date time records; after the first 144 classroom hours and 1,000 hours of related work, and then for each subsequent 1,000 hours, the standards entitle an apprentice to a pay increase.

Apprentices may not always receive classroom instruction and OJL at the same time. Apprentices with special job experience or other coursework may obtain credit toward their classroom requirements. This reduces the time they are required to spend in the classroom while meeting the 8,000-hour OJL requirement. These special cases depend on the type of program and its regulations and standards.

For those entering an apprenticeship program, a high school or technical school education is desirable, with courses in plumbing, shop, mechanical drawing, and general mathematics. Manual dexterity, good physical condition, and a good sense of balance are important. It is essential to be able to solve math problems quickly and accurately and to work closely with others.

If you want to become an apprentice, you must submit to the Apprenticeship Committee certain information, which may include the following (the actual requirements may vary by state):

- The results of your aptitude test (General Aptitude Test Battery [GATB] or GATB Form Test, normally administered by the local Employment Security Commission)
- Letters of reference and recommendation from your past employers or sponsors
- Proof of your age (e.g., driver's license)
- For veterans of the Armed Forces, a copy of your Form DD-214
- A record of any technical training you received that is related to the construction industry and/or a record of any training you received before applying for the apprenticeship
- Your high school diploma or General Educational Development (GED) credential

Once you are in an apprenticeship program, you must do the following:

- Wear proper safety equipment on the job.
- Buy and maintain tools of the trade as needed and as required by the contractor.
- Submit a monthly OJL report.
- Report any change in your employment status or change in sponsor.
- Attend classroom instruction and follow all regulations for the classroom, correspondence study, or distance learning.

2.2.4 What You Should Expect from Your Employer or Sponsor

Once the Apprenticeship Committee selects an applicant to be an apprentice, the employer or sponsor hires the apprentice. The apprentice's job has the same potential for career advancement as a non-apprentice employee. In return, the employer or sponsor requires the apprentice to make satisfactory progress in OJL and related classroom instruction throughout the duration of the apprenticeship program.

The employer or sponsor agrees not to employ the apprentice in a way that may violate the apprenticeship standards. The employer or sponsor may also pay a share of the cost of operating the apprenticeship program.

2.2.5 What You Should Expect from a Training Program

Employers and sponsors who take the time and initiative to provide quality training are willing to invest in their workforce and improve the abilities of their workers. Nevertheless, you should take the time to find a program that will train you well.

Select an employer who has an approved apprenticeship program and who is willing to act as your sponsor. The employer's training program should be comprehensive, standardized, and based on developing and demonstrating competencies, not just on the amount of time you spend in a classroom. Before you enroll in a training program, find out how many employers in the area use the same program. The training program should have a well-defined compensation ladder so that your pay will increase in recognition of your new skills and experience.

The program should be nationally recognized and should provide you with transcripts and completion credentials. As an employee in the construction profession, you may end up working for several different contractors over the course of your career. Contractors in other cities and states must be able to recognize the experience and qualifications that you obtained through your training program.

Finally, the training program's curriculum should be complete and up-to-date. The curriculum should feature the latest developments in plumbing techniques, materials, tools, and equipment. The program should also take advantage of modern interactive training technology, such as multimedia tools and the Internet. As a plumber, you know how important it is to stay current; the same applies to training programs.

2.2.6 What You Should Expect from the Apprenticeship Committee

The Apprenticeship Committee is the local administrative body to which apprentices are assigned. It is responsible for the appropriate training of apprentices. Every apprenticeship program, whether state or federal, is covered by the following standards that have been approved by appropriate government agencies:

- The Committee is responsible for enforcing standards and for making sure that training is conducted properly. This ensures that graduates are fully qualified in those areas of training designated by the standards.
- The Committee screens and selects individuals for apprenticeship and refers them to participating training programs.
- The Committee places apprentices under written agreement for participation in the program.
- The Committee establishes minimum standards for related classroom instruction and OJL and monitors the apprentice to see that these standards are followed during the training period.
- The Committee hears all complaints of violations of apprenticeship agreements, whether by employer/sponsor or apprentice, and takes action within the guidelines of the standards.
- The Committee notifies the registration agencies of all enrollments, completions, and terminations of apprentices.

2.0.0 Section Review

1. Which of the following is true with regard to professionalism?
    a. It is okay to speak poorly of your employer as long as you are being honest.
    b. It is acceptable to cheat if it benefits your employer.
    c. It is better to maintain the same procedures rather than adapt to new methods.
    d. You should do what is right even when there is pressure to do otherwise.

2. The Department of Labor's Office of Apprenticeship program requires _____.
    a. 500 hours of OJL per year
    b. 750 hours of OJL per year
    c. 1,000 hours of OJL per year
    d. 2,000 hours of OJL per year

SUMMARY

History has taught us that plumbing is vital to public health, safety, and comfort. Today, the plumbing profession is more important than ever. As the world's population grows, the built environment is expanding to accommodate it. The world needs more skilled and experienced plumbers to ensure that the public stays safe, healthy, and comfortable.

The plumbing profession relies upon a range of skills and a broad base of knowledge. Plumbers are responsible for installing and maintaining a wide variety of supply and drainage systems. Plumbers carefully follow local codes and regulations to ensure that systems are safe.

Because the plumbing profession is always evolving, successful plumbers never stop learning about new techniques, materials, tools, and standards. In addition to staying current with their technical skills and knowledge, successful plumbers also follow high ethical standards. They maintain a positive attitude, are honest and loyal, and are responsible for their actions. They cooperate with their teammates and supervisors, pay attention to rules and regulations, and are always prompt and reliable employees.

As an apprentice embarking on your plumbing career, you have the right to expect a quality education that serves you well. Your employer or sponsor and your Apprenticeship Committee should do their best to ensure that you are well prepared for the responsibilities you will take on. Ultimately, though, the decision is yours as to whether you will try hard to become the best professional plumber that you can be.

1. The modern word plumber is derived from the Latin word for _____.

   a. the water channels that supplied Rome
   b. a person who works with lead
   c. doctors who treated lead poisoning
   d. a person who repaired bathhouses

2. LEED applies to all areas of the plumbing profession, *except* _____.

   a. construction
   b. design
   c. maintenance
   d. renovation

3. Water that comes from laundry and bathtubs and reused in known as _____.

   a. blackwater
   b. stormwater
   c. graywater
   d. potable water

4. Suggested national plumbing standards are called _____.

   a. ordinances
   b. model codes
   c. regulations
   d. uniform codes

5. You are offered a job by another contractor before you have fulfilled your current obligations. The professional response is to _____.

   a. ask the contractor to wait until you have met your obligations
   b. notify your current employer of your intention to leave immediately
   c. ask your current employer to let you divide your time between both projects
   d. determine if your current employer is willing to negotiate better terms

6. Your supervisor points out that you forgot to complete a step in the assembly of a drain in a DWV system. One of the apprentices that you are supervising was the person installing the drain. The responsible action is to _____.

   a. explain that one of the apprentices was responsible
   b. ask the apprentice to explain to your supervisor
   c. explain to your supervisor why the apprentice did the work that way
   d. take responsibility for work done under your supervision

7. People who develop, design, manufacture, and test mechanical devices are called _____.

   a. contractors or owners
   b. inspectors
   c. mechanical engineers
   d. master estimators

8. Project managers are responsible for _____.

   a. assessing the materials and labor needed on projects
   b. ensuring that projects are in compliance with safety regulations
   c. controlling the scope and direction of a plumbing business
   d. instructing all new apprentices on a job site

9. The minimum standards for training programs in the United States are set by _____.

   a. NCCER
   b. NEMA
   c. OSHA
   d. Office of Apprenticeship

10. Participants in an NCCER training program receive all of the following, *except* _____.

    a. training transcripts for your employer
    b. a listing in the National Registry
    c. certification as a master plumber
    d. certificates for each completed training level

11. The non-profit _____ developed a system in 2000 that will certify a building as "green," meaning it is energy efficient.

    a. United States Green Building Council
    b. United States Energy Conservation Association
    c. United States Green Construction Council
    d. United States Energy and Environmental Design Association

12. Which of the following does not cause backflow?

    a. Failure of a pump
    b. Injection of air into the system
    c. Cuts in the water main
    d. Insufficient levels of water

13. A student must complete _____ hours of OJL per year, according to the Office of Apprenticeship.

    a. 200
    b. 2000
    c. 100
    d. 1000

14. _____ is the highest level of skill a craftsman can become in their craft profession.

    a. Master craftsman
    b. Elite craftsman
    c. Certified pro craftsman
    d. Site foreman

Trade Terms Quiz

Fill in the blank with the correct term that you learned from your study of this module.

1. During the _____ phase of a plumbing project, plumbers locate all supply and waste connections from the building systems to public utilities.

2. _____ is the process of cleansing water to remove chemicals and particles.

3. You can use _____ thermoplastic pipe as tubing for cold water systems.

4. Someone who installs or repairs plumbing systems and fixtures is called a(n) _____.

5. Water supply systems provide fresh water; _____ systems remove wastewater.

6. Always wear appropriate personal protective equipment when disinfecting water with _____, a heavy, greenish-yellow gas.

7. You are _____ water when you remove magnesium and sodium salts from it.

8. Office of Apprenticeship requires apprentices to have 2,000 hours of _____ per year.

9. Water that is safe to drink is referred to as _____.

10. Construction ordinances written by national construction organizations and based on suggested national plumbing standards are called _____.

11. During the _____ phase of a plumbing project, plumbers install fixtures, appliances, water purification systems, water heaters, and controls.

12. When a potable water system and a nonpotable water system are arranged so that an accidental pressure differential between the two systems would cause contaminated water to flow into the freshwater system, it is called a(n) _____.

13. Plumbers destroy harmful organisms in potable water through _____.

14. During the Roman Empire, a(n) _____ were used to carry water from rivers and lakes to cities.

15. In ancient Rome, a person who worked with lead was called a(n) _____.

16. As a construction professional, you must ensure that your conduct is in accordance with your profession's principles and values, which are called _____.

17. One of the most important safety concerns for plumbers is _____, which is the flow of contaminated water into the freshwater system through a cross-connection.

18. Plastic that becomes substantially infusible and insoluble when treated by heat or chemicals is called _____.

19. During the _____ phase of a plumbing project, plumbers cut holes in walls, ceilings, and floors to attached or hang pipe runs.

20. Plastic that is soft and pliable when heated and hard and rigid when cooled is called _____.

21. A(n) _____ is a legally binding requirement published by state and local governments to establish minimum standards for various types of construction.

22. Sinks, shower stalls, and toilets are common examples of _____.

23. The Latin word for lead is _____.

24. Plumbers install a _____ to keep nonpotable water from contaminating a potable supply system.

25. If you are installing piping for a system that is designed exclusively for environmental control, you are not engaged in _____.

26. To become a(n) _____, you must first complete an apprenticeship training program.

27. Plumbing _____ are types of apparatus used in such trades as installing drainage systems.

28. Heat that comes from below the surface of the earth is called _____.

29. A system that directly or indirectly uses the heat of the sun to heat water is called a(n) _____ system.

30. The process of _____ is the consumption of fresh water from underground sources faster than the water can be naturally replenished.

31. The non-profit construction trade organization _____ developed a system for certifying that buildings have been designed and constructed to green standards called _____ that includes between 10 and 14 points that are categorized as _____, which is the managed use of drinkable water to reduce waste.

32. Efforts to preserve natural resources include a number of water preservation techniques including _____ to collect and store rainwater, the reuse of _____ from laundries and bath/showers, artificial depressions in the ground that filter pollutants from stormwater called a _____, and the use of treated wastewater called _____ in flushing fixtures and other non-potable systems.

Trade Terms

Aboveground rough-in
Appurtenances
Aqueduct
Aquifer depletion
Backflow
Backflow preventer
Bioswale
Chlorine
Code
Cross-connection
Disinfection
Drain, waste, and vent (DWV)
Ethics

Filtration
Finish
Fixture
Geothermal
Graywater
Journey plumber
Leadership in Energy and
    Environmental Design (LEED)
Model code
On-the-job learning (OJL)
Plumbarius
Plumber
Plumbing

Plumbum
Polyvinyl chloride (PVC)
Potable
Rainwater harvesting
Reclaimed water
Solar hot water
Softening
Thermoplastic
Thermoset
Underground rough-in
United States Green Building
    Council (USGBC)
Water efficiency

Appendix A

ETHICAL PRINCIPLES FOR MEMBERS OF THE CONSTRUCTION TRADES

Honesty: Be honest and truthful in all dealings. Conduct business according to the highest professional standards. Faithfully fulfill all contracts and commitments. Do not deliberately mislead or deceive others.

Integrity: Demonstrate personal integrity and the courage of your convictions by doing what is right even where there is pressure to do otherwise. Do not sacrifice your principles because it seems easier.

Loyalty: Be worthy of trust. Demonstrate fidelity and loyalty to companies, employers, sponsors, co-workers, and trade institutions and organizations.

Fairness: Be fair and just in all dealings. Do not take undue advantage of another's mistakes or difficulties. Fair people are open-minded and committed to justice, equal treatment of individuals, and tolerance for and acceptance of diversity.

Respect for others: Be courteous and treat all people with equal respect and dignity.

Obedience: Abide by laws, rules, and regulations relating to all personal and business activities.

Commitment to excellence: Pursue excellence in performing your duties, be well informed and prepared, and constantly try to increase your proficiency by gaining new skills and knowledge.

Leadership: By your own conduct, seek to be a positive role model for others.

NCCER CREDENTIALS

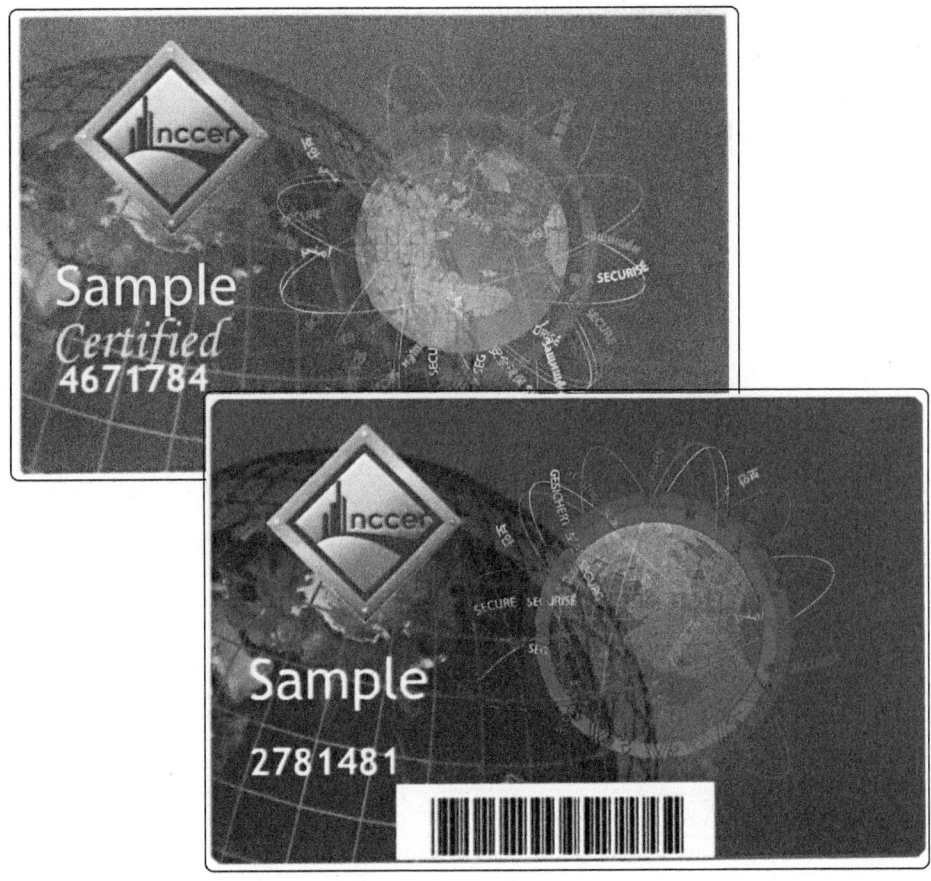

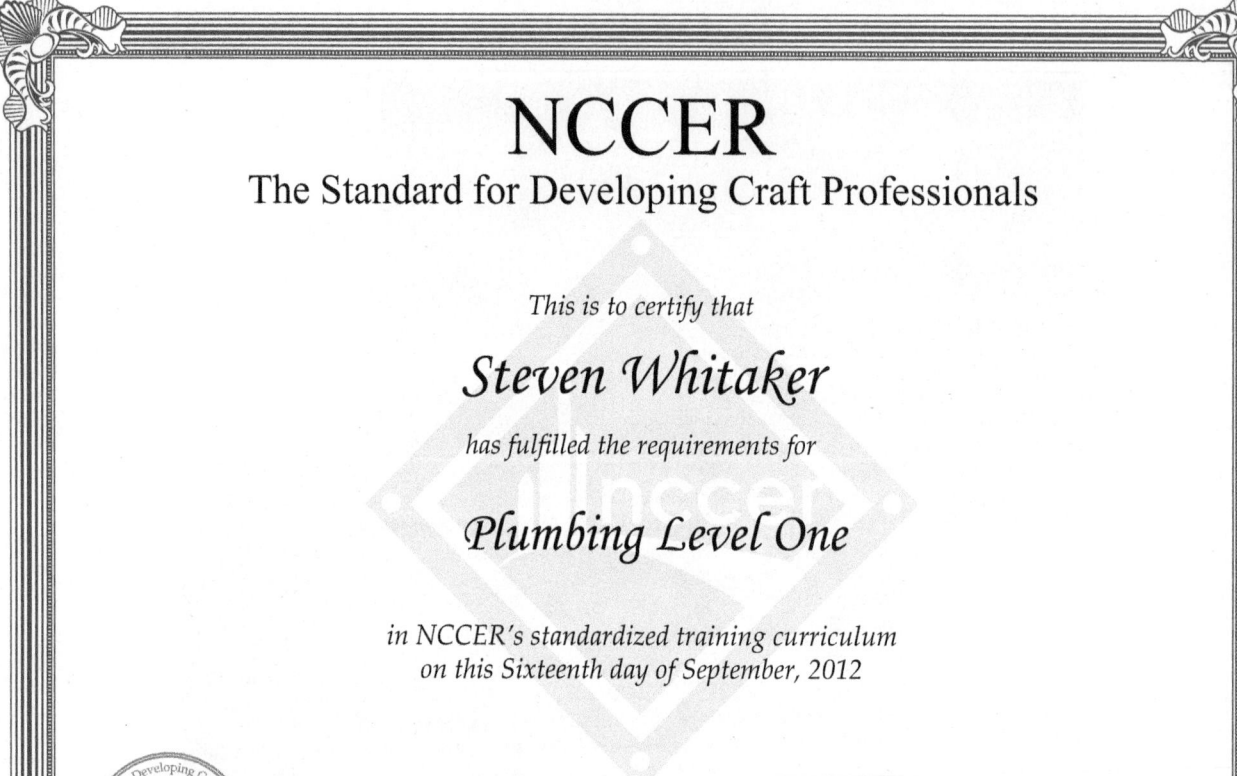

NCCER
The Standard for Developing Craft Professionals

This is to certify that

Steven Whitaker

has fulfilled the requirements for

Plumbing Level One

*in NCCER's standardized training curriculum
on this Sixteenth day of September, 2012*

Donald E. Whyte
President, NCCER

THE STANDARD FOR DEVELOPING CRAFT PROFESSIONALS

13614 Progress Boulevard, Alachua, Florida 32615 • p: 888.622.3720 f: 386.518.6255 • www.nccer.org

Official Transcript

January 17, 2012

NCCER Card #: 1720726
Trainee Name: John Q Smith
Sponsor: Austin Industrial Incorporated
Address: 2801 E 13th St
La Porte, TX 77571

Current Employer/School:
Solomon Plumbing Company

Module	Description	Instructor	Training Location	Date Completed
00101-04	Basic Safety	Kevin Jenkins	Solomon Plumbing Company	2/20/2008
00102-04	Introduction to Construction Math	Dave Buck	Building Trades Institute, LLC	8/8/2008
00103-04	Introduction to Hand Tools	Kevin Jenkins	Solomon Plumbing Company	1/1/2008
00104-04	Introduction to Power Tools	Dave Buck	Building Trades Institute, LLC	8/8/2008
00105-04	Introduction to Blueprints	Kevin Jenkins	Solomon Plumbing Company	3/20/2008
00106-04	Basic Rigging	Dave Buck	Building Trades Institute, LLC	8/8/2008
00108-04	Basic Employability Skills	Rod Blackburn	Utility Contractors, Inc.	3/15/2009
02101-05	Introduction to the Plumbing Profession	Kevin Jenkins	Solomon Plumbing Company	3/22/2008
26101-02	Electrical Safety	Don Whyte	National Center for Construction Education &	7/29/2002
26102-02	Hand Bending	Don Whyte	National Center for Construction Education &	7/29/2002
26103-02	Fasteners and Anchors	Don Whyte	National Center for Construction Education &	7/29/2002
26104-02	Electrical Theory One	Don Whyte	National Center for Construction Education &	7/29/2002
26105-02	Electrical Theory Two	Don Whyte	National Center for Construction Education &	7/29/2002
26106-02	Electrical Test Equipment	Don Whyte	National Center for Construction Education &	7/29/2002
26107-02	Introduction to the National Electrical Code	Don Whyte	National Center for Construction Education &	7/29/2002
26108-02	Raceways, Boxes, and Fittings	Don Whyte	National Center for Construction Education &	7/29/2002
26109-02	Conductors	Don Whyte	National Center for Construction Education &	7/29/2002

Page 1

Donald E. Whyte
President, NCCER

Aboveground rough-in: The second phase of a plumbing project. During this phase, holes are cut in walls, ceilings, and floors. Then, supply and waste pipes are attached or hung so they can be connected to fixtures. Also referred to as stack out, top out, or in-wall rough-in.

Appurtenances: Accessories or apparatus that require no demand from the water supply side and add no load to the waste side.

Aqueduct: A man-made channel used to carry water.

Aquifer depletion: The use of underground fresh water at a rate faster than it can be replenished.

Backflow: The flow of contaminated water into the freshwater system resulting from a cross-connection between potable and nonpotable water systems.

Backflow preventer: A device that prevents nonpotable water from entering a potable supply system.

Bioswale: A depression in the ground that filters pollutants from stormwater.

Chlorine: A heavy, greenish-yellow gas used as a disinfectant in water treatment. Chlorine should be handled only when wearing appropriate personal protective equipment.

Code: A requirement published by state and local governments to establish minimum standards for various types of construction. A code carries the force of law.

Cross-connection: An arrangement between a potable water system and a nonpotable water system in which an accidental pressure differential between the two systems causes backflow of contaminated water into the freshwater system.

Disinfection: The process of destroying harmful organisms in potable water.

Drain, waste, and vent (DWV): A piping system that combines sanitary drainage with venting.

Ethics: A set of principles and values that guide an individual's conduct.

Filtration: The process of cleansing water to remove particles and chemicals.

Finish: The third phase of a plumbing project. During the finish phase, plumbers install fixtures, appliances, water purification systems, water heaters, and controls. Also referred to as trim-out or trim finish.

Fixtures: Devices that receive water from a water supply line. Common fixtures include sinks, shower stalls, and toilets.

Geothermal: Heat that is generated below the earth's surface.

Graywater: Water that comes from baths and washing machines.

Journey plumber: A plumber who has successfully completed an apprenticeship-training program.

Leadership in Energy and Environmental Design (LEED): A system for certifying that buildings have been designed and constructed to environmental standards.

Model codes: Construction ordinances that are written by a national construction organization according to suggested national plumbing standards. Model codes that have not been adopted by a jurisdiction do not have the force of law

On-the-job learning (OJL): Field experience used in conjunction with classroom lessons in an apprenticeship program. Office of Apprenticeship requires 144 hours of classroom instruction per year and 2,000 hours of OJL per year.

Plumbarius: The Roman term for someone who works with lead. The root of the modern word plumber.

Plumber: One who installs or repairs plumbing systems and fixtures.

Plumbing: According to the National Standard Plumbing Code, plumbing is "the practice, materials, and fixtures within or adjacent to any building structure or conveyance, used in the installation, maintenance, extension, alteration, and removal of all piping, plumbing fixtures, plumbing appliances, and plumbing appurtenances... ."

Plumbum: Latin word for lead.

Polyvinyl chloride (PVC): A thermoplastic material frequently used in tubing for cold water systems and the first type of plastic approved for use in plumbing.

Potable: Water that is safe for cooking and drinking.

Rainwater harvesting: The collection and storage of rainwater for irrigation.

Reclaimed water: Wastewater that has had impurities and solids removed from it so that it can be reused for non-potable purposes.

Softening: The process of removing magnesium and sodium salts that cause scale on the inside of pipes and fittings.

Solar hot water: Water that has been directly or indirectly heated by sunlight.

Thermoplastic: A plastic material used in plumbing and sanitary systems that is soft and pliable when heated and hard and rigid when cooled.

Thermoset: A plastic material used in plumbing and sanitary systems that becomes substantially infusible and insoluble when treated by heat or chemicals.

Underground rough-in: The phase of a plumbing project during which the plumber locates all supply and waste connections from the building systems to public utilities, and establishes where these systems will enter or leave the building.

United States Green Building Council (USGBC): The non-profit construction trade organization responsible for the development of LEED.

Water efficiency: The managed use of drinkable water to reduce waste.

Additional Resources

This module presents thorough resources for task training. The following resource material is suggested for further study.

Plumbing a House, 1998. Peter A. Hemp. Newtown, CT: Taunton Press.

The National Standard Plumbing Code, current edition. Falls Church, VA: Plumbing-Heating-Cooling Contractors Association.

Figure Credits

Section Review Answer Key

SECTION 1.0.0

Answer	Section Reference	Objective
1. c	1.1.1	1a
2. a	1.2.3	1b

SECTION 2.0.0

Answer	Section Reference	Objective
1. d	2.1.9	2a
2. d	2.2.1	2b

This page is intentionally left blank.

NCCER CURRICULA — USER UPDATE

NCCER makes every effort to keep its textbooks up-to-date and free of technical errors. We appreciate your help in this process. If you find an error, a typographical mistake, or an inaccuracy in NCCER's curricula, please fill out this form (or a photocopy), or complete the on-line form at **www.nccer.org/olf**. Be sure to include the exact module ID number, page number, a detailed description, and your recommended correction. Your input will be brought to the attention of the Authoring Team. Thank you for your assistance.

Instructors – If you have an idea for improving this textbook, or have found that additional materials were necessary to teach this module effectively, please let us know so that we may present your suggestions to the Authoring Team.

NCCER Product Development and Revision
13614 Progress Blvd., Alachua, FL 32615

Email: curriculum@nccer.org
Online: www.nccer.org/olf

❏ Trainee Guide ❏ Lesson Plans ❏ Exam ❏ PowerPoints Other _____

Craft / Level: _____ Copyright Date: _____

Module ID Number / Title: _____

Section Number(s): _____

Description: _____

_____

_____

_____

_____

Recommended Correction: _____

_____

_____

_____

_____

Your Name: _____

Address: _____

_____

Email: _____ Phone: _____

This page is intentionally left blank.

Plumbing Safety

OVERVIEW

Unsafe acts and unsafe conditions lead to accidents, and while they may result in death or personal injury, accidents also cost companies time and money. The majority of accidents, however, can and should be prevented. There are many important safety measures that help plumbers remain safe on the job site.

Module 02102

Trainees with successful module completions may be eligible for credentialing through NCCER's National Registry. To learn more, go to *www.nccer.org* or contact us at 1.888.622.3720. Our website, *www.nccer.org*, has information on the latest product releases and training.

Your feedback is welcome. You may email your comments to *curriculum@nccer.org*, send general comments and inquiries to *info@nccer.org*, or fill in the User Update form at the back of this module.

02102 V4.5

Objectives

Successful completion of this module prepares trainees to:

1. Describe the causes and impacts of accidents.
   a. Identify the causes of accidents.
   b. Describe the costs and impacts of accidents.
2. Identify methods for preventing accidents.
   a. Explain the purpose of various types of personal protective equipment in preventing accidents.
   b. Explain the role of hazard communication in preventing accidents.
   c. Identify methods used to establish work zone safety.
3. Identify the safety precautions required when using hand and power tools.
   a. Describe the safety precautions associated with hand tools.
   b. Describe the safety precautions associated with power tools.
4. Identify the safety precautions associated with various work areas.
   a. Describe the safety precautions associated with work in trenches.
   b. Describe the safety precautions associated with confined spaces.
   c. Identify the safety precautions associated with underground work.
   d. Demonstrate a lockout/tagout procedure.
   e. Describe jobsite safeguards and emergency response procedures.

Performance Tasks

Under the supervision of your instructor, you should be able to do the following:

1. Inspect the following personal protective equipment:
   - Gloves
   - Body harness
   - Hard hat
   - Safety glasses
   - Safety shoes
   - Hearing protection
2. Put on the following personal protective equipment:
   - Hard hat
   - Body harness
   - Eye protection
   - Gloves
   - Hearing protection
   - Safety shoes
3. Demonstrate proper use of ladders.
4. Inspect power tools (corded and cordless) to ensure they are safe to use.
5. Inspect hand tools to ensure they are safe to use.
6. Demonstrate/simulate the proper methods of lockout/tagout for energy sources.

Trade Terms

Apparatus
Asbestos
Atmospheric hazards
Benching
Bladed tools
Combustible
Competent person
Confined spaces
Decibels (dB)
Electrically powered tools
Energy-isolating device
Energy sources
Fire watch
Gassy operations
Guards
Guy wires
Hazard Communication (HazCom) Standard
Hypothermia
Impact tools

Liquid-fuel tools
Lockout
Lockout devices
Lockout/tagout procedures
NFPA warning diamond
Nonpermit-required confined space
Occupational Safety and Health Administration
    (OSHA)
Oxygen-deficient atmosphere
Oxygen-enriched atmosphere
Permit-required confined space
Power tools
Protective system
Safety data sheet (SDS)
Shoring
Subsidence
Tagout devices
Trench shields

Industry Recognized Credentials

If you're training through an NCCER-accredited sponsor you may be eligible for credentials from NCCER's Registry. The ID number for this module is 02102. Note that this module may have been used in other NCCER curricula and may apply to other level completions. Contact NCCER's Registry at 888.622.3720 or go to *www.nccer.org* for more information.

CODE NOTE

Codes vary among jurisdictions. Because of the variations in code, consult the applicable code whenever regulations are in question. Referencing an incorrect set of codes can cause as much trouble as failing to reference codes altogether. Obtain, review, and familiarize yourself with your local adopted code. Safety codes are developed by the US Occupational Safety and Health Administration (OSHA).

Contents

Contents (continued)

Figures and Tables

This page is intentionally left blank.

1.0.0 CAUSES AND IMPACTS OF ACCIDENTS

Objective

Describe the causes and impacts of accidents.
a. Identify the causes of accidents.
b. Describe the costs and impacts of accidents.

Trade Terms

Competent person: An individual who is capable of identifying existing and predictable hazards or working conditions that are hazardous, unsanitary, or dangerous to employees, and who has authorization to take prompt, corrective measures to eliminate or control these hazards and conditions.

Hazard Communication (HazCom) Standard: A federal OSHA regulation requiring employers to educate and inform workers about chemical hazards on the job site (*29 CFR 1910.1200*).

Occupational Safety and Health Administration (OSHA): The division of the US Department of Labor mandated to ensure a safe and healthy environment in the workplace.

Most plumbing-related accidents and injuries are caused by worker carelessness, poor safety planning, lack of training, or failure of the employer or employee to follow safety regulations. Accidents do not only affect plumbers and their employers; they can also affect the health and safety of the public. Diseases, contamination, and flooding are just a few of the ways.

To help prevent accidents, your company must have a safety program. This program will provide you with the rules and safeguards you need to work safely. Safety must be part of all phases of the job and must involve employees at every level, including management. The US Department of Labor's Occupational Safety and Health Administration (OSHA) requires that a company-appointed competent person be on site before you start any job. OSHA's regulation *CFR 1926.32* defines a competent person as one who is capable of identifying existing and predictable hazards in the surroundings or working conditions that are unsanitary or dangerous to employees and who is authorized to take prompt, corrective measures to eliminate them. A competent

person has experience and training for the job and knows the job's hazards, as well as the rules and regulations associated with the job.

In addition to its own rules, your company must comply with many local, state, and national regulations. For example, OSHA requires the following:

- An employer "shall furnish to each of his employees employment and a place of employment which are free from recognized hazards that are causing or are likely to cause death or serious physical harm to his employees."
- An employer "shall comply with occupational safety and health standards" established in the federal Occupational Safety and Health Act.
- An employee "shall comply with occupational safety and health standards and all rules, regulations, and orders issued pursuant to this Act which are applicable to his own actions and conduct."

This is called OSHA's "general duty clause." OSHA also says that employees have a duty to fix a recognizable hazard when they see one. Safety regulations like these are intended to make work sites safe and accident-free. Safety policies and procedures are available to help you and your company comply with these regulations. Remember that there is a good reason for each regulation. Following good safety practices helps to save lives.

1.1.0 Causes of Safety Accidents

Accidents and injuries cause pain and suffering that could be avoided, as well as financial hardship for employees, their families, and their companies. All workers on a site, including plumbers, have a moral, legal, and financial obligation to prevent accidents and injuries. To do this, you must understand that unsafe acts and conditions cause accidents. Unsafe acts are things you do or do not do that can cause an accident. Unsafe conditions are factors that make the work area dangerous.

1.1.1 Unsafe Acts

Unsafe acts often lead to serious injury and sometimes death. You can prevent unsafe acts by changing your behavior. It is your responsibility to recognize unsafe acts and stop them immediately. This can mean telling your co-workers to stop what they are doing. If your co-workers do not stop acting in an unsafe manner, stop what you are doing and move as far away from them as possible. In some cases, you may need to inform your supervisor of the problem. Here are some examples of the most common unsafe acts:

- Operating equipment at improper speeds
- Operating equipment without authority
- Using defective equipment
- Disabling a safety device
- Servicing equipment while it is in motion or energized
- Using equipment improperly
- Failing to use personal protective equipment (PPE)
- Failing to warn co-workers of a dangerous or potentially dangerous situation
- Working in an improper position
- Working while impaired by alcohol or illegal drugs
- Operating tools or equipment when taking certain types of prescription drugs
- Lifting loads improperly
- Loading or placing equipment or supplies improperly
- Horseplay (see *Figure 1*)

An unsafe act can also be defined as work that is not done correctly. Workers who fail to use proper PPE, follow safety procedures, or warn co-workers of potentially hazardous conditions can cause or worsen accidents. Keep yourself safe, look out for the safety of others, always follow safety rules, and use the right equipment for the job.

1.1.2 Unsafe Conditions

Unsafe working conditions cause many accidents and injuries. Environmental factors such as noise, extreme heat or cold, poor lighting, and poor air circulation can create unsafe conditions by impairing your reactions or limiting your movements. Poor housekeeping is also a hazard. Clutter in walkways, spills, and improper waste disposal can make simply walking on a construction site dangerous.

To work safely, you must be able to hear, see, breathe, and keep your balance. Before you start a job, look for and correct any unsafe working conditions. Pick up and properly store loose pipes and fittings and keep your work area clean and free of debris that could cause accidents. Keep glue cans closed to reduce the amount of toxic fumes in the air. Check the ground for standing water that could be an electrical hazard. The toxic, biological, and electrical hazards affecting plumbers are covered later in this module.

1.2.0 Costs and Impacts of Accidents

Accidents are very costly. When they happen, everyone involved loses, including the company and its employees. You may not believe that all accidents will cost you money. Think about it this way: If there is an accident on a job site, the company will have to pay higher insurance rates, costs for medical care and/or repairs, downtime, and the cost of investigations or fines. If the company is paying for these things, it has less money to spend on raises and performance bonuses and to hire replacement workers. What that means to you is more work without an increase in pay. You are a part of the company (see *Figure 2*), and job-site accidents will affect you, even if you are not directly involved.

Accident costs can be classified as direct (insured) and indirect (uninsured). The costs associated with accidents can be compared to an iceberg (see *Figure 3*). The tip of the iceberg represents the direct costs, such as medical bills, compensation, and insurance premiums. The larger, indirect costs are unseen. They include property and equipment damage, production delays, and lost

Figure 1 Horseplay is dangerous.

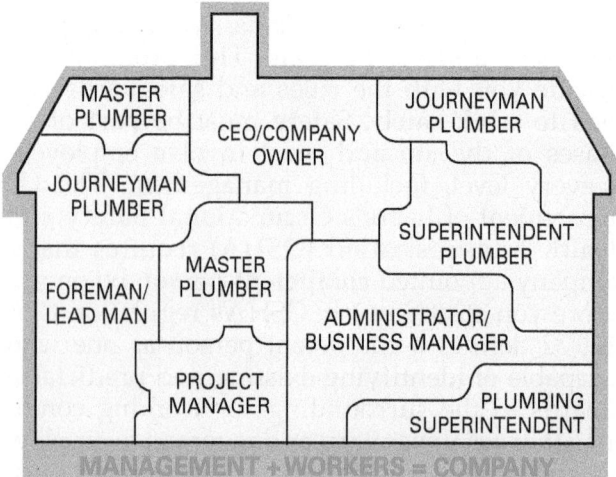

Figure 2 Companies are a combination of workers and management.

time. Injuries that occur off the job may equally delay production and involve other costs for the employers. A high incident rate can affect your company's ability to secure future work.

OSHA, part of the US Department of Labor, is a government agency that protects millions of workers each year. Its mission is to ensure a safe and healthy workplace. Since the agency was created in 1971, workplace deaths have been cut in half, and injury and illness rates have declined 40 percent. At the same time, the number of people working has increased. In fact, since 1971, the number of people working in the United States has doubled, from 56 million workers at 3.5 million work sites to 111 million workers at 7 million sites.

OSHA adopts and enforces safety regulations known as standards. Some, like the Hazard Communication (HazCom) Standard, apply to all industries. Others apply only to a specific industry. In addition to general industry standards, OSHA has a series of standards for plumbing construction. Twenty-six states have their own OSHA programs. The state regulations must be at least as strict as the national standards. Find out which regulations apply to your job site. The federal and state OSHA programs keep workplaces safe through accident investigation, inspections, and outreach.

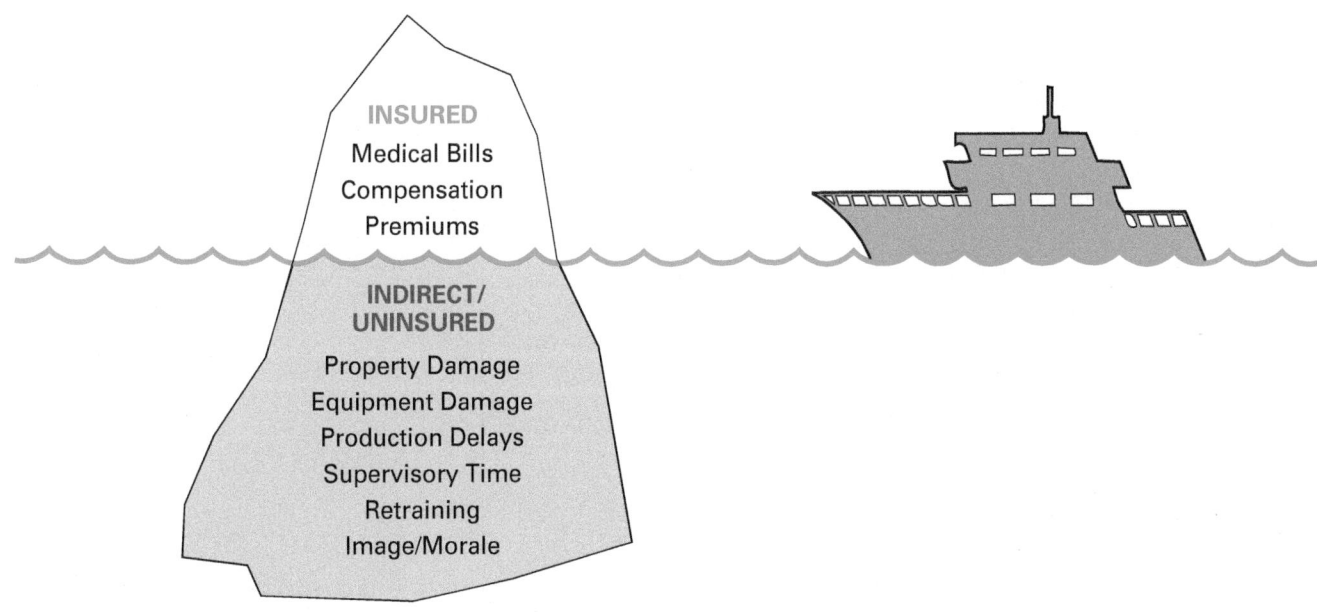

INSURED
Medical Bills
Compensation
Premiums

INDIRECT/
UNINSURED

Property Damage
Equipment Damage
Production Delays
Supervisory Time
Retraining
Image/Morale

Figure 3 Hidden costs of accidents.

Additional Resources

Environmental Protection Agency. *www.epa.gov.*

National Safety Council. *www.nsc.org.*

OSHA. *www.osha.gov.*

1.0.0 Section Review

1. You see a co-worker fooling around on the work site. Your first course of action is to _____.

    a. tell your supervisor
    b. tell your co-worker to stop
    c. move as far away from your co-worker as possible
    d. complain to other co-workers about this person's actions

2. OSHA's mission is to protect _____.

    a. the economy
    b. employers
    c. workers
    d. the whole workplace environment

2.0.0 PREVENTING ACCIDENTS

Objective

Identify methods for preventing accidents.

  a. Explain the purpose of various types of personal protective equipment in preventing accidents.
  b. Explain the role of hazard communication in preventing accidents.
  c. Identify methods used to establish work zone safety.

Performance Tasks

  1. Inspect the following personal protective equipment:
- Gloves
- Body harness
- Hard hat
- Safety glasses
- Safety shoes
- Hearing protection

  2. Put on the following personal protective equipment:
- Hard hat
- Body harness
- Eye protection
- Gloves
- Hearing protection
- Safety shoes

  3. Demonstrate proper use of ladders.

Trade Terms

Apparatus: One tool, which combines a variety of functions, to perform a particular job.

Asbestos: A fibrous, fire-resistant substance used in pipe insulation, shingles, wallboard, floor coverings, and certain types of insulation. Now banned by government regulation as a health hazard.

Confined spaces: Spaces that, by design and/or configuration, have limited openings for entry and exit, have unfavorable natural ventilation, may contain or produce hazardous substances, and are not intended for continuous employee occupancy.

Decibels (dB): A measure of sound intensity or loudness. The higher the decibel level, the louder and more potentially damaging the sound is.

Guards: Devices that protect tool operators from dangerous moving parts, such as blades, gears, and pulleys.

Guy wires: Ropes, chains, cables, or rods attached to something as a brace or guide.

Lockout/tagout procedures: Processes for identifying hazardous equipment, locking it so that no workers can use it until it is certified for safe use, and placing a tag on the equipment that describes the problem and warns against use.

NFPA warning diamond: A four-color diamond label placed on containers or doors to alert people to specific safety hazards in a product, room, or building.

Oxygen-deficient atmosphere: An atmosphere in which there is not enough oxygen to support life. Usually considered less than 19.5 percent oxygen by volume.

Power tools: Tools that require a power source, such as electricity, hydraulics, or pneumatics, to operate.

Safety data sheet (SDS): A document that must accompany any hazardous material. The SDS identifies the material and gives the exposure limits, the physical and chemical characteristics, the kind of hazard it presents, precautions for safe handling and use, and specific control measures.

There are many tools and safeguards that are used to prevent injury on a job site. The first line of defense is to remove the hazard itself. For example, do not work at elevation if the work can be done at ground level and the finished piece brought up to the installed location. If a hazard cannot be eliminated, appropriate personal protective equipment (PPE) must be worn. Workers must be advised of the hazard through effective communication, and a safe work zone must be established.

2.1.0 Personal Protective Equipment

PPE is designed to protect you from injury. Many plumbers are injured on the job because they are not using PPE.

You won't see all the potentially dangerous conditions on a job site just by looking around. Before doing any job, stop and consider what types of accidents could happen. Using common sense and PPE greatly reduces your chances of getting hurt. The best PPE is of no use unless you do the following four things:

- Inspect it regularly, and replace any PPE that is damaged or worn.

- Care for it properly.
- Use it correctly when it is needed.
- Avoid altering or modifying it in any way.

As a plumber, you will most commonly use the following types of PPE:

- Hard hats
- Eye and face protection
- Gloves
- Safety shoes
- Hearing protection
- Fall protection
- Respiratory protection
- Proper clothing

NOTE

The safety harness and lanyard are parts of a system that is known as the personal fall arrest system. Workers must know how to properly inspect, don, and maintain their system.

NOTE

Accidents and incidents are defined differently by OSHA. When reporting injuries, always use the word accident, not incident.

2.1.1 Hard Hats

Figure 4 shows a typical hard hat. The outer shell of the hat protects your head during a fall or from a flying object. The webbing inside the hat maintains a space between the shell and your head. When wearing a hard hat, adjust the headband, not the hard hat, so that the webbing fits your head and there is at least 1 inch of space between your head and the shell.

Never wear a baseball cap under a hard hat. The knob on the top of the cap will impede the hard hat's function if impacted. Additionally, pull back long hair (longer than 4 inches), which can easily get caught in machinery. You should not

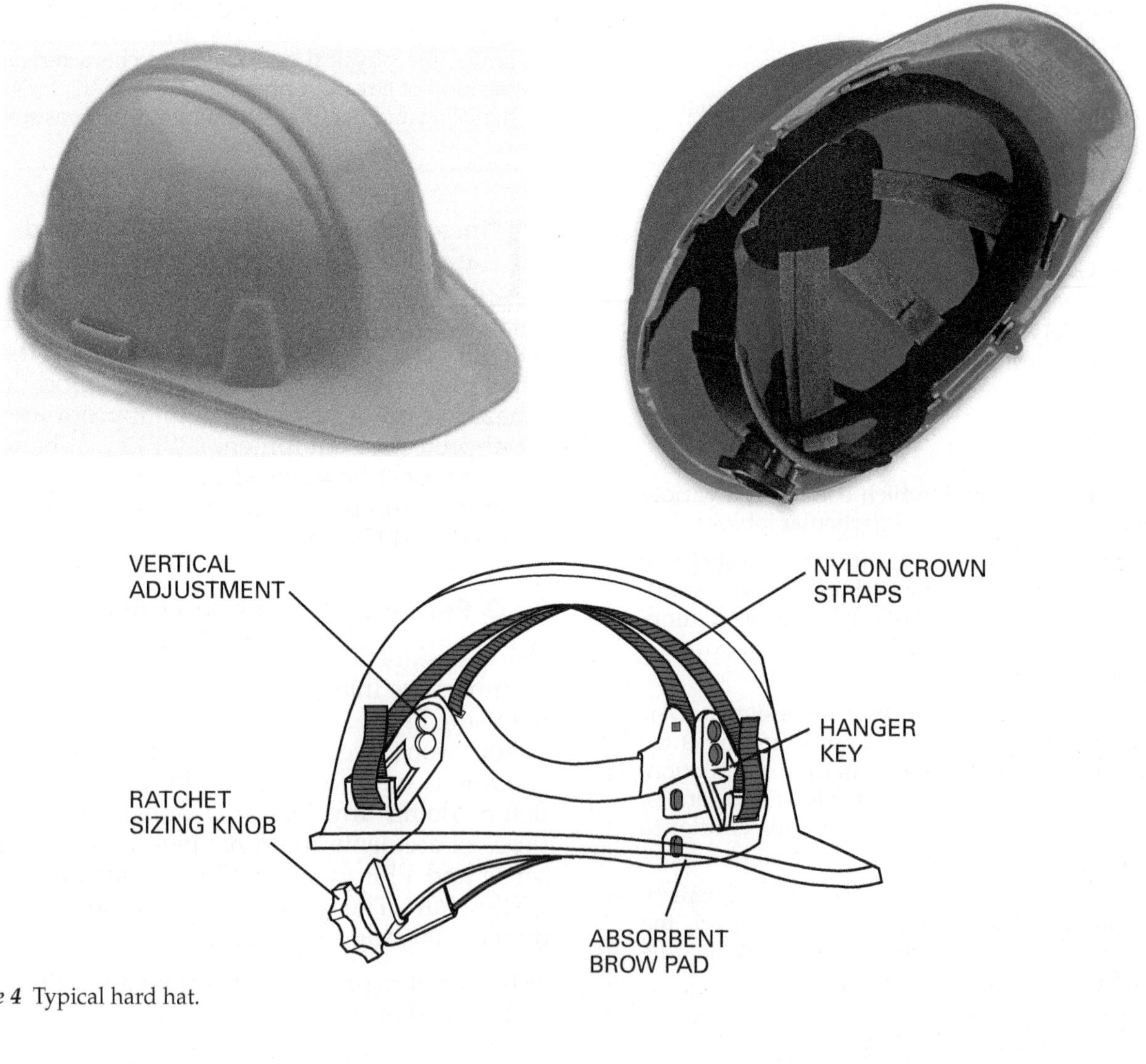

VERTICAL ADJUSTMENT

NYLON CROWN STRAPS

HANGER KEY

RATCHET SIZING KNOB

ABSORBENT BROW PAD

Figure 4 Typical hard hat.

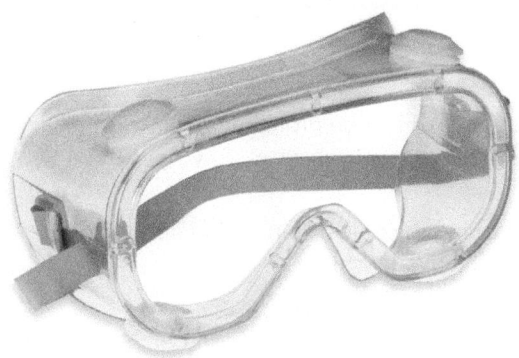

SAFETY GOGGLES

PRESCRIPTION GLASSES WITH SIDE SHIELDS

Figure 5 Typical safety goggles and glasses.

be required to cut your hair, but you may need to cover it with a bandana or hair net under your hard hat. Always ensure that the hard hat that you use has not exceeded its life expectancy.

Case History

Hard-Hat Safety

You won't get the protection that a hard hat is designed to provide if you do not wear it correctly. In one reported accident, a worker who was wearing his hard hat backwards was pulling down a piece of 6-inch pipe when the pipe began falling. As the pipe fell, and because the brim of his hat was turned backward couldn't deflect falling objects, the pipe hit him on the forehead. The worker suffered brain damage as a result and could not work again.

   The Bottom Line: Always wear your hard hat correctly.

You must wear your hat correctly. Notice that hard hats have a short, sturdy brim at the front. This brim is designed to deflect falling objects, so never wear the hat backward or pushed back from your face. Doing so eliminates the protection provided by the brim.

2.1.2 *Safety Glasses, Goggles, and Face Shields*

Wear eye protection (*Figure 5*) whenever there is even the slightest chance of an eye injury. Areas where there are potential eye hazards from falling or flying objects are usually identified, but you should always be on the lookout for other possible hazards, such as sewage and pressurized water.

   There are different types of eye protection. Safety goggles give your eyes the best protection from all directions. Regular safety glasses protect you from objects flying at you from the front, such as large chips, particles, sand, or dust, but you can add side shields for further protection. Do not wear contact lenses because the lenses can burn your eyes. You may, however, substitute prescription safety glasses if they provide the same protection as regular safety glasses. If welding is part of your job, you must use safety goggles with tinted lenses or a welding hood. Tinted lenses protect your eyes from the bright welding arc or flame.

CAUTION

Never wear tinted glasses indoors.

Case History

Eye Safety

In one reported accident, a plumber suffered a serious eye injury after debris from the work he was doing blew into his eyes. A supervisor had told this plumber several times to wear safety glasses. Had he listened to his supervisor, the plumber would not have been injured, nor would he have had to take time off from work to recover.

   The Bottom Line: Always wear eye protection while on the job site.

Eye Protection

If you must wear corrective glasses, OSHA regulations state that your eyes must be protected by one of the following ways:

- Spectacles whose protective lenses provide optical correction
- Goggles that can be worn over corrective spectacles without disturbing the adjustment of the spectacles
- Goggles that incorporate corrective lenses mounted behind the protective lenses

Your optometrist can provide the required eyewear.
Source: Occupational Safety and Health Administration *CFR 1926.102*

2.1.3 Gloves

On many jobs, you must wear heavy-duty gloves (*Figure 6*) to protect your hands from cuts, chemical burns, or exposure to raw sewage. Work gloves are usually made of cloth, canvas, or leather, but they may also be made of metal mesh or a material called Kevlar®, which protects against metal cuts. Never wear cloth gloves around rotating or moving equipment. Inspect your gloves every day to ensure that they are in good condition. Immediately replace gloves that are worn, torn, or no longer fit.

In addition to a visual test, an effective way to check rubber gloves is to conduct an air test. The following steps are outlined in *29 CFR 1910.137*:

Step 1 Hold the glove by its cuff, flip it several times to make a seal, and roll the glove toward its fingers. An air pocket will form inside the glove.

Step 2 Hold the rolled portion of the glove tightly. Inspect the inflated exterior of the glove for cracks or degradation of the insulating material surface. Forcing air into the glove will expose damage to the insulation that cannot be seen during visual inspections.

Step 3 Inspect the glove for holes in the insulating material. Hold the glove close to your ear. If you hear air escaping from the glove or if the glove does not hold pressure, the glove is damaged. Damaged gloves must be removed from service.

Figure 6 Work gloves.

CAUTION

Protective equipment that is good one day may be damaged the next. Always do a daily visual inspection of PPE to prevent injuries from defective or worn equipment.

WARNING!

Always thoroughly wash and disinfect your hands if you have been exposed to sewage, even if you were wearing gloves. Doing so can prevent disease or illness. In general, it is best to use an alcohol gel instead of an antibacterial cleaner. Recent research indicates that overuse of antibacterial cleaners may make bacteria more resistant.

2.1.4 Safety Shoes

To protect your feet from falling objects and punctures, wear steel-toed, steel-soled safety shoes (*Figure 7*). The steel toe protects your toes from falling objects. The steel sole keeps nails and other sharp objects from puncturing your foot. You are required to wear this type of safety shoe when using tamping equipment or jackhammers. If electrical hazards are present, wear metal-free shoes or boots. Welding jobs call for boots or other sturdy shoes without laces or eyelets to prevent hot metals or sparks from becoming trapped in the shoes. Do not wear oil-soaked shoes if you are welding. Doing so increases the risk of fire.

It is also important to consider how your shoes will be affected by water, especially the soles. Many injuries have occurred as a result of a slip on a wet surface. When working on wet or icy surfaces, wear rubber boots or shoes with skid-resistant soles. Always replace boots or shoes when the sole tread becomes worn or the shoes have holes, even if the holes are on top. To reduce the possibility of falls or injury to the feet, never wear open-toed shoes or open-heeled shoes because increases.

2.1.5 Hearing Protection

Damage to most parts of the body causes pain. Ear damage, however, does not always cause pain. Exposure to loud noise over a long period can cause hearing loss, even if the noise is not loud enough to cause pain. Most companies follow OSHA's rules for when workers must use hearing protection. Specially designed earplugs that fit into your ears and filter out noise are one type of hearing protection (*Figure 8*). You must clean earplugs regularly with soap and water to prevent ear infections. Because earplugs are made in different sizes, make sure you use earplugs that fit properly.

Another type of hearing protection is earmuffs, which are large, padded covers for the entire ear (*Figure 9*). You must adjust the headband on earmuffs for a snug fit. If the noise level is very high, you may need to wear both earplugs and earmuffs.

You can prevent noise-induced hearing loss by limiting exposure and using appropriate PPE. *Table 1* shows the maximum safe exposure to sound levels rated 90 decibels (dB) and higher.

Some kits also come with tourniquets, splints, safety whistles, safety streamers, safety pins, and garbage sacks. Always keep the first-aid kit stocked. That way, when someone needs to use the kit, the necessary items are available. It may also be helpful to keep an accident report and a pencil in your first-aid kit.

Never attempt to treat serious injuries yourself. Always call 911 in an emergency. Check your company's policy for administering medications. Some companies may prohibit the distribution of medications.

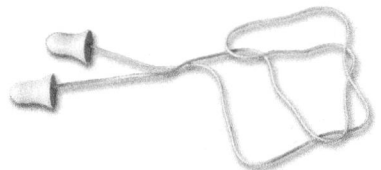

Figure 8 Earplugs.

Figure 9 Earmuffs.

Figure 7 Safety shoes.

Did You Know?

Statistics compiled by the National Institute of Occupational Safety and Health (NIOSH) indicate that 48 percent of plumbers report some hearing loss due to the use of noisy electric machinery in tight places.

Table 1 Maximum Noise Levels

Sound Level (decibels)	Maximum Hours of Continuous Exposure per Day	Examples
90	8	Power lawn mower
92	6	Belt sander
95	4	Tractor
97	3	Hand drill
100	2	Chop saw
102	1.5	Impact wrench
105	1	Spray painter
110	0.5	Power shovel
115	0.25 or less	Hammer drill

2.1.6 Fall Protection

OSHA requires workers to wear and be trained in the proper use of personal fall protection equipment in certain workplace situations. OSHA developed a Focused Inspection Program to target the four high-hazard areas. One of the four leading hazard groups is falls from elevation. Falls from elevation are accidents involving failure of, failure to provide, or failure to use appropriate fall protection. Wearing personal fall protection equipment can help to prevent these accidents. This equipment usually includes a safety harness and a lanyard. Refer to *Table 2* to determine the proper length of fall protection equipment for a specified height. For fall clearances less than 18-$\frac{1}{2}$ feet, use a self-retracting lifeline. For fall clearances over 18-$\frac{1}{2}$ feet, use a shock-absorbing lanyard or a self-retracting lifeline.

Fall protection equipment incorporates embedded red threads that will appear if the equipment has been stretched, torn, or otherwise damaged. If you see these red threads on your equipment, do not use it. Never write your name on the webbing of your fall protection equipment using a marker. The ink in the marker could react with the webbing material, causing it to corrode and degrade. Never attach a retractable lanyard to a fixed lanyard.

The safety harness (*Figure 10*) is an extra-heavy-duty harness that fits over your body and includes leg, shoulder, chest, and pelvic straps. The lanyard (*Figure 11*) is an engineered system that protects against free falls greater than 3 feet, with reinforced ends with D-rings snapped onto them. One D-ring attaches to the safety harness, and the other is attached to an anchor point above the work area. OSHA requires that anchors must be able to support 5,000 pounds per each person.

Table 2 Calculating Fall Distance

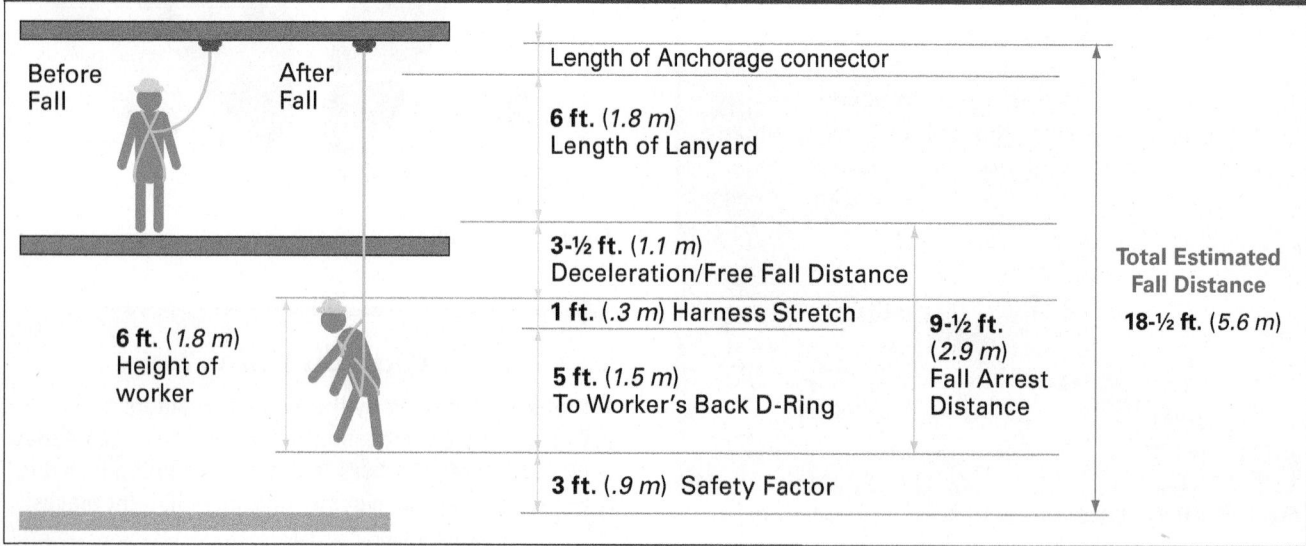

First-Aid Kits

Always keep a first-aid kit handy. Minor injuries can be treated easily when the necessary items are available. A complete first-aid kit should include the following items:

- Self-adhesive bandages
- Sterile dressing
- Gauze bandages
- First-aid tape
- Moleskin
- First-aid scissors
- Tweezers
- Surgical gloves
- Iodine and/or alcohol swabs
- Ibuprofen or other anti-inflammatory aids
- Antibiotic ointment
- Aspirin or other pain reliever

A qualified person will tell you where a strong anchor point is. The lanyard should be long enough to allow you to perform your tasks but short enough to bring you to a complete stop and limit the greatest distance that you can travel to 3.5 feet, and be strong enough to withstand twice the potential impact energy if you were to free fall 6 feet or the distance permitted by the manufacturer.

Fall protection includes covers or guardrail systems to prevent you from falling off any surface that is 6 feet or higher above a lower surface or into a hole or excavation that is 6 feet deep or more. Covers for holes also protect you from objects that may fall from above the work area. Use of fall protection equipment is required when working in these areas:

> **WARNING!**
>
> Never use a safety harness and lanyard for anything except their intended purpose. Always follow the manufacturer's instructions for hooking up a safety harness or lanyard. More than 70 percent of reported job-site accidents are caused by improper use of the lanyard and harness.

- More than 4 feet above the ground or the next lower surface, or 10 feet if on a scaffold (refer to your local applicable standards)
- Near openings in floors (for example, a stairwell, elevator shaft, chimney, or plumbing stack opening)
- Near holes or excavations deeper than 6 feet

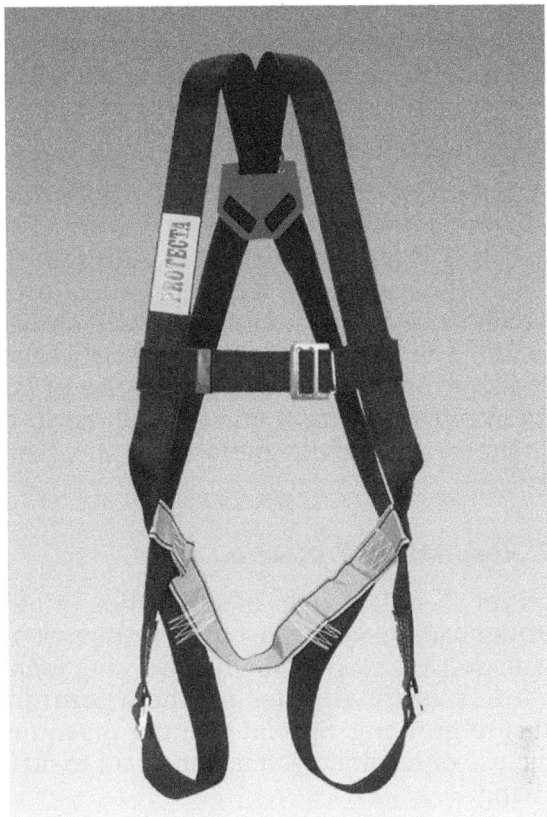

Figure 10 Typical safety harness.

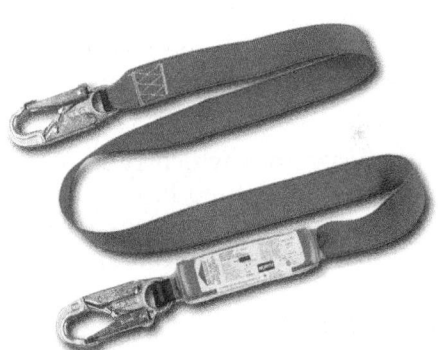

Figure 11 Lanyard.

Always tie off your fall protection equipment. Never tie off when leaning, as this will result in a miscalculation of the length required.

Keep your safety harnesses and lanyards in good working order. Before using this equipment, inspect it for damage, such as wear on the harness straps and buckles, or D-rings that are bent or deeply scratched. Check the harness for any cuts or rough spots. Do not use fall protection equipment that shows these signs of wear or damage. If you find any damage, turn in the harness for testing or replacement.

All companies are required by law to have a written Fall Protection Plan that addresses site-specific fall hazards and the steps taken to prevent each hazard. This information will normally be included in your training or regularly scheduled safety meetings. If you are unsure of your company's plan, or have questions about it, ask your supervisor before performing any aerial work.

2.1.7 Respiratory Protection

Wherever there is danger of suffocation or other breathing hazards, you must use a respirator. You must also wear a respirator when working with or near fire-resistant asbestos or where hazardous molds are growing. Special training is required for the use of respirators. It is important to do the following:

- Use the proper type of respiratory protection for the hazard.
- Pass a cardiopulmonary fitness test before using respiratory protection.
- Conduct a fit test with the respiratory protection equipment prior to use to ensure that it fits properly.

Federal law specifies which type of respirator to use for various breathing hazards. Respirators are grouped into three main types, based on how they protect the wearer from contaminants.

Air-purifying respirators—Air-purifying respirators provide the lowest level of protection. They are made for use only in atmospheres that have enough oxygen to sustain life (at least 19.5 percent). Air-purifying respirators are chemical-specific. They use special filters and cartridges to remove specific gases, vapors, and particles from the air. The respirator cartridges contain charcoal, which absorbs certain toxic or deadly vapors and gases. When the wearer detects any taste or smell, it indicates the charcoal's absorption capacity has been reached. This means the cartridge can no longer remove the contaminants. The respirator filters remove particles such as dust, mists, and metal fumes, by trapping them within the filter material. Filters should be changed when breathing becomes difficult.

Respirator manufacturers typically classify air-purifying respirators into four groups:

- No maintenance
- Low maintenance
- Reusable
- Powered air-purifying respirators (PAPRs)

Reusable respirators (*Figure 12*) are made in half-mask and full-facepiece styles. These respirators require the replacement of cartridges, filters, and respirator parts. Their use also requires a complete respirator maintenance program. Half-mask air-purifying respirators have several limitations, so be sure to refer to the manufacturer's instructions before using.

Powered air-purifying respirators (PAPRs) (*Figure 13*) are made in half-mask, full-facepiece, and hood styles. They use battery-operated blowers to pull air through the cartridges and filters attached to the respirator. The blower motors can be either mask- or belt-mounted. Depending on the cartridges used, PAPRs can filter particles, dust, fumes, and mists

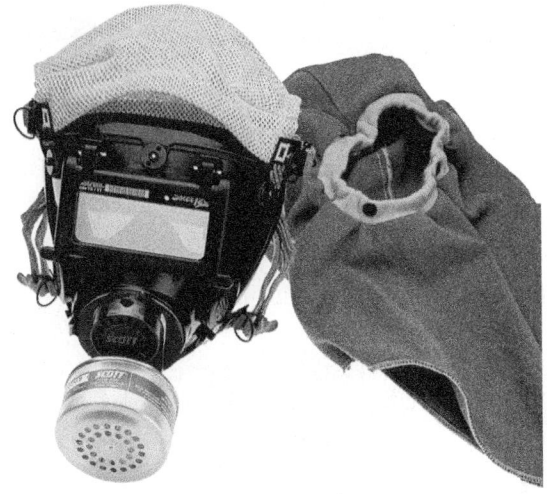

Figure 12 Half-face respirator.

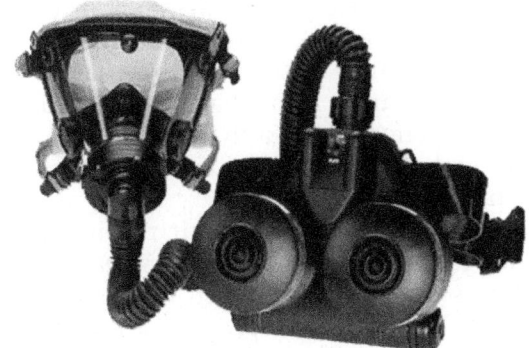

Figure 13 Full-face powered air-purifying respirator.

models have an audible and visible alarm that is activated as soon as the airflow falls below the required minimum level. This feature alerts the user to a loaded filter or low battery charge.

Supplied-air respirators—Supplied-air respirators (*Figure 14*) provide air for extended periods through a high-pressure hose connected to an external source of air, such as a compressor, compressed air cylinder, or pump. They provide protection in atmospheres where air-purifying respirators are not adequate. Supplied-air respirators are typically used in toxic atmospheres. Some can be used in atmospheres that are immediately dangerous to life and health (IDLH),

along with certain gases and vapors. Other than units with the blower mounted in the mask, PAPRs have a belt-mounted, powered air-purifier unit connected to the mask by a breathing tube. Many

Asbestos

Asbestos is a hazardous, fibrous substance that causes lung diseases, including cancer. It was once used regularly in construction. Pipe insulation, shingles, wallboard, floor covering, and blown-in insulation are just a few products that may contain asbestos.

The federal government stopped production of most asbestos products in the early 1970s. However, installation of these products continued through the late 1970s and early 1980s. Today, asbestos fibers can be released during renovations of older buildings. Breathing asbestos dust can have chronic and lasting effects.

Source: Agency for Toxic Substances and Disease Registry

Figure 14 Supplied-air respirator.

as long as they are equipped with an air cylinder for emergency escape. An atmosphere is considered IDLH if it poses an immediate hazard to life or produces immediate, irreversible, and debilitating effects on health. There are two types of supplied-air respirators: continuous flow and pressure demand.

Continuous-flow, supplied-air respirators provide air to the user in a constant stream. One or two hoses deliver the air from the source to the facepiece. Unless the compressor or pump is specially designed to filter the air, or a portable air-filtering system is used, the unit must be located where there is breathable air. Pressure-demand, supplied-air respirators are similar to continuous-flow respirators except that they supply air to the user's facepiece via a pressure-demand valve as the user inhales.

Self-contained breathing apparatus (SCBA)— SCBAs provide the highest level of respiratory protection. They can be used in an **oxygen-deficient atmosphere** (below 19.5 percent oxygen), in poorly ventilated or **confined spaces**, which have limited openings, and in IDLH atmospheres. These respirators supply air for about 30 to 60 minutes from a compressed-air cylinder worn on the user's back. An emergency escape breathing **apparatus** (EEBA) is a smaller version of an SCBA cylinder (see *Figure 15*). EEBAs are used to escape from hazardous environments and generally provide a 5- to 10-minute air supply.

Figure 15 Emergency escape breathing apparatus.

> **NOTE**
>
> You must be certified to wear an SCBA. Training for certification covers maintaining and checking the equipment, as well as correctly putting it on and removing it.

Respirator training requirements—An OSHA Letter of Interpretation clarified the annual training requirement for respiratory protection. "Annual means that training and fit testing must be conducted every year, before or on the anniversary date of the employee's previous training and fit test; for example, if the employee is trained or fit tested on February 1, 2011, then the employee must be trained or fit tested before or on February 1, 2012." If the anniversary date is missed, the reason for the delay and the expected date of completion should be noted in the employee's file.

Trained employees must be able to demonstrate knowledge in the following areas:

- Why a respirator is necessary, and the consequences of improper fit, use, or maintenance
- The limitations and capabilities of respirators
- How to use the respirator effectively in emergency situations, including when the respirator malfunctions
- How to inspect, put on, remove, use, and check the seals of the respirator
- The procedures for maintaining and storing respirators

Respirators

OSHA regulations (*29 CFR 1910.134*) require employees who use respirators in the workplace to be trained on an annual basis. Employees must complete the training before they are required to use a respirator. According to OSHA, "The training must be comprehensive, understandable, and recur annually, and more often if necessary." Additional training may be required if earlier training becomes obsolete or if the employee has not mastered the use of a respirator and demonstrates difficulty using a respirator.

- How to recognize medical signs and symptoms that may limit or prevent the effective use of respirators
- The information in *OSHA Standard 29 CFR 1910.134*

Respirator selection—A respirator must be selected based on the contaminant present and its concentration level. It must be fitted and used properly, in accordance with the manufacturer's instructions. It must be worn during all times of exposure.

Employers must have a respiratory protection program with the following components:

- Standard operating procedures for selection and use
- Employee training
- Regular cleaning and disinfecting
- Sanitary storage
- Regular inspection
- Annual fit testing
- Pulmonary function testing

Always check the cartridge on your respirator to ensure that it is the correct type for the air conditions and contaminants on your job site. In certain concentrations, vapors or fumes can be eliminated by the use of air-purifying devices as long as the oxygen levels are acceptable. Smoke billowing from a fire and the fumes generated when welding are examples.

When selecting a respirator to wear while working with specific materials, you must first determine the hazardous ingredients in the material and their exposure levels. Always read the safety data sheet (SDS), which is often located in a binder in the project manager's or supervisor's office. SDSs identify the hazardous ingredients and should list the type of respirator and cartridge recommended for use with the material. You will learn more about SDSs later in this module.

> **WARNING!**
> If breathing becomes difficult, if you become dizzy or nauseated, if you smell or taste the chemical, or if you have other noticeable effects of exposure, leave the area immediately. Get to an area with fresh air and seek any necessary assistance. You must have at least one other worker with you when working in a confined space. You will learn more about working in confined spaces later in this module.

Did You Know?

The United Nations Globally Harmonized System of Classification and Labeling of Chemicals (GHS) is an international standard for safety data sheets (SDSs). Its purpose is to define and communicate the health, physical, and environmental hazards of chemicals as well as the necessary protective measures on labels and SDSs.

Respirator fit testing—Respirators are useless unless properly fit tested to each wearer. To obtain the best protection from your respirator, you must perform positive and negative fit checks each time you wear it. Repeat these fit checks until you have obtained a good face seal.

Use the following steps to perform a positive fit check:

Step 1 Adjust the facepiece for the best fit, then adjust the head and neck straps to ensure good fit and comfort.

Step 2 Block the exhalation valve with your hand or other material.

Step 3 Breathe out into the mask.

Step 4 Check for air leakage around the edges of the facepiece.

Step 5 If the facepiece puffs out slightly for a few seconds, you have a good face seal.

Perform a negative fit check using the following steps:

Step 1 Block the inhalation valve with your hand or other material.

Step 2 Attempt to inhale.

Step 3 Check for air leakage around the edges of the facepiece.

Step 4 If the facepiece caves in slightly for a few seconds, you have a good face seal.

Ensure that your respirator is clean and in good condition and that all of its parts are in place. Otherwise, it will not protect you. Respirators must be cleaned every day. Failure to do so will limit their effectiveness, and they will offer little or no protection. For example, suppose you wore a respirator for two weeks and did not clean it. The bacteria that accumulated as you breathed into the respirator, plus the airborne contaminants that managed to enter the facepiece, would make the inside of your respirator very unsanitary. Continued use could do you more harm than good. Remember, only a clean and complete respirator will give you the necessary protection. Use the following general guidelines for taking care of your respirator:

- Inspect your respirator before and after each use.
- Do not wear a respirator if the facepiece is distorted or if it is worn and cracked. You will not be able to get a proper face seal.
- Do not wear a respirator if any part is missing. Replace worn straps or missing parts before use.
- Do not expose respirators to excessive heat or cold, chemicals, or sunlight.
- Clean and wash your respirator each day.
- Sanitize your respirator each week.

Clean your respirator daily. OSHA's Respirator Cleaning Procedures (Mandatory) are located in *Appendix B* of *CFR 1910.134*. Cleaning your respirator is an easy task that helps to ensure its safe use. Briefly, the procedures include the following steps:

Step 1 Remove filters, cartridges, or canisters. Disassemble facepieces and discard or repair any defective parts.

Step 2 Wash components in warm water with a mild detergent. Use a stiff bristle (not wire) brush to remove dirt.

Step 3 Rinse components thoroughly in clean, warm water. Drain.

Step 4 If your cleaner does not contain a disinfecting agent, immerse components for two minutes in one of the following: hypochlorite solution, aqueous solution of iodine, or other commercially available cleansers of equivalent disinfectant quality.

Step 5 Rinse components thoroughly in clean, warm water.

Step 6 Air-dry components or hand-dry them with a clean, lint-free cloth.

Step 7 Reassemble facepiece, replacing filters, cartridges, and canisters where necessary.

Step 8 Test the respirator to ensure that all components work properly.

2.1.8 *Proper Clothing and Grooming*

Except for PPE, OSHA has not set a standard for proper workplace clothing and personal grooming. However, proper clothing and good grooming can be as important to your safety as PPE. Treat your work clothing the same way you treat your PPE: inspect it often for signs of wear that could cause problems and keep it clean.

Avoid wearing clothing that is too tight or too loose. Tight clothing can restrict your ability to move quickly; loose clothing can catch in machinery or equipment. Make sure that your clothing does not have flaps, strings, or ragged edges that could be tangled in power tools, or tools that require a power source to operate. Don't wear pants that drag on the floor or shirts or jackets with sleeves that are too long. Leg chaps and kneepads can also provide increased protection to your legs. When welding, you must not wear pants with cuffs into which bits of molten metal or sparks may fall.

Do not wear jewelry, which can catch in machinery. If your hair or beard is long, be sure to keep it out of your way while working. Long hair falling into your eyes can affect your ability to see clearly. Like jewelry, long hair can get caught in machinery. Tie back, wrap, or cover long hair. Dressing and grooming in a way that is appropriate for your job will not only help to keep you safe but will also reflect well on you as an employee.

Beards and Respirators

If you wear a beard, you may not be able to seal a respirator facepiece properly. There are, however, loose-fitting respirators that are available for routine or emergency use. According to OSHA, these respirators can be used by bearded workers because facial hair does not interfere with the facepiece seal of these units. Your employer may determine whether the respirator is acceptable for your use.

2.2.0 Hazard Communication

Exposure to hazardous chemicals and materials, such as sealants, asbestos, cleansers, and compressed gas, can cause both environmental and health problems. Health problems can include skin or eye irritation, breathing difficulty, allergic reactions, and cancer. Environmental hazards can include fire, corrosion, and reactivity (explosions).

The types of hazardous materials on a site can range from chemicals to radiation. Sewage is also considered hazardous. Radiation is probably the least thought-about hazard because it cannot be seen or tasted. Radiation is present during radiographic testing of welds in piping, vessels, medical equipment, or pumps.

Chemicals present a significant danger because they exist in many different forms. Chemicals are not only liquids. They can also be solids, gases, fumes, and mists. Many common products contain several chemicals. For example, some paints contain cadmium and lead. Many chemicals pose health hazards, such as disease or burns. Others pose physical hazards, including fire or explosion. Some pose both health and physical hazards. A strong HazCom program that includes SDSs will help you identify the hazards and understand how to protect yourself. You have the right to know the hazards of all the chemicals you will be exposed to on the job.

2.2.1 Right to Know

OSHA directs employers to tell workers about hazardous materials on the job site through HazCom. You may have heard it called the worker right-to-know program. Everyone on site must be educated about the hazardous materials they might use on the job. This is done through a written HazCom program and training. In addition, all materials must have proper labels and SDSs. Labels and SDSs provide information about health hazards, safety precautions, and emergency responses. You need to understand this information in order to protect yourself. The final responsibility for your safety rests with you. Plumbers have the following responsibilities when it comes to HazCom:

- Learn to recognize hazardous materials labels.
- Know where SDSs are kept on your job site.
- Report any hazards you spot on the job site to your supervisor.
- Know the physical and health hazards of the materials you use.

- Know how to protect yourself from hazards.
- Know what to do in an emergency.
- Understand your employer's HazCom program.

2.2.2 Labels

On a construction site, all materials in containers must have a label. Labels describe what is in a container. They also warn you of chemical hazards.

The HazCom Standard states that hazardous material containers must be labeled, tagged, or marked. The label must include the name of the material, the appropriate hazard warnings, and the name and address of the manufacturer.

OSHA does not require specific labels. Label information can be any type of message using words, pictures, or symbols. However, labels must describe the hazards present. Labels must also be readable and easy to see.

Common HazCom labels come from the National Fire Protection Association (NFPA), the National Paint and Coatings Association (NPCA), and the US Department of Transportation (DOT). The NFPA's hazardous materials classification system is often referred to as the NFPA warning diamond (*Figure 16*). The four-color diamond can be a container label. It is also used on doors to note the hazards in a room or building to aid firefighters and emergency responders. Each section and color represents a hazard: health, flammability, stability, and specific hazards. Numbers from zero to four indicate increasing hazards.

The ACA's Hazardous Materials Information System (HMIS®) (*Figure 17*), like the NFPA warning diamond, uses a color coding system to identify the hazards associated with a particular product and assigns numbers to each section to reflect the degree of hazard. HMIS® is meant for alerting employees to potential hazards.

The DOT requires labels for hazardous materials shipments (*Figure 17*). Either of these labels can be part of your company's HazCom labeling program. You may see these labels at your job site.

Hazardous materials at the site must be properly labeled. If the material is transferred from a labeled container, the new container must be labeled with all of the information from the original label. Make sure that any materials you work with are labeled. Be sure that you understand your company's labels.

2.2.3 Safety Data Sheets

Each product on a construction site must have an SDS, a facts sheet prepared by the product's manufacturer or importer. An SDS describes the substance and its hazards, safe handling requirements, first-aid needs, and emergency spill procedures. OSHA does not have a mandatory form, but it does require inclusion of specific information. The Chemical Manufacturers Association has developed a standard form that meets national and international standards. Most chemical manufacturers use this form. The sections of the form include the following:

- Chemical product and company information
- Composition/information on ingredients
- Hazard identification
- First-aid measures
- Firefighting information
- Accidental release measures

- Handling and storage
- Exposure controls/personal protection
- Physical and chemical properties
- Stability and reactivity
- Toxicological properties
- Ecological properties
- Disposal considerations
- Transportation information
- Regulatory information
- Other information related to the chemical

An SDS can be difficult to read. The scientific information is fairly technical. *Figure 18A* and *Figure 18B* show an SDS for a PVC cement. Look for the following important information on an SDS:

- Specific hazards
- Personal protection requirements
- Handling procedures
- First-aid information
- The 24-hour emergency-response telephone number

Using *Figure 18A* and *Figure 18B*, try to find the information you would need to use the chemical described on the SDS. First locate the hazards. Section 2 of the SDS shows that this cement poses a hazard to the skin and eyes. Next, find out how to minimize these hazards. Section 8 explains the appropriate eye, skin, and respiratory protection to use. To find out how to handle and store this adhesive, check Section 7. Now you have the facts you need to protect your health and that of your co-workers.

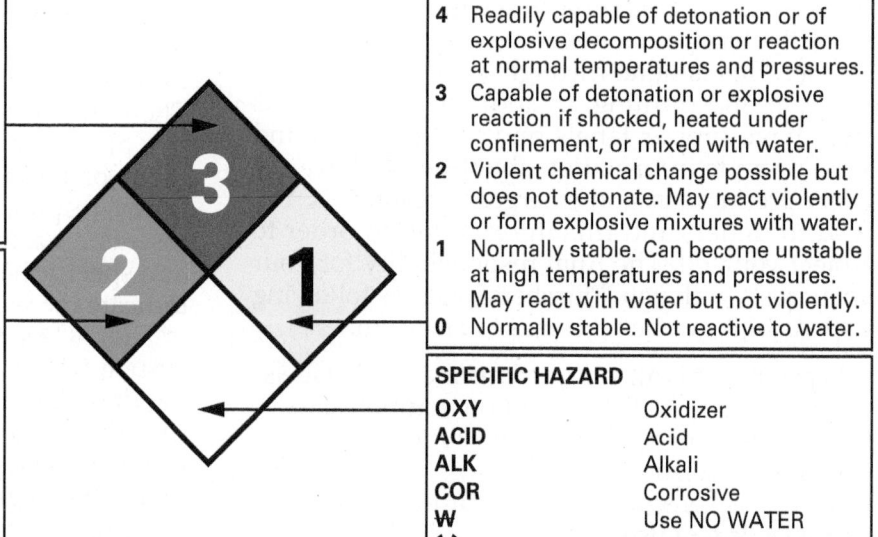

FLAMMABILITY HAZARD

4 Extremely flammable – Rapidly vaporizes at normal pressure and temperature or is readily dispersed in air and will burn readily.
3 Flammable – Ignites at normal temperatures.
2 Ignites when moderately heated.
1 Ignites when preheated.
0 Will not burn.

HEALTH HAZARD

4 Extreme – Fatal with very short exposure. Wear special full protective suit and breathing apparatus.
3 Serious – Serious injury with short exposure. Wear full protective suit and breathing apparatus.
2 Moderate – Continued exposure can cause injury. Use breathing apparatus.
1 Slight – Exposure can cause irritation. Breathing apparatus may be worn.
0 Normal – No hazard.

INSTABILITY HAZARD

4 Readily capable of detonation or of explosive decomposition or reaction at normal temperatures and pressures.
3 Capable of detonation or explosive reaction if shocked, heated under confinement, or mixed with water.
2 Violent chemical change possible but does not detonate. May react violently or form explosive mixtures with water.
1 Normally stable. Can become unstable at high temperatures and pressures. May react with water but not violently.
0 Normally stable. Not reactive to water.

SPECIFIC HAZARD

OXY	Oxidizer
ACID	Acid
ALK	Alkali
COR	Corrosive
W	Use NO WATER
☢	Radioactive

Figure 16 NFPA warning diamond.

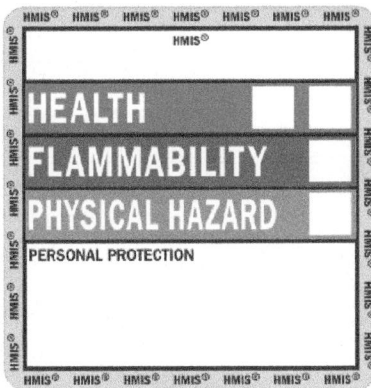

NPCA HAZARDOUS MATERIALS
INFORMATION SYSTEM (HMIS®)

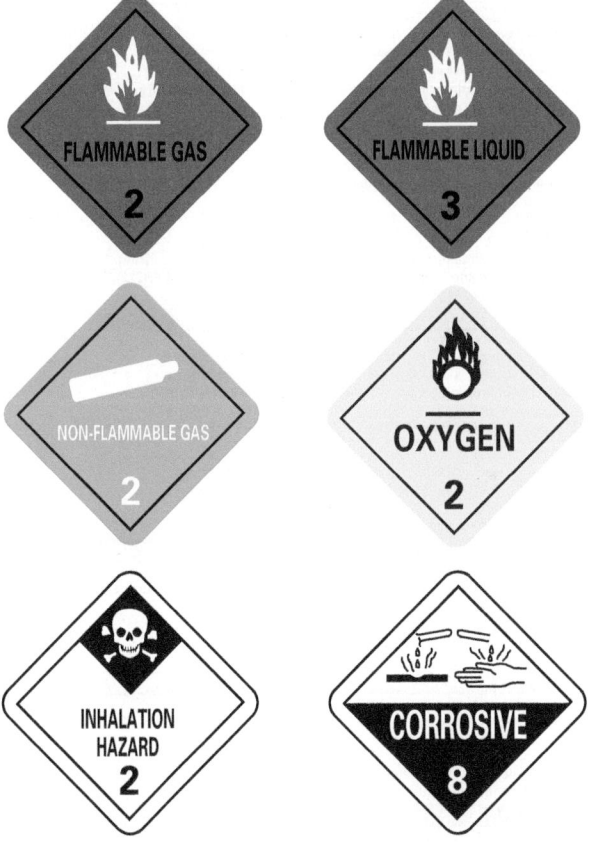

DOT HAZARDOUS MATERIALS LABELS

Figure 17 HazCom labels.

Section 4 lists the first-aid measures for inhalation, eye contact, skin contact, and ingestion. Section 5 explains fire hazards and firefighting measures. Section 6 provides accidental release measures. You will find the emergency telephone number in Section 1. Now you have the information you need in case of an emergency. All SDSs must be kept on site. Ask your supervisor to tell you where the SDSs are and to point out the sections that relate to your job. Your health and safety and that of your co-workers depend on it.

2.2.4 *Responding to Emergencies*

Every job site should have an emergency-response plan. Planning should be coordinated well and communicated to everyone involved. If you are told by your supervisor to evacuate a work site, go to a safe location and wait until you are notified that conditions are safe. Being prepared can reduce the severity of an emergency. Planning is especially important if the accident happens in a remote area that does not have a telephone. Be aware of the location of nearby eye wash stations, including portable stations in vehicles. Eye wash stations are marked with signs (*Figure 19*).

Make sure you know your employer's emergency plan. Find out whom to call in an emergency and what you need to do to protect yourself. Know where your emergency hospital is and where you and your co-workers should go in an emergency. Response plans will vary from company to company. At a minimum, all plans should include the name of the nearest emergency hospital, safety meeting points, and an emergency point of contact. If your company does not have a plan, see to it that one is made.

2.3.0 Work Zones

Plumbing work is often done at construction sites in or near public areas. Creating a clear work zone is an important part of working safely. Barricades, fencing, caution tape, signs, and cones mark a construction work zone. To ensure everyone's safety, you must keep the public and their vehicles away from your work area.

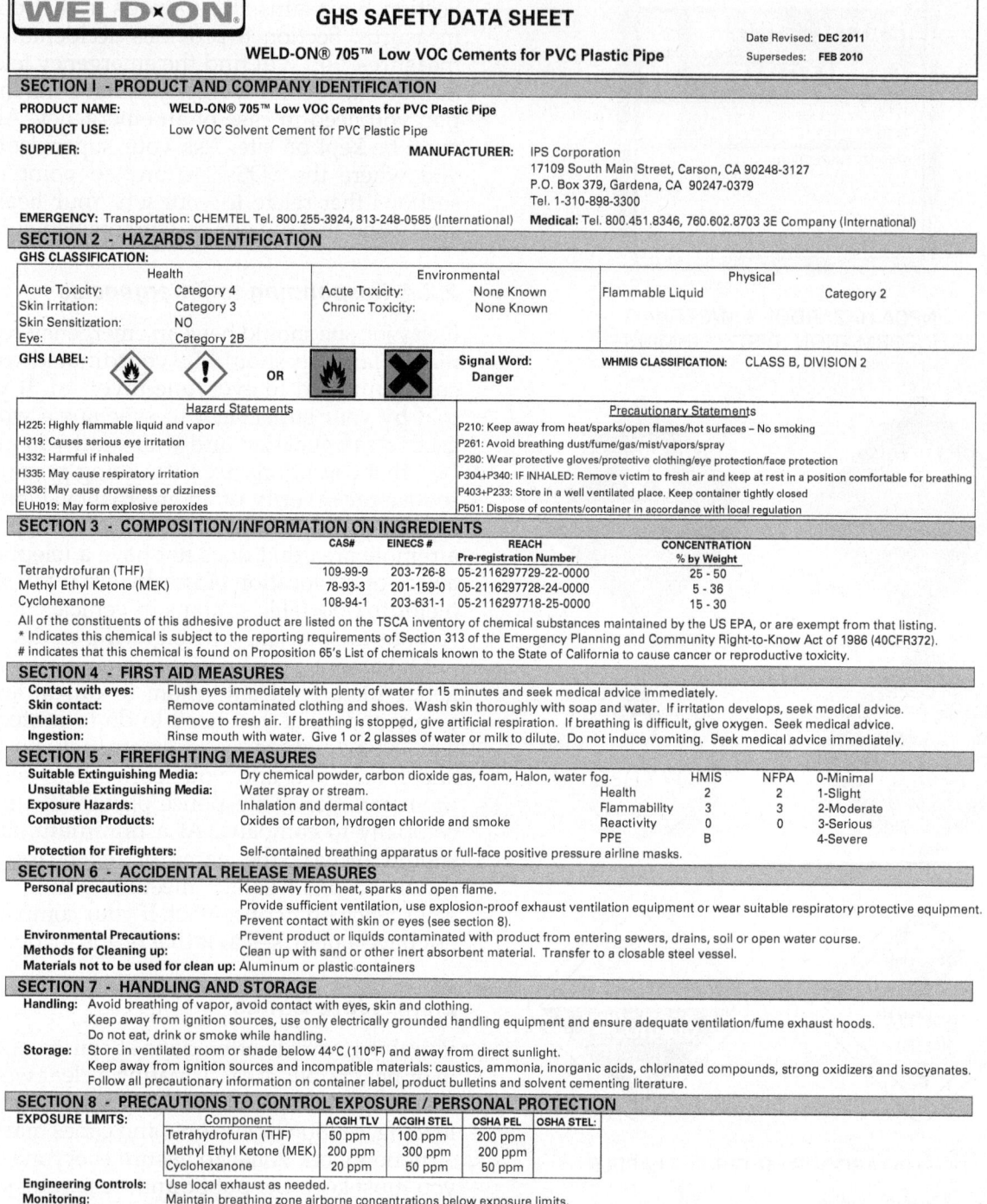

GHS SAFETY DATA SHEET

WELD-ON

WELD-ON® 705™ Low VOC Cements for PVC Plastic Pipe

Date Revised: **DEC 2011**
Supersedes: **FEB 2010**

SECTION I - PRODUCT AND COMPANY IDENTIFICATION

PRODUCT NAME: WELD-ON® 705™ Low VOC Cements for PVC Plastic Pipe
PRODUCT USE: Low VOC Solvent Cement for PVC Plastic Pipe
SUPPLIER:

MANUFACTURER: IPS Corporation
17109 South Main Street, Carson, CA 90248-3127
P.O. Box 379, Gardena, CA 90247-0379
Tel. 1-310-898-3300

EMERGENCY: Transportation: CHEMTEL Tel. 800.255.3924, 813-248-0585 (International) **Medical:** Tel. 800.451.8346, 760.602.8703 3E Company (International)

SECTION 2 - HAZARDS IDENTIFICATION

GHS CLASSIFICATION:

Health		Environmental		Physical	
Acute Toxicity:	Category 4	Acute Toxicity:	None Known	Flammable Liquid	Category 2
Skin Irritation:	Category 3	Chronic Toxicity:	None Known		
Skin Sensitization:	NO				
Eye:	Category 2B				

GHS LABEL: OR

Signal Word: Danger

WHMIS CLASSIFICATION: CLASS B, DIVISION 2

Hazard Statements	Precautionary Statements
H225: Highly flammable liquid and vapor	P210: Keep away from heat/sparks/open flames/hot surfaces – No smoking
H319: Causes serious eye irritation	P261: Avoid breathing dust/fume/gas/mist/vapors/spray
H332: Harmful if inhaled	P280: Wear protective gloves/protective clothing/eye protection/face protection
H335: May cause respiratory irritation	P304+P340: IF INHALED: Remove victim to fresh air and keep at rest in a position comfortable for breathing
H336: May cause drowsiness or dizziness	P403+P233: Store in a well ventilated place. Keep container tightly closed
EUH019: May form explosive peroxides	P501: Dispose of contents/container in accordance with local regulation

SECTION 3 - COMPOSITION/INFORMATION ON INGREDIENTS

	CAS#	EINECS #	REACH Pre-registration Number	CONCENTRATION % by Weight
Tetrahydrofuran (THF)	109-99-9	203-726-8	05-2116297729-22-0000	25 - 50
Methyl Ethyl Ketone (MEK)	78-93-3	201-159-0	05-2116297728-24-0000	5 - 36
Cyclohexanone	108-94-1	203-631-1	05-2116297718-25-0000	15 - 30

All of the constituents of this adhesive product are listed on the TSCA inventory of chemical substances maintained by the US EPA, or are exempt from that listing.
* Indicates this chemical is subject to the reporting requirements of Section 313 of the Emergency Planning and Community Right-to-Know Act of 1986 (40CFR372).
\# indicates that this chemical is found on Proposition 65's List of chemicals known to the State of California to cause cancer or reproductive toxicity.

SECTION 4 - FIRST AID MEASURES

Contact with eyes: Flush eyes immediately with plenty of water for 15 minutes and seek medical advice immediately.
Skin contact: Remove contaminated clothing and shoes. Wash skin thoroughly with soap and water. If irritation develops, seek medical advice.
Inhalation: Remove to fresh air. If breathing is stopped, give artificial respiration. If breathing is difficult, give oxygen. Seek medical advice.
Ingestion: Rinse mouth with water. Give 1 or 2 glasses of water or milk to dilute. Do not induce vomiting. Seek medical advice immediately.

SECTION 5 - FIREFIGHTING MEASURES

			HMIS	NFPA	
Suitable Extinguishing Media:	Dry chemical powder, carbon dioxide gas, foam, Halon, water fog.				0-Minimal
Unsuitable Extinguishing Media:	Water spray or stream.	Health	2	2	1-Slight
Exposure Hazards:	Inhalation and dermal contact	Flammability	3	3	2-Moderate
Combustion Products:	Oxides of carbon, hydrogen chloride and smoke	Reactivity	0	0	3-Serious
		PPE	B		4-Severe
Protection for Firefighters:	Self-contained breathing apparatus or full-face positive pressure airline masks.				

SECTION 6 - ACCIDENTAL RELEASE MEASURES

Personal precautions: Keep away from heat, sparks and open flame.
Provide sufficient ventilation, use explosion-proof exhaust ventilation equipment or wear suitable respiratory protective equipment. Prevent contact with skin or eyes (see section 8).
Environmental Precautions: Prevent product or liquids contaminated with product from entering sewers, drains, soil or open water course.
Methods for Cleaning up: Clean up with sand or other inert absorbent material. Transfer to a closable steel vessel.
Materials not to be used for clean up: Aluminum or plastic containers

SECTION 7 - HANDLING AND STORAGE

Handling: Avoid breathing of vapor, avoid contact with eyes, skin and clothing.
Keep away from ignition sources, use only electrically grounded handling equipment and ensure adequate ventilation/fume exhaust hoods.
Do not eat, drink or smoke while handling.
Storage: Store in ventilated room or shade below 44°C (110°F) and away from direct sunlight.
Keep away from ignition sources and incompatible materials: caustics, ammonia, inorganic acids, chlorinated compounds, strong oxidizers and isocyanates.
Follow all precautionary information on container label, product bulletins and solvent cementing literature.

SECTION 8 - PRECAUTIONS TO CONTROL EXPOSURE / PERSONAL PROTECTION

EXPOSURE LIMITS:

Component	ACGIH TLV	ACGIH STEL	OSHA PEL	OSHA STEL:
Tetrahydrofuran (THF)	50 ppm	100 ppm	200 ppm	
Methyl Ethyl Ketone (MEK)	200 ppm	300 ppm	200 ppm	
Cyclohexanone	20 ppm	50 ppm	50 ppm	

Engineering Controls: Use local exhaust as needed.
Monitoring: Maintain breathing zone airborne concentrations below exposure limits.
Personal Protective Equipment (PPE):
Eye Protection: Avoid contact with eyes, wear splash-proof chemical goggles, face shield, safety glasses (spectacles) with brow guards and side shields, etc. as may be appropriate for the exposure.
Skin Protection: Prevent contact with the skin as much as possible. Butyl rubber gloves should be used for frequent immersion.
Use of solvent-resistant gloves or solvent-resistant barrier cream should provide adequate protection when normal adhesive application practices and procedures are used for making structural bonds.
Respiratory Protection: Prevent inhalation of the solvents. Use in a well-ventilated room. Open doors and/or windows to ensure airflow and air changes. Use local exhaust ventilation to remove airborne contaminants from employee breathing zone and to keep contaminants below levels listed above. With normal use, the Exposure Limit Value will not

Figure 18A Typical SDS.

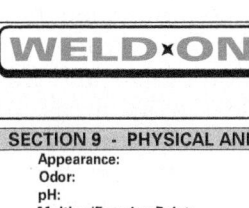

GHS SAFETY DATA SHEET

WELD-ON® 705™ Low VOC Cements for PVC Plastic Pipe

Date Revised: **DEC 2011**
Supersedes: **FEB 2010**

SECTION 9 · PHYSICAL AND CHEMICAL PROPERTIES

Appearance:	Clear or gray, medium syrupy liquid		
Odor:	Ketone	**Odor Threshold:**	0.88 ppm (Cyclohexanone)
pH:	Not Applicable		
Melting/Freezing Point:	-108.5°C (-163.3°F) Based on first melting component: THF	**Boiling Range:**	66°C (151°F) to 156°C (313°F)
Boiling Point:	66℃ (151°F) Based on first boiling component: THF	**Evaporation Rate:**	> 1.0 (BUAC = 1)
Flash Point:	-20℃ (-4°F) TCC based on THF	**Flammability:**	Category 2
Specific Gravity:	0.9611 @23°C (73°F)	**Flammability Limits:**	**LEL:** 1.1% based on Cyclohexanone
Solubility:	Solvent portion soluble in water. Resin portion separates out.		**UEL:** 11.8% based on THF
Partition Coefficient n-octanol/water:	Not Available	**Vapor Pressure:**	129 mm Hg @ 20°C (68°F)based on THF
Auto-ignition Temperature:	321°C (610°F) based on THF	**Vapor Density:**	>2 (Air = 1)
Decomposition Temperature:	Not Applicable	**Other Data: Viscosity:**	Medium bodied
VOC Content:	When applied as directed, per SCAQMD Rule 1168, Test Method 316A,VOC content is: < 510 g/l.		

SECTION 10 · STABILITY AND REACTIVITY

Stability:	Stable
Hazardous decomposition products:	None in normal use. When forced to burn, this product gives off oxides of carbon, hydrogen chloride and smoke.
Conditions to avoid:	Keep away from heat, sparks, open flame and other ignition sources.
Incompatible Materials:	Oxidizers, strong acids and bases, amines, ammonia

SECTION 11 · TOXICOLOGICAL INFORMATION

Likely Routes of Exposure: Inhalation, Eye and Skin Contact

Acute symptoms and effects:

Inhalation:	Severe overexposure may result in nausea, dizziness, headache. Can cause drowsiness, irritation of eyes and nasal passages.
Eye Contact:	Vapors slightly uncomfortable. Overexposure may result in severe eye injury with corneal or conjunctival inflammation on contact with the liquid.
Skin Contact:	Liquid contact may remove natural skin oils resulting in skin irritation. Dermatitis may occur with prolonged contact.
Ingestion:	May cause nausea, vomiting, diarrhea and mental sluggishness.

Chronic (long-term) effects: None known to humans

Toxicity:

	LD_{50}	LC_{50}
Tetrahydrofuran (THF)	Oral: 2842 mg/kg (rat)	Inhalation 3 hrs. 21,000 mg/m³ (rat)
Methyl Ethyl Ketone (MEK)	Oral: 2737 mg/kg (rat), Dermal: 6480 mg/kg (rabbit)	Inhalation 8 hrs. 23,500 mg/m³ (rat)
Cyclohexanone	Oral: 1535 mg/kg (rat), Dermal: 948 mg/kg (rabbit)	Inhalation 4 hrs. 8,000 PPM (rat)

Reproductive Effects	Teratogenicity	Mutagenicity	Embryotoxicity	Sensitization to Product	Synergistic Products
Not Established	Not Established	Not Established	Not Established	Not Established	Not Established

SECTION 12 · ECOLOGICAL INFORMATION

Ecotoxicity:	None Known
Mobility:	In normal use, emission of volatile organic compounds (VOC's) to the air takes place, typically at a rate of ≤ 510 g/l.
Degradability:	Biodegradable
Bioaccumulation:	Minimal to none.

SECTION 13 · WASTE DISPOSAL CONSIDERATIONS

Follow local and national regulations. Consult disposal expert.

SECTION 14 · TRANSPORT INFORMATION

Proper Shipping Name:	Adhesives
Hazard Class:	3
Secondary Risk:	None
Identification Number:	UN 1133
Packing Group:	PG II
Label Required:	Class 3 Flammable Liquid
Marine Pollutant:	NO

EXCEPTION for Ground Shipping
DOT Limited Quantity: Up to 5L per inner packaging, 30 kg gross weight per package.
Consumer Commodity: Depending on packaging, these quantities may qualify under DOT as "ORM-D" .

TDG INFORMATION
TDG CLASS:	FLAMMABLE LIQUID 3
SHIPPING NAME:	ADHESIVES
UN NUMBER/PACKING GROUP:	UN 1133, PG II

SECTION 15 · REGULATORY INFORMATION

Precautionary Label Information:	Highly Flammable, Irritant
Symbols:	F, Xi

Ingredient Listings: USA TSCA, Europe EINECS, Canada DSL, Australia AICS, Korea ECL/TCCL, Japan MITI (ENCS)

Risk Phrases:
R11: Highly flammable.
R20: Harmful by inhalation.
R36/37: Irritating to eyes and respiratory system.
R66: Repeated exposure may cause skin dryness or cr
R67: Vapors may cause drowsiness and dizziness

Safety Phrases:
S9: Keep container in a well-ventilated place.
S16: Keep away from sources of ignition - No smoking.
S25: Avoid contact with eyes.
S26: In case of contact with eyes, rinse immediately with plenty of water and seek medical advice.
S33: Take precautionary measures against static discharges.
S46: If swallowed, seek medical advise immediately and show this container or label.

SECTION 16 · OTHER INFORMATION

Specification Information:
Department issuing data sheet:	IPS, Safety Health & Environmental Affairs
E-mail address:	<EHSinfo@ipscorp.com>
Training necessary:	Yes, training in practices and procedures contained in product literature.
Reissue date / reason for reissue:	12/14/2011 / Updated GHS Standard Format
Intended Use of Product:	Solvent Cement for PVC Plastic Pipe

All ingredients are compliant with the requirements of the European Directive on RoHS (Restriction of Hazardous Substances).

This product is intended for use by skilled individuals at their own risk. The information contained herein is based on data considered accurate based on current state of knowledge and experience. However, no warranty is expressed or implied regarding the accuracy of this data or the results to be obtained from the use thereof.

Figure 18B Typical SDS.

Figure 19 Eye wash station sign.

Signs, tags, and color codes in the workplace protect workers from hazardous conditions and help them respond to emergencies (*Figure 20*). For signs, tags, and color codes to be effective, all workers must understand what they mean and know what action they are required to take. This reduces confusion and ensures their effectiveness.

Emergency Preparedness

Your company expects you to be prepared in case of an emergency. Therefore, you must read and understand the company's emergency-response plan. Many companies have guidelines for employees to follow during an emergency. These guidelines are designed to prevent panic and to protect people, property, and the environment:

- Stay calm, and quickly evaluate the situation.
- Notify affected personnel.
- Follow company safety procedures to protect personnel, property, and the environment.
- Submit any required reports.

OSHA requires reports for certain emergency situations. These requirements are detailed in *29 CFR 1904*. Although you may not be the person submitting the report, you may be asked to provide information to the reporting authority.

Signals, such as alarms, bells, buzzers, whistles, and horns, also communicate hazards to workers. For example, backup alarms are used on forklifts, construction equipment, and trucks. Fire alarms are used to clear work areas. Conveyer belt lines have buzzers, bells, or whistles to let workers know they are about to be started.

Barricades are another way to warn of danger. They are used on construction sites to keep out unauthorized personnel and control traffic (*Figure 21*). Barricades are also used to control pedestrian traffic outside, or in rooms or hallways that have been recently washed or waxed.

It's important to recognize all of the signs and signals on your job site and make sure they are placed properly and working correctly. Doing this can save a life.

2.3.1 Signs

All work sites have specific markings and signs to identify hazards. Signs can also provide emergency information (*Figure 22*). Common signs on a site include danger signs, caution signs, informational signs, and safety instruction signs or tags.

Danger signs—Danger signs are usually red, black, and white. They tell workers that an immediate hazard exists, such as high voltage or flammable materials. Danger signs also have specific precautions that must be observed to avoid an accident (*Figure 23*). Examples of danger signs include No Smoking and Keep Out.

Caution signs—Caution signs are yellow with black letters. Caution signs warn workers about potential hazards or unsafe practices (*Figure 24*). When you see a caution sign, take action to protect yourself. Common examples include Do Not Operate, Keep Aisles Clear, and Electric Fence.

Yellow is the basic color used for caution. It identifies places where physical hazards may be caused by striking against objects; stumbling, falling, tripping; or being caught between obstacles. Solid yellow, yellow and black stripes, or yellow and black checkers caution workers against these hazards.

Caution signs for piping systems that contain dangerous materials are also yellow. Yellow warns workers against starting machinery that is under repair. Painted barriers and flags should be at the starting point of the work area or the power source. Signs and barriers should be displayed so that workers notice them easily on such things as electrical controls, ladders, scaffolds, vaults, valves, dryers, boilers, elevators, and tanks.

INFORMATION SIGN

SAFETY SIGN

CAUTION SIGN

DANGER SIGN

Figure 20 Work zone signs and tags.

Figure 21 Typical uses of barricades.

Figure 22 Work zone signs.

Figure 23 Typical danger sign.

Figure 24 Typical caution sign.

Informational signs—Informational signs provide general information that is not related to safety (*Figure 25*). The standard color is blue. The background, the entire sign, or just a panel may be blue. Common examples include No Admittance, No Trespassing, and Employees Only.

Informational signs can also be black and white. These signs are used as traffic and house-keeping markers. They identify information such as the following:

- Dead ends of aisles or passageways
- Location of trash cans
- Location and width of aisles
- Rooms or passageways
- Stairways (risers, direction, borders)
- Drinking fountains and food-dispensing machines

Safety instruction signs—Safety instruction signs are used for general instructions and suggestions related to safety measures (*Figure 26*). The background and lettering on these signs are often white and green, but they can vary depending on the message and the location of the sign. Any letters used against the white background are black. Common examples include Report All Unsafe Conditions To Your Supervisor; Walk, Don't Run; and Help Keep This Plant Safe and Clean.

Safety tags—Safety tags are used as a temporary warning about immediate and potential hazards (*Figure 27*). They are similar to signs, but they are not designed to be used in place of signs or as a permanent means of protection. For example, an Out of Order tag may be used on damaged equipment until it can be disposed of or repaired.

Figure 25 Common informational sign.

Figure 26 Common safety sign.

A Do Not Start tag may be placed on machinery during lockout/tagout procedures, which identify hazardous equipment and prevent their use until they have been certified as safe. Tags and the devices used to attach them must meet specific requirements to ensure their durability and effectiveness.

2.3.2 Signals

Signals are used to inform workers of potential dangers. Types of signals include alarms, bells, buzzers, whistles, horns, and hand signals. Hand signals control vehicle traffic, guide the handling of materials, and assist equipment operators. All affected workers must know what each hand signal means before it is used. The meaning should be confirmed between the equipment operator and spotter or person giving the operator the signals before a task is started.

Figure 27 Examples of safety tags.

2.3.3 Barricades and Barriers

Any opening in a wall, floor, or the ground is a safety hazard. There are two types of protection for these openings: guarded or covered. You should cover a hole whenever possible, but when it is not practical to cover a hole, use barricades. If the bottom edge of a wall opening is less than 3 feet above the floor and would allow someone to fall 4 feet or more, place guards around the opening. For instance, you would place guards around a window opening several stories up to prevent a worker from sitting on the opening and falling out the window. The two most commonly used guard methods are railings and warning barricades. Typical warning barricades are made of plastic tape or rope strung from wire or between posts. The tape or rope is color-coded red, yellow, or yellow and purple.

Red means danger. No one may enter an area with a red warning barricade. A red barricade is used when there is danger from falling objects or when a load is suspended over an area.

Yellow means caution. You may enter an area with a yellow barricade, but you must know what the hazard is and be careful. Yellow barricades are used around wet areas or areas containing loose dust. Yellow with black lettering warns of physical hazards, such as bumping into something, stumbling, or falling.

Yellow and purple indicates a radiation warning. No one may pass a yellow and purple barricade. These barricades are often used where piping welds are being X-rayed.

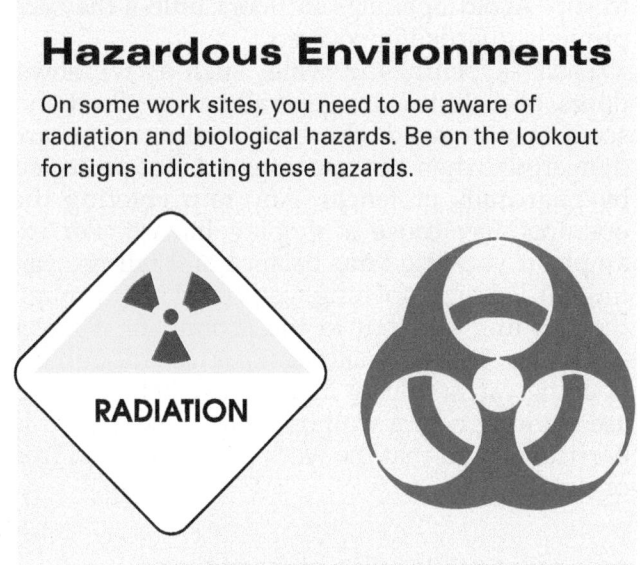

Hazardous Environments

On some work sites, you need to be aware of radiation and biological hazards. Be on the lookout for signs indicating these hazards.

Protective barricades provide both a visual warning and protection from injury. They can be posts, rails, chain, or cable. People cannot get past protective barricades. Blinking lights are placed on barricades so they can be seen at night.

2.3.4 Walking and Working Surfaces

Slips, trips, and falls on walking and working surfaces cause 15 percent of all accidental deaths in the construction industry. Some accidents occur because of environmental conditions, such as snow, ice, or wet surfaces. Others happen because of poor housekeeping and careless behavior, such as leaving tools, materials, and equipment out and unattended. Such accidents can be avoided if workers are aware of their surroundings and follow the rules on the site. It is your responsibility to keep all walking and working surfaces clean and dry.

Floors—Floors can be a hazard in a number of ways. Ice, grease, oil, or wet processes can make them slippery. Tools, equipment, materials, or litter can clutter them. Unguarded openings in the floor or ground can cause fatal falls. Some of the types of openings to be aware of on a work site include stairs, hatches, chutes, trapdoors, and maintenance access holes (manholes). You can avoid slips, trips, and falls on floors by making sure the surface is free of ice, snow, moisture, and clutter. If you cannot remove the ice, wear shoes with skid-resistant cleats. If wet processes are used in the work area, make sure there is proper drainage, grating, and mats. Keep the floor clear of tools, equipment, materials, or litter that you could trip over or would cause you to slip. Avoid openings in floors unless they are properly guarded or covered.

Walls—Openings in walls, such as windows, doors, and chutes, are generally at eye level and seem easy to avoid. However, wall openings are dangerous when the openings are not protected by guardrails or fences. Any rain entering the opening may cause a slipping hazard. For example, if you lose your balance and fall near an unguarded wall opening, you could slip through the opening and fall to the ground or a lower work area. Tools or materials that fall through the opening can seriously injure those below. If you are working near a wall opening, make sure it is barricaded and that the work area is dry and free of clutter.

Platforms—Platforms are work areas elevated above the floor or ground. Platforms may be located above dangerous equipment, such as galvanizing tanks or degreasing units. Many platforms do not have guardrails. These platforms, called open-sided platforms, are hazardous because they do not protect workers from falling over the edge. To help prevent accidents, make sure platforms are dry and clear of materials and debris before stepping onto them.

Ramps and runways—The hazards of ramps and runways are similar to those of floors. Workers can trip on tools and equipment or slip on wet or icy surfaces. Ramps and runways can be more dangerous, however, because they are sloped. When slips, trips, and falls happen at a downhill angle, the worker slides or rolls down the ramp or runway more quickly and hits the ground harder. The resulting injuries are often more serious. Imagine if a worker were carrying a tool with a sharp blade or edge during such an accident. It is likely that the worker would be injured from the blade or edge of the tool, as well as the fall. You can avoid slips, trips, and falls on ramps and runways by following these guidelines:

- Check the surface before using it. If the ramp or runway is icy or wet, don't use it until it is dry and free of ice. If the ice cannot be removed, wear shoes with skid-resistant cleats.
- Make sure the ramp is clear of tools, equipment, materials, or debris.
- Make sure any tools or equipment you are carrying are turned off and secured. This will help prevent injuries if you fall.

Stairs—Workers use stairs to travel between levels, in and out of pits, and on and off platforms. Stairs can be wet, icy, or slippery depending on the location. They can also be damaged or cluttered with tools and equipment. All of these conditions can cause workers to slip, trip, or fall. Don't use stairs if you notice any of these conditions or if the stairs do not have a guardrail.

Ladders—The greatest hazard of using a ladder is falling. Workers can fall from the ladder, or the ladder can slip out from under them. Most falls occur when the ground or ladder is wet (*Figure 28*) or icy, or the ladder is not properly secured. Serious injuries or death can result. These safe practices can help prevent slips and falls from ladders:

- Use appropriate fall protection.

- Wear safe, strong work boots that are in good condition.
- Watch where you step. Be sure your footing is secure.
- Maintain clean, smooth walking and working surfaces. Fill holes, ruts, and cracks.
- Clean up slippery material.
- Pick up litter.
- If you must climb to reach something, use a sound ladder that has been safely set up and properly secured at the top and bottom (*Figure 29*).
- When climbing a ladder, always face the ladder and use both hands.
- Always maintain three-point contact with the ladder: one hand and two feet or two hands and one foot.
- Don't overreach from a ladder. Climb down and move the ladder to the desired position.
- Don't walk a ladder.

WARNING!

Metal ladders conduct electricity. Never use metal ladders around electrical equipment. Although wooden ladders do not conduct electricity, you must not use them around electrical equipment if the wood is wet or damp. Even a small amount of water will act as a conductor.

Figure 28 Liquid on a stepladder.

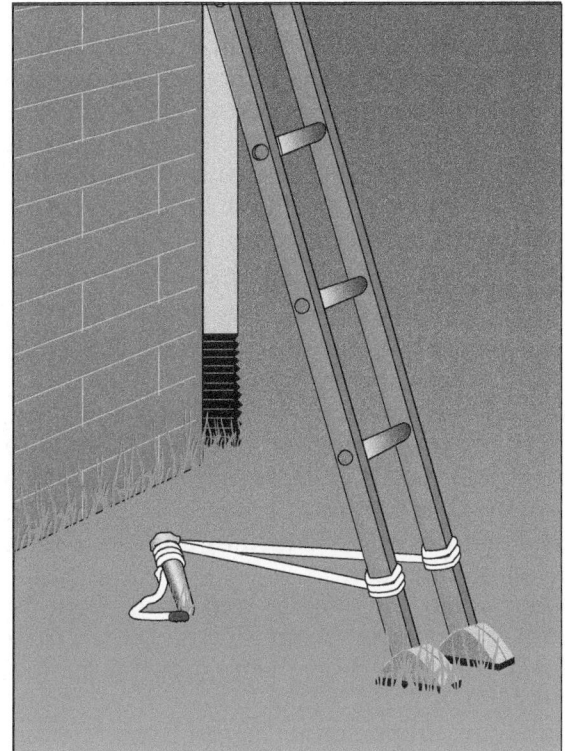

BOTTOM SECURED

TOP SECURED

Figure 29 Properly secured ladders.

Scaffolding—Scaffolding supports workers and materials on elevated platforms. It is generally used on multistory buildings or structures. Weather conditions, poor housekeeping, and carelessness can make working on scaffolding very dangerous. Always refer to *29 CFR* for detailed instructions before setting up scaffolding. Always pay attention and follow these guidelines to avoid injury or death:

- Use appropriate fall protection.
- Keep scaffolding planks clear of extra tools and materials. Clean up any slippery substances that get spilled on scaffolding.
- Anchor freestanding scaffolding with **guy wires** to prevent tipping or sliding. Guy wires are ropes, chains, cables, or rods attached to something as a brace or guide.
- Read the posted safety rules and regulations for scaffolding use.
- When removing objects and tools from the work area, lower them down with a rope. Never throw or drop them from the scaffolding.
- Make sure all personnel are off the scaffolding before moving it.
- Keep tools and materials back from the edge of the scaffolding.
- Use only approved scaffolding.
- Inspect the scaffolding regularly.
- Install all braces and accessories according to the manufacturer's recommendations.
- Ensure that the scaffolding is plumb and level.
- Install all safety rails according to regulations.
- Make sure the wheels are locked before climbing or disassembling rolling scaffolding.
- Do not work on scaffolding during storms or high winds or when the scaffolding is covered with ice or snow.
- Do not work on scaffolding that is wobbly or bouncy or can be pulled down easily; it is not safe. Scaffolding must have a sound structure and include toeboards and handrails. Wood floor planks must be free of knots. Broken members must be repaired immediately.

Roofs—Walking and working on roofs presents two different kinds of risks: falling through an opening, such as a skylight or vent, and falling off the roof. Weather hazards, poor housekeeping, or carelessness can cause both to happen. The following guidelines will help ensure your safety when working on a roof:

- Wear appropriate fall protection, even on shallow-pitch roofs.
- Wear boots or shoes with rubber or crepe soles that are in good condition.

- Rain, frost, and snow are all dangerous because they make a roof slippery. If possible, wait until the roof is dry. Otherwise, wear special roof shoes with skid-resistant cleats in addition to wearing fall protection.
- Brush or sweep the roof periodically to remove any accumulated dirt or debris.
- Remove any unused tools, cords, and other loose items from the roof. They can be a serious hazard.
- Be alert to any other potential hazards, such as live power lines.
- Use common sense. Taking chances can lead to injury or death.

> **WARNING!**
> When working 6 feet or more above ground, you must tie off and wear appropriate fall protection. Failure to do so can result in falls that lead to severe injury or even death.

Always use caution when walking or working on or near these surfaces, especially if there are other hazards in the area.

2.3.5 Motorized Vehicles

Your ability to get and keep a job in the plumbing industry depends partly on your ability to maintain an acceptable driving record. According to experts, motor vehicle accidents are a top issue for plumbing contractors. Auto liability losses account for 43 percent of all contractor insurance losses and 30 percent of the dollars paid to settle these losses. One of your responsibilities is to operate motorized vehicles safely on construction sites and on the road.

Case History

Pay Attention to Warnings

A 21-year-old laborer had been throwing old roofing materials off a roof with six unguarded skylights. During a work break, he sat down on one of the skylights, which began to break under his weight. He tried to raise himself from the skylight with his arms, but the plastic dome collapsed. He was killed when he fell 27 feet to the concrete floor below. The victim had been warned by his supervisor and co-workers not to sit on the skylights.

The Bottom Line: Ignoring safety warnings can be fatal.

Source: National Institute of Occupational Safety and Health

Motorized vehicles include trucks, vans, fork-lifts, backhoes, and cranes. Motorized equipment includes portable equipment, such as generators, compressors, and pumps, and larger equipment, such as earth-moving equipment and personnel platform. A piece of motorized equipment is considered portable if it can be transported from job site to job site or to different areas on the same job site. Whether you are operating one of these vehicles on the construction site, working nearby, or driving a vehicle on the roadways, you must adhere to the safety precautions outlined in this section.

Many motor vehicle accidents can be prevented. It is your responsibility to be aware of the necessary safety measures to protect yourself and your co-workers. The following are some personal safety guidelines:

- Do not work if you are taking a prescription medicine that could impair your motor or thinking skills.
- Never drink alcohol prior to work or during work hours. Never drink alcohol before or when driving.
- Never, under any circumstances, use illegal drugs.
- Buckle up. Always wear your safety belt and ensure that any passengers wear theirs.
- Ensure that all passengers are seated in a firmly secured seat. Do not allow anyone to ride in a truck bed.

Just as you routinely check your PPE, you should also routinely check any vehicle you are responsible for operating. Use the following guidelines to ensure vehicle safety:

- Ensure that safety devices such as the horn, backup alarm, mirrors, and brakes are in good working order. (Brakes include trailer brake connections, parking system or hand brakes, and emergency brakes.)
- Keep windshields, side windows, mirrors, and lights clean and functional. Keep windshield washer wells supplied with cleaning fluid.
- Always shut off the engine before fueling.
- Never run motorized vehicles in an enclosed, unventilated area.
- Ensure that the area behind your vehicle is clear before backing up. Safe backing tips are presented later in this section.
- Carry safety equipment, such as flares and fire extinguishers.
- Tie down or secure truck bed loads, and secure rear gates.
- Never remain in a truck that is being loaded by excavating equipment. Stand clear of the area until the loading is completed and the load is secured.

Reckless Driving

Reckless driving is a common cause of accidents. It is also one of the most preventable. The following are some examples of reckless driving:

- Driving too fast for conditions
- Not looking in the direction of travel
- Stopping, starting, or turning suddenly

To avoid these and other hazards, always observe the following rules:

- Adjust speed for driving conditions. Slow down on wet, bumpy, or slippery surfaces.
- Drive with the load as low as possible and tilted back.
- Look in the direction of travel; this includes backing up.
- Make smooth, deliberate starts, stops, and turns. Sudden movements can cause load shifts or vehicle instability.
- Watch for people and other vehicles.

- Ensure that levers used to control hoisting or dumping are latched when not in use to prevent accidental starting or tripping of the mechanism.
- Stand clear of levers used to control hoisting or dumping while these devices are in operation.
- Always turn off the engine and set the brakes before leaving the vehicle.
- Immediately report any unsafe conditions on vehicles to your supervisor.

When you drive a company-owned vehicle, you must use it only for your assigned tasks. Most companies have a written policy governing the use of company-owned vehicles. Employees who do not follow the policy may be, and should be, disciplined or fired. Although specific rules vary from one company to another, in general, they cover the following areas:

- If you operate a licensed vehicle owned or controlled by the company, you must maintain a current driver's license as required by federal or state regulations.
- Do not transport non-employee passengers. Do not allow non-employees or unqualified employees to use company vehicles, unless an authorized official of the company has given permission.
- Inspect your company vehicles at the beginning of each workday and keep them clean.

- Obey all traffic laws. All fines are your responsibility. Report any traffic citations to your supervisor in writing. Repeated violations may result in suspension or dismissal.
- Wear your seat belt at all times, and be sure that others in the vehicle wear theirs.
- When you leave a vehicle unattended, remove the keys, set the brakes, roll up the windows, and lock the doors.
- Consumption of alcohol or nonprescription drugs is grounds for immediate dismissal, whether ingested before work or while on the job. If you are taking prescribed medication that may affect your ability to perform your duties safely, you must notify your supervisors when you report to work.
- Tell your safety director or supervisor immediately about any incidents involving damage to company property, property of others, personal injury, or injury to others. Failure to report accidents involving a company vehicle is grounds for termination.
- Do not use radar-detection equipment in any company vehicle.
- Be courteous to other motorists. You and the vehicle are a rolling billboard for your company.
- Be trained in and use good defensive driving techniques while operating company vehicles. Employees should follow all insurance requirements.
- Remember that you are also responsible for all tools and equipment assigned to your company vehicle.
- Be sure your vehicle is equipped with an appropriate fire extinguisher and a first-aid kit.

You must take extra precautions when backing up construction vehicles. Whenever possible, avoid backing to avoid the increased risks that come with reduced visibility behind any vehicle. Find a parking spot that allows you to leave the site without backing up. However, many sites have tight or constricted turning areas, and you are not always able to avoid backing up. In those cases, follow these backing safety guidelines:

- Avoid blocking the rearward inside view with equipment and stock.

- Walk completely around the vehicle first, to check for hazards you may encounter when backing up. Don't forget to look up for overhangs and electrical wires.
- Before backing up, roll down the window and turn off the radio. Check all mirrors and look over both shoulders. Sound the horn twice or activate the backup alarm to warn others. Back up slowly.
- Whenever possible, have a co-worker guide you. The use of a radio can also be helpful when communicating with the guide. The guide should stand at the rear driver's side of the vehicle (if there is room) and use full-motion arm signals, not hand signals. If you lose sight of your guide, stop immediately. Do not start again until your guide is visible.
- When parking in an area from which you will have to back up, place orange traffic cones behind the vehicle to ensure that this area remains clear.

WARNING!

Operating a motorized vehicle in an indoor location without proper ventilation can sicken or even kill you. All motorized vehicles emit carbon monoxide as part of their exhaust. You cannot see, smell, or taste carbon monoxide, and you will not know when it is in the air. Never leave your vehicle running when it is not being used. Do not leave keys in an unattended vehicle. Only use approved vehicles indoors, and do so under the supervision of a competent person.

WARNING!

Construction sites can be extremely noisy. Noise on the site may prevent you from hearing a back-up alarm. Use your eyes as well as your ears when operating or working around motorized vehicles to prevent being struck by a motorized vehicle, lever, crane, or load.

Defensive Driving

Your best protection against a motor vehicle accident is to become a defensive driver. Defensive drivers stay focused on the road, on road conditions, and on the other drivers around them. Here are some tips (in addition to the safety guidelines included in this section) to help you stay focused and be a good defensive driver:

- Avoid long cell phone calls while driving. Even with a hands-free unit, you can become distracted. Pull over to complete your calls.
- Get in the habit of checking your blind spots frequently.
- Avoid eating, drinking, talking on the cell phone, or smoking while driving. Hot burning ashes or spills of hot or cold food or liquids can all distract you momentarily, which is enough time for an accident to happen.
- Clean off the dashboard and seats. Papers, clipboards, empty paper cups, small tools, and other debris on your dashboard and seats will slide around while you are driving and can distract you. If you have to make a sudden stop, those items will become airborne and could cause injuries.
- Know where you are going before you start your vehicle. If you must consult a map, pull over. Do not try to navigate and drive at the same time.
- Become a weather-wise driver. When the weather includes ice, snow, fog, or wet pavements, change your driving pattern to accommodate the conditions. Slow down and be especially watchful of the cars around you. Be prepared to take defensive actions to protect yourself.

Remember, most accidents happen while the vehicle is being backed up. Always check your rearview and sideview mirrors, use backup alarms, and be sure that your path is clear.

Additional Resources

Environmental Protection Agency. *www.epa.gov.*

National Safety Council. *www.nsc.org.*

OSHA. *www.osha.gov.*

2.0.0 Section Review

1. When welding, you must wear _____.
   a. safety glasses
   b. safety glasses with side shields
   c. tinted goggles or a hood
   d. safety goggles with a wrap-around strap

2. The color blue on an NFPA warning diamond indicates a(n) _____.
   a. fire hazard
   b. health hazard
   c. explosion hazard
   d. radioactive hazard

3. The sign or tag that is used to tell workers that an immediate hazard exists and that specific precautions must be observed to avoid an accident is a(n) _____.
   a. danger sign
   b. caution sign
   c. informational sign
   d. safety instruction sign

3.0.0 SAFETY PRECAUTIONS FOR USING TOOLS

Objective

Identify the safety precautions required when using hand and power tools.

    a. Describe the safety precautions associated with hand tools.

    b. Describe the safety precautions associated with power tools.

Performance Tasks

    4. Inspect power tools (corded and cordless) to ensure they are safe to use.

    5. Inspect hand tools to ensure they are safe to use.

Trade Terms

Bladed tools: Tools that use sharp edges to accomplish their tasks. Bladed tools include saws, knives, scissors, tin snips, and wire cutters.

Electrically powered tools: Tools that use electrical current to operate.

Fire watch: One or more people who are responsible for preventing and extinguishing fires and notifying the fire department and/or occupants in the event of a fire emergency.

Impact tools: Tools that must strike or be struck to accomplish their task. They include hammers, chisels, and taps.

Liquid-fuel tools: Tools that use a liquid fuel, such as gasoline or liquid propane, to operate.

Tools are used every day on a job site, and it is easy to forget that they can pose serious safety risks. The types of injuries that occur vary. Some are the result of workers tripping on tools, electrical shock, sharp edges, or fire. Others are due to carelessness or lack of training. The following list demonstrates the types of injuries that occur as a result of hand and power tool accidents:

- Burns
- Cuts
- Sprains
- Electrical shock
- Eye injuries
- Hearing loss
- Broken bones

3.1.0 Hand Tools

Hand tools are nonpowered tools that require the worker's strength and force to operate. Hand tools become dangerous when they are misused or improperly maintained. Bladed tools (tools with sharp edges) and impact tools (tools that must be hit or struck) present a great risk of injury because of the force needed to operate them. The best way to reduce hand-tool accidents is to inspect and maintain tools regularly and always wear appropriate PPE.

3.1.1 Bladed Tools

Bladed tools (*Figure 30*), such as saws and utility knives, present the risks of cutting, slicing, snipping, and stabbing. These injuries can be as minor as a small cut or as severe as an amputation or stab wound. To minimize these risks, adhere to the following:

- Use bladed tools with the blades and points aimed away from yourself and others.
- Direct bladed tools away from aisles.
- Store bladed tools properly; use the sheath or protective covering if there is one.
- Keep blades sharp and inspect them regularly. Dull blades are hard to use and control and can be far more dangerous than sharp blades.

3.1.2 Impact Tools

Impact tools (*Figure 31*), such as chisels and punches, work by striking them with a hammer. The most common types of injuries caused by impact tools are hammer strikes and eye injuries from flying tool fragments. Getting struck by a hammer is usually the result of carelessness or not paying attention. If you are using a hammer, keep your eyes and your mind on what you're doing.

You can prevent injuries caused by tool fragments flying off damaged tools by inspecting the tool before using it. Look for mushroomed heads, cracks, chips, or other signs of damage or weakness. Wear proper eye protection while using impact tools.

3.1.3 Dust and Suspended Particles

Some hand tools create a lot of airborne dust and particles. Sandpaper, planes, files, and saws, for example, may create a lot of sawdust or drywall powder. Short-term exposure to these particles may cause minor irritation of the eyes, nose, or sinuses. In some people, exposure may trigger a

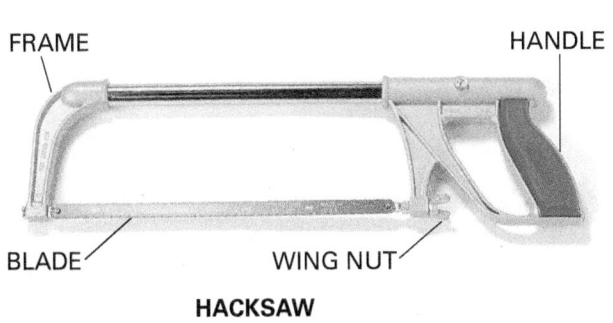

HACKSAW

TUBE CUTTER

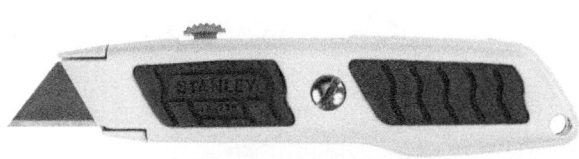

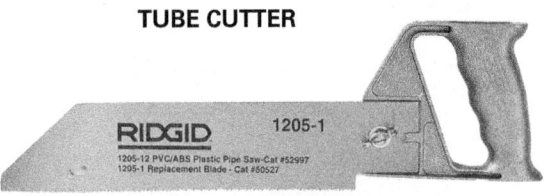

UTILITY KNIFE

ALL PURPOSE (PVC) SAW

Figure 30 Bladed tools.

more serious reaction, such as an asthma attack. Long-term exposure over the course of a career can affect your lungs, possibly causing lung disorders. To help prevent these conditions, wear filtering masks and eye protection when you are using a tool that creates dust.

3.2.0 Power Tools

Power tools, as explained above, are powered by electricity, pressurized air, or fuel (*Figure 32*). They can be hazardous when they are improperly used or poorly maintained. Most of the risks associated with hand tools are also associated with power tools. Adding a power source to a tool increases the risk factors. For example, a radial saw arm is far more dangerous than a hand-powered saw.

Power tools are powered by different sources. Power sources for those used in plumbing include electricity and fuel, such as propane gas.

You should know the safety rules and operating procedures and be properly trained for each tool you use. The user's manual supplied by the tool's manufacturer provides specific operating procedures and safety rules. Before operating any power tool for the first time, always read the manual to become familiar with the tool. If the manual is missing, contact the manufacturer for a replacement.

CHISELS

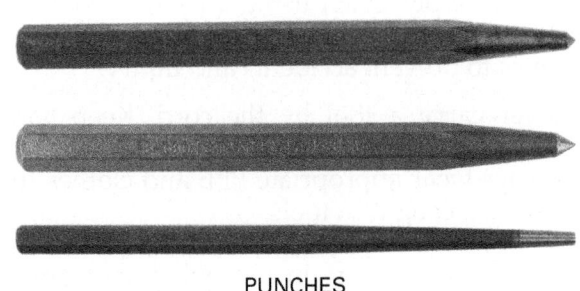

PUNCHES

Figure 31 Impact tools.

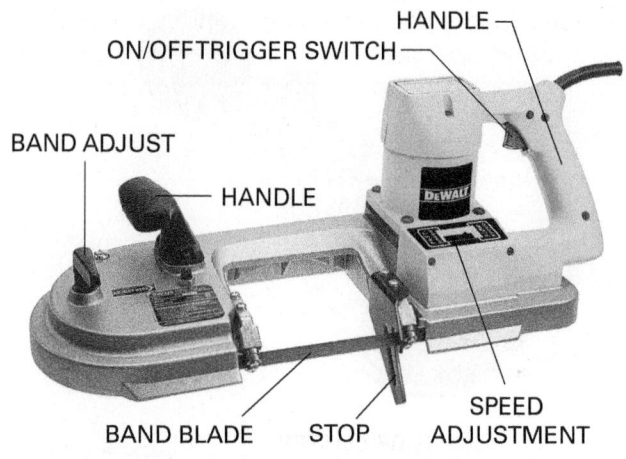

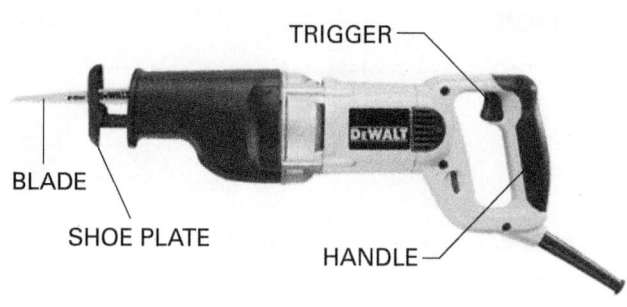

(A) PORTABLE BAND SAW

(B) RECIPROCATING SAW

(C) HAMMER DRILL

(D) ROTARY HAMMER DRILL

Figure 32 Power tools.

It's important to understand the general safety rules that apply when using all power tools, regardless of type. Use the following safeguards at all times to prevent accidents and injury:

- Never carry a tool by the cord. Keep cords away from heat, oil, and sharp edges.
- Always wear appropriate PPE and clothes that won't get caught in tools.
- Do not distract others or let anyone distract you; do not engage in horseplay; and do not run or throw objects.
- Never leave a power tool running unattended.
- Secure work with clamps or a vise, leaving both hands free to operate the tool.
- Be sure that an electrical power tool is properly grounded and connected to a ground fault circuit interrupter (GFCI) before using it (see *Figure 33*).

- Do not use dull or broken tools or accessories. Never use tools with frayed cords.
- Never use a power tool with guards or safety devices removed or disabled. Use electric extension cords of sufficient size to service the power tool you are using.
- Never operate a power tool if your hands or feet are wet. Keep the work area clean at all times.
- Keep a firm grip on the power tool at all times.
- Report unsafe conditions to your instructor or supervisor. Remove damaged tools from use and tag out with Do Not Use tags.

Figure 33 Ground fault circuit interrupter.

3.2.1 Electrically Powered Tools

A serious danger in using **electrically powered tools** is electrocution. The electricity that powers the tool can cause burns, shocks, explosions, electrocution, and fires. Electrical shocks can be minor and uncomfortable or they can be severe, causing burns or death. Even a small amount of current can cause the heart to stop pumping in rhythm. If not corrected, this condition will result in death. Electrical shock can also cause a loss of balance, muscle control, or consciousness, which could then cause the victim to fall or drop a tool. A fall from a ladder or scaffolding can be quite serious. To prevent electrical shock, tools must provide at least one of the following types of protection:

- *Double insulated*—Double insulation is more convenient than three-wire cords. The user and tools are protected in two ways: by normal insulation on the wires inside and by a housing that cannot conduct electricity to the user in the event of a malfunction.
- *Powered by a low-voltage isolation transformer*—If your electrically powered tools do not have either a ground plug or double insulation, check with your supervisor to make sure that you are protected by a low-voltage isolation transformer.

- *Grounded with a three-wire cord*—Three-wire cords have two current-carrying conductors and one grounding conductor. You are probably familiar with the three-prong plug common on electrically powered tools. When there is a three-prong cord, that means the tool is powered by a grounded three-wire cord, and it should only be plugged into a three-prong, grounded receptacle. Never remove the third prong (grounding conductor) from a plug. If you are using a three-prong extension cord, make sure that it is properly grounded at its source. The tool being used must be protected by a GFCI or a generator.

- *Twist locks*—Twist locks on tool plugs ensure the power source matches the tool's voltage rating. The arrangement of the plug conductors can only fit into receptacles that are wired to the appropriate voltage. A second advantage is that the plugs, once inserted, don't pull out. Twist locks have a high impact resistance.

You can avoid accidents and injury when using electrically powered tools. Follow these guidelines to protect yourself and your co-workers:

- Use the right tools for the job, and use them the right way.
- Wear all appropriate PPE, such as gloves and safety footwear.
- Store power tools in a dry place when not in use.
- Never use electrically powered tools in damp or wet places.
- Work only in well-lit work areas.
- Consult local codes before working with electrical equipment.

Because malfunctioning electrically powered tools can cause sparks, these tools can cause fires and explosions. Make sure you are aware of any fire hazards in your work area. Avoid using electrically powered tools around flammable materials, fumes, and gases.

Finally, electrical cords and extension cords pose a tripping hazard. Extension cords should be brightly colored to make them more visible. Cords and cables should be run somewhere other than walkways, or at least along a wall, rather than in the middle of a walkway or across a walkway. Avoid running cables and cords across elevated work areas and scaffolding. Occasionally, it may be necessary to run a cord or cable across a walkway. If so, either tape the cord down and put a carpet over it, or place it in a cord runner designed to minimize the tripping hazard. When electrical and extension cords are no longer in use, always hang the cords according to OSHA standards for keeping floors clear. To avoid creating tripping hazards, workspaces, walkways, and similar paths must be clear of cords. Never use worn or frayed cables (*Figure 34*). Turn in any frayed cords that you encounter to your supervisor.

CAUTION

Do not run a cord through doorways or through holes in ceilings, walls, and floors, which might pinch the cord. Check to see whether there are sharp corners along the cord's path. Any of these situations can lead to cord damage. Extension cords are a tripping hazard. They should never be left unattended and should always be put away when not in use.

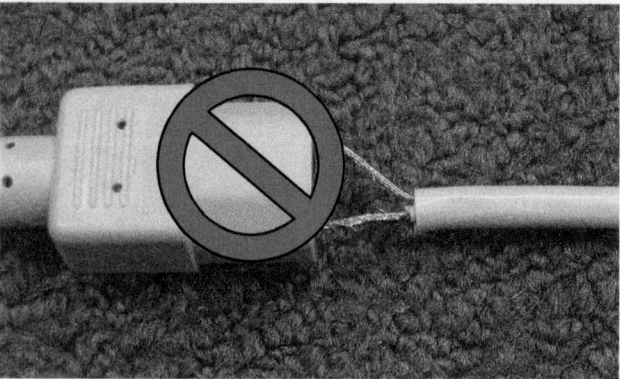

Figure 34 Never use damaged cords.

CAUTION

Frayed electrical cords can cause electrocution. Never use a power tool if the cord is frayed or damaged. Never use frayed or damaged extension cords. Follow your company's procedures for removing equipment with damaged or frayed cords from service.

3.2.2 Liquid-Fuel Tools

Some power tools, like torches (see *Figure 35*), are powered by a liquid fuel, such as propane. Whenever a torch is used, a fire extinguisher must be nearby. The most serious hazard with fuel-powered tools comes from fuel vapors that can burn or explode. Burning liquid fuel also gives off exhaust fumes, which can be dangerous. Here are a few tips on using liquid-fuel tools safely:

- Always wear the appropriate PPE, including eye protection, gloves, and respirators, if necessary.
- Handle, transport, and store the fuel only in approved flammable-liquid containers.

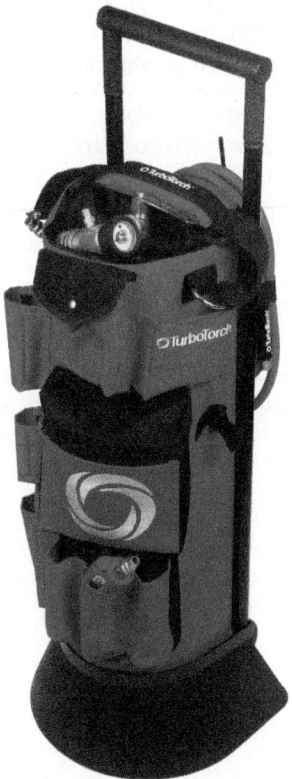

Figure 35 Acetylene torch kit and tank on rolling cart.

- Before refilling the tank for a liquid-fuel powered tool, shut down the engine and allow it to cool. This reduces the risk of a hot tool igniting fuel vapors.
- If you are using a liquid-fueled tool inside a closed area, there must be adequate ventilation and/or respirators in use so you can avoid breathing dangerous exhaust fumes.
- When you are using a liquid-fuel powered tool, make sure fire extinguishers are available nearby.
- When using torches, a fire watch (people who are responsible for preventing and putting out fires) should be maintained.

Case History

Added Oxygen Causes Death

A welder entered a 24-inch-diameter steel pipe to grind a bad weld on a valve about 30 feet from the entry point. Before he entered, other crew members decided to add oxygen to the pipe near the bad weld to make sure the air was safe. The welder had been grinding off and on for about five minutes when a fire broke out. The fire covered his clothing. He was pulled from the pipe, and the fire was put out. The burns were so serious that the welder died the next day.

   The Bottom Line: This accident could have been avoided. It happened because of poor communication among workers and unsafe work practices.

   Source: Occupational Safety and Health Administration

Additional Resources

OSHA website. *www.osha.gov.*

3.0.0 Section Review

1. A bladed hand tool is safest if it is _____.

    a. slightly dull to prevent serious cuts
    b. inspected and sharpened regularly
    c. well oiled
    d. kept warm to prevent shattering

2. Liquid-fuel tools could start a fire when the _____.

    a. fuel vapor comes into contact with a hot tool
    b. power cord comes into contact with fuel vapor
    c. operator is using the tool while standing in water
    d. air hose is not grounded

4.0.0 SAFETY PRECAUTIONS FOR SPECIFIC WORK AREAS

Objective

Identify the safety precautions associated with various work areas.

a. Describe the safety precautions associated with work in trenches.
b. Describe the safety precautions associated with confined spaces.
c. Identify the safety precautions associated with underground work.
d. Demonstrate a lockout/tagout procedure.
e. Describe jobsite safeguards and emergency response procedures.

Performance Tasks

6. Demonstrate/simulate the proper methods of lockout/tagout for energy sources.

Trade Terms

Atmospheric hazards: Potential dangers in the air or conditions of poor air quality.

Benching: A method of protecting workers from cave-ins by excavating the sides of an excavation to form one or a series of horizontal levels or steps, usually with vertical or near-vertical surfaces between levels.

Combustible: Air or materials that can explode and cause a fire.

Energy-isolating device: Any mechanical device that physically prevents the transmission or release of energy. Can include manually operated electrical circuit breakers, disconnect switches, line valves, and blocks.

Energy sources: Any sources of electrical, mechanical, hydraulic, pneumatic, chemical, thermal, or other energy.

Gassy operations: Working conditions in which one or more of the following conditions exist: higher than minimum levels of methane or explosive gases are present; a gas ignition has previously occurred there; or the area is connected to an underground area designated a gassy operation.

Hypothermia: A life-threatening condition caused by exposure to very cold temperatures.

Lockout: The placement of a lockout device on an energy-isolating device, in accordance with an established procedure, ensuring that the energy-isolating device and the equipment being controlled cannot be operated until the lockout device is removed.

Lockout devices: Any devices that use positive means such as a lock to hold an energy-isolating device in a safe position, thereby preventing the energizing of machinery or equipment.

Nonpermit-required confined space: A confined workspace free of any atmospheric, physical, electrical, and mechanical hazards that can cause injury or death.

Oxygen-enriched atmosphere: An atmosphere in which there is too much oxygen. Usually considered more than 23.5 percent oxygen by volume.

Permit-required confined space: A confined space that has actual or possible hazards. These hazards can be atmospheric, physical, electrical, or mechanical.

Protective system: A method of protecting employees from cave-ins, from material that could fall or roll from an excavation face or into an excavation, or from the collapse of adjacent structures. Protective systems include support systems, sloping and benching systems, and shielding systems.

Shoring: A structure such as a metal hydraulic, mechanical, or timber system that supports the sides of an excavation and is designed to prevent cave-ins.

Subsidence: A depression in the earth that is caused by unbalanced stresses in the soil surrounding an excavation.

Tagout devices: Any prominent warning devices, such as a tag and a means of attachment that can be fastened securely to an energy-isolating device in accordance with an established procedure. The tag indicates that the machine or equipment to which it is attached is not to be operated until the tagout device is removed in accordance with the energy-control procedure.

Trench shields: Structures that are able to withstand the forces imposed on them by a cave-in and can thereby protect employees within the excavation. Shields can be permanent structures or portable and moved along as work progresses. Shields can be either premanufactured or job-built in accordance with *29 CFR 1926.652 (c)(3) or (c)(4)*.

Certain work areas have hazards that require special safety precautions. These areas include work in trenches, confined spaces, and underground areas. In addition, work in areas of operating equipment or stored energy systems will require the use of appropriate lockout/tagout procedures. All workers must be aware of the specific safeguards and emergency response procedures for each job site.

4.1.0 Trenching

Safety is crucial during any excavation job. Safety precautions must be exercised at all times to prevent injury to yourself or other workers. Always use the most recent edition of the OSHA manual as the governing standards for excavation safety.

Excavations are done for a number of reasons, including laying pipe and locating utility and sewage lines. During an excavation, earth is removed from the ground, creating a narrow excavation made below the surface of the ground called a trench. The trench width, and usually the depth, are limited to 15 feet. The soil that is removed from the ground is called spoil. When soil is removed from the ground, extreme pressures may be generated on the trench walls. If the walls are not properly secured by shoring (a structure that supports the sides of an excavation), sloping, or shielding, they will collapse (*Figure 36*). The collapse of unsupported trench walls can instantly crush and bury workers. This type of collapse happens because not enough material is available to support the walls of an excavation.

Figure 36 Trench shield.

Case History
Ditch Collapse

A plumber was laying sewer lines in a ditch 2-½ feet wide and 8 feet deep. When he stood up to stretch his legs, the ditch collapsed. Dirt from the inadequately supported ditch buried him past his hat. Luckily, another plumber was nearby to uncover his head quickly. A backhoe was then used to free him.

The plumber was seriously injured. He suffered a skull fracture, broken jaw, damaged hips, and minor brain damage. He recovered from the accident but now has two artificial hips and a neurological disorder.

The Bottom Line: This plumber would not have been injured if he had followed OSHA's requirements for trenching. Be aware of your surroundings and follow OSHA regulations.

Excavation Safety

Federal and state safety regulations have established standards for protecting those who work in excavations. *OSHA 29 CFR 1926* defines the trench protective devices that are acceptable. OSHA mandates how the protective devices must be used. In accordance with OSHA regulations, using appropriate manufactured and engineered trench-shielding and -shoring devices or sloping trench walls to angles that eliminate the risk of cave-ins other than in a stable location is required for excavations 5 feet or deeper.

4.1.1 Trenching Hazards

Working in and around excavations is one of the most hazardous jobs you will ever do. The design of the excavation may not be your responsibility, but you should be aware of the safety hazards involved in the placement and design of the excavation. You must take safety precautions at all times to prevent injury to yourself and others. Some of the hazards you may encounter during an excavation include the following:

- Surface encumbrances (buildings, vegetation, rocks, or other objects along the surface area of the trench that can hinder operations and block sightlines)
- Underground installations (sewer, telephone, fuel, electric, water lines)

- Flooding from broken water or sewer mains
- Hazardous atmospheres (toxic liquid or gas leaks)
- Cave-ins due to trench failure
- Falls from employees working too close to the trench edge
- Electrical shock from striking electrical cable in the trench or striking overhead lines
- Auto traffic, if the excavation site is near a highway
- Exposure to falling loads
- Collapse of walls or buildings adjacent to the trench
- Loose soil or rock inside the trench
- Exposure to hazardous atmospheres

The type of soil in and around a trench contributes to the collapse of trench walls. Soil type is a major factor to consider in trenching operations. Although you should be aware that soil type plays a role in the safe construction of trenches, only a competent person has enough experience, training, and education to determine whether the soil in and around a trench is safe and stable.

4.1.2 Guidelines for Working in and Around Trenches

Working in a trench exposes a worker to many potential hazards. The most obvious hazard is a trench failure. However, there are other situations you should be aware of when working in a trench. You must always use caution when working near the edge of a trench. A trench wall could suddenly give way, sending you to the bottom of the trench.

When you are working in a trench, flooding can be a concern. If a water or sewer main is ruptured while excavating, the trench can quickly fill with water. A little less obvious but just as dangerous are those conditions where a natural water supply has been dammed off. Tons of water may be held back, and if the dam fails, the trench could be quickly flooded.

Electrical shock is also a potential hazard in the trench. While excavating you could accidentally pierce the insulation of an electrical cable in the trench. If overhead electrical lines are near the trench, there is always the danger that the excavator will come into contact with the electrical lines.

When working near highways, vehicle traffic is a potential hazard. Always check for traffic as you move about the job site. Wear appropriate PPE at all times. Sometimes, trenches can fill with dangerous gases, or oxygen can be displaced from the bottom of the trench. This can happen if the trench is near facilities that store large volumes of chemicals. The liquids or gases could flow or leak into the trench.

Finally, you must be aware of the possibility of injury from objects falling into the trench. Chunks of dirt from the excavator bucket, parts of the bucket, or dirt and stone from the excavation can unexpectedly fall into the trench. Never enter the trench without appropriate PPE.

OSHA requires that a competent person inspect excavations, the adjacent areas, and fall protection and shielding devices daily. The competent person must perform these inspections at the following times:

- Before the start of work
- As needed throughout the shift
- After every rainstorm or other occurrence that increases hazards to workers

When working in or around any excavation or trench, you are responsible for your personal safety. You are also responsible for the safety of others in the trench. According to OSHA, the primary hazard in trenching and excavation is injury resulting from collapse. To prevent that, soil analysis is conducted to determine the appropriate sloping, benching (stepped excavation), and shoring. Be aware of the potential for additional hazard when working with heavy machinery, manually handling materials, and working in proximity to traffic. Overhead and underground power lines and underground utilities such as natural gas represent hazards as well.

Workers need a safe and reliable way of entering and exiting excavations. OSHA refers to this as access (getting into a trench) and egress (getting back out). Several methods are available, including ramps, stairways, and ladders (see *Figure 37*).

Did You Know?

Violations of scaffolding and trenching regulations are the two most frequent citations issued by OSHA. These violations are fined per person.

When structural ramps are used, a competent person must design them. The ramps must be constructed so that each section fits together well to prevent displacement of the soil in the excavation. *Table 3* shows the categories of rock and soil as determined by OSHA. In addition to proper displacement of soil, the surface of ramps must provide enough traction so that workers will not slip. Guardrails on ramps and stairways must be

Figure 37 Ladder in a trench.

smooth enough to protect workers from punctures, lacerations, and snagging of clothing.

The following are additional safety measures for access/egress:

- Ladders, ramps, or stairways are required in trench excavations that are more than 4 feet deep.
- Ladders, ramps, or stairways used as exits must be located every 25 feet in any trench that is more than 4 feet deep.
- Ladder side rails must extend a minimum of 3 feet above the landing or top of the trench.
- Ladders must have nonconductive side rails if work will be performed near equipment or systems using electricity.
- Two or more ladders must be used when 25 or more workers are working in an excavation in which ladders are the primary means of entry and exit or are used for two-way traffic in and out of the trench.
- All ladders must be inspected before each use for signs of damage or defects.
- Damaged ladders should be labeled Do Not Use and removed from service until repaired.
- Use ladders only on stable or level surfaces.
- Secure ladders when they are used in any location where they could be displaced by excavation activities or traffic.
- While on a ladder, do not carry any object or load that could cause you to lose your balance.
- Exercise caution whenever using a trench ladder.

4.1.3 Indications of an Unstable Trench

A number of stresses and weaknesses can occur in an open trench or excavation. For example, increases or decreases in moisture content can affect the stability of a trench or excavation. The following sections discuss some of the more frequent causes of trench failure. These conditions are illustrated in *Figure 38*.

Tension cracks usually occur from one-quarter to half the distance from the top of a trench. Sliding or slipping may occur as a result of tension cracks. In addition to sliding, tension cracks can cause toppling. Toppling occurs when the trench's vertical face shears along the tension crack line and topples into the excavation. An unsupported excavation can create an unbalanced stress in the soil, which in turn causes subsidence at the surface and bulging of the vertical face of the trench. If uncorrected, this condition can cause wall failure and trap workers in the trench or greatly stress the protective system used to protect workers from harm in the event of a cave-in, collapses, or other occurrence. Bottom heaving is caused by downward pressure created by the weight of adjoining soil. This pressure causes a bulge in the bottom of the cut. Heaving and squeezing can occur even when shoring and shielding are properly installed.

Another indication of an unstable trench is boiling. Boiling is when water flows upward into the bottom of the cut. A high water table is one cause of boiling. Boiling can happen quickly and can occur even when shoring or trench boxes are used. If boiling starts, leave the trench immediately.

4.1.4 Trench Failure

The most common hazard during an excavation is trench failure or cave-in. Using common sense and following all applicable safety precautions makes the trench a safer place to work.

To understand the seriousness of trench failure, consider what can happen when there is a shift in the earth that surrounds an unsupported trench. Workers could be buried when any of the following events happen:

- One or both edges of the trench cave in
- One or both walls slide in
- One or both walls shear away and collapse

Failure of unsupported trench walls is not the only cause of burial. Tons of dirt can be dumped on workers if the spoil pile or excavated earth slides into the trench. Such slides occur when the pile is placed too close to the edge of the trench or when the ground beneath the pile gives way.

Table 3 Determination of Soil Type

Category/Type	Description
Stable rock	Natural solid mineral matter that can be excavated with vertical sides and remain intact while exposed. It is usually identified by a rock name such as granite or sandstone. Determining whether a deposit is of this type may be difficult unless it is known whether cracks exist and whether or not the cracks run into or away from the excavation.
Type A soils TYPE A SOIL Supported of shielded Vertically sided lower portion SUPPORT OR SHIELD SYSTEM 20' MAXIMUM 1 / 3/4 18" MINIMUM	Cohesive soils with an unconfined compressive strength of 1.5 tons per square foot (tsf) (144 kPa) or greater. Examples of Type A cohesive soils are often clay, silty clay, sandy clay, clay loam, and, in some cases, silty clay loam and sandy clay loam. (No soil is Type A if it is fissured; is subject to vibration of any type; has previously been disturbed; is part of a sloped, layered system where the layers dip into the excavation on aslope of four horizontal to one vertical [4H:1V] or greater; or has seeping water.)
Type B soils TYPE B SOIL Supported of shielded Vertically sided lower portion SUPPORT OR SHIELD SYSTEM 20' MAXIMUM 1 / 1 18" MINIMUM	Cohesive soils with an unconfined compressive strength greater than 0.5 tsf (48 kPa) but less than 1.5 tsf (144 kPa). Examples of other Type B soils are angular gravel; silt; silt loam; previously disturbed soils unless otherwise classified as Type C; soils that meet the unconfined compressive strength or cementation requirements of Type A soils but are fissured or subject to vibration; dry unstable rock; and layered systems sloping into the trench at a slope less than 4H:1V (only if the material would be classified as a Type B soil).
Type C soils TYPE C SOIL Supported of shielded Vertically sided lower portion SUPPORT OR SHIELD SYSTEM 20' MAXIMUM 1 / 1-1/2 18" MINIMUM	Cohesive soils with an unconfined compressive strength of 0.5 tsf (48 kPa) or less. Other Type C soils include granular soils such as gravel, sand and loamy sand, submerged soil, soil from which water is freely seeping, and submerged rock that is not stable. Also included in this classification is material in a sloped, layered system where the layers dip into the excavation or have a slope of four horizontal to one vertical (4H:1V) or greater.
Layered geological strata	Where soils are configured in layers (i.e., where a layered geologic structure exists), the soil must be classified on the basis of the soil classification of the weakest soil layer. Each layer may be classified individually if a more stable layer lies below a less stable layer (i.e., where a Type C soil rests on top of stable rock).

Source: Occupational Safety and Health Administration website. "OSHA Technical Manual, Section V, Chapter 2. Excavations: Hazard Recognition in Trenching and Shoring," *www.osha.gov*, reviewed February 23, 2004.

There must be a minimum of 2 feet between the trench wall and the spoil pile. This area must also be kept free of any tools and materials.

The following conditions will likely lead to a trench cave-in. If you notice any of these conditions, immediately inform your supervisor. The conditions are listed in order of seriousness:

- Disturbed soil from previously excavated ground

- Trench intersections where large corners of earth can break away
- A narrow right-of-way, causing heavy equipment to be too close to the edge of the trench
- Vibrations from construction equipment, nearby traffic, or trains
- Increased subsurface water that causes soil to become saturated and therefore unstable

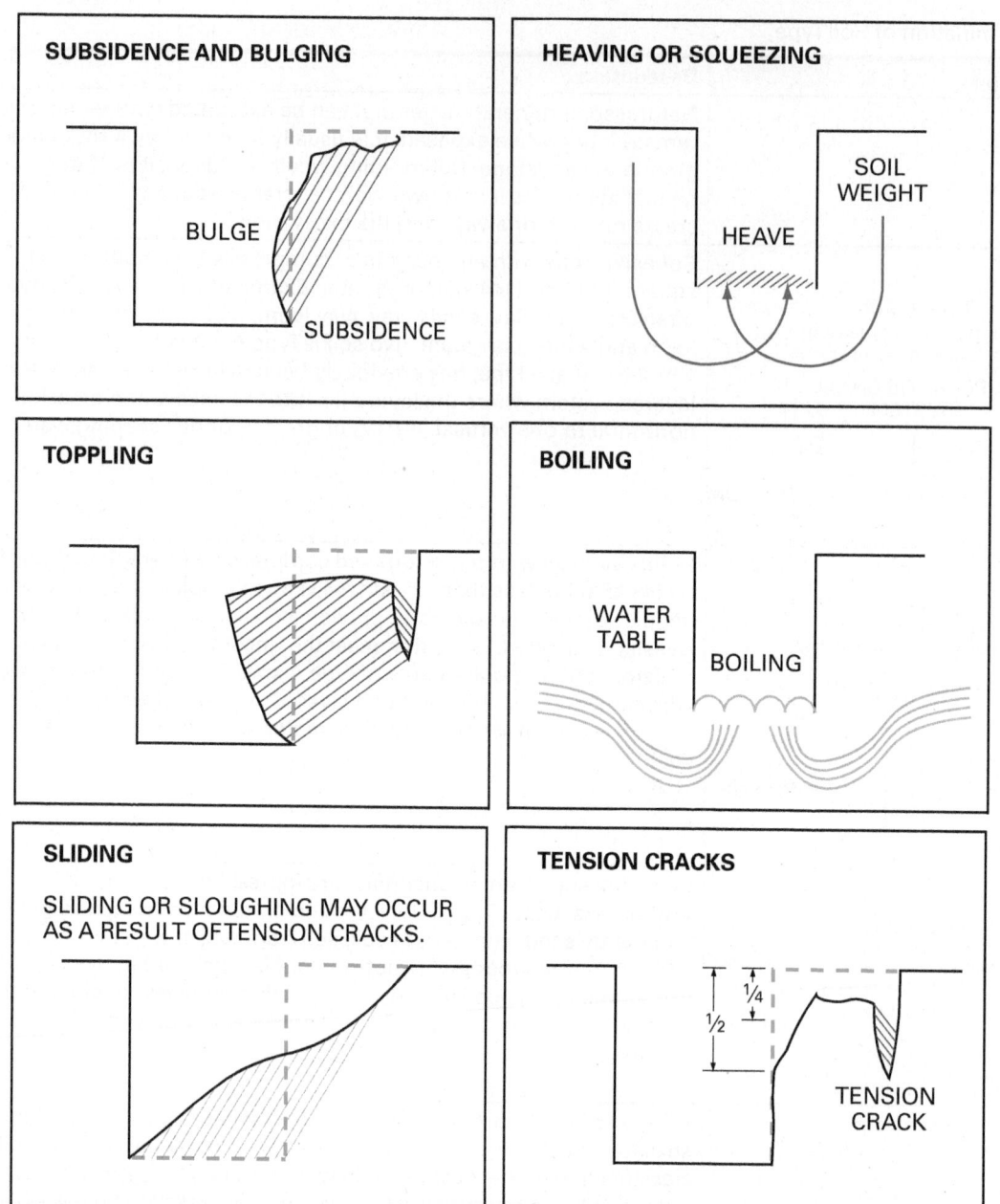

Figure 38 Indications of an unstable trench.

- Drying of exposed trench walls, which causes the natural moisture that binds together soil particles to be lost
- Inclined layers of soil dipping into the trench, causing layers of different types of soil to slide on each other and cause the trench walls to collapse

4.1.5 *Making the Trench Safer*

There are several ways to make the trench a safer place to work (see *Figure 39*). Trench shoring, shielding, and sloping are used to protect workers and equipment. It is important that you recognize the differences among them.

- *Shoring*—Shoring supports the walls of a trench and prevents wall movement and collapse. Shoring does more than provide a safe environment for workers in a trench. Because it restrains the movement of trench walls, shoring also stops the shifting of adjacent soil formations containing buried utilities or on which sidewalks, streets, building foundations, or other structures are built.
- *Trench shields*—Trench shields, also called trench boxes, are placed in unshored excavations to protect workers from wall collapse.

One-Call Notification

States operate one-call notification centers to notify the owners of underground utilities prior to digging activities. These "Call Before You Dig" programs provide a toll-free telephone number that can be called to request utility owners to visit the site and mark all underground lines using paint, chalk, flags or whiskers according to a standardized color code. One-call centers help prevent damage to lines and injury to workers as a result of digging into a buried utility line.

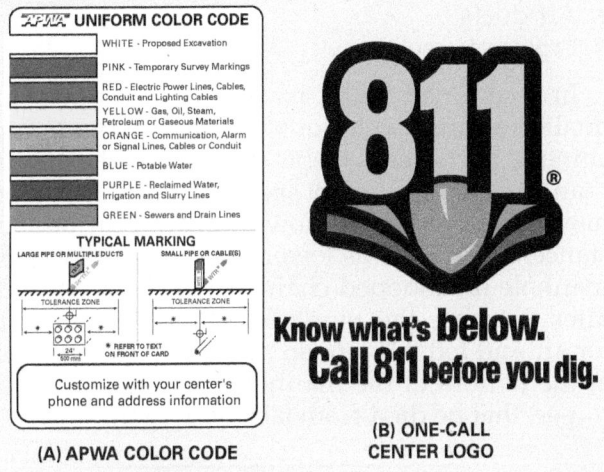

(A) APWA COLOR CODE

(B) ONE-CALL CENTER LOGO

They provide no support to trench walls or surrounding soil, but for specific depths and soil conditions, they withstand the side weight of a collapsing trench wall.

* *Sloping*—Sloping an excavation means cutting its walls back at an angle to its floor. OSHA regulations (*CFR 1926*) give companies two options for setting a safe and appropriate angle: Companies may use the tables and charts recommended by OSHA, which are included in *CFR 1926*, or they may use a sloping design that has been developed by a registered, professional engineer.

Many large excavations require the use of walkways so that workers can cross from one side of the excavation to the other. Where these walkways are 6 feet or more above the ground or lower work levels, OSHA requires the installation of guardrails and, when necessary, safety nets. *CFR 1926.502* outlines the specifications for the type of guardrail systems to be used for various types of projects. In general, guardrails must comply with the following provisions:

* The top rail must be 42 inches (plus or minus 3 inches) above the walking/working level.
* When midrails are used, they must be installed midway between the top rail and the walking/working level.
* When screens and mesh are used, they must extend from the top rail to the walking/working level and along the entire opening between rail supports.
* Balusters between posts must not be more than 19 inches apart.

(A) BENCHING

(B) SLOPING

(C) SHORING

Figure 39 Shoring methods.

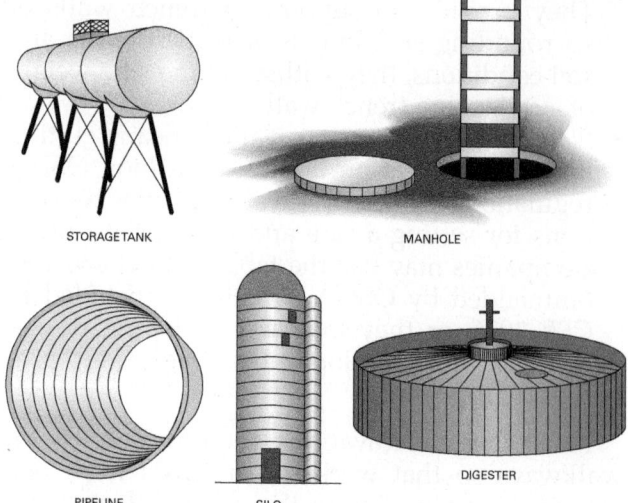
Figure 40 Examples of confined spaces.

- The surface of guardrail systems should be smooth to prevent punctures, lacerations, and snagging of clothing.
- Where safety nets are required, the nets must be inspected at least once a week for wear or damage and must meet the requirements set forth in *CFR 1926.502*.

4.2.0 Confined Spaces

Spaces on a job site are considered confined when their size and shape restrict the movement of anyone who must enter, work in, and exit the space (*Figure 40*). Confined spaces often have poor ventilation and are difficult to enter and exit. For example, employees who work in process vessels generally must squeeze in and out through narrow openings and perform their tasks in a cramped or awkward position. In some cases, confinement itself creates a hazard.

Case History

The Dangers of Gas in Confined Spaces

At the end of a workday, a worker put away his Presto-Lite acetylene torch without completely shutting the regulator valve. The torch remained sealed in its box, as the gas slowly leaked through the night. The next morning, as the worker lit a cigarette, he also opened the box. When the lid opened, the gas escaped and caught the flame, causing an explosion. Fortunately, the worker was only singed, but he was lucky.

The Bottom Line: Always check gas valves to ensure that they are completely closed. Failure to do so can result in severe burns or death.

Confined spaces are entered for inspection, equipment testing, repair, cleaning, or emergencies. They should be entered only for short periods of time. Some of the confined spaces you may work in include the following:

- Manholes
- Boilers
- Trenches
- Tunnels
- Sewers
- Underground utility vaults
- Pipelines
- Pits
- Air ducts
- Process vessels

In a confined space, hazards such as poor air quality, toxins, explosions, fire, and moving machinery parts tend to be even more dangerous than they are in open spaces. Confined spaces may also contain unknown hazards. In one instance, a worker was lowered into a 21-foot-deep manhole on a looped chain seat. Twenty seconds after entering the manhole, he started gasping for air and fell. He landed face down in the water at the bottom of the manhole. An autopsy determined that he died from lack of oxygen.

WARNING!

Often more than one worker is killed in manhole accidents. Never enter a manhole to rescue a co-worker. You may also be overcome or killed by toxic gases. To keep safe when working in or around manholes, follow these rules:

- The worker inside the manhole should wear a safety harness so that co-workers can quickly pull him or her out if necessary.
- The worker inside the manhole should whistle, sing, or keep up a steady stream of conversation so that co-workers know that person is all right.
- If the worker inside the manhole falls silent, co-workers should immediately pull that worker out.
- Always test for gases before going into a manhole.

Most confined spaces have restricted entrances and exits. Workers are often injured as they enter or exit through small doors and hatches. It can also be difficult to move around in a confined space, and workers can be struck by moving equipment. Escapes and rescues are much more difficult in confined spaces.

Case History

Confined Space Ventilation

A worker was using a liquid propane blowtorch in the bilge area of a construction barge. The torch flamed out, and the worker left the area without turning off the flow of gas. The worker was killed when he returned and lit the torch, igniting the accumulated gas.

The Bottom Line: Always provide adequate ventilation in confined spaces. Never leave a liquid-fuel tool without checking that the flow of fuel has been turned off.

Source: Occupational Safety and Health Administration

Figure 41 Permit-required confined space.

Management typically develops written confined-space entry programs to protect workers. These programs identify the hazards and specify the equipment or support that is needed to avoid injury. All industrial and some construction sites have written confined-space entry programs. It is your responsibility to know and follow your company's program. Never work alone in a confined space. OSHA requires an attendant to remain outside a permit-required confined space, which is a space that has potential or actual hazards. The attendant monitors entry, work, and exit (*Figure 41*). Refer to the OSHA flowchart in *Figure 42* when determining whether a space requires permitting.

4.2.1 Confined-Space Classification

All confined spaces must be inspected before work can begin. Inspection helps to identify possible hazards. After an inspection by a company-authorized person, the confined space is classified based on any hazards that are present. The two classifications are nonpermit required and permit required (*CFR 1910.146*).

A nonpermit-required confined space is a workspace free of any mechanical, physical, electrical, or atmospheric hazards (airborne danger or air quality problem) that can cause death or injury. After a space has been classified as nonpermit required, workers can enter using the appropriate PPE for the type of work to be performed. Always check with your supervisor if you are not sure which PPE is required.

A permit-required confined space, as noted above, has actual or possible hazards. These hazards can be atmospheric, physical, electrical, or mechanical. *OSHA CFR 1910.146* defines a permit-required confined space as having one or more of the following characteristics:

- Contains or has the potential to contain a hazardous atmosphere
- Contains a material that has the potential for engulfing (overwhelming) an employee who enters the confined space
- Has an internal configuration such that an employee entering the space could be trapped or asphyxiated by inwardly converging walls or by a floor that slopes downward and tapers to a small cross-section
- Contains any other recognized serious safety or health hazard

The job-site supervisor must issue and sign an entry permit before anyone enters the confined space. No one is allowed to enter a confined space without a valid entry permit. The permit is to be kept at the confined space while work is being done. Always check with your supervisor if you are not sure whether you need a permit to enter a confined space.

4.2.2 Entry Permits

Confined spaces can be extremely dangerous. Entry into the space begins when any part of your body passes the entrance or opening of the space. Before entering a permit-required confined space, you must have an entry permit (*Figure 43A* and *Figure 43B*).

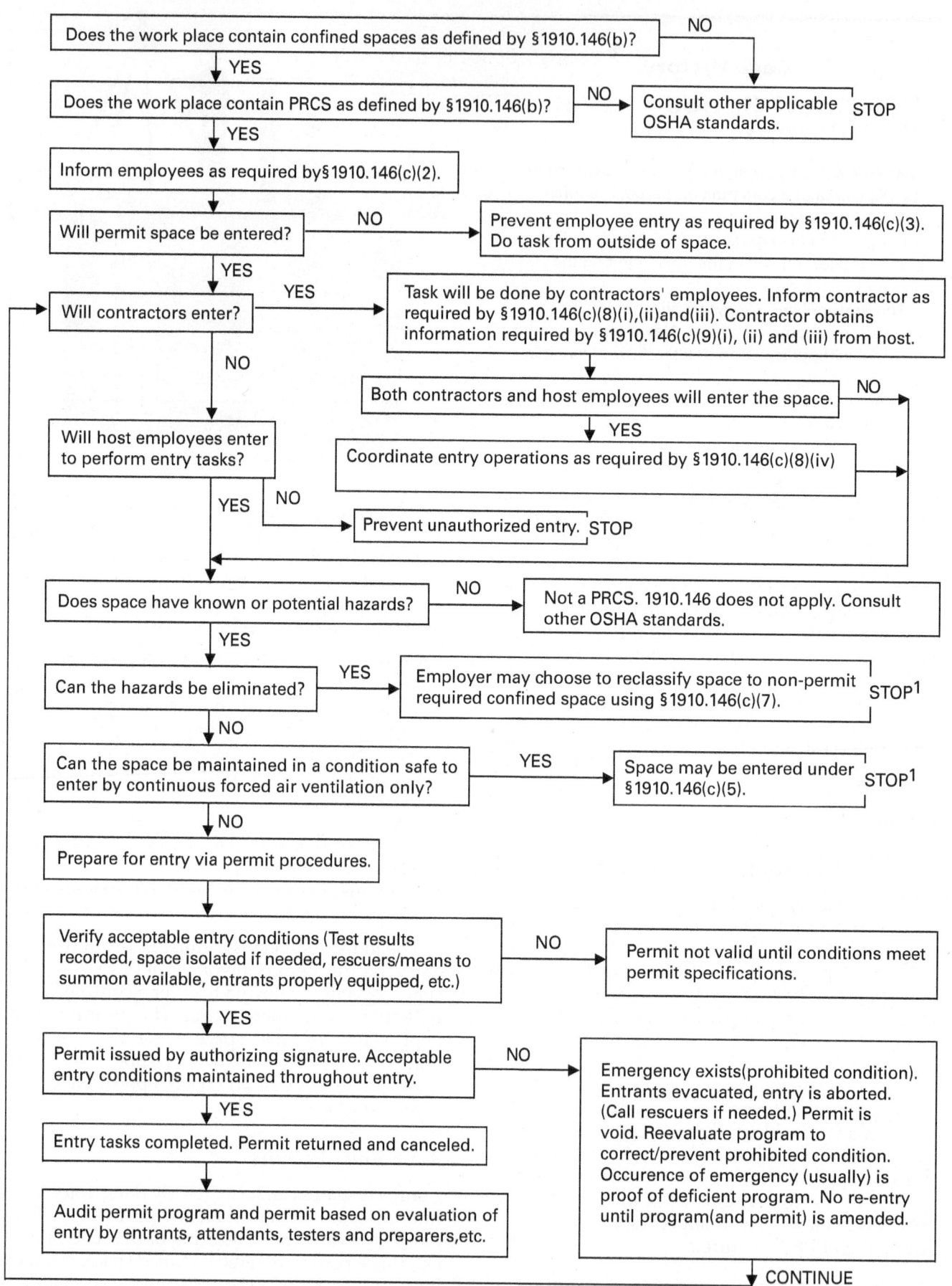

Does the work place contain confined spaces as defined by §1910.146(b)? — NO →

↓ YES

Does the work place contain PRCS as defined by §1910.146(b)? — NO → Consult other applicable OSHA standards. STOP

↓ YES

Inform employees as required by§1910.146(c)(2).

↓

Will permit space be entered? — NO → Prevent employee entry as required by §1910.146(c)(3). Do task from outside of space.

↓ YES

Will contractors enter? — YES → Task will be done by contractors' employees. Inform contractor as required by §1910.146(c)(8)(i),(ii)and(iii). Contractor obtains information required by §1910.146(c)(9)(i), (ii) and (iii) from host.

↓ NO

Both contractors and host employees will enter the space. — NO →

↓ YES

Coordinate entry operations as required by §1910.146(c)(8)(iv) →

Will host employees enter to perform entry tasks?

↓ NO → Prevent unauthorized entry. STOP

↓ YES

Does space have known or potential hazards? — NO → Not a PRCS. 1910.146 does not apply. Consult other OSHA standards.

↓ YES

Can the hazards be eliminated? — YES → Employer may choose to reclassify space to non-permit required confined space using §1910.146(c)(7). STOP¹

↓ NO

Can the space be maintained in a condition safe to enter by continuous forced air ventilation only? — YES → Space may be entered under §1910.146(c)(5). STOP¹

↓ NO

Prepare for entry via permit procedures.

↓

Verify acceptable entry conditions (Test results recorded, space isolated if needed, rescuers/means to summon available, entrants properly equipped, etc.) — NO → Permit not valid until conditions meet permit specifications.

↓ YES

Permit issued by authorizing signature. Acceptable entry conditions maintained throughout entry. — NO → Emergency exists(prohibited condition). Entrants evacuated, entry is aborted. (Call rescuers if needed.) Permit is void. Reevaluate program to correct/prevent prohibited condition. Occurence of emergency (usually) is proof of deficient program. No re-entry until program(and permit) is amended.

↓ YES

Entry tasks completed. Permit returned and canceled.

↓

Audit permit program and permit based on evaluation of entry by entrants, attendants, testers and preparers,etc.

CONTINUE

¹

Figure 42 Permit-required confined space decision flow chart.

Attachment 16 – 2
Confined-Space Entry Permit
Master Card / Safe Work Ticket No. _____

1. **Work Description:** _____
   Equip. Name / Number & Location or Area _____
   Purpose of Entry _____
   Valid Start Date _____   **Duration Time** _____ to _____

2. **Hazardous Materials:**
   What did the equipment last contain? _____
   Will the work generate a hazardous atmosphere? ☐ Yes ☐ No If yes, specify hazards and controls.
   _____

3. **Rescue Requirements:**
   ☐ External, by attendant   ☐ Complex Rescue, by rescue team at point of entry
   ☐ Non-IDLH and/or Simple Rescue, by rescue team on-site ☐ IDLH, by rescue team at point of entry
   Has the rescue team been notified of the entry? ☐ Yes   ☐ N/A Time of notification _____
   How will the rescue team be summonsed for an emergency? ☐ Radio Channel: ____ ☐ Other: _____

4. **Gas Test Requirements:**
   LEL/0₂ - Instrument Mfg./No. _____/_____   Bump Check Time/Gas Tester - _____/_____
   Toxicity - Instrument Mfg./No. _____/_____   Bump Check Time/Gas Tester - _____/_____
   Frequency of Testing: ☐ Continuous ☐ Other - Specify - _____
   • Continuous monitoring results must be recorded every three hours

Acceptable Levels	Results	Time	Initials	Results	Time	Initials	Results	Time	Initials
Oxygen: 19.5%-23.5%									
Combustible Gas: %LEL - <10%									
Other _____ < PEL* _____									
Other _____ < PEL* _____									
Other _____ < PEL* _____									

   • Entry in excess of the PEL will require appropriate PPE.

5. **Ventilation / Exhaust Equipment:**
   ☐ None required, natural ventilation adequate ☐ Forced air ventilation ☐ Exhaust ventilation
   Equipment Type: ☐ Air powered horn ☐ Electric blower   Volume Required - _____ cfm

6. **Personal Protection:**
   ☐ Gloves (type) _____   ☐ Respirator (type) _____
   ☐ Goggles or face shield   ☐ Self Contained Breathing Equipment
   ☐ Lifelines Attached to Harness   ☐ Other, specify: _____
   ☐ Chemical Resistant Suit, Specify Type _____

7. **Fire Protection:** ☐ None required ☐ Portable Fire Extinguisher – type and size:_____
   ☐ Fire Watch   ☐ Other, specify: _____

Figure 43A Entry permit.

8. Condition of Area and Equipment:

Required Yes	N/A		THESE KEY POINTS MUST BE CHECKED
		a.	Equipment locked and tagged out?
		b.	Piping is disconnected, capped or plugged and/or blinded.
		c.	Equipment emptied, washed, purged and ventilated?
		d.	Low voltage or GFCI protected equipment provided?
		e.	Explosion proof electrical equipment provided?
		f.	Provisions are made to barricade or post signs at entry points when attendant is not on duty.

Other Requirements:

9. Special Instructions: ☐ None   ☐ Check with issuer before starting work

10. Approval

10. Approval	Permit			Permit Acceptance	
	Supt. / Area Supv.	Date	Time	Maint. Supv. / Engineer / Contractor Supv.	Date
Issued by					
Endorsed by					
Endorsed by					

11. Individual Review / Entrant Roster: I have been instructed in the proper Work Permit, Confined Space Entry, Lockout/Tagout Procedures, associated physical and atmospheric hazards and have reviewed the gas-testing results

Entrants	Date	Time In / Out	Time In / Out	Time In / Out	Time In / Out	Time In / Out

I have been informed of the duties and responsibilities for an attendant, the associated physical and atmospheric hazards, and have reviewed the gas-testing results.

Attendants	Date	Time On / Off	Time On / Off	Time On / Off

12. Job Completion:
☐ Yes ☐ N/A   Has the rescue team been notified?
☐ Yes ☐ No   Is the work on equipment complete and the confined space ready to return to service?
☐ Yes ☐ No   Has the worksite been cleaned and made safe?
Workers answering above questions: _____

13. Post Job Review: Were any hazards encountered or created during entry operations?
☐ Yes ☐ No   If yes, describe: _____
Possible solutions: _____

Forward to job file within 7 days of job completion.

Figure 43B Entry permit.

An entry permit is a job checklist that verifies that the space has been inspected. It also tells everyone on the site about the hazards of the job. The supervisor must fill out and sign all entry permits before anyone enters the space. The permit must be posted at the entrance to the confined space and be available for workers to review. Entry permits must include the following information:

- A description of the space and the type of work to be done
- The date the permit is valid and how long it lasts
- Test results of all atmospheric testing, including oxygen, toxin, and flammable material levels
- The name and signature of the person who did the tests
- The name and signature of the entry supervisor
- A list of all workers, including supervisors, who are authorized to enter the site
- The means by which workers and supervisors will communicate with each other
- Special equipment and procedures that are to be used during the job
- Other permits needed for work done in the space, such as welding
- The contact information for the emergency-response rescue team

4.2.3 Atmospheric Hazards

Atmospheric hazards are the most common type of hazard in a confined space. In a hazardous atmosphere, the air can have either too little or too much oxygen, be explosive or flammable, or contain toxic gases. The atmosphere in a confined space must be monitored constantly. Special meters are used to detect atmospheric hazards (see *Figure 44*). These meters must be calibrated and operated according to the manufacturer's instructions. In addition, meters must be able to detect oxygen and combustible gases at the levels specified in *OSHA 29 CFR 1910.146*.

A confined space that does not have enough oxygen is called an oxygen-deficient atmosphere. A confined space with too much oxygen is called an oxygen-enriched atmosphere. For safe working conditions, the oxygen level in a confined space must range between 19.5 and 23.5 percent by volume, with 21 percent considered the normal level. Oxygen concentrations below 19.5 percent by volume are considered oxygen-deficient; those above 23.5 percent are considered oxygen-enriched. Portable, battery-operated gas meters can be used to measure oxygen levels.

Many of the processes that occur in a confined space use oxygen and may reduce the percentage of oxygen to an unsafe level. These processes include the following:

- Burning
- Rusting of metal
- Breaking down of plants or garbage
- Oxygen mixing with other gases

When the oxygen in a confined space is reduced, it becomes harder for a worker to breathe. The symptoms of insufficient oxygen happen in this order:

1. Fast breathing and heartbeat
2. Impaired mental judgment
3. Extreme emotional reaction
4. Unusual fatigue
5. Nausea and vomiting
6. Inability to move your body freely
7. Loss of consciousness
8. Death

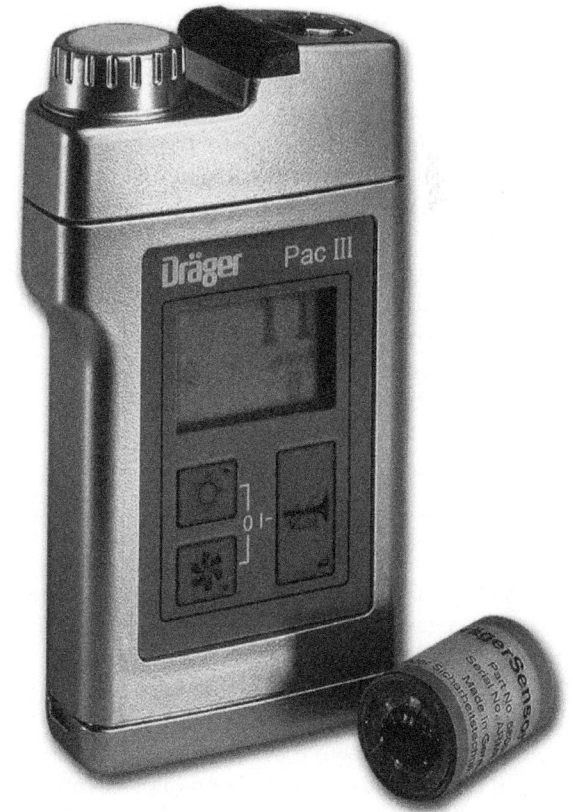

Figure 44 Detection meter.

4.2.4 Explosion Hazards

Too much oxygen in a confined space is a fire hazard and can cause explosions. Materials like clothing and hair are highly flammable and burn rapidly in oxygen-enriched atmospheres. Fires can start easily in a confined space with an oxygen-enriched atmosphere. Air in a confined space becomes combustible when chemicals or gases reach a certain concentration. Flammable gases can be trapped in confined spaces. Flammable gases include, but are not limited to, acetylene, butane, propane, and methane. Dust and work by-products from spray painting or welding can also form a combustible atmosphere.

Some flammable gases are lighter than air and concentrate at the top of a confined space. Vapors from fuels are generally heavier than air and will concentrate at the bottom of the space. A spark or flame will cause an explosion in a combustible atmosphere.

4.2.5 Toxic Atmospheres

Toxic gases and vapors come from many sources. They can be deadly when they are inhaled or absorbed through the skin above certain concentration levels. In spaces with no ventilation, high concentrations can gather and quickly become toxic. Even in lesser amounts, some chemicals can seriously affect your breathing and brain functions.

The effects of toxic gases and vapors vary. Many toxic gases, such as carbon monoxide, cannot be seen or smelled. Some toxic gases have harmful effects that may not show up until years after contact. Others, such as nitric oxide, can kill quickly.

Special meters, which are calibrated to detect toxic gases and vapors, sound an alarm to alert workers of a toxic atmosphere. These meters must be set and operated according to the manufacturer's instructions.

Did You Know?

Generally, a mixture of flammable gas and air must be 5 to 15 percent gas and 85 to 95 percent air for the gas to explode. A mixture of gas and air that does not have enough gas to explode is below the lower explosive limit. This mixture is too lean to explode. A mixture with more than the maximum amount of gas is above the upper explosive limit. This mixture is too rich to explode.

4.2.6 Additional Hazards

In addition to atmospheric hazards, there are several other physical and environmental hazards in confined spaces. Be aware of and take precautions against the following dangers:

- *Electric shock*—Electric shock can occur when power tools and line cords are used where there are wet floors or surfaces. Always ground tools and equipment, or use a GFCI when working in a confined space.
- *Purging*—Purging happens when toxic, corrosive, or natural gases enter and mix with the air in a confined space. Purging most often occurs in pipelines. Purging gases are used to clean pipelines. These gases can create an oxygen imbalance in the space that can suffocate workers almost immediately. Once purging is complete, appropriate ventilation must be established to render the atmosphere safe. Air monitoring is necessary to verify air quality.
- *Falling objects*—Materials or equipment can fall into confined spaces and strike workers. This usually happens when another worker enters or exits a confined space with a topside opening, such as a manhole. Vibrations can also cause materials or tools to fall.
- *Engulfment*—Engulfment occurs when a worker is buried alive by a liquid or material that enters a confined space. Small, loose material stored in bins and hoppers, such as grain, sand, or coal, can engulf and suffocate a worker. Heavier materials or liquids that enter a confined space can crush or strangle a worker. Even when a worker is engulfed only up to the waist, death may result from the massive pressure of the materials and constriction of the body.
- *Extreme temperatures*—Confined spaces that are too hot or too cold can be hazardous. Spaces that are too hot can cause heat stroke or heat exhaustion. Spaces that are too cold can cause hypothermia, a condition in which your body temperature is too low. Another temperature-related hazard in a confined space is steam. Steam is extremely hot and can cause severe or fatal burns.
- *Noise*—Noise in a confined space can be very loud. This is because the sound bounces off the walls of a small space. Too much noise, or noise that is too loud, can permanently damage hearing. It can also prevent workers from communicating. If this happens, you could miss an evacuation warning.
- *Slick or wet surfaces*—Workers can be seriously hurt by slips and falls on slick or wet surfaces. Wet surfaces also add to the chance

---

of electrocution from electrical circuits, equipment, and power tools.

- *Moving parts*—Workers can get struck or trapped by moving parts, such as augers or belts. This usually happens when a worker slips or falls. It also can happen if the operator of the moving machinery is wearing loose clothing or jewelry.

It is important to understand how to protect yourself and your co-workers in a confined space. Everyone must be aware of what is happening on the site and understand how to work safely. These are the most common safeguards to follow during confined-space operations:

- Monitoring and testing
- Ventilation
- PPE
- Communications
- Training

4.2.7 Monitoring and Testing

The air in a confined space must be tested before any workers enter and continuously monitored while workers are in the space. Testing must be done by a properly trained, qualified person, called the confined-space attendant. This person must be company-approved or otherwise designated. The confined-space attendant tests the air for oxygen content, explosive gases or vapors, and toxic gases or vapors.

The atmosphere in a confined space may need to be monitored during the entire job. This is done by attaching monitors to workers entering the confined space or by using outside devices. When the atmosphere is monitored, workers can be assured that the air quality is good and that they will know immediately about changes in the atmosphere that would require them to leave. The confined-space attendant always remains outside the workspace to monitor the workers and their environment. The attendant does not enter the space even to perform rescues but is responsible for calling for evacuation when needed.

Workers who are trained and authorized to monitor the air in a confined space must do the following:

- Read and understand the operating and maintenance manual provided by the monitor's manufacturer.
- Maintain monitors according to the manufacturer's instructions.
- Inspect, test, and calibrate monitors according to the manufacturer's instructions.

- Verify that batteries for battery-operated monitors are working properly.
- Test the atmosphere in a confined space before employees enter the space.
- Continuously monitor the air in the confined space while employees are in the space.
- Carry out the company's safety procedures regarding hazardous atmospheres or emergency situations that may arise.
- Complete any reports required by company policy or OSHA standards (*CFR 1904*).
- Seek retraining or update training as necessary.

4.2.8 Ventilation

If the air in a confined space is hazardous or may become dangerous, the space must be ventilated immediately to remove toxic gases or vapors and to replace lost oxygen. Ventilators blow clean air into the space. They must stay on as long as workers are in the space.

Just because air is being blown into or out of the space, it doesn't mean that the space is being ventilated. Toxic gases can hide in confined spaces. Make sure the attendant has carefully tested the entire space before you enter it.

4.2.9 Personal Protective Equipment

Every job requires some type of PPE. Standard PPE includes hard hats, safety goggles and glasses, boots, and gloves. In a confined space, the following items may also be needed:

- Full-body harness
- Lifelines
- Air-purifying respirator
- Air-supplying respirator

4.2.10 Communication

All workers on a site must be able to communicate with one another. It is especially important for attendants and workers entering the site to be able to communicate. This allows attendants to warn workers about dangers and order an evacuation when necessary. Communication among workers is another way to monitor the confined space.

> **WARNING!**
> When combustible gases or vapors are present, sparks from communication devices can cause an explosion. Therefore, you must not operate electrical equipment, including telephones or cell phones that are not classified as intrinsically safe/explosion-proof.

4.2.11 Training

Entering a confined space requires specialized training. No one is allowed to enter a confined space without the proper training and authorization from the site supervisor. Training gives workers the knowledge to complete their jobs safely and efficiently. If you have not been properly trained, do not enter any confined space.

Explosion-Proof Equipment

Some tasks require workers to use intrinsically safe/explosion-proof electrical equipment. An intrinsically safe (I-safe) communication device is one that has been certified for use in hazardous environments. This equipment is designed to protect workers from explosions caused by sparks. Such equipment includes certified I-safe phones, mobile computers, and portable data terminals. It must be certified to meet the following requirements of the National Electrical Code:

- Outside air cannot penetrate the casing.
- The casing must be strong enough to contain an explosion inside the equipment without igniting the surrounding air.
- The interior must be filled with nonflammable powder or inert gas to prevent ignition.

4.3.0 Underground Safety

Tunnels, shafts, chambers, and passages located underground have their own special safety requirements. Because underground workspaces are completely enclosed except for access and equipment tunnels, the results of an accidental release of gas, fire, and other hazards are magnified accordingly. Plumbers installing lines in underground locations should follow the OSHA requirements outlined in *29 CFR 1926.800* very closely.

Before going underground, ensure that the proper safety precautions are in place both aboveground and in the workspace. Once underground, familiarize yourself with the location of safety equipment, potential hazards, and escape routes. Be prepared, so that in an emergency you can take the correct action immediately to protect yourself and your co-workers from injury and death.

Employers are responsible for ensuring that their employees who work underground receive proper training on underground hazards and safety, including the following elements when appropriate:

- Access
- Air monitoring
- Ventilation
- Illumination
- Communications
- PPE
- Explosives
- Fire prevention and protection
- Emergency procedures

4.3.1 Access

OSHA regulations specify the number and placement of access shafts for personnel and equipment. Hoisting mechanisms are used in shafts to provide access for workers. These hoists must be able to operate safely and effectively in an emergency. They must be able to operate even if the power fails in the rest of the underground workplace. Independent power sources for personnel hoists should meet all safety requirements for the underground workspace.

4.3.2 Air Monitoring and Ventilation

Fresh air must be supplied to underground work areas in sufficient quantities to prevent the build-up of harmful dust, fumes, mist, vapors, or gases. OSHA regulations specify the use of mechanical ventilation in cases where natural ventilation is insufficient to provide fresh air to the workers. Refer to your local applicable code for the minimum safe airflow requirements. Ensure that all machines used in an underground space are properly vented. OSHA regulations require a minimum of 200 cubic feet of fresh air per minute for each employee underground.

The site supervisor is responsible for designating a competent person to perform air-monitoring tests in an underground work site. The atmosphere in an underground workplace must be the same content and pressure as normal surface air. Air-monitoring tests check for oxygen content and contaminants. If flammable or noxious gases or other contaminants are present at higher-than-permitted levels, OSHA regulations specify the correct response.

Working conditions are classified as gassy operations when air monitoring shows one or more of the following:

- The level of methane or explosive gases in the air exceeds minimum acceptable levels.
- A gas ignition has occurred previously in the underground space.

- The underground work area is connected to another area that has been classified as a gassy operation.

Vent all fan and compressor exhausts to ensure that they do not contaminate the underground air. If air monitoring shows that conditions are dangerous to life, evacuate the underground work area, post signs by the entrances, and take all other necessary precautions as specified in OSHA regulations and in your emergency plan.

> **WARNING!**
>
> If the levels of methane or other explosive gases reach 5 percent of the lower explosive limit (LEL), increase ventilation to remove the gases. If the levels reach 10 percent of the LEL, immediately suspend all hot work in an underground workspace. If the levels reach 20 percent of the LEL, immediately evacuate all workers who are not required to deal with the situation, and disconnect all non-emergency electrical equipment. Otherwise, a fire or explosion could result, causing injury or death.

4.3.3 Illumination

Refer to your local code and the OSHA regulations for proper lighting requirements. Use only approved lighting equipment when working underground. All employees working underground should have a portable lamp (either handheld or attached to a hard hat) for emergency use, unless there is sufficient natural light or an emergency lighting system is in place.

4.3.4 Communications

Use radios, walkie-talkies, or other approved communications equipment when voice communications are not effective within the workspace or between the workspace and the surface. OSHA regulations require at least two communication methods, one of which must be voice, for use near shafts when hoisting or transferring personnel. Ensure that powered communications systems are able to operate even if one of the units breaks down. Test radios and other communication equipment before going into the underground workplace.

A specially designated person must be on duty on the surface whenever a worker or workers are in the underground workplace. That person is responsible for calling in aid if needed and keeping an accurate count of all employees working underground.

At the end of your shift, notify the incoming shift about any safety occurrences on your shift,

such as cave-ins, gas leaks, or equipment failures. Ensure that people on the rest of the job site coordinate their activities with the underground work so as to avoid conflicts.

4.3.5 Personal Protective Equipment

Refer to your local applicable code and site safety regulations to determine the appropriate PPE to be used. Use only NIOSH-approved respirators when working underground. Ensure that all respirators and other safety devices are properly maintained according to the manufacturer's specifications. Employers are responsible for training all workers in the proper use of PPE. PPE and safety equipment that is not worn, such as respirators, must be located in easily accessed areas near workstations where smoke or gas is a risk.

4.3.6 Explosion and Fire Hazards

Open flames and fires are not permitted underground. Smoking is also prohibited, and all ignition sources, such as matches and lighters, must be collected before workers are allowed to descend underground. Ensure that appropriate warning signs are posted where fire and explosion hazards are present. Never use or store gasoline underground. Diesel fuel can be used, though you must be sure to follow local code and the manufacturer's specifications when pumping diesel fuel to and from an underground space.

When using gas such as acetylene and liquid petroleum gas (LPG) for hot work, refer to the local applicable code and the OSHA regulations for proper storage and handling. When fuels and gases are not in use, ensure that all fuel containers are sealed and gas tank valves are shut tight. On the surface, flammable and explosive materials must be stored at least 100 feet away from underground entrances.

Clearly mark all entrances and ventilation shafts with signs or other approved indicators. When hot work is being performed underground, post fire watches according to OSHA regulations.

4.3.7 Emergency Procedures

Employers must ensure that rescue teams are on site or available to respond to an emergency within 30 minutes. According to OSHA regulations, when 25 or more people are working underground, two 5-person rescue teams must be available. One team must be on site or no more than 30 minutes away. The other team cannot be more than two hours away. With fewer than

25 workers, only one team is required on site or within 30 minutes of the work site.

Emergency team members must be qualified in rescue procedures, in the use of breathing apparatuses, and in firefighting. Employers are responsible for ensuring that rescue teams are familiar with the layout of the work site.

4.4.0 Lockout/Tagout

Lockout devices and tagout devices protect workers from all possible sources of energy, including water or other liquids. In a lockout, an energy-isolating device that prevents the transmission or release of energy, such as a stop valve, a disconnect switch, or a circuit breaker, is placed in the Off position and locked. Lockout devices can be used with key or combination locks. Multiple lockout devices (*Figure 45*) are used when more than one person has access to the equipment. In some instances, as with work involving valves, a chain and lock can be used to hold the valve in place and keep it from being turned.

In a tagout, components that allow the flow of water or power equipment and machinery, such as switches or valves, are set in a safe position, and a written warning or tagout device is attached to them (*Figure 46*). An effective lockout/tagout program should include the following:

- An inspection of equipment by a trained person who is thoroughly familiar with the equipment operation and associated hazards
- Identification and labeling of lockout devices
- The purchase of locks, tags, and blocks
- A standard written operating procedure that all employees follow

The exact procedures for lockout/tagout may vary at different companies and job sites. Ask your supervisor to explain the lockout/tagout procedure on your job site. You must know and follow this procedure. This is for your safety and

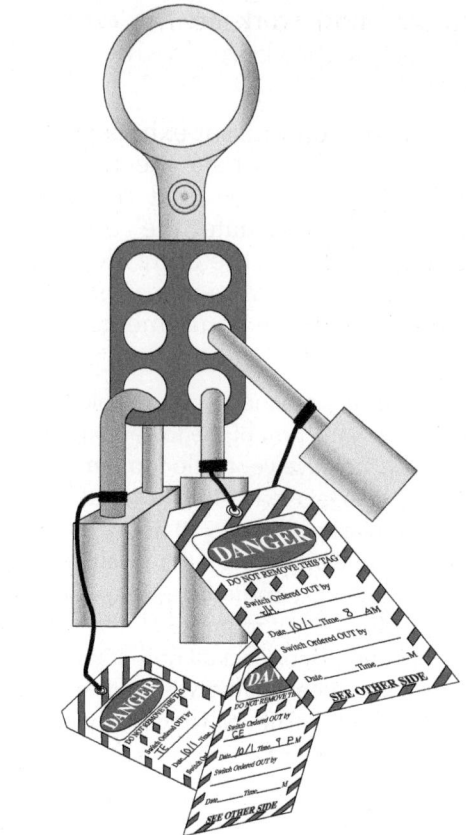

Figure 45 Multiple lockout device.

the safety of your co-workers. If you have questions about lockout/tagout procedures, ask your supervisor. Each foreman should have a kit.

A typical lockout/tagout procedure is made up of these three steps:

- Sequence for lockout/tagout
- Restoring energy
- Emergency removal authorization

Case History

Follow Lockout/Tagout Procedures

The threaded male end of a pressure-relief valve on a sprinkler system riser broke off during an attempt to re-pipe the discharge side of the valve. This caused approximately 550 gallons of water to be discharged into a utility closet and surrounding office areas. There were no injuries as a result of this accident, but there was significant property damage.

The Bottom Line: A lockout/tagout should have been performed on the sprinkler system.

4.4.1 Sequence for Lockout/Tagout

A typical lockout/tagout program sets out a sequence of events that must occur during the procedure. A typical lockout/tagout follows these steps:

Step 1 Check the procedures to ensure that no changes have been made since you last used a lockout/tagout.

Step 2 Identify all authorized and affected employees involved with the pending lockout/tagout.

Step 3 Notify all authorized and affected personnel that a lockout/tagout is to be used and explain why it is necessary.

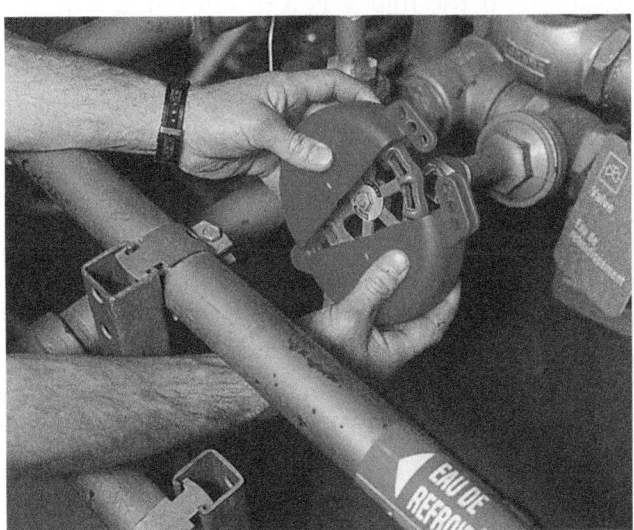

Figure 46 Placing a lockout/tagout device.

Step 4 Shut down the equipment or system using the normal Off or Stop procedures.

Step 5 Lock out all energy sources, which provide power to the device, and test disconnects to be sure they cannot be moved to the On position. Next, open the control cutout switch. If there is no cutout switch, block the magnet in the switch Open position before working on electrically operated equipment or apparatus such as motors or relays. Remove the control wire.

Step 6 Lock and tag the required switches or valves in the Open position. Each employee who is authorized to use the equipment has to have a lock.

Step 7 Dissipate any stored energy.

Step 8 Verify that the test equipment is functional using a known power source.

Step 9 Confirm that all switches are in the Open position, and use test equipment to verify that all components are de-energized.

Step 10 If you have to leave the area temporarily, upon returning retest to ensure that the equipment or system is still de-energized.

4.4.2 Restoration of Energy

Once work is done on the machinery or equipment, energy can be restored. The following steps are typically used to restore energy:

Step 1 Completely reassemble and secure the equipment or system.

Step 2 Confirm that all equipment and tools, including shorting probes, are accounted for and removed from the equipment or system.

Step 3 Replace and reactivate all of the safety controls.

Step 4 Remove the locks and tags from the isolation switches. All employees must remove only their own lock and tag. Notify all affected personnel that the lockout/tagout has ended and that the equipment or system will be re-energized.

Step 5 Operate or close the isolation switches to restore energy.

4.4.3 Emergency Removal Authorization

OSHA regulations (*CFR 1019.147*) state that only the employee who applied a lockout/tagout device is authorized to remove it. However, there are times when emergency removal of a lockout/tagout device is required. For example, the employee who applied the device may be absent. Therefore, OSHA allows an exception to this rule, provided the company follows these additional rules:

- Verify that the authorized employee who applied the lockout/tagout device is not at the facility.
- Make a reasonable effort to inform the authorized employee that the lockout/tagout device has been removed.
- Ensure that the authorized employee knows this before resuming work at the facility.

OSHA regulations require employers to have specific procedures and training in place to handle emergency removal. This training and procedures should be part of the company's energy control program. Some companies may have additional requirements. For example, a company may require that a written record of the emergency removal procedure be made and signed by authorized personnel.

Did You Know?

OSHA provides a basic lockout procedure form that is available for free on its website. The form is designed to help employers develop their lockout/tagout procedures so that they will meet the OSHA requirements. The website address is *www.oshacfr. com*. The form is titled *Typical Minimal Lockout Procedure, 1910.147, Appendix A.*

4.5.0 Safeguards and Emergency Response

It is important to know how to protect yourself and your co-workers on the job. Everyone must be aware of the activities taking place on the site and understand how to perform them safely. Some common safeguards to keep the job site safe include the following:

- Never operate any device, valve, switch, or piece of equipment that has a lock or a tag attached to it.
- Use only tags that have been approved for your job site.

- If a device, valve, switch, or piece of equipment is locked out, make sure the proper tag is attached.
- Lock out and tag all electrical systems.
- Lock out and tag pipelines containing acids, explosive fluids, and high-pressure steam.
- Lock and tag motorized vehicles and equipment when they require repair or are being replaced. Disconnect or disable any starting devices.

You can avoid accidents, injuries, or contamination by making sure you are properly trained, by working safely, and by following the rules. When accidents do happen, it is important to stay calm and tell your supervisor immediately. A worker who is trained in first aid should evaluate the injury and decide on the best treatment plan. For example, if the injury is a minor cut, a bandage can be applied. If the cut is deep and requires stitches, the worker should be taken to the hospital. If the cut is severe and an extreme loss of blood or amputation is involved, immediately call 911 or your local emergency response number.

Many companies, regardless of their size, have an emergency action plan (*Figure 47*). An emergency action plan describes the actions employees should take to ensure their safety in an emergency situation. A company appointed competent person or emergency action plan coordinator is responsible for creating an emergency action plan. Plans vary by company. In general, most plans include some or all of the following items:

- Trainees must not perform emergency actions that they are not trained to do
- Conditions under which an evacuation would be necessary
- A clear chain of command
- Person authorized to order an evacuation
- Evacuation procedures and emergency escape route assignments (*Figure 48*)
- Designated meeting area outside of the building
- Procedures for helping visitors to evacuate
- Procedures for helping those with disabilities or difficulty understanding English
- Procedures for workers who remain to operate critical plant operations before evacuating
- Procedures for accounting for workers after the evacuation
- Designated rescue and first-aid personnel
- The location and telephone numbers of local emergency agencies, such as hospitals and fire and police departments (generally, calling 911 is the most appropriate response)

If you are unsure about the action plan at your job site, ask your supervisor to explain it to you. It can mean the difference between life and death.

EMERGENCY ACTION PLAN

In the event of a fire or other emergency situation on the project that warrants emergency response or evacuation, the following procedures will be strictly adhered to:

1. **SOUND THE ALARM** by shouting loudly and repeatedly the description of the emergency situation you have observed, for example:

   • FIRE, FIRE, FIRE

   • EMPLOYEE INJURED, EMPLOYEE INJURED, EMPLOYEE INJURED

   • NEED PARAMEDICS, NEED PARAMEDICS, NEED PARAMEDICS

   • EVACUATE, EVACUATE, EVACUATE

   Make sure that at least one other employee has heard and understood the alarm.

2. **CALL 911 AND NOTIFY THE EMERGENCY OPERATOR.** The employee first observing the emergency will travel to the closest telephone or cellular phone, **DIAL 911** and **REPORT THE EMERGENCY**. State the following:

   THERE IS AN EMERGENCY SITUATION AT _____

   IN THE EVENT OF A REQUEST FOR PARAMEDIC RESPONSE, THE EMPLOYEE REPORTING THE EMERGENCY SHALL TRAVEL TO THE PROJECT ENTRY POINT AND STAND BY TO DIRECT EMERGENCY RESPONSE PERSONNEL.

3. In the event of an **INJURED EMPLOYEE ALARM**, personnel trained, capable, and prepared to render first aid may respond to the scene of the accident and render the assistance required to sustain the injured employee, until such time as the paramedics arrive. Assist if and ONLY when your assistance may be rendered without personal risk to you, the first-aid provider, due to hazards present in the accident area.

4. In the event of an uncontrollable fire or other emergency that warrants evacuation, **EVACUATE!** Select the closest, safest route to exit the building and proceed in an orderly and expeditious manner outside. While en route to the exit, assist in the notification of other employees by re-sounding the alarm to evacuate. Yell loudly and repeatedly:

   FIRE, EVACUATE! FIRE, EVACUATE! FIRE, EVACUATE!

5. **ASSEMBLE:** As you exit the building, select the closest and safest route, and proceed in an orderly manner to the DESIGNATED ASSEMBLY AREA, where a head count will be taken to ensure that everyone has safely evacuated. Remain in the assembly area until released to return to work or instructed otherwise.

   DESIGNATED ASSEMBLY AREA FOR THE JOB SITE:

   _____

Note: Each individual subcontractor is responsible for taking head counts for his or her employees and for reporting to emergency response personnel if or when employees are unaccounted for.

Figure 47 Sample emergency action plan.

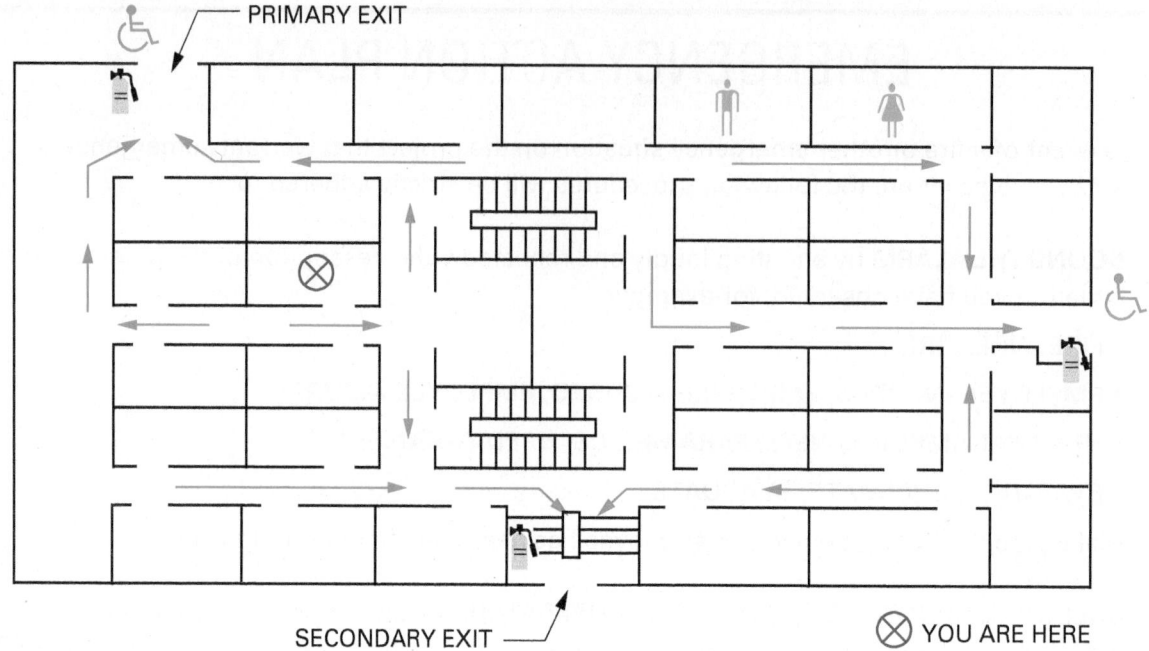

PRIMARY EXIT

SECONDARY EXIT

⊗ YOU ARE HERE

Figure 48 Typical emergency escape plan.

If you are working in a facility where escape maps are necessary, study the map carefully. This is especially important if you are working in a large building. Memorize the route. It is best, of course, to use an exit nearest your location. However, the closest exit may be blocked by fire or debris. Therefore, always know alternate escape routes so that you have more than one way to evacuate.

Alarms

At most job sites, the person responsible for coordinating the company's emergency action plan may have to consider the needs of workers with disabilities or workers for whom English is a second language. For example, sound alarms may also have bright flashing lights, and warning and exit signs may be written in more than one language.

Additional Resources

Environmental Protection Agency. *www.epa.gov.*

Managing Electrical Hazards, Latest edition. NCCER. New York, NY: Pearson.

OSHA. *www.osha.gov.*

4.0.0 Section Review

1. You see construction workers cutting back the walls of an excavation at an angle to its floor. This is called _____.

   a. sloping
   b. shoring
   c. shielding
   d. trenching

2. For working conditions in a confined space to be considered safe, the oxygen level by volume must range between _____.

   a. 6.5 and 10.5 percent
   b. 11.5 and 14.5 percent
   c. 19.5 and 23.5 percent
   d. 24.5 and 28.5 percent

3. The minimum amount of fresh air for each employee working underground required by OSHA is _____.

   a. 100 cubic feet per minute
   b. 100 cubic feet per hour
   c. 200 cubic feet per hour
   d. 200 cubic feet per minute

4. Stop valves, disconnect switches, and circuit breakers are examples of _____.

   a. energy-isolating devices
   b. energy removal devices
   c. lockout/tagout devices
   d. multiple lockout devices

5. If you are unsure about the emergency action plan at your job site, the person you should ask to explain it to you is your _____.

   a. OSHA representative
   b. local fire department
   c. co-worker
   d. supervisor

SUMMARY

Serious accidents and injuries can occur on a work site. Whether you work for a large contractor or you are a subcontractor, you have a responsibility to work safely and follow the rules and regulations of the site. You also have a responsibility to be aware of what's going on around you and to notice the behavior of your co-workers. Personal injury and damage to equipment is less likely to happen when you do so. It is also important that everyone on a job site knows and follows the company's emergency-response plan.

1. According to OSHA, a competent person is _____.

   a. a worker who has been trained to use tools properly
   b. a worker who can identify hazards
   c. a worker who can identify hazards and is authorized to take prompt corrective measures
   d. any foreman or supervisor on the job site

2. OSHA stands for _____.

   a. Occupational Standards for Health Act
   b. Occupational Services and Health Administration
   c. Office of Safety and Health Administration
   d. Occupational Safety and Health Administration

3. One simple way to help ensure a safe work site is to _____.

   a. check the job site periodically and remove any clutter
   b. purchase your own first-aid kit
   c. keep foul weather gear stored in your truck
   d. avoid working in cold weather

4. When you wear a hard hat, the space between your head and the shell of the hard hat should be _____.

   a. $\frac{1}{4}$ inch
   b. $\frac{1}{2}$ inch
   c. 1 inch
   d. 2 inches

5. Air-purifying respirator cartridges absorb certain toxic vapors from the air through _____.

   a. a treated fiber
   b. a baffle chamber
   c. filters filled with baking soda
   d. charcoal

6. The highest level of respiratory protection is provided by _____.

   a. air-purifying respirators
   b. supplied-air respirators
   c. self-contained breathing apparatus
   d. exhaust hoods

7. To obtain the best protection from a respirator, you must perform a positive and negative fit check _____.

   a. the first time you wear it
   b. every time you wear it
   c. after you have worn it for at least 20 minutes
   d. once or twice a week

8. The SDS is a fact sheet prepared by _____.

   a. OSHA
   b. the product's manufacturer or importer
   c. the National Fire Protection Agency
   d. local authorities according to local codes

9. The most important things to look for on an SDS are the _____.

   a. specific hazards, personal protection requirements, handling procedures, first-aid information, and the emergency-response telephone number
   b. chemical hazards, explosive properties, personal protection requirements, and ecological information
   c. chemical composition, toxicological properties, and disposal and transportation information
   d. accidental-release measures, handling and storage requirements, transportation and regulatory information, and the emergency-response telephone number

10. Section 4 of an SDS _____.

    a. provides the manufacturer's contact information
    b. lists firefighting measures
    c. provides general handling and storage information
    d. lists first-aid measures

11. If your supervisor tells you to evacuate the work site, you should _____.

    a. return to work the next day
    b. go to a safe location and wait until you are notified that conditions are safe
    c. walk at least 50 feet away from the site and wait there for further instructions
    d. gather up all of your belongings before leaving the work site

12. The sign or tag that is used to tell workers about potential hazards or unsafe practices is a(n) _____.

    a. danger sign
    b. caution sign
    c. informational sign
    d. safety instruction sign

13. The sign or tag that is used to tell workers about general information not related to safety is a(n) _____.

    a. danger sign
    b. caution sign
    c. informational sign
    d. safety instruction sign

14. The sign or tag that is used for general instructions and suggestions related to safety measures is a(n) _____.

    a. danger sign
    b. caution sign
    c. informational sign
    d. safety instruction sign

15. The two types of protection for openings in walls, floors, or the ground are _____.

    a. guards and covers
    b. poles and screens
    c. ropes and planks
    d. blocks and berms

16. The percent of all accidental deaths in the construction industry caused by slips, trips, and falls is _____.

    a. 10
    b. 15
    c. 20
    d. 25

17. When job site surfaces are wet or icy, you should _____.

    a. delay the job until conditions improve
    b. wear shoes with cleats
    c. spread gravel or other granular material
    d. move to another work area

18. Walking and working surfaces that are located aboveground and sometimes over large equipment are called _____.

    a. ceilings
    b. floors
    c. platforms
    d. ramps

19. To remove tools from scaffolding, it is best to _____.

    a. lower them to the ground using a rope
    b. lower the height of the scaffolding to hip level and then remove the tools
    c. use the buddy system, and pass each tool from one person to another
    d. toss them gently into a strong net set up below the scaffold

20. When working on a roof, it is necessary to _____.

    a. build a flat holding pen to store necessary tools and equipment
    b. erect a wooden wind break
    c. periodically sweep the roof to remove dirt and debris
    d. mount temporary lightning rods on all gable ends

21. Chisels, wedges, and punches are designed to be used with a _____.

    a. wedge
    b. spud wrench
    c. hammer
    d. burring tool

22. When inspecting an impact tool, an indication that the tool may be dangerous is _____.

    a. a mushroomed head
    b. a missing guard
    c. a two-prong plug
    d. no insulation

23. A trench is a narrow excavation in which the depth is greater than the width and the width does not exceed _____.

    a. 10 feet
    b. 15 feet
    c. 20 feet
    d. 25 feet

24. To protect workers from the collapse of an excavation wall in an unshored excavation, install a(n) _____.

    a. cantilever
    b. ladder
    c. trench shield
    d. brace

25. Lockout and tagout devices are used to _____.

    a. prevent the theft of construction tools and equipment
    b. help keep inventory of construction tools and equipment
    c. protect workers from all possible sources of energy
    d. allow only inspectors to gain access to energy sources

Trade Terms Quiz

Fill in the blank with the correct term that you learned from your study of this module.

1. _____ is a condition that occurs when body temperature is too low.

2. A(n) _____ provides information about hazardous chemicals and materials.

3. _____ are structures used in unshored trenches to protect workers from cave-ins.

4. The _____ is a four-color label placed on containers or doors to alert people to specific safety hazards in a product, room, or building.

5. _____ operate using a power source.

6. Confined workspaces that are free of any mechanical, physical, electrical, or atmospheric hazards are _____.

7. When working with or near _____ or hazardous mold, you must always wear a respirator.

8. _____ restrains the movement of trench walls.

9. Acetylene, butane, propane, and methane are considered _____ gases.

10. Torches are considered _____.

11. An unbalanced stress in the soil causes _____ at the surface.

12. Electrocution is the most serious hazard of using _____.

13. _____ is a method of cave-in protection that uses horizontal steps in an excavation.

14. _____ are written warnings to protect workers from energy sources.

15. Safety standards are enforced by the _____.

16. Hazards in a(n) _____ can include chemical, mechanical, electrical, and physical hazards.

17. Subsidence can cause great stress on a trench's _____.

18. _____ tools have sharp edges.

19. When an atmosphere has less than 19.5 percent oxygen, it is considered to be a(n) _____.

20. Tags are placed on machinery and equipment during a(n) _____.

21. A stop valve is considered a(n) _____.

22. A(n) _____ has the training, experience, and authority to stop a job if it is not being done safely.

23. A confined space containing too much oxygen is referred to as a(n) _____.

24. Hammer strikes are a common injury when using _____.

25. The _____ requires employers to tell workers about hazardous materials on the job site.

26. _____ can be used with key or combination locks.

27. Too much or too little oxygen in a confined space creates a(n) _____.

28. Ventilation is essential if you are working in _____.

29. _____ should be locked out and disconnected to make sure they are not powered on.

30. SCBA stands for self-contained breathing _____.

31. _____ are also called barricades and barriers.

32. When air monitoring shows that an underground work area contains higher-than-minimum levels of methane or explosive gases, the work areas are classified as _____.

33. Sound is measured in _____.

34. Freestanding scaffolding must be anchored with _____.

35. A stop valve, a disconnect, or a circuit breaker is placed in the Off position during a(n) _____.

36. Always post a(n) _____ when using torches.

Trade Terms

Apparatus
Asbestos
Atmospheric hazards
Benching
Bladed tools
Combustible
Competent person
Confined spaces
Decibels (dB)
Electrically powered tools
Energy-isolating device

Energy sources
Fire watch
Gassy operations
Guards
Guy wires
Hazard Communication
  (HazCom) Standard
Hypothermia
Impact tools
Liquid-fuel tools
Lockout

Lockout devices
Lockout/tagout procedures
NFPA warning diamond
Nonpermit-required con-
  fined space
Occupational Safety and
  Health Administration
  (OSHA)
Oxygen-deficient atmo-
  sphere
Oxygen-enriched atmo-

sphere
Permit-required confined
  space
Power tools
Protective system
Safety data sheet (SDS)
Shoring
Subsidence
Tagout devices
Trench shields

Section Review Answer Key

Answer	Section Reference	Objective
1. b	1.1.1	1a
2. d	1.2.0	1b

SECTION 2.0.0

Answer	Section Reference	Objective
1. c	2.1.2	2a
2. b	2.2.2; Figure 16	2b
3. a	2.3.1	2c

SECTION 3.0.0

Answer	Section Reference	Objective
1. b	3.1.1	3a
2. a	3.2.2	3b

SECTION 4.0.0

Answer	Section Reference	Objective
1. a	4.1.5	4a
2. c	4.2.3	4b
3. d	4.3.2	4c
4. a	4.4.0	4d
5. d	4.5.0	4e

NCCER CURRICULA — USER UPDATE

NCCER makes every effort to keep its textbooks up-to-date and free of technical errors. We appreciate your help in this process. If you find an error, a typographical mistake, or an inaccuracy in NCCER's curricula, please fill out this form (or a photocopy), or complete the online form at **www.nccer.org/olf**. Be sure to include the exact module ID number, page number, a detailed description, and your recommended correction. Your input will be brought to the attention of the Authoring Team. Thank you for your assistance.

Instructors – If you have an idea for improving this textbook, or have found that additional materials were necessary to teach this module effectively, please let us know so that we may present your suggestions to the Authoring Team.

NCCER Product Development and Revision

13614 Progress Blvd., Alachua, FL 32615

Email: curriculum@nccer.org
Online: www.nccer.org/olf

❏ Trainee Guide   ❏ Lesson Plans   ❏ Exam   ❏ PowerPoints   Other _____

Craft / Level: _____   Copyright Date: _____

Module ID Number / Title: _____

Section Number(s): _____

Description: _____

_____

_____

_____

_____

Recommended Correction: _____

_____

_____

_____

_____

Your Name: _____

Address: _____

_____

Email: _____   Phone: _____

This page is intentionally left blank.

Tools of the Plumbing Trade

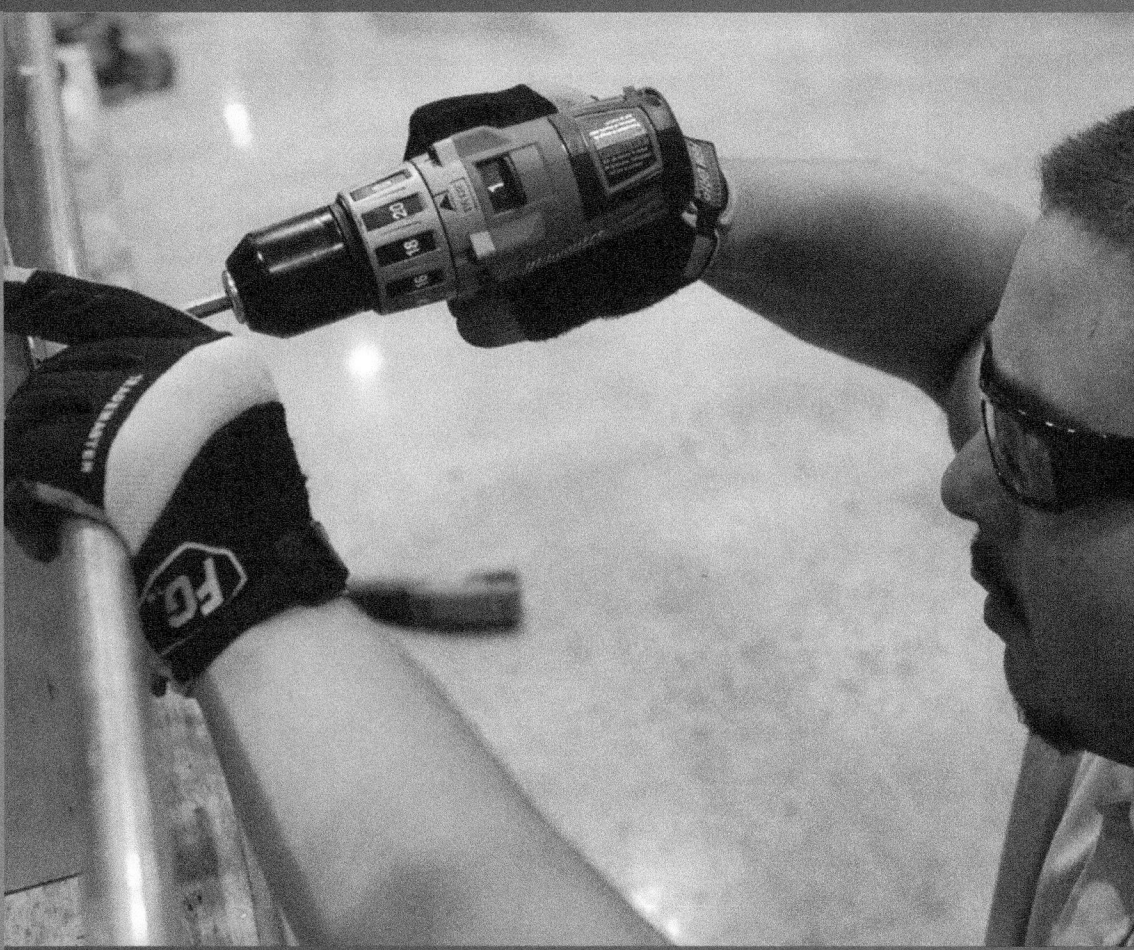

OVERVIEW

A plumber's tools are a plumber's livelihood. Learning to use them safely and maintain them properly is essential to the trade. As plumbers advance in their trade, it is important to invest in quality tools that pertain to the work they perform and skills they develop. Apprenticeship training is a great way to learn how to properly use hand and power tools with expertise and precision.

Module 02103

Trainees with successful module completions may be eligible for credentialing through NCCER's National Registry. To learn more, go to *www.nccer.org* or contact us at 1.888.622.3720. Our website, *www.nccer.org*, has information on the latest product releases and training.

Your feedback is welcome. You may email your comments to *curriculum@nccer.org*, send general comments and inquiries to *info@nccer.org*, or fill in the User Update form at the back of this module.

02103 V4.5

Objectives

Successful completion of this module prepares trainees to:

1. Describe the hand tools used in the plumbing trade.
   a. Identify the basic safety requirements when using hand tools.
   b. Identify the measuring and layout tools used in the plumbing trade.
   c. Identify the leveling tools used in the plumbing trade.
   d. Identify the hand cutting tools used in the plumbing trade.
   e. Describe the tools used to hold and assemble pipe.
   f. Describe the hammers used by plumbers.
   g. Describe the screwdrivers used by plumbers.
2. Describe the power tools used in the plumbing trade.
   a. Identify the basic safety requirements when using power tools.
   b. Identify the power cutting tools used in the plumbing trade.
   c. Identify the drilling and boring tools used in the plumbing trade.
   d. Identify the power threading tools used in the plumbing trade.
   e. Identify the soldering tools used in the plumbing trade.

Performance Tasks

Under the supervision of your instructor, you should be able to do the following:

1. Identify plumbing tools.
2. Properly use plumbing tools.
3. Demonstrate proper maintenance and storage of hand and power tools.

Trade Terms

Amperage
Bar stock
Burr
Diameter
Electrical ground
Ferrous
Flux
Joist
Keel crayon
Kerf
Level
Miter
Nonferrous

Offset
Plumb
Reaming
Scribe
Soapstone
Solder
Soldering
Spirit level
Temper
Tolerance
Torque
Tube stock
Voltage drop

Industry Recognized Credentials

If you're training through an NCCER-accredited sponsor you may be eligible for credentials from NCCER's Registry. The ID number for this module is 02103. Note that this module may have been used in other NCCER curricula and may apply to other level completions. Contact NCCER's Registry at 888.622.3720 or go to *www.nccer.org* for more information.

CODE NOTE

Codes vary among jurisdictions. Because of the variations in code, consult the applicable code whenever regulations are in question. Referencing an incorrect set of codes can cause as much trouble as failing to reference codes altogether. Obtain, review, and familiarize yourself with your local adopted code. Safety codes are developed by the US Occupational Safety and Health Administration (OSHA).

Contents

Contents (continued)

Figures and Tables

This page is intentionally left blank.

1.0.0 HAND TOOLS USED IN PLUMBING

Objectives

Describe the hand tools used in the plumbing trade.

  a. Identify the basic safety requirements when using hand tools.
  b. Identify the measuring and layout tools used in the plumbing trade.
  c. Identify the leveling tools used in the plumbing trade.
  d. Identify the hand cutting tools used in the plumbing trade.
  e. Describe the tools used to hold and assemble pipe.
  f. Describe the hammers used by plumbers.
  g. Describe the screwdrivers used by plumbers.

Performance Tasks

  1. Identify plumbing tools.
  2. Properly use plumbing tools.
  3. Demonstrate proper maintenance and storage of hand and power tools.

Trade Terms

Burr: Uneven or jagged edge left on metal by certain cutting tools.

Diameter: The distance across the center of a circle.

Keel crayon: A waxy crayon used to mark a cutting line on the surface of a tube. Also called soapstone.

Kerf: The cut or groove made by a saw blade, determined by the way the teeth are set on the blade.

Level: Straight on a horizontal plane.

Miter: A surface forming the beveled end or edge of a piece where a joint is made by cutting two pieces at an angle and then fitting these pieces together.

Offset: A combination of elbows or bends that brings one section of the pipe out of line but into a line parallel with the other section.

Plumb: Straight on a vertical plane.

Reaming: A process that removes burrs from pipe after it has been cut.

Scribe: A sharply pointed and hardened steel tool used for marking a surface to be cut by etching a line or a point into the surface.

Soapstone: Another term for keel crayon.

Solder: An alloy (tin plus antimony, copper, and silver) with a low melting point used to join metals or seal joints.

Soldering: A method of joining metals or sealing joints using solder and heat.

Spirit level: A level in which the adjustment to the horizon is shown by the position of a bubble in liquid contained in a nearly horizontal glass tube or a circular box with a glass cover.

Straightedge: A length of wood or metal that does not bow or twist along its length.

Tolerance: Allowable variation in a given measurement or quantity.

Torque: Twisting or turning force applied in a rotating motion. Measurements are given in either inch-pounds or foot-pounds.

Plumbing, like any other skilled profession, has its own tools. You need to become familiar with the wide range of hand and power tools that are specific to plumbing, as well as the many other general-use tools necessary in the plumbing profession. You need to know how to use tools properly, how to select the proper tools for the task, and how to prepare surfaces for tool use. Most important, you must practice all the safety and maintenance rules that apply to the tools you are using. Your tools are your livelihood; they are an investment in your future just like your training. As you begin your plumbing career, you will acquire the tools of your profession, such as instruments for measuring, laying out, leveling, cutting, drilling, boring, and soldering, which joins pipe with solder (a soft metal alloy filler) and heat. Your toolbox will also include tools most construction professionals use: wrenches, pliers, hammers, screwdrivers, and vises.

As you advance in your plumbing career, the tools you will want to purchase will depend in large part on the projects you work on most frequently and on any specialties you develop. Whether you are using your employer's tools or your own, you will want to make sure you have all the tools you need. When purchasing tools, you should consider your career goals and choose tools for the jobs that will help you achieve your goals.

To get the most out of your tools and extend the longevity of their use, you should regularly inspect and maintain them. You should purchase the best-quality tools you can afford but remember that even the highest-quality tools require maintenance; therefore, you should know how to take care of all your tools. Care for tools includes knowing how often to inspect your tools for damage and how often to perform regular maintenance. Well-maintained tools will last longer and save you money since you won't have to replace tools as frequently. The money saved allows you to buy more tools and become a better-equipped plumber. No one expects you to be fully equipped just starting out, but you should always be looking to add to your toolbox as your personal finances allow. Additionally, the care you take of your own tools should be applied to any tools your company lends you.

As an apprentice plumber, you should have the basic tools listed in the *Appendix* of this module. The plumbing tasks you take on as an apprentice require supervision by more experienced workers. Their direction will include how to use and care for tools properly and safely. Use this as an opportunity to learn more about your craft. Learning how to use your tools goes hand in hand with learning how to install, repair, and maintain plumbing systems and fixtures. This learning should become habitual.

1.1.0 Hand Tool Safety

Safe work habits are important to all workers. Plumbing professionals must know how to use tools safely. Many of the tools you use as a plumber can be dangerous if used improperly or if you are unfamiliar with the tool. This is one reason why your apprenticeship training is so important: it teaches you proper use of the tools you need to do your job.

Safety on the job site protects everyone from injury and possible death. All workers need to follow safety procedures not only for their own sake, but also for the sake of their fellow workers. Developing safe work habits is more than just knowing specific dos and don'ts. You should develop an attitude wherein safety is always the foremost consideration. You will see signs posted on job sites that urge workers to "Think Safety." These signs are a constant reminder of the importance of safety to you, your co-workers, and the job. In general, safe operation of tools means careful operation of tools. If you are using a tool safely, you are likely doing a top-quality job as well.

As a plumber, you should follow the same safety precautions in your use of tools as other construction professionals do. That means always using the appropriate personal protective equipment for the tools you use. When using a potentially dangerous tool, you must never skip safety preparations in order to save a few minutes or to speed up the work. An injury that results from your failure to take those safety precautions will cost you far more time and money than a few minutes of preparation. As an apprentice plumber, create a work ethic in which safety is of utmost importance.

Listing every possible safety precaution would be impractical. Yet you can start by becoming familiar with and adhering to the basic safety principles outlined in the following sections. You may also want to refer back to the *Plumbing Safety* module.

Using personal protective equipment (PPE) is the first step in ensuring a safe work environment:

- Wear the PPE that is appropriate for the task.
- Keep your PPE clean and in good condition.
- Inspect your PPE often for signs of wear or damage.

Following good work habits is essential to preventing accidents and injuries on the job. Exercise these guidelines as you develop safe and professional work habits:

- Keep your work area neat. Short lengths of pipe and other debris scattered around can cause you or someone else to fall. It also makes the job site look unprofessional.
- Concentrate on your work. Avoid distracting fellow workers. Give your undivided attention to what you are doing. That way, you will not make careless mistakes that could hurt you or your co-workers or that could slow down the job.
- Do not hold or contort your body in such a way that a tool's operation might hurt you. Avoid using hand tools with your wrist bent, if possible. Always have a firm grasp on tools, especially power tools.
- Know which tools are available and what each tool has been designed to do. Always follow the manufacturer's recommended specifications when using tools. Use each tool as it is supposed to be used (for instance, do not use wrenches as hammers or screwdrivers as chisels).
- Inspect tools for defects before each use, and repair or replace defective tools.
- Do not apply excessive force or pressure when using tools.

- Do not cut toward yourself when using cutting tools. Always aim the blade away from your body.
- Carry hand tools to and from the work site in a strong, secure toolbox. Never carry a sharp tool or blade in your pockets.
- Do not throw hand tools to co-workers; hand them (handle first, if the tool has a handle) to co-workers.
- Point the tips of sharp tools, such as saws, chisels, and knives, that are lying on benches or tables away from aisles or places where people are walking. Make sure tool handles do not extend over the edge of the workbench or worktable. A job site should not be littered with unattended or unused tools.
- Store tools properly after use.

Maintenance and safety go hand in hand. Ensure the safe use of hand tools by caring for them and maintaining them properly:

- Check to see that cutting edges are sharp, and keep them sharp by using a flat file followed by a whetstone.
- Keep tools clean and lubricated (as appropriate).
- Prevent tools from rusting by applying light oil to metal parts, especially when the tool is in storage for long periods.
- Replace steel blades as needed.
- Check to see that tool handles are secure.
- Never use tools that are in poor condition. A well-maintained tool is far less likely to cause an injury than a tool in poor condition.

1.2.0 Measuring and Layout Tools

A plumber's basic instruments include measuring and layout tools. You need to take precise measurements before installing plumbing piping, systems, and fixtures. Measuring tools allow you to determine the length, width, height, and diameter (distance across the center) of objects such as pipe. These tools also allow you to determine if a surface is level (straight on the horizontal plane) or plumb (straight on the vertical plane). Layout tools are used to make accurate lines, circles, and curves.

Accurate measurements are fundamental to plumbing. Inaccurate measurements lead to poorly designed or useless plumbing systems. Getting the right measures at the beginning is essential to productivity in plumbing work. To get the right measures, you need to have the right tools (rules, tape measures, squares) and know how to use them.

1.2.1 Folding Rule

Folding rules are usually made of wood and come in 6-foot and 8-foot lengths (*Figure 1*). They often come equipped with a 6-inch-long brass extension that can be used to take inside measurements between two walls or to measure the depth of an opening. The brass extension allows you to measure in tight places or to measure short distances beyond the end of the folding rule.

You can extend the folding rule above your head to measure heights. This allows one person to take measurements that would otherwise require two people and a ladder. To maintain a folding rule, lubricate the joints lightly with oil or silicone. Be careful when opening or closing the rule because it is easy to break a folding rule at its joints. Try not to drop it, because that might loosen the joints enough to give inaccurate measurements. Small inaccuracies in measuring inches can lead to much larger errors over a span of many feet.

1.2.2 Plumber's Rule

A variation of the wooden folding rule is the plumber's rule (*Figure 2*). It has measurements on one side and a 45-degree scale on the other. You use it to measure offsets when you are installing a pipe or vent around an obstacle. Using elbows or bends, an offset changes the direction of a run of pipe (for example, to clear an obstacle) and then returns the pipe to a path that is parallel to its original direction.

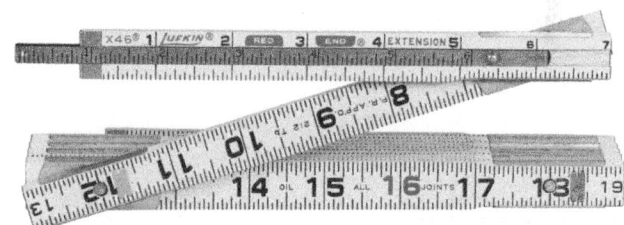

Figure 1 Folding rule.

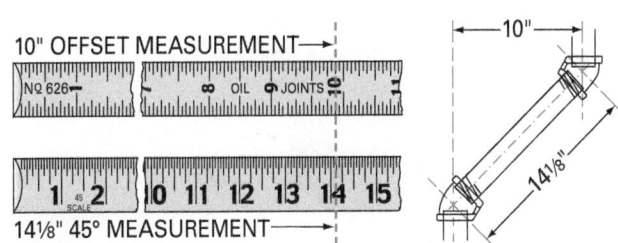

Figure 2 Plumber's rule with 45-degree scale.

1.2.3 Tape Measure

Steel tape measures (*Figure 3*) are valuable in many professions, including plumbing. Steel tape measures are compact, flexible, and easy to carry. They can clip onto a belt or fit into special leather holders. Retractable steel tape measures are available in lengths up to 30 feet. Plumbers most often use steel tape measures that retract into a casing at the push of a button.

When measuring long pipe runs, a 25-foot, 50-foot, or 100-foot-long steel tape can be used (*Figure 4*). These lengths of tape are wound back into their cases by a hand crank.

Steel tape measures come in various widths. Most plumbers prefer a steel tape that is 1-inch wide, because it is stiffer and easier to work with. You can use it to take longer measurements without having to call on a co-worker to help. A 1-inch tape is bulkier, but its stiffness lets it extend farther without kinking or bending than, say, a $\frac{1}{2}$-inch tape.

All steel tape measures must be kept clean, dry, and free of kinks. Dirt and sand wear away the markings on the tape. Water causes the tape to rust, especially if it gets into the casing that holds the rolled-up tape. Kinks in the tape that are not straightened out prevent it from rewinding. Wipe away water and dirt before rewinding a steel tape. Apply a light machine oil or penetrating oil periodically to keep the tape lubricated and operating smoothly.

Plumbers also use cloth tape measures. When using a cloth tape measure, be sure to avoid stretching it. Otherwise, you will get an inaccurate measurement.

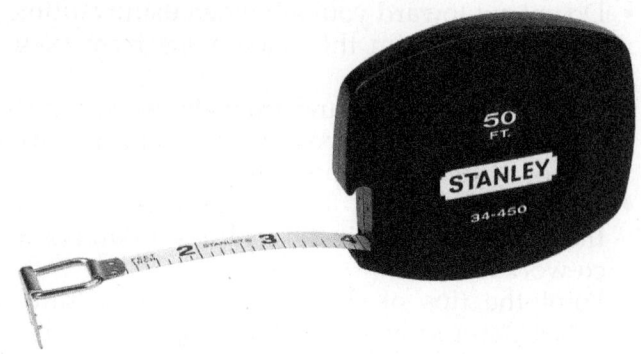

Figure 4 Steel tape measure.

1.2.4 Squares

Straight, accurate lines and angles are necessary for any construction work, including plumbing. A man named Laroy Starrett recognized the importance of straight lines and true angles (usually 90 degrees, with one edge perpendicular to another) and invented the combination square in 1877. Because of its versatility, the square is an essential tool for tradespeople.

Square tools are used to make sure lines are straight and angles are true. Squares are used for marking, testing, and measuring. The type of square to be used depends on the job and the plumber's preference. Square tools are not square-shaped at all; squares that you will use most often are the speed square (shaped like a triangle), the combination square, and the framing square.

Speed square—The speed square, also known as a rafter angle, magic square, or protractor square, is used for marking and measuring. As shown in *Figure 5*, the speed square is actually shaped like a triangle, but it functions as a square. The speed square is made of cast aluminum. It combines a try miter, protractor, and framing square in one tool.

Figure 3 Retractable steel tape measure.

Figure 5 Speed square.

One leg of the triangle has degree gradations marked on its face, allowing for fast and easy layout. The other leg has a raised edge that allows it to fit into the workspace. The legs of a speed square range from 6 to 12 inches in length. The longest side of the triangle, also called the hypotenuse, has degrees marked on it (from 0 to 90) to help in measuring and in marking miter cuts. A speed square with 12-inch-long legs will have a hypotenuse of just less than 17 inches (16.97 inches, to be exact).

Use a speed square to see if a cut or joint is square, to mark cutoff lines, or as a straightedge, to see if a surface has been warped or cupped. You can use a speed square as a straightedge because it does not bow, twist, or warp along its length. You can also use it as a cutting guide when using a handheld circular saw. The speed square is small, making it easy to store and carry on the job site. Smaller speed squares, in fact, fit into the pocket of most tool belts.

Combination square—The combination square (*Figure 6*) is used to establish whether two sides of an object join in a true 90-degree angle. The combination square is adjustable; it has a rigid steel blade 12 inches long with a slot in the middle of it that allows the head to slide along the length of the blade and be fixed to any point along it. The head has both a 45-degree and a 90-degree angle measure. Some combination squares have a small spirit level in their handle for leveling and a carbide scribe for marking metal.

The combination square is useful in marking and checking 90-degree crosscuts and miter cuts. When the head is set at the end of the rule, the combination square can measure height and can also be adjusted to measure depth. Like the speed square, the combination square can be used as a saw guide. Combination squares are precision tools and are very useful for work that requires extremely accurate measurements. The best combination squares will not lose their accuracy even when used a lot. To maintain combination squares, keep them clean and dry, avoid dropping them, and apply a little machine oil to them periodically.

Framing square—The framing square (*Figure 7*) is not square-shaped either. This tool is L-shaped, made of flat steel or aluminum, with a 24-inch long, 2-inch wide blade that is perpendicular to a 16-inch long, 1-$\frac{1}{2}$-inch wide tongue. The framing square has a variety of uses: marking and laying out patterns, testing for squareness, and measuring. The framing square can also be used as a straightedge for testing the flatness of a surface. Most framing squares are marked in inches and fractions of an inch, down to $\frac{1}{8}$ inch.

> **NOTE**
>
> Be careful not to drop a square or strike it hard enough to change the angle between the blade and head (of a combination square) or tongue (of a framing square). Keep squares dry to prevent them from rusting. A periodic coat of light machine oil or penetrating oil helps these tools last longer.

1.3.0 Leveling Tools

Squares are useful for measuring small rectangular objects. For larger, fixed structures, however, a square is not large enough to determine true horizontal and vertical lines. This is where leveling tools become useful.

The term level describes the straightness of horizontal members. The term plumb describes the straightness of vertical members. Level and plumb members intersect at 90-degree angles, and level and plumb tools are used to establish precise horizontal and vertical lines, respectively. There are many of these tools, ranging from simple spirit levels to more sophisticated electronic and laser instruments.

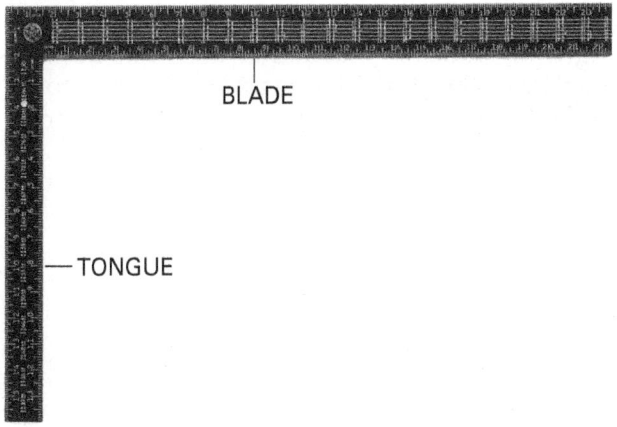

BLADE

TONGUE

Figure 7 Framing square.

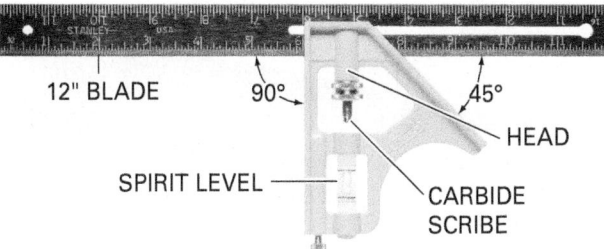

12" BLADE   90°   45°

HEAD

SPIRIT LEVEL   CARBIDE SCRIBE

Figure 6 Combination square.

1.3.1 Spirit Level

The most common tool used to check both level and plumb is the spirit level (*Figure 8*). Spirit levels are made of wood, aluminum, or a lightweight alloy. Plumbers generally buy aluminum or magnesium spirit levels because they are resistant to moisture.

The essential element of the spirit level is a sealed glass or plastic tube containing a specific amount of alcohol. This tube, or vial, is slightly curved, and the outside of each vial is marked with two parallel lines in its center. This vial is mounted into the body of the level tool. The tube is almost completely filled with the liquid, except for an air bubble, upon which the entire function of the instrument depends. When the frame of the tube is precisely level horizontally or precisely plumb vertically, the air bubble is centered between the two parallel lines.

Spirit levels generally contain three bubble vials. The vial mounted at the center of the tool is used to establish true horizontal (and verify level). One vial is mounted transversely at each end for establishing true vertical (and verifying plumb). Vials are replaceable. Spirit levels come in a variety of sizes. The 24-inch-long spirit level is standard, but spirit levels also come in 28-inch, 48-inch, 72-inch, and 78-inch-long sizes.

Spirit levels are accurate only up to their lengths. The longer the level tool is, the greater its accuracy. For accuracy over greater lengths, you must use a longer level. But if you are working in a cramped space, as plumbers often do, a level tool that is too long is useless. These considerations—the length of plumb or level to be measured and the amount of room you have when taking these measures—should guide your choice of spirit level for a job.

One way to increase the accuracy of a spirit level is with a straightedge. A straightedge is a length of wood or metal that does not bow or twist. Place the straightedge across the span to be checked, and put the spirit level in the center of the straightedge.

1.3.2 Plumber's Level

The engineer's level or plumber's level (*Figure 9*) can measure slope (angle) as well as plumb and level. One of the vials is adjustable to different angles. When you adjust the vial and hold the level at the desired angle, the bubble in the vial is centered. Check one level against another to ensure the reading is correct.

You can check the accuracy of your level by placing the level on your work. Note the position of the bubble. Place the level end to end and check the bubble again. The readings should be identical.

1.3.3 Torpedo Level

The torpedo level (*Figure 10*) is a small spirit level that is useful in confined areas or for leveling short runs of pipe. Most torpedo levels have a vial that indicates 45-degree angles. Plumbers often use a torpedo level with a magnetic base. This is handy because the magnetic base sticks to iron pipe, making maneuvering in a confined space easier. Torpedo levels are usually 9 inches long and can fit comfortably into a pants pocket. They are also tapered at the ends.

Did You Know?

When referring to a spirit level, the word spirit refers to alcohol, the fluid used in the level.

OPENING FOR TOP-READING

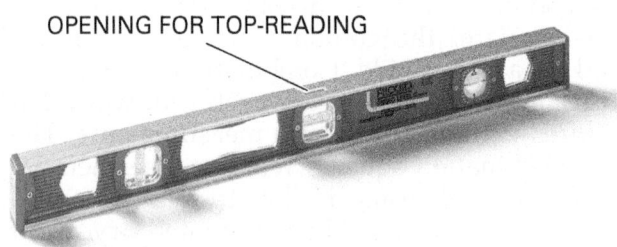

Figure 9 Plumber's level.

45°-ANGLE VIAL

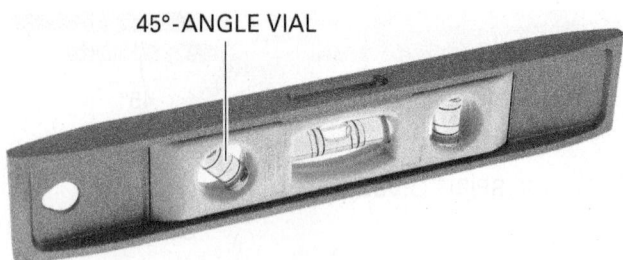

Figure 10 Torpedo level.

OPENING FOR
TOP-READING

Figure 8 Spirit level.

1.3.4 Line Level

Plumbers use a line level by hooking it on a tightly stretched, nylon string or line between two points to be leveled with each other (*Figure 11*). Hooks at either end of the line level attach to the line at about the midpoint of the reach. When the air bubble is centered between the markings on the vial, this indicates the trueness of the line to which the line level is attached. This is how you make sure that pipe laid between two distant points is level along its entire distance. By using a line level, you can transfer (determine or calculate) vertical dimensions, such as height or elevation, over long distances without using the larger leveling instrument called a transit. Laser levels have largely superseded line levels, but you still might be called on to use a line level in the field.

When using a line level, make sure the string is taut. A sagging string leads to an inaccurate measurement. Even the most tightly stretched string, however, can sag a little, which means the line level, for all its convenience, is not completely accurate. It should therefore not be used in situations where precise measures are required.

Top Reading

As a plumber, you will take most of your readings at eye level, by looking at the spirit level as it lies on the surface of the material being leveled. Sometimes, however, you need to place the spirit level on a pipe below you, and you are only able to see the spirit level by looking down on it. In these cases, you need to use a spirit level that allows the user to top-read. This type of spirit level has an opening on top that allows you to see the vial and the air bubble from above.

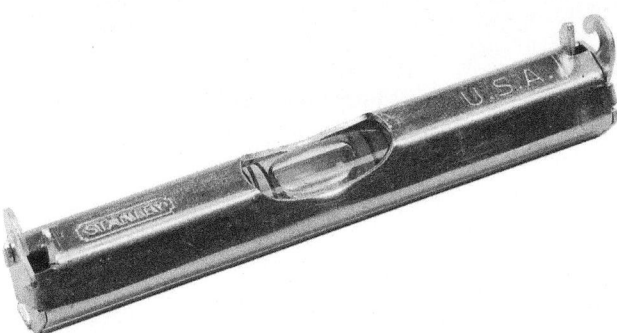

Figure 11 Line level.

1.3.5 Precision and Distance Leveling Instruments

If distances are too great to be measured accurately by the levels mentioned above, or if more accuracy is required, you will need to use a builder's level (also called a surveyor's level) or a transit level with a leveling rod (*Figure 12*). For these instruments to work, they must be calibrated properly. You will learn more about instrument calibration later in this module.

Both the builder's level and the transit level work on the principle that a line of sight is perfectly straight. A telescope is mounted on a tripod and rotated in a complete circle without changing its horizontal position. You then look, or sight, through the telescope to a stadia (or graduated) rod mounted or held vertically by a co-worker some distance away. The stadia rod is a long, brightly painted rod with graduated measurements (usually in feet and tenths of a foot). The rod is located at a point of interest and sighted through the telescope to determine vertical dimensions.

The main difference between a builder's level and a transit level is that the telescope with the builder's level is fixed, but the telescope with a transit level can move up and down. A builder's level, then, can work only on the horizontal plane; the transit level can work in either the horizontal or the vertical plane. You can use either the builder's level or the transit level to measure differences in elevation between two distant points.

The transit level is a precision instrument, calibrated to indicate not only true horizontal but also the angle of inclination in degrees, minutes, or seconds. The telescope is mounted on a horizontal circular plate; attached to this plate are tools that balance and orient the telescope. A compass permits the user of the transit to approximate the bearing of the scope. Once the transit is set up on a tripod, the legs of the tripod must not be moved, because doing so affects the accuracy of the measurement. The telescope is then pivoted on a horizontal axis to point in any direction. When a co-worker positions the stadia rod, you view the rod through the telescope and then determine the relative height of the grade (also known as slope) or the height of the object upon which the rod stands.

Did You Know?

How can you be sure your level is level? Check one level against another periodically to make sure you get accurate readings.

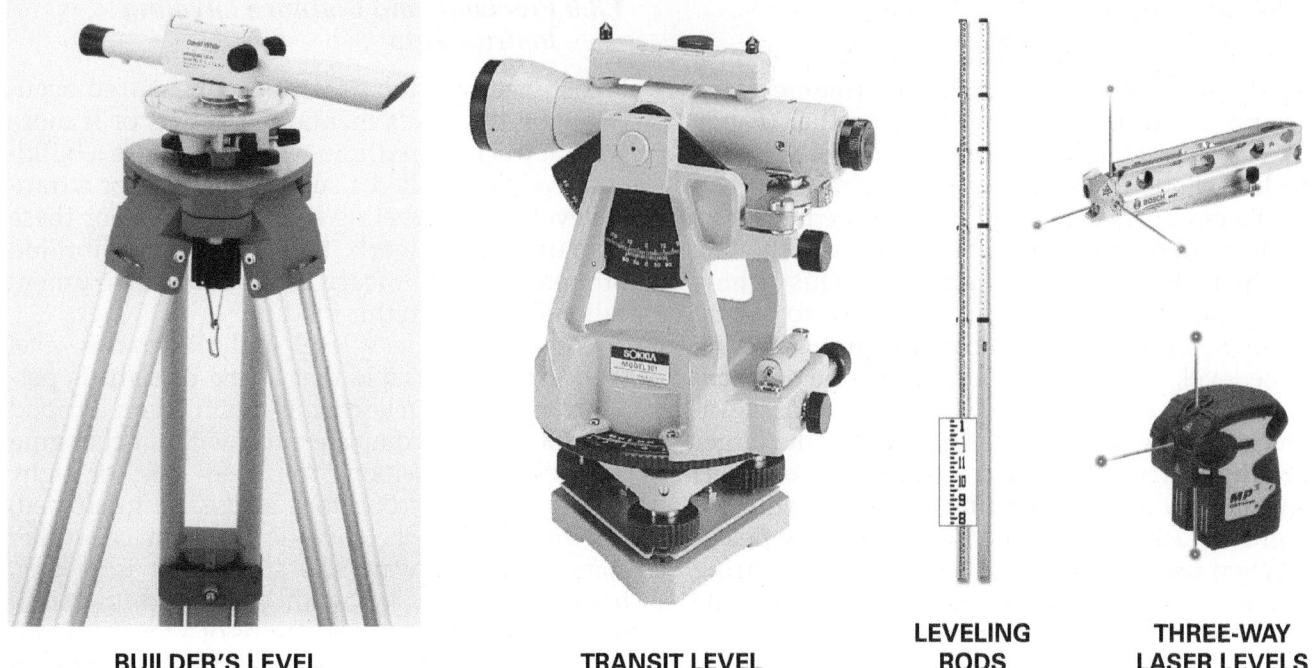

| BUILDER'S LEVEL | TRANSIT LEVEL | LEVELING RODS | THREE-WAY LASER LEVELS |

Figure 12 Precision levels.

1.3.6 Laser Level Tools

Another tool you will use is a cold beam laser (*Figure 13*). This leveling device projects a thin, 90-degree beam of low-wattage light. Laser leveling instruments are more expensive than other leveling tools (although they are becoming more affordable), but they offer several advantages. Lasers are more accurate over longer distances. Lasers are very easy to use when establishing vertical or horizontal lines and are used for measuring or laying out openings.

> **CAUTION**
> Use cold beam lasers with extreme care. Although low-power lasers can be used, arcs can still be unpredictable. Always wear appropriate personal protective equipment and refer to the manufacturer's specifications.

> **CAUTION**
> Don't lean on your level. These precision tools can lose their calibrated accuracy if you do so and will need to be reset, which wastes time on the job.

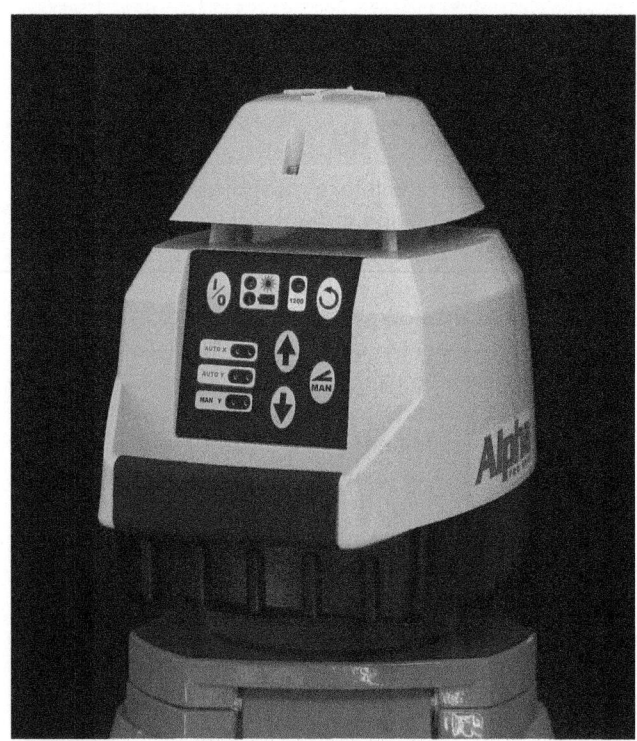

Figure 13 Cold beam laser leveling device.

Some laser level tools can project beams up to 1,500 feet to hit a specific point. They can have a variety of components designed to get the most accurate possible readings, including the following:

- Large vertical and horizontal indicator bubbles
- Lens refractors that show line or dot images
- Dual diodes for increased beam visibility
- Wall, floor, or tripod mounts, brackets, or clamps
- Built-in bump sensors that automatically notify the user, or shut off the device, if the instrument has been bumped hard enough to affect the accuracy of its readings
- Electronic level vials that dampen vibrations from conditions at the job site, to ensure the stability of the laser beam over distance
- Remote control capability
- Optional receivers that increase the laser's operating range even more

Using this tool, you can tell immediately if something has been aligned. If you are installing gravity-flow pipelines, for instance, you can use this tool as an alignment guide and lay the pipe along the beam of light emitted by the laser. Other applications of cold beam lasers include installing and aligning such building elements as walls, partitions, and access floors.

Portable, handheld lasers that feature plumb (both up and down), level, and square capabilities can greatly enhance productivity, which saves time and money. Lasers can be set up more quickly than other leveling devices and, because of their accuracy, require fewer repeat measures.

Calibrating a Builder's Level

A leveling vial is used on the builder's level. It is attached directly to the telescope. To make sure the telescope measures a grade accurately, you must level it by using the attached leveling vial. The vial is a spirit level tool design, but it has been manufactured to be much more accurate.

Four leveling screws are used to level the vial. You rotate the telescope until it lies over two opposing leveling screws. You then adjust the two screws in opposite directions until the bubble is centered in the leveling vial. After rotating the telescope 90 degrees, you adjust the other two screws. This process should be repeated at least once to make sure that the level is accurately calibrated. Once the level is properly set, you are ready to measure grade.

The time saved when using a laser leads to cost savings. The initial expense of a laser is quickly made up in increased productivity.

These devices require relatively little maintenance. Like other levels, lasers should be cleaned off after use and stored in a safe, dry place. Once a year, the unit should be cleaned, checked, and adjusted by a professional. Heat, cold, and moisture—conditions common to plumbing sites—can affect a laser's operating performance. The degree to which these conditions affect the laser's stability and accuracy depends on the quality of the instrument itself. Higher-quality lasers are generally more resilient to adverse environmental conditions. Some lasers are specifically designed and tested to endure shock, moisture, and temperature changes. The performance of all lasers, however, begins to deteriorate rapidly at temperatures of 110°F or higher.

The light emanating from a cold beam laser will not instantly injure humans, but long-term exposure could injure the eyes. Therefore, never point a laser level at another person. Only qualified and trained plumbers should use this equipment. Cold beam lasers must meet all of the Electronic Product Radiation Control provisions of the federal Food, Drug, and Cosmetic Act.

WARNING!

Never look directly into the laser beam or point the beam at a co-worker. Over time, exposure to the beam can injure eyes.

NOTE

Levels and transits are precision instruments, and they are very expensive. Take care not to drop them or bump them against other materials, because loose or broken vials make these tools useless. You also need to keep levels and transits clean and dry so they retain their true measure.

Did You Know?

In some situations, such as in bright sunlight, it may be difficult to see the laser beam. Enhancement goggles are available that allow the user to see the laser beam more easily. Some laser level instruments are specifically designed for outdoor use and have beams that are easily visible in sunlight.

1.3.7 Plumb Bob

A plumb bob (*Figure 14*) is a precise, yet simple, measuring tool used to align points vertically, from one floor to another floor below. To establish this exact vertical line, the plumb bob uses the law of gravity. A plumb bob generally consists of a specially designed weight attached to one end of a nylon thread. Nylon is used because it is more resistant to dampness. A precision plumb bob has a pointed tip.

When you are setting a vertical run of pipe, you need to know the center point on that run of pipe. To find this center point, dangle the plumb bob from a point in the top plate of the framed ceiling. The bob is allowed to swing freely on the string. When the bob stops swinging and is hanging straight and free, the point of the bob will be directly below the point where the string has been affixed above. The point on the level plane directly below the tip of the bob, then, will line up exactly vertically with the point above, and the level plane will also be exactly perpendicular to the string holding the bob. You can then pass a segment of pipe through the level planes at precisely the same point from floor to floor.

Although the plumb bob is not an intricate tool, it needs to be handled carefully. Be sure not to drop the plumb bob. If the tip of the plumb bob point becomes bent or rounded, it provides inaccurate readings. If the string is not properly attached to the center point of the plumb bob, it will give inaccurate readings because the bob hangs at a slight angle to the string (*Figure 15*). If the tip becomes damaged or out of alignment, replace the plumb bob. If the string is not properly attached, replace it. Before using a plumb bob, always check the condition of the tip and the alignment

Figure 15 String attached to a plumb bob.

of the string. Laser plumb bobs create an accurate and highly visible plumb reference instantly. Laser plumb bobs have magnetically dampened stands that allow the plumb bob to settle and be used much faster than the traditional plumb bob with string.

> **WARNING!**
> The point on a plumb bob is sharp. It is particularly easy to stab your toe with this instrument. Handle this tool with care to avoid hurting yourself or a co-worker.

1.3.8 Chalkline

The chalkline (*Figure 16*) is used to lay out long, straight lines of chalk on smooth or semi-smooth surfaces. You pull the chalkline taut between two points, then carefully stretch the line away from the surface and let it snap back to produce a straight line of chalk. To maintain a chalkline, add powdered chalk, and replace the string line when it is worn.

Because of the way most chalklines are designed, they can be used instead of a plumb bob. When you hang the chalkline overhead from the metal clip and allow the box to hang free and straight, the pointed end at the base of the chalkline indicates the exact point below on the vertical plane. You then transfer the mark to the floor to match the mark on the overhead framing member.

Figure 14 Plumb bob.

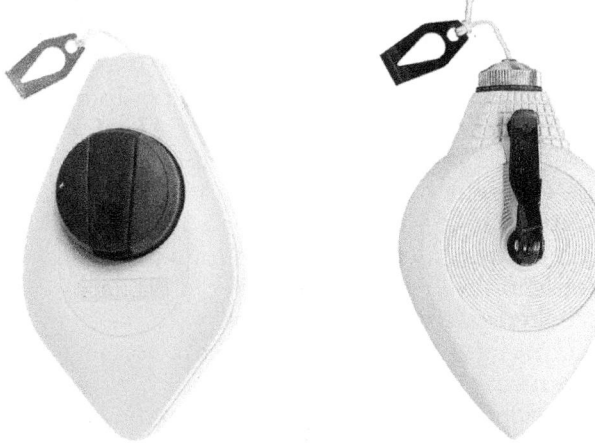

Figure 16 Chalklines.

1.4.0 Hand Cutting Tools

Sometimes plumbers need to make changes to a commercial or residential structure before installing plumbing materials and fixtures. Plumbers use tooth-edged cutting tools such as hacksaws to make these changes.

Smooth-edged cutting tools are also used by plumbers. These tools have a smooth cutting edge instead of teeth. Chisels, manual pipe cutters, and shovels are the smooth-edged tools most commonly used by plumbers.

> **WARNING!**
>
> Rusty or dull cutting tools not only perform poorly but can also cause serious accidents if they slip.

Preparation is essential to making accurate and safe cuts. Before starting to cut, determine what kind of saw and what kind of blade is most appropriate for the material you are cutting. Carefully measure and mark the dimensions of the cut (its distance, angle, and depth), and then adjust the saw controls to match those measurements. Whatever saw you choose, make sure it is in good working condition, with sharp teeth and a tight handle.

> **NOTE**
>
> When using personal protective equipment (PPE), be sure to follow the manufacturer's instructions carefully.

> **CAUTION**
>
> Walls and floors may conceal wiring, pipes, conduits, or supporting beams. Before you saw into a wall or floor, check to find out what is behind the wall or below the surface of the floor.

1.4.1 Hacksaws

The hacksaw is a multipurpose cutting tool. Hacksaw types (*Figure 17*) include common hacksaws, which can be used to cut metal and other materials, PVC saws, and keyhole saws. Keyhole saws are used for cutting openings in drywall and plywood. Mini-hacksaws often come with 6-inch-long blades called Tiny Tims, which are commonly used by plumbers (*Figure 18*).

Hacksaw blades come in different styles for different uses. Blades differ in length, flexibility, the set (or bend) of their teeth, and coarseness (number of teeth per inch), depending on the material they are made to cut. The hacksaw blade most often used to cut pipe is a flexible-back blade. Its teeth are set, which means each tooth is bent to alternating sides so the blade cuts a kerf (the groove made by a saw blade). The kerf is slightly wider than the blade is thick (*Figure 19*). This prevents the blade from binding, and it reduces the friction that results from cutting. Blades are typically very coarse (*Figure 20*). Keep the blades sharp to make a good cut.

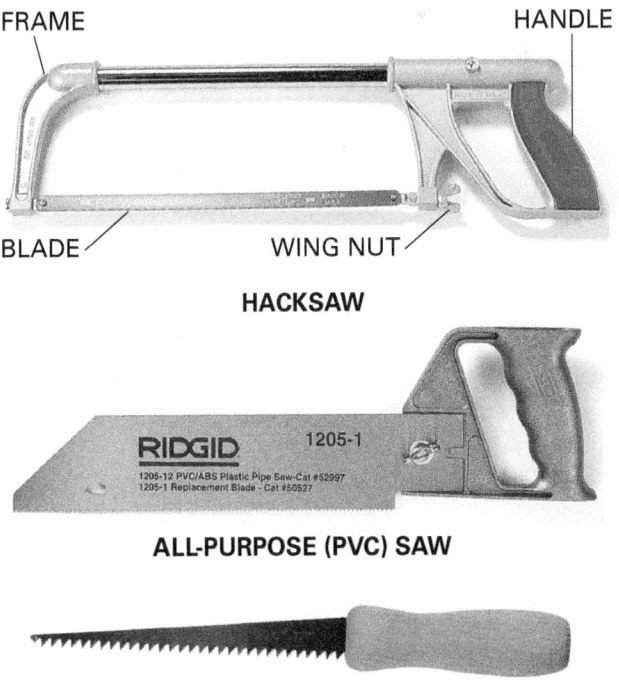

Figure 17 Hacksaws.

Tools of the Plumbing Trade

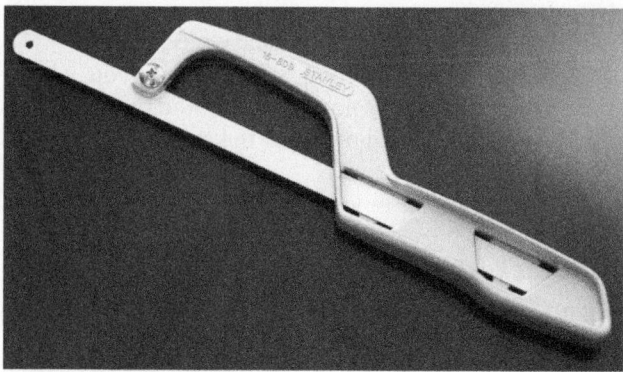
Figure 18 Mini-hacksaw.

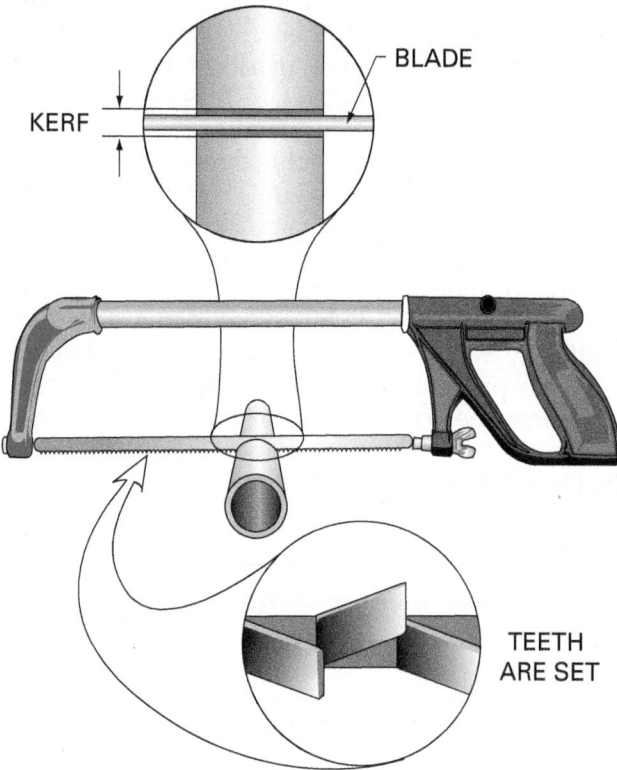
Figure 19 Hacksaw blade kerf cut.

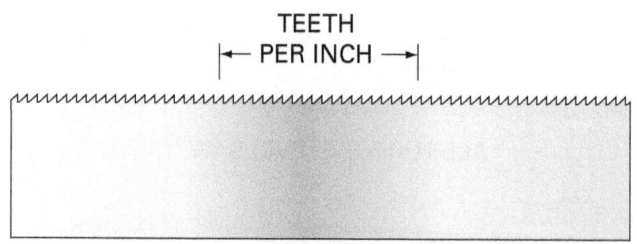
Figure 20 Blade coarseness.

The more teeth per inch of the blade, the finer the cut that the blade will make. Generally, a blade with 18, 24, or 32 teeth per inch is used for cutting pipe. A saw with more teeth per inch generates more debris, which can clog the kerf. When choosing a saw, remember that a larger saw, while it might have a larger blade that is better for cutting tougher material, is also less portable, which is always a consideration for a plumber.

You should not use a hacksaw for square cuts. A crooked cut can cause problems later during threading and other operations that require accurate measurements. Close tolerance (the allowable variations in a given measurement or quantity) means there is little room for extra space. If a plumbing fixture must be installed to meet close tolerances, a crooked cut can make it nearly impossible to fit the fixture in the prepared space.

When using a hacksaw, follow these safety and maintenance instructions:

- Select a saw blade suitable to the material you are cutting.
- Place the blade in the hacksaw frame so that the blade's teeth point toward the end of the frame and away from the handle (and from you). Tighten the blade firmly, and make sure the hacksaw frame is properly aligned.
- Do not try to make straight cuts with crooked hacksaw frames or loose hacksaw blades; this might cause the blade to buckle, twist, or break, which is dangerous.
- Do not apply too much pressure or twisting when sawing; this might cause the hacksaw blade to break, which could result in injury.
- Replace dull hacksaw blades. Worn and dull blades are more likely to cause accidents.
- Make the cuts away from you, and use long, straight, and steady strokes, using almost the entire length of the blade, when cutting.
- Ease pressure on the hacksaw on the backward stroke to avoid dulling the blade's teeth.

Tolerances

A tolerance is the allowable variation from a given specified dimension or quantity. Tolerances are expressed as "±" (plus or minus). For example, if you are told that a pipe can be installed in a location "±$\frac{1}{8}$ inch," that means the measurement can vary by no more than one-eighth of an inch and still be within acceptable specifications.

1.4.2 Chisels

Plumbers use wood chisels and cold chisels. Both types of chisels are made from heat-treated steel to increase the hardness of their cutting edges. You force the chisel's cutting edge into and through the material you are working on by hitting the chisel's handle with a hammer. Plumbers use digger bars, also called spud bars, to pry or cut through obstructions such as rocks, hardened soil, or roots (*Figure 21*).

> **WARNING!**
>
> To avoid injury to your hands, face, and eyes, always wear PPE when sharpening chisels. Appropriate eye protection is of the utmost importance.

The wood chisel (*Figure 22*) is used to make openings or notches in wooden structural material so that pipes can be installed through these openings. For the wood chisel to cut well, its blade needs to be beveled at a precise 25-degree angle (*Figure 23*). To keep it keen, hone the chisel's cutting edge on an oilstone.

The cold chisel (*Figure 24*) is used for cutting, shaping, and removing cold metal softer than the chisel's cutting edge, such as cast iron, wrought iron, bronze, copper, and steel. It is also used for enlarging holes. The blade of a cold chisel is beveled on both sides to a 60-degree angle (*Figure 25*). If the blade of the cold chisel is dull, do not use the tool.

Mushrooming of the chisel head (or handle) is a common problem with cold chisels. Mushrooming of the chisel head results from hammering it. Hammering on a mushroomed head can be dangerous because it can cause metal fragments to fly off the chisel head, seriously injuring the user's eyes or hands. Never use a chisel that has a mushroomed head. When the handle end of a chisel mushrooms, grind off or trim the mushroomed part of the chisel at a slight bevel (*Figure 26*).

Figure 21 Digger bar.

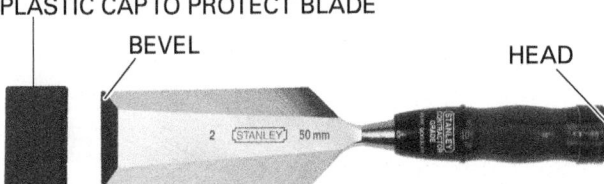

PLASTIC CAP TO PROTECT BLADE
BEVEL
HEAD

Figure 22 Wood chisel.

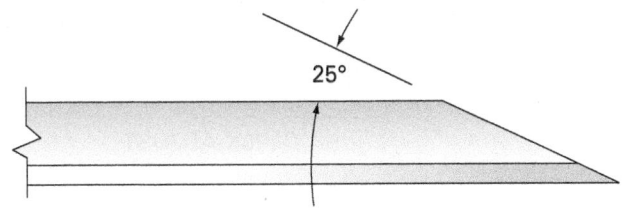

25°

Figure 23 25-degree angle cutting edge.

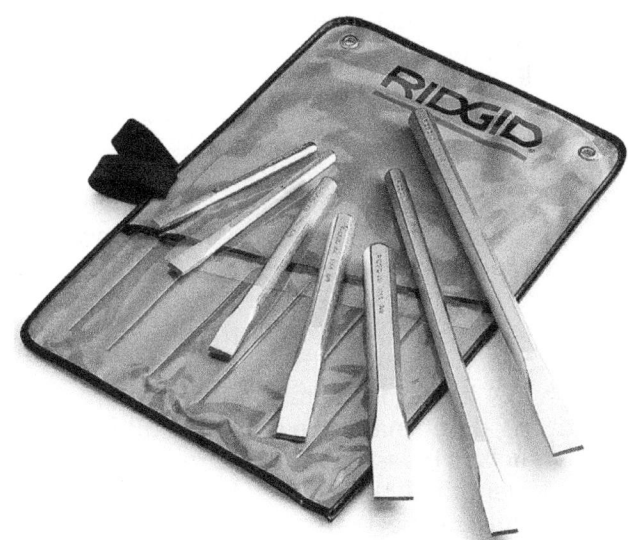

Figure 24 Cold chisels.

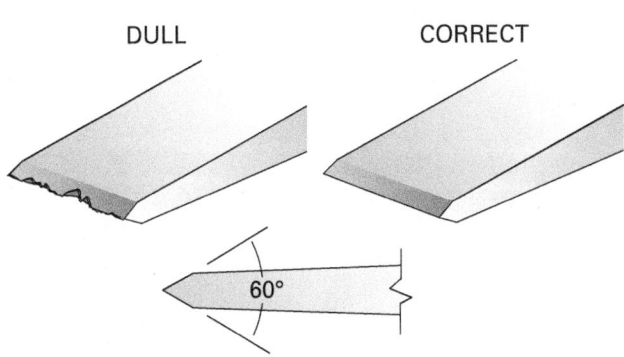

DULL   CORRECT

60°

Figure 25 60-degree angle cutting edge and dull versus correct chisel blades.

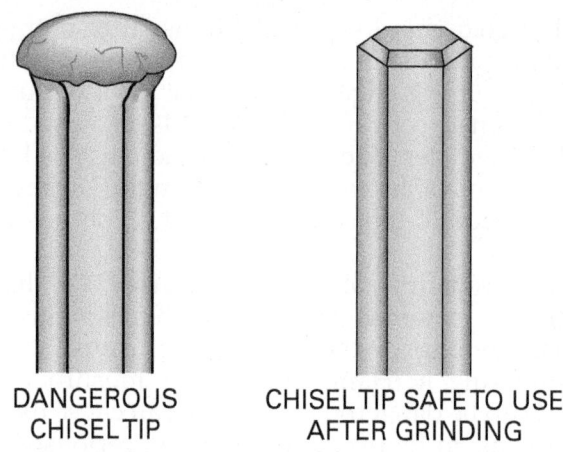

DANGEROUS
CHISEL TIP

CHISEL TIP SAFE TO USE
AFTER GRINDING

Figure 26 Repairing a mushroomed chisel head.

Cold chisels can also be used for trimming concrete, but never use these tools to cut or split stone or concrete. When striking a chisel head, point the tool away from your body. Do not use a chisel as a prying or wedging tool, as its brittle steel can break and cause injury. Wear gloves and a face shield or goggles when using chisels.

1.4.3 Manual Pipe Cutters

Pipe-cutting tools come in a variety of sizes, shapes, and designs. Common sizes range from $\frac{1}{8}$ inch to 4 inches. The three most commonly used manual pipe cutters are the tube cutter, the pipe cutter, and the soil pipe cutter.

Tube cutter—The tube cutter (*Figure 27*) is used to cut copper tubing or thin-wall conduit that is $\frac{1}{8}$ inch to 4 inches in diameter. It also is effective for cutting other malleable (pliable) building materials, such as brass and aluminum.

A conventional tube cutter has four movable parts: a cutter wheel, an adjusting screw, and at least two guiding wheels. The blade on the cutter

wheel slices into the pipe as you make revolutions around the pipe, gradually tightening the cutter.

When doing this, you should keep the blade in the same groove and to make steady and repetitive revolutions with the cutter, in order to obtain a clean cut without damaging the pipe. The tube cutter also has a reamer feature, which allows you to remove any metal burrs from the inside of the pipe once the cut is completed.

A specialized midget tube cutter is used for cutting copper tubing $\frac{1}{8}$ inch to $\frac{7}{8}$ inch in diameter and is handy when working in tight places. Another special tool of this type, a quick-adjust tube cutter, has a sliding adjustment instead of an adjusting screw. You will often use a waxy crayon, also called a keel crayon, to mark a cutting line on the surface of the tube. In other cases, you will use soapstone (a soft, talc-based rock) to mark the pipe. Keel produces a narrower line than soapstone. You will then set the cutting wheel of the tube cutter on the cutting line and follow that line when making the cut.

Pipe cutter—The pipe cutter (*Figure 28*) is used to cut steel pipe. This tool is basically the same as the tube cutter, but it is heavier and it often has more than one cutter wheel. It is used to cut large-diameter pipe.

Soil pipe cutter—Soil pipe cutters are used to cut cast iron or clay pipe. There are two types, the ratchet cutter and the snap cutter (*Figure 29*). Soil pipe cutters are especially appropriate to use in confined spaces or when cutting hard-to-reach piping.

The soil pipe ratchet cutter has a length of chain that wraps around the pipe. The chain is attached to a handle so that ratcheting the handle tightens the chain, which in turn pushes cutters into the pipe and creates a clean break.

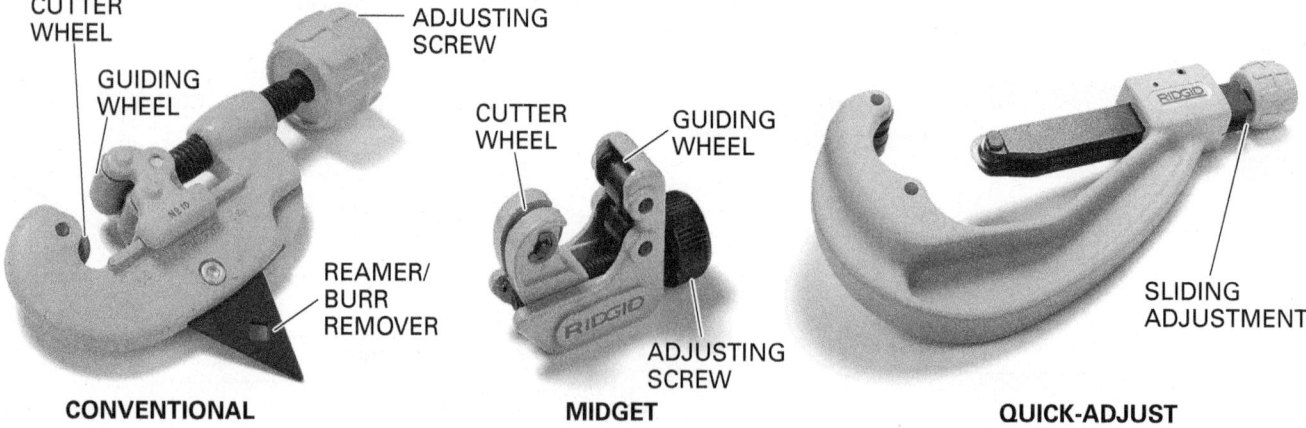

CUTTER WHEEL

GUIDING WHEEL

ADJUSTING SCREW

REAMER/ BURR REMOVER

CONVENTIONAL

CUTTER WHEEL

GUIDING WHEEL

ADJUSTING SCREW

MIDGET

SLIDING ADJUSTMENT

QUICK-ADJUST

Figure 27 Tube cutters.

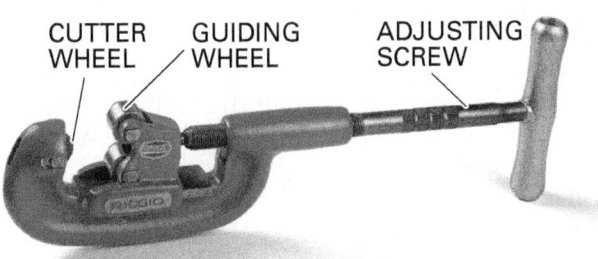

CUTTER WHEEL   GUIDING WHEEL   ADJUSTING SCREW

CONVENTIONAL

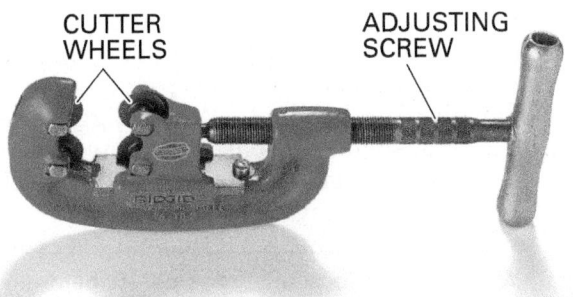

CUTTER WHEELS   ADJUSTING SCREW

FOUR-WHEELED

Figure 28 Pipe cutters.

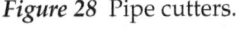

SOIL PIPE RATCHET CUTTER

SOIL PIPE SNAP CUTTER

Figure 29 Soil pipe cutters.

The soil pipe snap cutter, which is more common, cuts with a scissors action. Two long handles provide leverage to the cutting chain wrapped around the pipe. The chain is set with cutting rollers at each link. Tighten the chain, and turn it around the pipe to be cut. Because the chain tightens and cuts evenly, the pipe breaks cleanly, with a snap. Electrically powered chain snaps that attach to power tools are also available (*Figure 30*).

You will sometimes mark a line on the surface of the pipe with soapstone to guide the cut. Make sure that the line is perpendicular to the pipe, otherwise you will shatter the pipe when you make the cut.

Care of cutters—If the tube or pipe you are cutting looks as if it has been mashed after you cut it, you probably need to replace the cutting wheel or wheels. Replace the cutting wheels that are nicked or otherwise damaged. Apply lubricating oil regularly to all movable parts of the pipe cutter so that it operates smoothly.

Figure 30 Electrically operated soil pipe cutter.

Reaming—After you cut a pipe, you will find a burr on the inside of the pipe. A pipe reamer is used to remove the burr (*Figure 31*) in a process called **reaming**. If the burr is left in the pipe, deposits collect inside the pipe and clog it, slowing the flow of liquid within the pipe (*Figure 32*). Because reamers are tapered, one reamer can deburr many sizes of pipe. Be sure, however, not to overream the inside of a pipe. One type of reamer is a deburr tool (sometimes called a pencil reamer because it is about the size of a pencil), which is used to remove burrs from copper tubing (*Figure 33*).

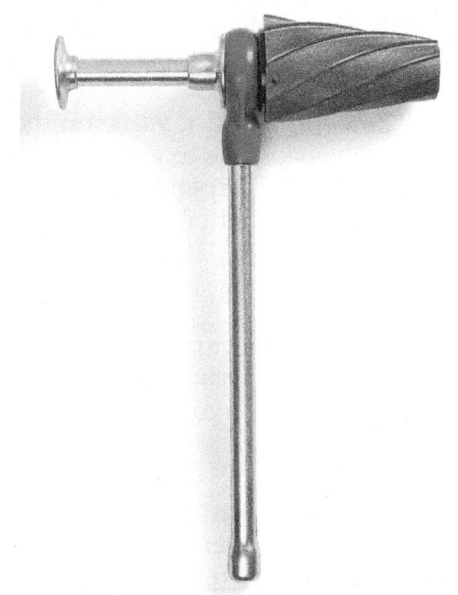

Figure 31 Pipe reamer.

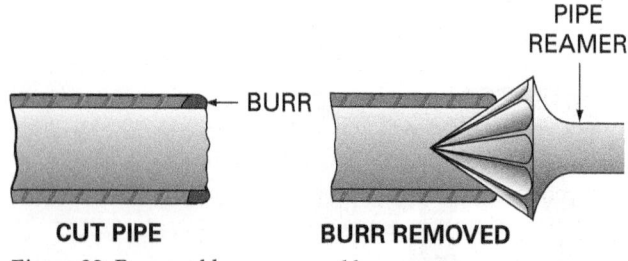

CUT PIPE **BURR REMOVED**

Figure 32 Burr and burr removal by reaming.

Figure 33 Deburr tool.

1.4.4 Shovels

Plumbers use a variety of shovels, including handheld trench shovels, to dig trenches for water pipes. Before digging, you must always identify the location of all nearby gas, electrical, water, and waste and sewage lines and systems. Shovels hitting underground utility wires and pipes can cause a number of injuries and expensive damage. When digging around tree roots, use shovels with shock-absorbing handles. The size of the shovel you use should be dictated by the task and by your own size. Shovels with long handles, for instance, are easier for taller plumbers to use than short-handled shovels. When digging in a confined area, such as a soak well, however, you will need to use shovels with extra-short handles, whatever your size may be.

> **WARNING!**
>
> You must alert local utility companies before doing any digging to avoid damaging underground cables and piping. The utility company will mark the area so that you can dig safely. In addition, all states maintain a One-Call notification system. Construction companies can call this number to have pipelines that carry petroleum or natural gas marked before any digging takes place.

1.5.0 Tools for Assembly and Holding

Holding tools include wrenches, vises, and pliers. These tools hold plumbing materials and workpieces in place so that you can do something else, like drill a hole, without the material or workpiece moving around. These holding tools help you to be precise in your work.

1.5.1 Wrenches

The wrench is a crucial assembly tool. As a plumber, you will use a variety of wrenches for turning pipe, fittings, and fasteners. The physics of a wrench allow it to work as a lever, with a jaw, strap, or chain that grips or fits around the object needing to be tightened or loosened. Some wrenches, such as a strap wrench, are designed to leave no marks. They are used for turning finish fasteners, where the slightest mark will ruin the appearance of the finish. Other wrenches are used to tightly grip materials that are hidden in a building's structure, where marks left behind are not a concern. Plumbers use a wide variety of wrenches.

Pipe wrenches—The pipe wrench is used to grip and turn pipe and tubing. Types include regular heavy-duty pipe wrenches, 45-degree and 90-degree offset pipe wrenches, and heavy-duty end wrenches (*Figure 34*). Wrenches come in a number of lengths, such as 6, 12, 14, 18, 24, 36, and 48 inches. The wrench's length determines the size of pipe that it can handle.

The straight pipe wrench is the most common type. When working in tight quarters, use offset wrenches; their angles (45 or 90 degrees) make it easier to reach pipe in cramped areas. The jaws of pipe wrenches always leave marks on the pipe; therefore, do not use them on pieces where appearance is important. When applying force to a wrench, always direct it toward the open side of the jaws (*Figure 35*). This technique gives you the best grip and leverage.

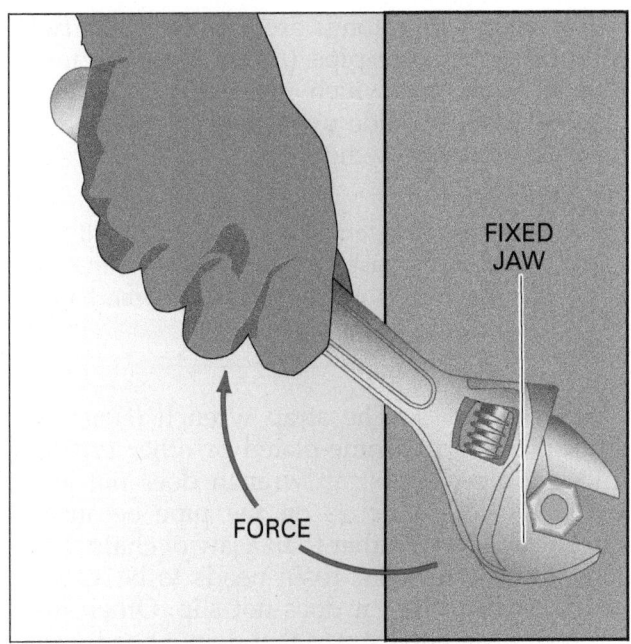

Figure 35 Correct method of applying force to a wrench.

Another variation of the pipe wrench is the chain wrench (*Figure 36*). This tool has a length of chain permanently attached to the wrench handle at one end. The chain is looped around the pipe to grip and secure it, and the other end of the chain can be secured to various positions on the wrench's handle. Oil the chain frequently to prevent it from becoming stiff or rusty.

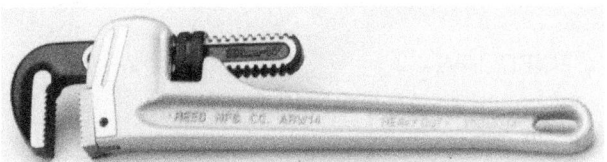

STRAIGHT ALUMINUM PIPE WRENCH

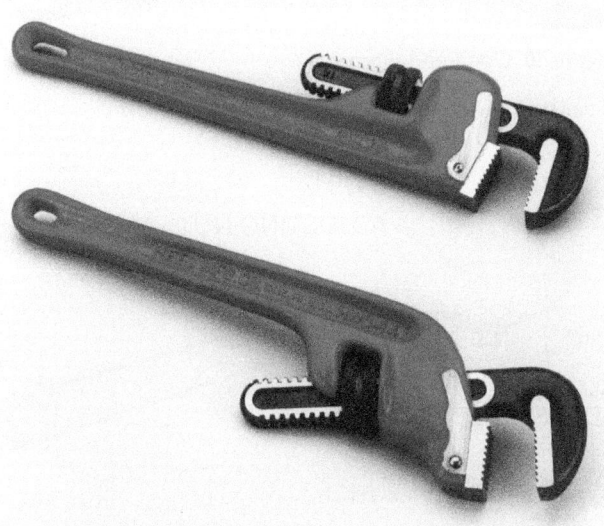

HEAVY-DUTY CAST STEEL PIPE WRENCH
45-DEGREE CAST STEEL OFFSET WRENCH

Did You Know?

The pipe wrench was invented by a steamboat firefighter named Daniel Stillson. Stillson had suggested to Walworth, a heating and piping firm, that it design a wrench that could screw pipes together. The firm's owner told Stillson to take a crack at making a prototype that would twist off the pipe. To the everlasting gratitude of future plumbers, Stillson was successful and his prototype twisted off the pipe. In 1870, Stillson's design was patented, and Walworth began manufacturing the wrench. Even today, pipe wrenches are sometimes called Stillson wrenches.

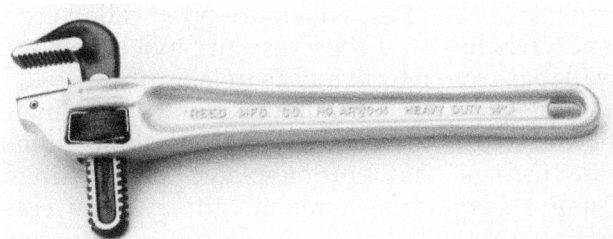

90-DEGREE OFFSET ALUMINUM PIPE WRENCH
Figure 34 Pipe wrenches.

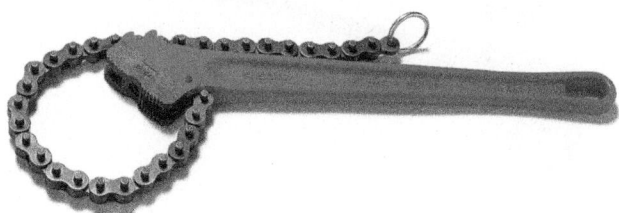

Figure 36 Chain wrench.

Pipe tongs—Pipe tongs are the wrenches typically used on large pipe (*Figure 37*). Pipe tongs also have chains, which need to be oiled frequently. They provide extra leverage needed for tougher, more heavy-duty jobs.

Strap wrenches—The strap wrench (*Figure 38*) is used to hold chrome-plated or other types of finished pipe. The strap wrench does not leave jaw marks or scratches on the pipe because it grips with a strap rather than a jaw or chain. With some strap wrenches, rosin needs to be applied to the strap so that it does not slip. Other strap wrenches use vinyl straps that do not need rosin.

Spud wrenches—The spud wrench is similar to a pipe wrench, except that it has no teeth in its jaws (*Figure 39*). Spud wrenches loosen and tighten fittings on drain traps, sink strainers, and toilet connections and are also used on large, oddly shaped nuts. The wrench's narrow jaws allow it to fit into tight places. The spud wrench is difficult to adjust, though, and plumbers often use adjustable wrenches and open-end wrenches instead.

Open-end wrenches—The open-end wrench (*Figure 40*) usually has two different-sized openings, each on one end of the tool. The size of each opening is stamped on the handle. This wrench is usually used for assembling fittings and fasteners that are less than 1 inch across.

Adjustable wrenches—The adjustable wrench in *Figure 41*, is similar to the open-end wrench, except that it has an adjustable jaw. A roller at the stem of the jaw adjusts the size of the opening.

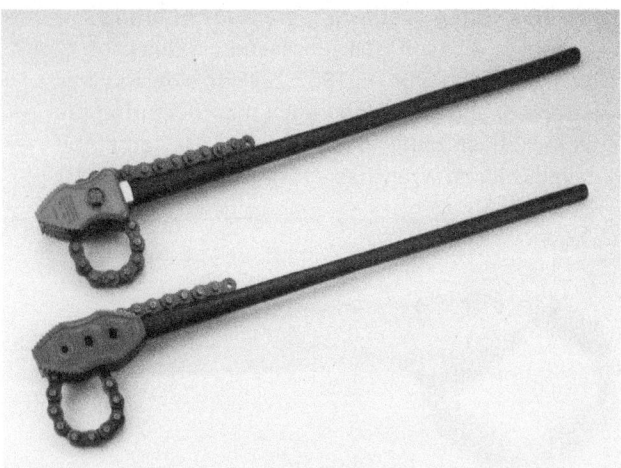

Figure 37 Pipe tongs.

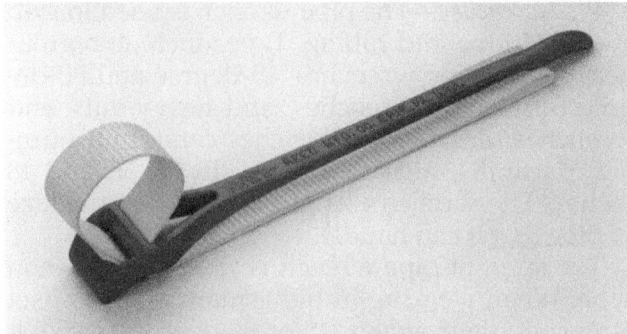

Figure 38 Strap wrench.

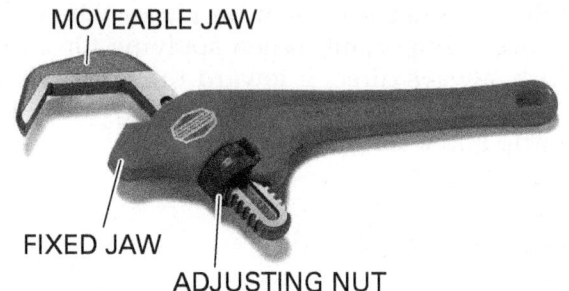

MOVEABLE JAW

FIXED JAW

ADJUSTING NUT

Figure 39 Spud wrench.

Figure 40 Open-end wrench.

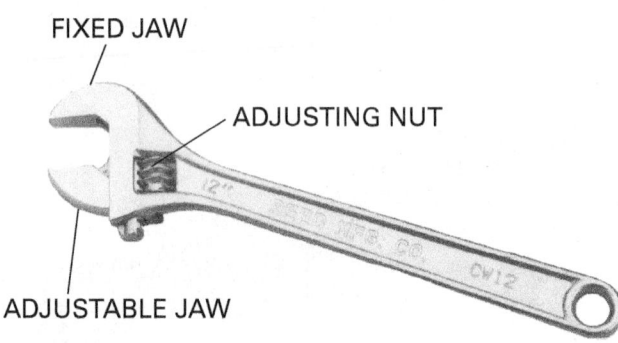

FIXED JAW

ADJUSTING NUT

ADJUSTABLE JAW

Figure 41 Adjustable wrench.

Because it is so easy to adjust, this wrench is found in virtually every toolbox. Various sizes of adjustable wrenches, which are often called crescent wrenches after the Crescent® tool brand, are available, from 6 to 18 inches in length.

Adjustable wrenches have smooth jaws, making these tools ideal for turning nuts, bolts, small pipe fittings, and chrome-plated pipe fittings. Using an adjustable wrench can save you considerable time, because you do not have to constantly switch wrench sizes as you work with different sizes of nuts and bolts.

Basin wrenches—The basin wrench (*Figure 42*) was developed to work in tight, hard-to-reach areas, such as recesses where sink and lavatory faucets are secured. This specialty wrench, with its offset jaws, permits you to turn a nut or a fastener that you could not reach with other wrenches.

Angle stop wrenches—The angle stop wrench is used to remove angle stop valves and compression couplings. Some designs use a ratchet mechanism. This type of wrench is designed for ease of use in tight locations such as under bathroom sinks.

Torque wrenches—The torque wrench has a gauge attached to it, which indicates the torque (the force that is being applied to the workpiece). This wrench is known for its accuracy and reliability. There are several types of torque wrenches. The no-hub design does not have a gauge; instead, it is designed to disengage at a preset torque limit. The clicker-type wrench makes adjustments easy because it has a sliding scale with a locking mechanism. Other torque wrenches have ratcheting heads that can be removed. These wrenches allow you to remove the head and re-insert it to torque in the left-hand direction. Because of the low-torque range, this wrench is popular with transmission mechanics.

> **WARNING!**
> When using a torque wrench, be careful not to force the tool. Battery-operated wrenches can cause just as much damage as electric ones. Using excess torque on either model can break your wrist.

> **CAUTION**
> When using a torque wrench, you must install fasteners in a sequence so that all fasteners are tightened at the same rate. If you tighten one fastener more than another, you can distort and damage the workpiece.

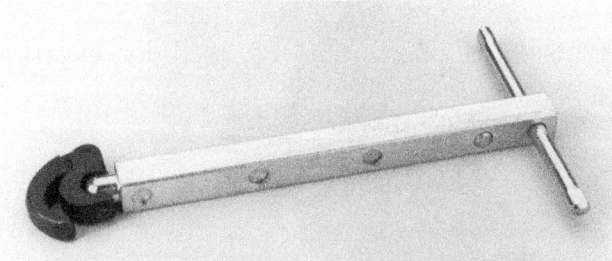

Figure 42 Basin wrench.

Wrench safety—You should develop some specific safety habits when using wrenches, including the following:

- Use wrenches that are the right type and size for the job.
- Do not use loose wrenches. They will slip and can cause serious injury. A loose wrench can also ruin the corners of a cap screw or nut.
- Take care when working in a tight space that your grip on the wrench will not injure your hand as you turn the wrench.
- Pull up on the wrench; do not push it. If you push on a wrench, slippage can cause injury to your hand, or even to your face and body. If you cannot avoid pushing a wrench, do so with a stiff arm, holding your face and body back.
- Do not place too much strain on a small wrench.
- Do not use a wrench as a hammer.
- Switch to a larger wrench if you need more leverage.

Did You Know?

The monkey wrench is a general-purpose wrench with an adjustable jaw for turning nuts of various sizes. Charles Moncky invented this tool around 1858. "Monkey wrench" also has an informal meaning: to disrupt, as in, "He threw a monkey wrench into our plans." Monkey comes from the last name of the wrench's inventor. Wrench was a term that meant trick or deception long before the tool of the same name was developed.

Do not use a monkey wrench as a hammer. Doing so can damage the wrench, causing its jaws to loosen and making it difficult to adjust. A loose-fitting wrench can easily slip off the nut and cause injuries.

Torque

Torque is a measure of the force exerted by a rotating motion. Torque wrenches measure this force, or resistance to turning. You need torque wrenches when you are installing fasteners that must be tightened in sequence so that the workpiece is not distorted. You will use a torque wrench only when a torque setting is specified for a particular bolt.

1.5.2 Pliers

Pliers are a type of adjustable wrench and are used for bending, gripping, and holding. Some pliers can also be used to cut wires. Pliers come in many different head styles, depending on their use. Do not use pliers on exposed pipe or fasteners, because they leave jaw marks.

Adjustable pliers—Adjustable pliers, also called slip-joint pliers, are very popular (*Figure 43*). They are used for bending wire and for gripping and holding during assembly operations. The adjustable jaws make it easy to change the distance between them. There are two jaw settings: one for small materials and one for larger materials.

Lineman's pliers—Lineman's pliers (*Figure 44*), also known as side cutters, have wider jaws than slip-joint pliers. Lineman's pliers are used mainly for cutting heavy or large-gauge wire and for holding work. The shape of the jaws reduces the chances that wires will slip, and the hook bend in the handles provides a better grip.

Slip-lock pliers—Slip-lock pliers work on the same principle as slip-joint pliers, but their jaws have five or more adjustments (*Figure 45*). Slip-lock pliers can hold much larger materials than channel-lock pliers. Do not use slip-lock pliers to hold small materials because the jaws cannot be adjusted flat against each other. Slip-lock pliers are often referred to by the trade name Channellock®.

Locking pliers—Locking pliers are handy and adjustable pliers that can be used both as pliers and as a handheld vise or clamp (*Figure 46*). They are also known as plier wrenches, lever-wrench pliers, and by their proprietary name, Vise-Grips®. The jaws of these pliers can be adjusted by turning a screw attached to one of the handles, to lock firmly onto a workpiece (usually one made of metal) in a vise-like grip. A trigger on the other handle loosens the grip. The standard design for locking pliers has serrated, straight jaws, but

Figure 43 Adjustable pliers.

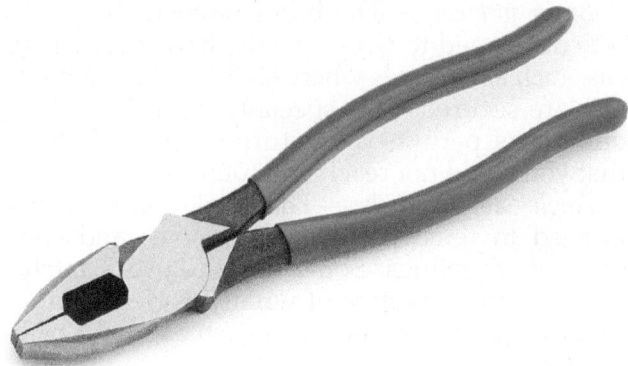

Figure 44 Lineman's pliers.

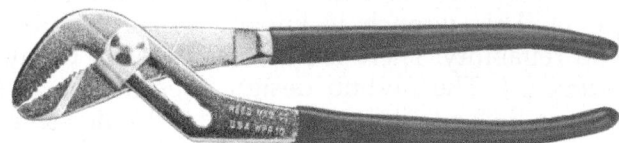

Figure 45 Slip-lock pliers.

these pliers also have other designs (long-nosed, flat-jawed, C-clamp jaws). Because of their fine jaw adjustment and tremendous clamping power, locking pliers are an excellent tool for removing bolts with heads that have been rounded off.

Care of pliers—Pliers need very little preventive maintenance. A little oil on all movable joints and some oil on the locking pliers' adjustment screw prevents rust and keeps them working smoothly. Do not expose pliers to extreme heat, which can melt their plastic handles. To ensure the longevity of your pliers, use them only for their intended purpose. Do not use pliers to hammer nails or to turn nuts or bolts.

Plier safety—The following are the specific safety guidelines for pliers:

- Pliers must not be used as a substitute for wrenches, because pliers cannot hold a workpiece as securely as wrenches. Using pliers as wrenches leads to slippage, which could cause injury.
- Insulated pliers must be used for electrical work.

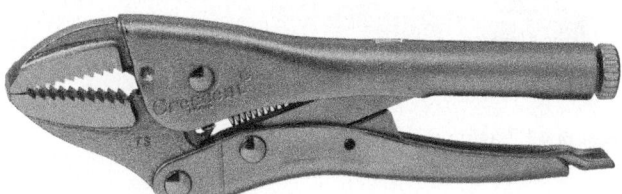

Figure 46 Locking pliers.

1.5.3 Vises

Vises are holding tools. Plumbing jobs, such as cutting, threading, and reaming, would be difficult without a vise to hold the workpiece secure. The vise permits you to do work that otherwise would require two people (one to hold the workpiece and the other to cut, thread, or ream it).

Standard yoke vises—The standard yoke vise is probably the most common one used by plumbers (*Figure 47*). Its jaws hold pipe firmly and prevent it from turning. Because the teeth of the yoke vise usually leave marks on the material, use it only with pipe that will not be exposed.

Chain pipe vises—The chain pipe vise is used in the same way as the standard yoke vise, but the chain vise can hold much larger pieces of pipe (*Figure 48*). The chain must be kept oiled or it will become stiff. A stiff chain degrades the operation of the chain vise. Chain pipe vises are available in bench-mounted and free-standing models.

Bench vises—The bench vise has two sets of jaws: one to hold flat work and another to hold pipe (*Figure 49*). Rotating the T-handle screw tightens or loosens the vise. You will not see this vise very often on the job site, but you might find it mounted in some plumbing trucks.

TRISTAND® vises—The RIDGID® TRISTAND® vise has a stand with built-in folding legs and a tray (*Figure 50*). The vise base extends over the front legs of the stand to provide a clear tool swing. Some models include additional tool slots, pipe benders, and a ceiling brace screw. When used with the proper sized pipe, the ceiling brace screw allows you to secure the vise by wedging it between the floor and ceiling.

Figure 48 Chain pipe vise.

1.5.4 Ratchet Threaders

One method of joining pieces of pipe together is to screw one pipe end into another screwed fitting, such as male and female fittings and couplings. For pipe segments to be joined in this way, the ends of the segments must be prepared by threading. Threading entails passing something through or around a pipe segment to form grooves that allow another segment to be screwed onto or into it. Galvanized pipes and black-iron (cast-iron) pipes need to be threaded on the outside. To make clean cuts of these threads, die tools

Figure 47 Yoke vise.

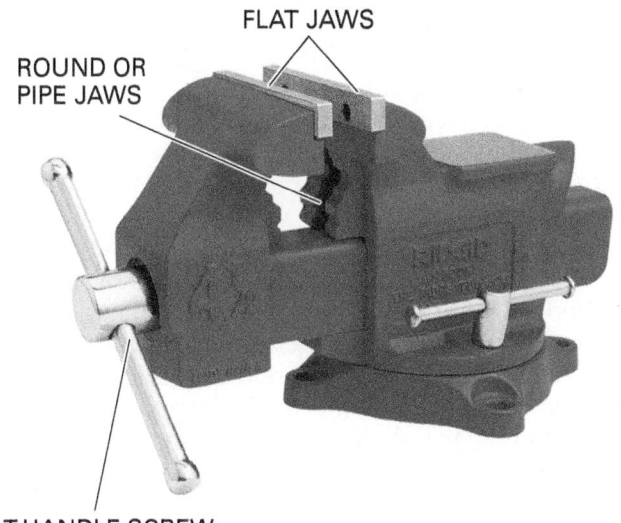
ROUND OR PIPE JAWS

FLAT JAWS

T-HANDLE SCREW

Figure 49 Bench vise.

CEILING BRACE SCREW

Figure 50 RIDGID® TRISTAND® vise.

(*Figure 51*) are used along with ratchet threaders. Dies must be sharp so that the threads they cut will be correctly tapered. Power threaders are also available.

Applying cutting oil to the die tool reduces the friction and heat caused by cutting. If there is too much friction or heat, the die tool pushes the metal that is being threaded instead of making a clean cut. Heat and friction can also break the teeth off the die.

Die tools are made to fit all standard pipe sizes and piping materials. Dies cut threads in a variety of sizes, typically in $\frac{1}{2}$ inch, $\frac{3}{4}$ inch, 1 inch, and $1\frac{1}{4}$ inch sizes. Be sure the die tools are appropriate for use with the material of the pipe to be threaded. Never use steel-pipe die tools to thread stainless-steel pipe.

The die tool is locked or clamped into a ratchet stock, where the ratcheting mechanism is attached to a handle (*Figure 52*). The ratchet stock helps to turn the die by increasing leverage on it. The ratcheting mechanism makes it easy to operate the stock. To cut the thread, apply pressure in a downward direction and then ratchet the handle upward to prepare for another downward stroke (*Figure 53*).

1.6.0 Hammers

Plumbers, like all other skilled workers in the construction profession, use hammers. There are many types of hammers, including claw hammers, ball peen hammers, sledgehammers, and mauls.

Hammers come in many sizes. The size of a hammer is determined by the weight of its head. Usually a 14- or 16-ounce hammer is considered medium weight (the weight is stamped on the hammer head). Some claw hammers are as heavy as 32 ounces, and, of course, sledgehammers can weigh several pounds.

Hammers also come with different types of handles. Hammers with wooden (usually hickory wood) handles are the least expensive, and the handles are comfortable, but the hammer head can easily work loose from the handle and may need periodic reshimming. Wooden handles can also break from vigorous nail pulling or when subjected to repeated overstriking. Steel handles are very strong (the head and the handle are one forged tool) but they transfer more of the shock of the hammer blow to your hand than hammers with wooden or fiberglass handles do.

DIE-TOOL SET

PIPE WITH THREADED ENDS

Figure 51 Die-tool set and threaded pipe.

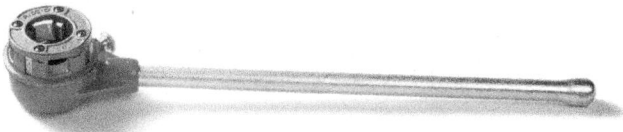

Figure 52 Die and ratchet stock.

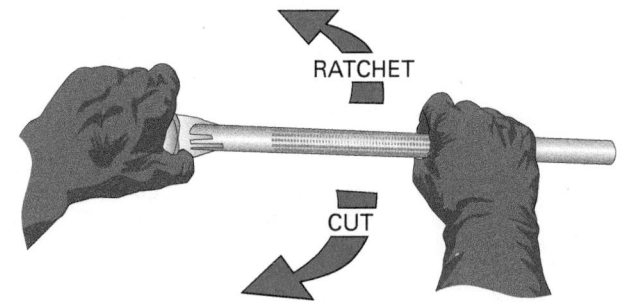

Figure 53 Operating a die and ratchet stock.

Fiberglass-reinforced plastic handles can be a good compromise. Fiberglass handles are almost as comfortable as wooden handles, and they are also durable and shock absorbent. Fiberglass handles can splinter from repeated overstriking.

Always wear eye protection when using a hammer. Check to make sure handles are securely wedged onto the hammer head before use, and do not use hammers that have cracked, splintered, or badly worn handles. Never strike one hammer face with or against another hammer or a hatchet. Never strike nail-pullers, steel chisels, or other hardened objects with a nail hammer, because the face of the hammer may chip. Nail hammers are intended for driving or pulling common, unhardened nails only.

1.6.1 Claw Hammer

The most common type of hammer is the carpenter's curved claw hammer (*Figure 54*). The hammer's head has a face for striking nails and, on the opposite side of the hammer head, a two-pronged claw for pulling nails. The head is steel. The handle can be fiberglass, wood, or steel. Fiberglass and steel hammers usually have shock-absorbing rubber or vinyl coatings on the handle to improve the grip. Curved-claw hammers can be 7 or 16 ounces.

To pull a nail, slip the claw of the hammer under the nail head and pull until the handle is nearly straight up (vertical) and the nail is partly drawn out of the wood. Then, pull the nail straight up from the wood (*Figure 55*). To gain added leverage when pulling out long nails, firmly secure a block underneath the hammer's head and on the same surface from which you are pulling the nail.

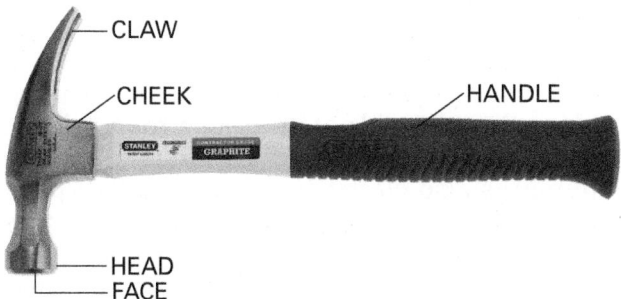

Figure 54 Claw hammer.

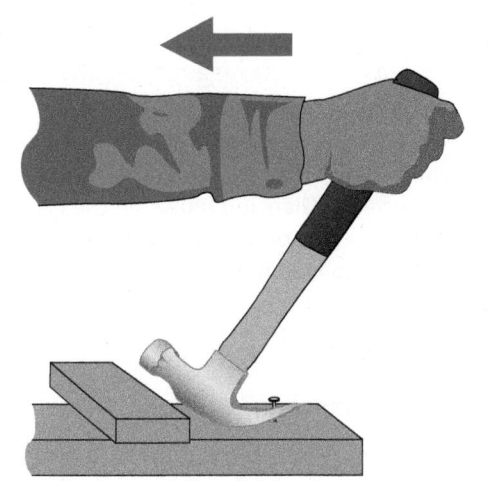

Figure 55 Using a claw hammer to pull a nail.

Variations on the claw hammer are straight-claw hammers and rip-claw hammers (*Figure 56*). Straight-claw hammers are better at prying out nails, but rip-claw hammers are more compact and better at pulling nails. Rip-claw hammers are 16 or 20 ounces, and the handles can be made of jacketed graphite, jacketed fiberglass, or tubular steel.

WARNING!

When working in close quarters, it is best to use a ball peen hammer instead of a claw hammer. In a tight space, you could accidentally claw yourself with the curved end of the hammer.

Figure 56 Stanley® jacketed graphite nail rip-claw hammer.

1.6.2 Ball Peen Hammer

The ball peen hammer (*Figure 57*), also known as the plumber's hammer, has a flat face for striking and, on the other side of its head, a rounded face that can align brackets and drive out bolts. This hammer is also classified by weight. Ball peen hammer heads range in weight from 6 ounces to $2\frac{1}{2}$ pounds. Ball peen hammers are designed for striking chisels and for riveting, shaping, and straightening unhardened metal. When using a ball peen hammer to strike a chisel or other tools that are designed to be struck, make sure the hammer's face that is doing the striking has a diameter at least $\frac{3}{8}$ inch larger than the face of the tool that is being struck.

> **WARNING!**
>
> When using a ball peen hammer, strike squarely at the chisel. Glancing blows off the chisel can cause the edge of the hammer's face to chip, which could lead to serious injury, especially to the eyes.

1.6.3 Sledgehammer

A sledgehammer is a particularly heavy hammer (*Figure 58*). The sledgehammer is a heavy-duty tool used for driving posts or other large stakes. It can also be used for breaking up cast iron or concrete. The head of the sledgehammer is made of high-carbon steel and weighs from 2 to 20 pounds. The shape of the sledgehammer's head and the length of its handle depend on the job it is designed to do.

Always wear eye protection when using a sledgehammer, and make sure the handle is securely wedged to the sledgehammer's head. Never attempt to strike an object at or above shoulder level with a sledgehammer. Instead, use a platform to raise yourself above the object you want to strike. If a co-worker is holding a stake, nail, pin, or other object to be driven by the sledgehammer, that person should stand at a right angle to the direction of the sledgehammer, grip the item being driven with a holding device, and wear gloves.

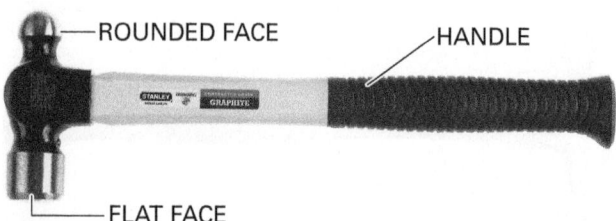

Figure 57 Ball peen hammer.

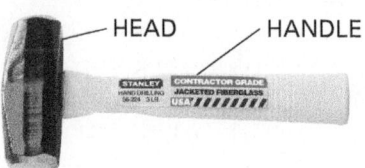

DOUBLE-FACE SHORT-HANDLED SLEDGEHAMMER

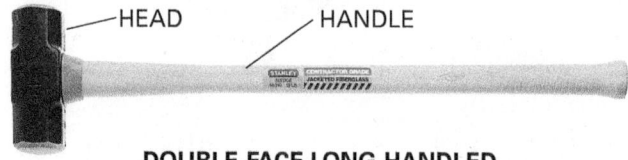

DOUBLE-FACE LONG-HANDLED SLEDGEHAMMER

Figure 58 Sledgehammers.

1.6.4 Metal Stud Punch

Metal stud punches use a steel head to punch through steel studs by squeezing a lever handle. Metal stud punches *are designed to automatically center the hole on* a standard stud. They also smooth the hole edges, preventing snagging or cutting when running anchor wire through the hole. Many metal studs come pre-punched (*Figure 59*).

1.7.0 Screwdrivers

Screwdrivers are used to install and remove a wide variety of screws: slotted, Phillips, clutch-drive, and Allen. Screwdrivers have generally the same design: a steel shank, the screwdriver blade at the tip of the shank, and a handle. Because of the wide variety of screws available, there are many types of screwdriver blades and handles.

Figure 59 Pre-punched metal stud.

The most common types are the standard screwdriver and the Phillips head screwdriver (*Figure 60*). The screwdriver's handle is most often bulbshaped and made of wood or plastic. The length of its blade and the width of its tip determine a screwdriver's size. Screwdrivers range in length from $2^1/_2$ inches to as much as 2 feet. A correct fit between the screwdriver tip and the screw head prevents the marring of the screw head. Use the correct screwdriver size to prevent the tip from slipping out of the screw head and possibly stripping the screw.

> **WARNING!**
>
> Do not use a screwdriver as a substitute for a hammer, wedge, nail puller, lever, chisel, or punch. To protect screws, always replace worn screwdriver tips. When using a screwdriver, secure the workpiece in a vise. Do not hold the workpiece in your hand unless the screw turns easily. Use only screwdrivers with insulated handles for electrical work. Do not carry exposed screwdrivers in your pockets.

1.7.1 Nut Drivers

A nut driver (*Figure 61*) is similar to a screwdriver. Instead of a blade, the nut driver has a socket end for tightening hex-head screws and bolts. It comes in all bolt sizes.

1.7.2 Multiple-Tip Screwdrivers

Some screwdrivers come with interchangeable bits (*Figure 62*). You simply remove one bit from the end of the shank and replace it with another bit. The bits are stored in the handle of the screwdriver. The bits typically have a flat blade on one end and a Phillips head on the other. Specialty bits are also included.

Calculator

You will need to make quick calculations when doing various jobs, so you should always carry a calculator. The traditional battery-powered, handheld calculator that helps you do basic math problems is one option. Specialized calculators can help you perform many types of construction calculations.

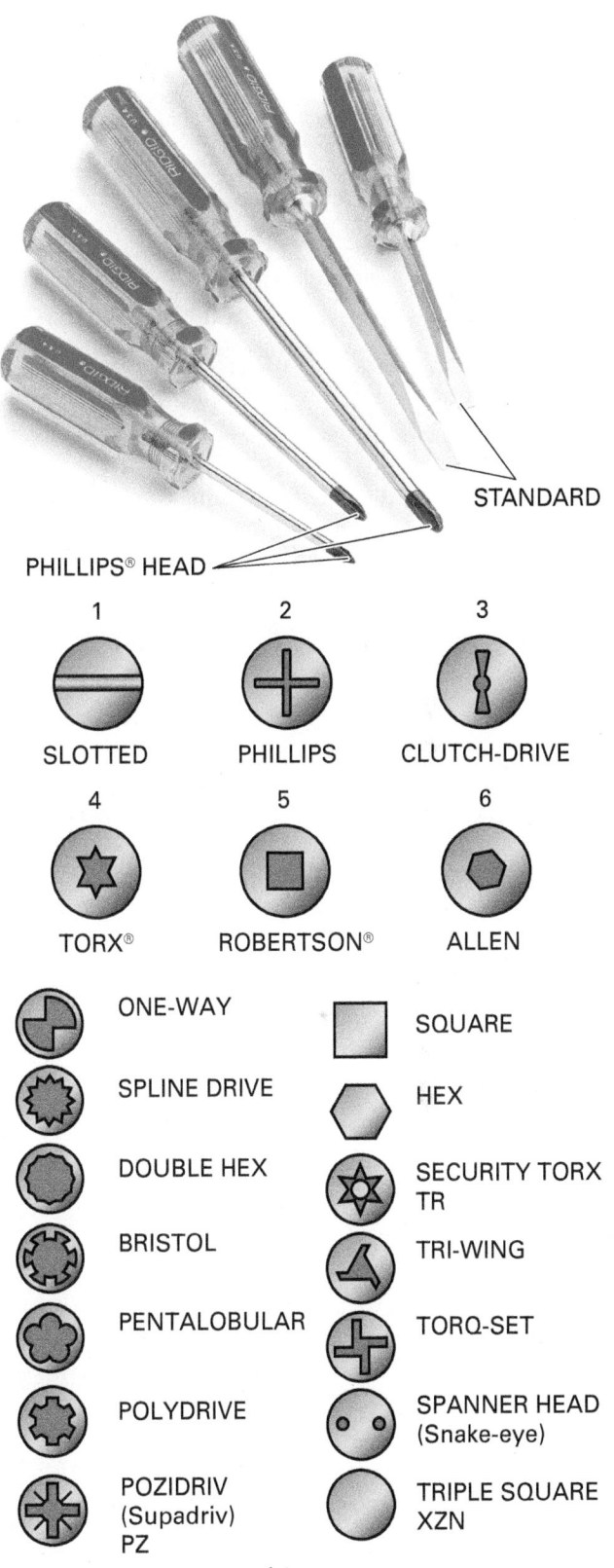

Figure 60 Common screwdrivers.

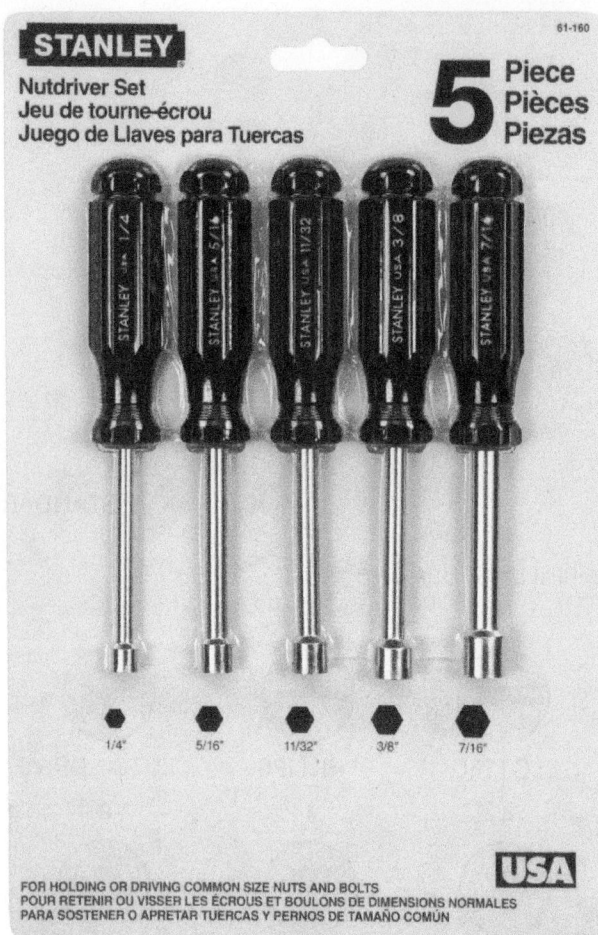

Figure 61 Nut driver set.

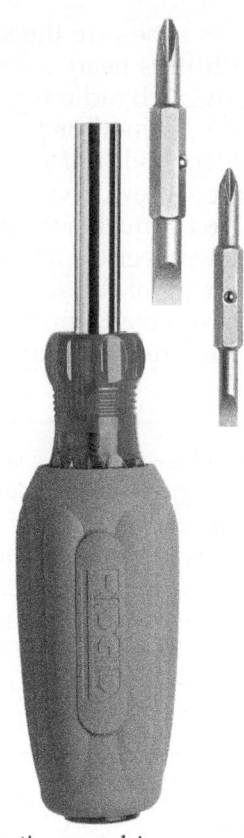

Figure 62 Multiple-tip screwdriver.

Additional Resources

Core Curriculum, Latest Edition. NCCER. New York, NY: Pearson.

Council Tool website. *www.counciltool.com*.

The Stanley Tools website. *www.stanleytools.com*.

1.0.0 Section Review

1. Prevent rust on the metal parts of tools by applying _____.

   a. water
   b. grease
   c. light oil
   d. petrol

2. To measure offsets, the plumber's rule has markings for _____.

   a. 30 degrees
   b. 45 degrees
   c. 60 degrees
   d. 90 degrees

3. Level and plumb members intersect at _____.

   a. 30-degree angles
   b. 45-degree angles
   c. 90-degree angles
   d. 180-degree angles

4. An opening in drywall is typically cut using a _____.

   a. band saw
   b. hacksaw
   c. keyhole saw
   d. mini-hacksaw

5. The type of pliers used mainly for cutting heavy or large-gauge wire and for holding work are called _____.

   a. locking pliers
   b. slip-joint pliers
   c. lineman's pliers
   d. slip-lock pliers

6. The best tool for prying out a nail is a(n) _____.

   a. straight-claw hammer
   b. rip-claw hammer
   c. flat-blade screwdriver
   d. auger

7. An Allen-head screw has a _____.

   a. round opening
   b. hex-shaped opening
   c. square opening
   d. slotted opening

2.0.0 POWER TOOLS

Objectives

Describe the power tools used in the plumbing trade.

  a. Identify the basic safety requirements when using power tools.
  b. Identify the power cutting tools used in the plumbing trade.
  c. Identify the drilling and boring tools used in the plumbing trade.
  d. Identify the power threading tools used in the plumbing trade.
  e. Identify the soldering tools used in the plumbing trade.

Performance Tasks

  1. Identify plumbing tools.
  2. Properly use plumbing tools.
  3. Demonstrate proper maintenance and storage of hand and power tools.

Trade Terms

Amperage: A measure of electrical current.

Bar stock: Uncut bar material on hand for use on a project.

Electrical ground: A conductive connection that provides a path for electrical current to pass from an electrical component into the earth.

Ferrous: Containing iron.

Flux: A water-soluble substance that facilitates the fusion (joining) of metals and helps prevent surface oxidation (rusting, tarnishing) during welding, brazing, and soldering. Also called soldering paste.

Joist: A piece of lumber used horizontally as a support for a ceiling or a floor.

Nonferrous: Not containing iron and therefore not magnetic.

Temper: The strength and resilience of a metal.

Tube stock: Uncut tubing material on hand for use on a project.

Voltage drop: The tendency of electricity traveling through the length of an extension cord to lose voltage.

Plumbers use a wide variety of power tools. Common power tools used by plumbers include various power saws, grinders, and drilling/boring tools. These tools present additional safety hazards, and it is essential that you be trained in the proper use of any tool before operating it.

2.1.0 Power Tool Safety

OSHA and your company have specific policies and procedures to keep the workplace safe from electrical hazards. You can do many things to reduce the chance of an electrical accident. If you ever have any questions about electrical safety on the job site, ask your supervisor. Adhere to these basic, job-site electrical safety guidelines:

- Use three-wire extension cords, and protect them from damage. Never fasten them with staples, hang them from nails, or suspend them from wires. Never use damaged cords.
- Make sure that panels, switches, outlets, and plugs are grounded.
- Never use bare electrical wire.
- Never use metal ladders near any source of electricity.
- Never wear a metal hard hat.
- Always inspect electrical power tools before you use them.
- Never operate any piece of electrical equipment that has a danger tag or lockout device attached to it.
- Use three-wire cords for portable power tools, and make sure they are properly connected.
- Never use worn or frayed cables.
- Check electrical tools periodically for damage to power cords. Also, test three-prong electrical plugs with a meter to make sure the safety ground is properly connected. Three-prong electrical plugs are effective only if properly grounded. A supervisor should oversee any repairs.
- Do not use electrical tools in damp or wet locations, even if the tools are equipped with a safety ground.
- Do not use electrical tools near flammable liquids; one spark is all it takes to ignite these liquids and start a fire.
- Never operate tools with the safety guards removed, missing, or disabled.

2.1.1 Ground Fault Circuit Interrupter (GFCI) Protection

All power tools must be either double-insulated or protected by a ground fault circuit interrupter (GFCI). Ground faults are the most common form of electrical-shock hazard at a job site. GFCIs are fast-acting circuit breakers designed to shut off electric power almost instantaneously in the event of a ground fault.

2.1.2 Extension Cords

Electrical cords are frequently seen on construction sites, yet they are often overlooked. Use the following safety guidelines to ensure your safety and the safety of other workers:

- Every electrical cord should have an Underwriters Laboratories (UL) label attached to it. Check the UL label for specific wattage. Do not plug more than the specified number of watts into an electrical cord.
- A cord set not marked for outdoor use is to be used indoors only. Check the UL label on the cord for an outdoor marking.
- Do not remove, bend, or modify any metal prongs or pins of an electrical cord.
- Extension cords used with portable tools and equipment must be three-wire type and designated for hard or extra-hard use. Check the UL label for the cord's use designation.
- Avoid overheating an electrical cord. Make sure the cord is uncoiled, and that it does not run under any covering materials, such as tarps, insulation rolls, or lumber.
- Do not run a cord through doorways or through holes in ceilings, walls, and floors, which might pinch the cord. Also, check to see that there are no sharp corners along the cord's path. Any of these situations will lead to cord damage.
- Extension cords are a tripping hazard. They should never be left unattended and should always be put away when not in use.
- To function properly, a power tool must receive enough electrical energy. As electricity travels the length of an extension cord, it tends to lose voltage. This loss is called voltage drop. When voltage drops, the tool has to draw more amperage (a measure of electric current), causing the motor to run hotter. If such unusual amperage draw continues for a prolonged period, the motor can overheat and be ruined. *Table 1* shows the wire gauge required for extension cords at given lengths and amperage draws. The lower the gauge number, the heavier the wire. Be sure that the extension cords you use, especially long ones, are heavy enough to minimize voltage drop.

- Receptacles on the ends of extension cords are not part of the permanent wiring of the building. Therefore, receptacles must be protected by GFCIs, according to OSHA regulations, whether or not the extension cord is plugged into the permanent wiring. Even when GFCIs are in use, always protect extension cords from excessive moisture. Never daisy-chain extension cords together. Manufacturers and OSHA do not allow it because it is dangerous.

2.2.0 Power Cutting Tools

Power cutting tools are used to save time and produce uniform cuts. They are also required when working with heavier materials. The most common power cutting tools used in plumbing include reciprocating saws, portable band saws, abrasive saws, and angle grinders.

2.2.1 Reciprocating Saws

The reciprocating saw (*Figure 63*) is electrically powered and can be used with a variety of blades. The choice of blade depends upon the material to be cut. A shaft pushes the blade forward and back. On the backstroke, it also lifts the blade to clear the cut and keep the saw's teeth from being worn. Single-speed, two-speed, and variable-speed models are available. The strokes per minute made by a reciprocating saw range from 1,700 (the slowest speed on two-speed models) to 2,400 (the highest speed on most variable-speed models).

> **WARNING!**
>
> Never reverse a saw while it is running. If the blade is pinched, it can cause a powerful kickback, which can jerk the saw toward the operator. Severe injuries, lacerations, and even death may result. Always use a safety chain and read the manufacturer's instructions.

> **Did You Know?**
>
> Amperage is the strength of an electrical current, which is expressed in amperes (commonly called amps). An ampere is a unit of electrical current. The higher the amps, the more current is flowing through an electrical wire. All electrical wire is rated by the number of amps the wire is designed to carry. For example, there are 20-amp, 60-amp, and 80-amp wires. Always be sure that you use a wire that can handle the amperage drawn by the tool you are using. Too many amps flowing through a wire not designed for that current will cause the wire to heat up and start a fire.

Table 1 Amperage Ratings

Amps Rating (or nameplate)	0-2.0	2.1-3.4	3.5-5.0	5.1-7.0	7.1-12.0	12.1-16.0	
Extension Cord Length				**Wire Size**			
25'	18	18	18	18	16	14	Not normally available as flexible extension cords
50'	18	18	18	16	14	12	
75'	18	18	16	14	12	10	
100'	18	16	14	12	10	8	
150'	16	14	12	12	8	8	
200'	16	14	12	10	8	6	
300'	14	12	10	8	6	4	
400'	12	10	8	5	4	4	
500'	12	10	8	5	4	2	
600'	10	8	6	4	2	2	
800'	10	8	6	4	2	1	
1000'	8	6	4	2	1	0	

The reciprocating saw brings brute force to bear on a task and is designed for use on hard-to-cut materials. It is used in practically every construction-related profession. You will use this powerful saw to cut cast iron, metal, plastic, and wood. It can saw through walls or ceilings to create openings for plumbing lines. The reciprocating saw is a necessity in any demolition work.

At the front of every reciprocating saw is a shoe plate. Never use a reciprocating saw without the shoe plate because, for the saw to make a good cut, the shoe plate must rest against the work surface. Offset blade adapters can be used to move the blade to the top of the shoe plate for cutting flush or even with the material. Always be sure to use the types of blades specified by the manufacturer for the kind of work you are doing with the reciprocating saw.

> **WARNING!**
> Always unplug any saw and/or remove the battery before you change the blades.

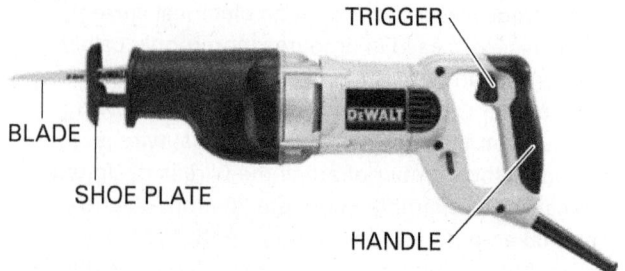

TRIGGER

BLADE

SHOE PLATE

HANDLE

Figure 63 Reciprocating saw.

The reciprocating saw, like all electrical tools, must be grounded with an 3, which is a conductive connection that provides a path for an electric current to pass from an electrical appliance to the earth. Inspect the body and the cord for damage before you operate the saw. Do not operate any electric tools when you are in contact with water or near flammable gases.

> **WARNING!**
> Do not remove the safe-ground pin on the plug of the reciprocating saw.

Some manufacturers make a cordless reciprocating saw. This saw works the same way as a regular reciprocating saw, but it uses battery power instead of an electrical current. The lack of a cord can be an advantage in tight spaces. You must use all the same precautions, however, when operating a cordless tool as you would with an electric tool. When you are using any battery-powered tool, make sure you have a spare battery handy so that you do not have to waste time looking for one if the battery you are using dies or needs recharging.

> **WARNING!**
> Be sure not to stand in water when using any power tool. Doing so greatly increases the chances of electrocution, which can cause serious injuries and even death.

2.2.2 Portable Band Saws

The portable band saw (*Figure 64*) is used to cut a variety of materials, including pipe both with and without iron. These types of pipe are referred to as ferrous and nonferrous pipe, respectively. Other types of pipe and materials that are cut with a portable band saw are bar stock (uncut bar-shaped material) and tube stock (uncut tube-shaped material). Plastics and materials with irregular shapes are also cut using a portable band saw. This saw's blade, which is tooth-edged like a hacksaw, revolves continuously around guides located at both ends of the saw. For best results, lubricate the blade to reduce friction and prolong the life of the saw.

2.2.3 Abrasive Saws

You might also use abrasive saws to cut pipe and other materials. These saws use a special wheel that can slice through either metal or masonry. The most common abrasive saws are the demolition saw and the chop saw. The main difference between the two is that the demolition saw is not mounted on a base. Use respiratory protection when using an abrasive saw. Before using an abrasive saw, ensure that the blade is correctly sized for the safety rating of the saw's rpm range. Never exceed the saw's safe rpm range.

Demolition saws—Demolition saws (*Figure 65*) run on either electricity or gasoline. These saws are effective for cutting through most materials found at a construction site. Although all saws are potentially hazardous, the demolition saw is a particularly dangerous tool that requires the full attention and concentration of the operator. Before operating a gasoline powered saw, check the belt and other moving components to ensure that they are not damaged.

CAUTION

When sawing, be sure to use the appropriate PPE for your eyes, ears, and hands. If you have long hair, be sure to tie it back or cover it properly.

CAUTION

OSHA 29 CFR 1910.106 regulates the amount of flammable or combustible liquids that may be located outside of or inside of a gas storage cabinet or in any one fire-area of a building. The standard prohibits more than 60 gallons of Class I or Class II liquids or more than 120 gallons of Class III liquids in one storage cabinet.

WARNING!

Gas-powered demolition saws must be handled with caution. Using a gas-powered saw where there is little ventilation can cause carbon monoxide poisoning. Carbon monoxide gas is hard to detect, but it is deadly and can kill you! Use fans to circulate air, and have a trained person monitor air quality.

WARNING!

Because a demolition saw is handheld, there is a tendency to use it in positions that can easily compromise your balance and footing. Be very careful not to use the saw in any awkward positions.

Chop saws—The chop saw is a versatile and accurate tool. It is a lightweight circular saw mounted on a spring-loaded arm that pivots and is supported by a metal base (*Figure 66*). Because of its size and weight, it is often portable. This is a good saw to use to get exact square or angled cuts. It uses a wheel to cut pipe, channel, tubing, conduit, or other light-gauge materials. The most common wheel houses an abrasive metal blade, but blades made for other materials such as PVC or wood can also be used. A vise on the base of the chop saw holds the material securely and pivots to allow miter cuts. Some can be pivoted past 45 degrees in either direction.

Choosing the Proper Saw Blade

There are several different types of saw blades. Steel blades are generally the least expensive, but they dull faster. If you do not use your power tools frequently, these blades might be good enough. If you use your power saws a lot, however, carbide-tipped blades are a better choice. These blades are more expensive than steel ones, but they also last longer.

Power Tool Safety

Before operating a reciprocating saw or any other corded or cordless electrical tool, always check to ensure the tool is in proper working condition. If it is a corded tool, check the entire length of the cord to ensure that it is free of cuts, tears, or stretching. If it is a cordless tool, check the battery to ensure that it is free of dents, punctures, or other damage. Always disconnect the power source before removing blades from a corded or cordless saw.

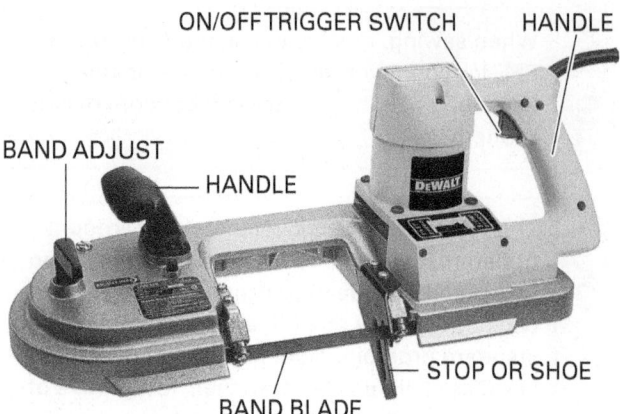

Figure 64 Portable band saw.

Chop saws are sized according to the diameter of the largest abrasive wheel they accept. Two common sizes are 12-inch and 14-inch wheels. Each wheel has a maximum safe speed. Never exceed that speed. Typically, the maximum safe speeds are 5,000 rpm for 12-inch wheels and 4,350 rpm for 14-inch wheels.

Abrasive saw safety—Abrasive saws, such as demolition and chop saws, require extreme care during use. To avoid injuring yourself, follow these guidelines:

- Wear PPE to protect your eyes, ears, and hands, including a respirator or a dust mask.
- Use abrasive wheels that are rated for higher rpms than the tool can produce. This way, you will never exceed the maximum safe speed for the wheel you are using. The blades are matched to a safety rating based on rpm.

- Be sure the wheel is secured on the arbor before you start using a demolition saw. Point the wheel of a demolition saw away from yourself and others before you start the saw.
- Use two hands when operating a demolition saw.
- Secure the materials you are cutting in a vise. The jaws of the vise hold the pipe, tube, or other material firmly and prevent it from turning while you cut. (Vises are discussed later in this module.)
- Be sure the guard is in place and the adjustable shoe is secured before using a demolition saw.
- Keep the saw and the blades clean.
- Inspect the saw each time you use it. Never operate a damaged saw. Ask your supervisor if you have a question about the condition of a saw.
- Inspect the blade before you use the saw. If you see damage, throw the wheel away (*Figure 67*).

2.2.4 Angle Grinders

Angle grinders, also called side grinders, are handheld power tools used to grind away hard, heavy materials and surfaces, such as pipes, plates, or welds (*Figure 68*). Angle grinders have a rotating grinding disc set at a right angle to the motor shaft. The grinding disc on the end grinder rotates in line with the motor shaft. Grinding is also done with the outside of the grinding disc. Always perform a ring test on the grinding wheel before using an angle grinder. This involves mounting the wheel on a rod and tapping it. A clear ring indicates the wheel is in good condition; a dull thud means the wheel is in poor condition and should be discarded.

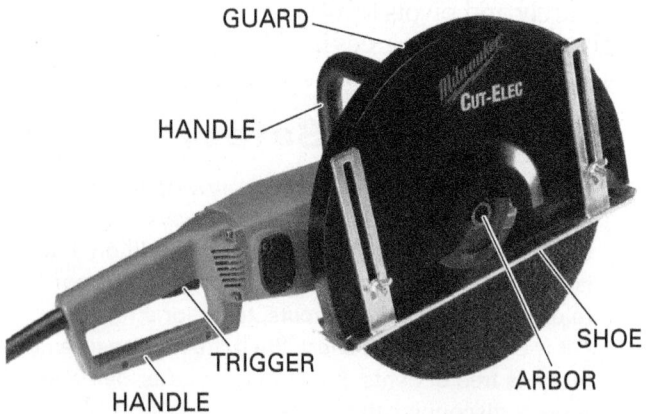

ELECTRIC

GAS-POWERED

Figure 65 Demolition saws.

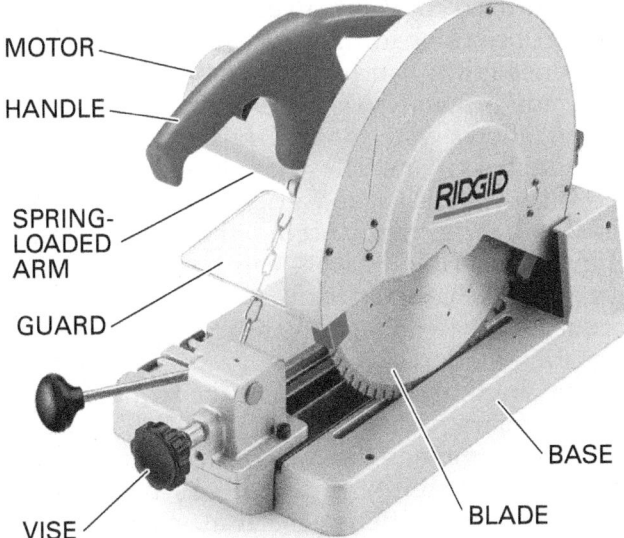

Figure 66 Chop saw.

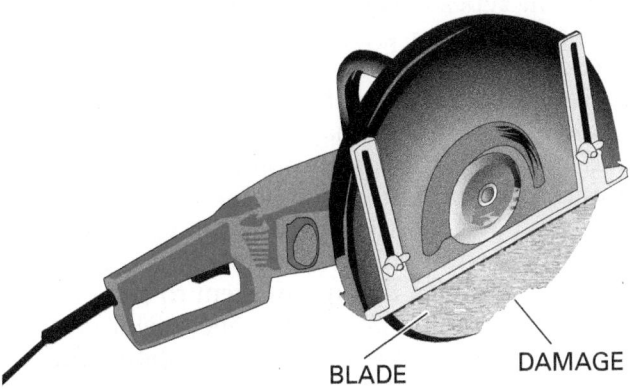

Figure 67 Damaged saw blade.

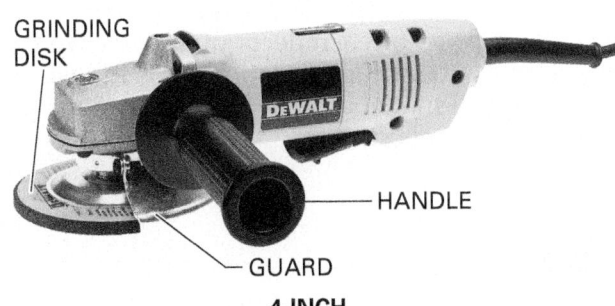

4-INCH

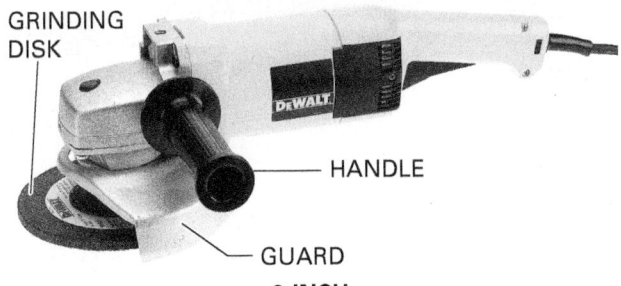

9-INCH

Figure 68 Angle grinders.

2.3.0 Drilling and Boring Tools

Electric drilling and boring tools are versatile instruments that have changed the plumbing profession by greatly shortening the amount of time it takes to install pipe. Few things have enhanced the plumber's productivity more than the electric drill, and die tools have made piecing pipe together easier. Drilling tools allow you to make holes in a building structure quickly.

> **CAUTION**
>
> Never assume you know how to use a power tool. Always thoroughly read and understand the operator's instructions before using and starting power tools.

Picking up a drill and pressing the trigger is not difficult, but using a drill with skill and care requires training and experience. To match a drill bit to a task, you need to know both the capabilities of the drill bits and the properties of the materials into which you are drilling. While drills are less dangerous than some tools, like saws, using or maintaining drills improperly can lead to injury.

Successful drilling depends on preparation. If possible, place the workpiece you plan on drilling in a clamp or a vise so that both your hands are free to operate the drill. This also prevents the workpiece from moving around as you drill. Before you drill a hole in metal, you need to establish and mark a precise center point on the metal's surface. If you are drilling a large hole, you might want to drill a pilot hole with a smaller drill bit first.

When you start drilling, place the drill bit, which is the part of the drill that does the cutting, on the center point you have marked on the workpiece. First run the drill at a slow speed so that the drill bit does not stray from your center point. Never force a drill. If you apply too much pressure to a drill, you are likely to break the drill bit. If you are drilling all the way through a workpiece, slow the drill speed as the drill bit approaches the other side so that when the bit breaks through, it leaves a smooth exit hole. If the drill speed is too high or you are applying pressure to the drill, the exit hole will be rough.

If the drill seems to jam or threatens to jam, withdraw the drill. Doing so allows the drill bit to clear any debris from the hole that you have made. Then you can resume where you left off.

Unlike other cutting tools, drill bits do not allow for an exit path for the debris they create; the sidewall of the hole blocks the debris caused by the drilling. It is usually necessary, especially when drilling deep holes, to back the drill bit out of the hole in order to clear out the debris. This also helps to dissipate the heat that is generated by the constant friction between the cutting edge of the drill bit and the workpiece.

Drilling is something of an art form. To do it well, you have to consider several factors, including the drill's rotational speed (which is defined by its rpm), the type of drill bit, the diameter and depth of the hole being drilled, and the material properties of the workpiece. These factors must be balanced in order to avoid too much heat buildup, which could lead to tool failure. These factors are discussed in the following sections.

2.3.1 Portable Electric Drills

The portable electric drill (*Figure 69*), sometimes called a pistol drill because of its shape, is used to drill holes in structural materials so that pipe can be run and fixtures can be installed. This basic drill has a motor built into the pistol-shaped body. The chuck is a gripping mechanism that houses the drill bit. After the drill bit is inserted into the jaws of the chuck, the bit is locked in place by tightening the chuck. Always unplug the drill when tightening and locking a drill bit into place.

If you need to drill between joists (horizontal support beam) or in tight corners where a pistol drill does not fit, an offset portable electric drill can come in handy (*Figure 70*). This drill's offset design allows you to drill holes where space is limited. Drill models with extended handles allow you to get a good grip so you can better guide the drill when boring through material. Milwaukee's Hole Hawg® can be used in tight places. This brand of drill is legendary for its torque, or the twisting or turning force applied in a rotating motion.

Two other kinds of drills are the hammer drill and the rotary hammer drill (*Figure 71*). The hammer drill combines the force of a hammer with the boring capability of a drill. Its hammering action penetrates dense material, such as concrete or masonry, and the rotating bit removes the chips. A variety of drill bits are available for different types of jobs. The rotary hammer drill is used for larger jobs; the regular hammer drill is used for drilling holes to set small anchors that secure piping to a structure.

Pistol drills are perhaps the most widely used and most versatile of all power tools. Their primary function is to drill holes in steel, wood, concrete, and plastic—in fact, in virtually any material used in building. Pistol drills come with many accessories that allow them to function as grinders, sanders, polishers, sheet-metal cutters, and, with the addition of wire or flapper wheels, metal finishers.

Many types of pistol drills are available, made by a variety of manufacturers. Some are durable, industrial-quality tools meant for heavy use, while others are better suited to occasional use. Naturally, price often reflects quality as well as adaptability to various functions. An industrial-strength drill model designed for continuous use can be a wise investment.

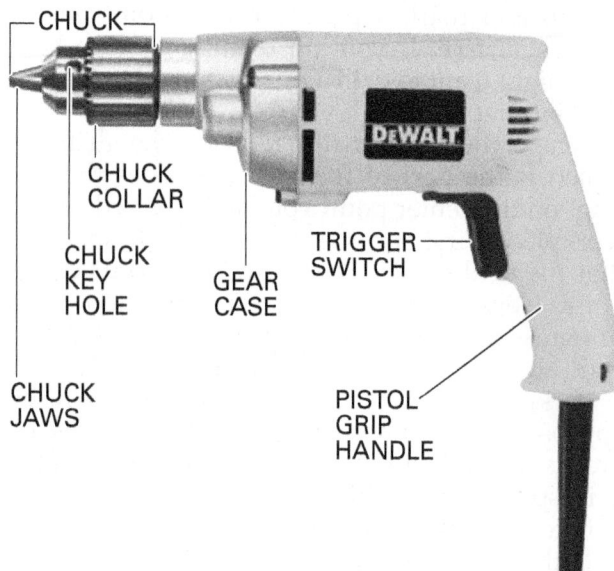

Figure 69 Portable electric drill (pistol drill).

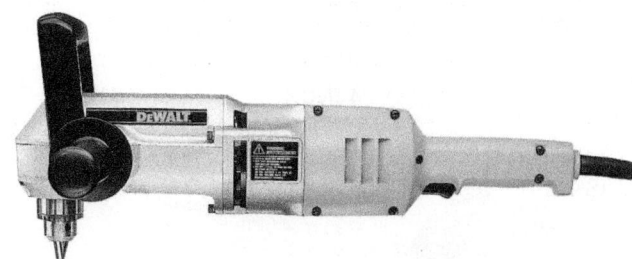

Figure 70 Offset drill.

HAMMER DRILL

ROTARY HAMMER DRILL

Figure 71 Hammer drills.

Though there are many different styles of pistol drills, each one has a specified capacity (the maximum size of the chuck cavity) that dictates the size of the drill bit the chuck cavity can hold. Common capacities are $\frac{1}{4}$ inch, $\frac{3}{8}$ inch, and $\frac{1}{2}$ inch.

Drill chucks usually contain three jaws that open and close simultaneously to release or grip a drill bit or other accessory (*Figure 72*). The chuck mechanism is worked with a chuck key or, in newer models, by hand. The chuck key is a gear that meshes with another gear at the base of the drill chuck. The key is attached to the drill cord with a piece of rubber.

The size of the pistol drill is stated in terms of the diameter of the chuck cavity when the chuck jaws are fully open. For instance, if a pistol drill's chuck cavity measures $\frac{1}{4}$ inch in diameter when the jaws are fully open, that drill is called a $\frac{1}{4}$-inch drill. It uses drill bits with shanks that are up to $\frac{1}{4}$ inch in diameter.

Capacity—Each drill motor has a maximum no-load rating, which is stated in rpm. "No-load" means the pistol drill is not doing any actual drilling and the motor is running freely at its maximum speed. No-load ratings vary depending on the drill's capacity. On all drills, as drill capacities increase, drill speeds decrease. Simply stated, larger drill bits turn more slowly. For example, a $\frac{1}{4}$-inch drill may have a no-load rating between 2,000 and 3,500 rpm. A $\frac{3}{8}$-inch drill may have a rating between 650 and 1,200 rpm. A $\frac{1}{2}$-inch drill may have a rating between 350 and 850 rpm. These ratings vary slightly by manufacturer. The rpm of the Hole Hawg® drill, for instance, is geared down to allow it to use larger, self-feeding drill bits.

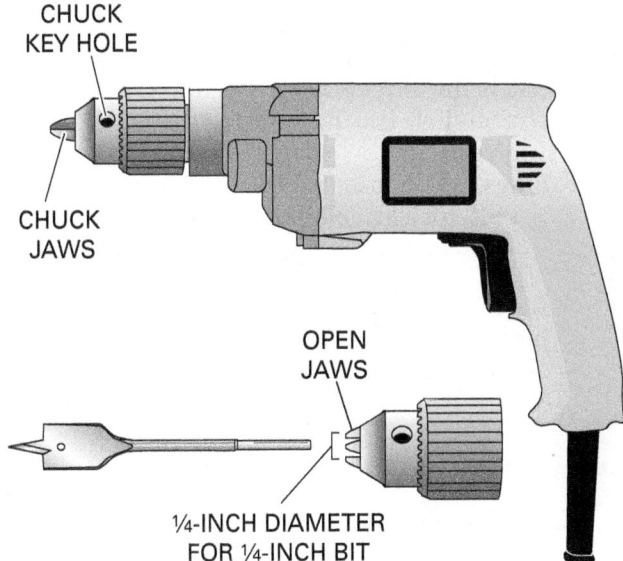

CHUCK KEY HOLE

CHUCK JAWS

OPEN JAWS

¼-INCH DIAMETER FOR ¼-INCH BIT

Figure 72 Drill chuck.

The reason an increase in drill capacity means a decrease in drill speed is that, as drill capacity increases, so does the drill's torque. Torque is a measure of a tool's ability to do work. It requires more torque, for example, to drill a $\frac{1}{2}$-inch hole through an I-beam than it does to drill a $\frac{1}{8}$-inch hole through a metal stud. Therefore, you would choose a larger drill for the I-beam than for the metal stud.

A larger-capacity drill motor turns more slowly than a smaller-capacity motor. A larger drill bit cannot tolerate the high speeds of a smaller-capacity drill motor without overheating and losing its **temper**. Temper is the drill bit's resilience and strength. This is one reason why, when drilling through thick steel, you should periodically withdraw the drill bit to prevent it from overheating.

Rotation—Variable-speed drills can operate in a wide range of rotational speeds, from slightly above zero to the maximum speed of the drill's motor. The pressure applied to the drill's trigger determines the rotational speed: the more pressure, the higher the speed. Manufacturers realize that drilling applications are not standard and that different applications require different speeds. Holes in some materials are best drilled at high speeds, while holes in other materials are best drilled at lower speeds.

A variation on the basic drill motor is the reversing motor. These drill motors enable either clockwise or counterclockwise rotation of the drill bit. This is useful if you need to back the drill bit out of a hole when it has become pinched. You switch the direction of rotation by toggling a lever or switch on the body of the drill. Most reversing drill motors also have variable speed capabilities. Most manufacturers designate these tools as VSR, for variable-speed reversing.

> **WARNING!**
>
> Never reverse drill motors while they are running. You can damage the drill or the workpiece or injure yourself.

Tempering

Certain metals and special glass are tempered (heated and cooled) to enhance their hardness and resilience. Alternately heating and cooling a metal, such as steel, pre-stresses the metal, greatly increasing its strength. Drill bits are often tempered to make them stronger than the materials on which they're used. Overheating a tempered metal, however, causes it to lose its enhanced hardness.

2.3.2 Cordless Drills

Cordless drills are becoming more popular because of their versatility and convenience over the traditional portable drill. A cordless drill is especially useful when working in areas where a power source is difficult to find or where it would be awkward or hazardous to run an electrical cord.

In cordless drills, a rechargeable battery pack runs the motor (*Figure 73*). The removable pack is attached to the drill during use and can be plugged into a battery charger when the drill is not in use. Some battery chargers can recharge the pack in an hour; others require more time. If you use a cordless drill a lot, always have a spare battery pack on hand so you can continue to use your cordless drill while your other battery pack recharges.

All the features of traditional pistol drills are also found on the cordless models. Many cordless drills, for instance, have a VSR capability. Some also have adjustable clutches that shift the force (torque) to allow the drill to double as a power screwdriver. Cordless drills generally cost more than traditional drills, but that expense may be more than offset by their flexibility.

2.3.3 Bits

Drill bits are the cutting tools used in electric drills, traditional and cordless alike. Drill bits are mounted in the drill chuck cavity and then locked into place. There are several different types of drill bits: twist, spade, hole saw, self-feeding, auger, and masonry. Each is shaped differently, and each has special properties or special uses (*Figure 74*).

Figure 73 Cordless drill.

Twist bits—The twist bit is a versatile drill bit that can be used to drill into almost any material. Twist bits are most commonly used for drilling small holes before installing plumbing fixtures. Twist bits are commonly sized at $\frac{1}{2}$ inch. Two types of twist bits are available: those made from high-speed steel and those made from carbon steel. Twist bits made from high-speed steel are of higher quality, can better withstand high temperatures without "burning up," and are used to drill metal (they can also drill wood, plastics, and composites). These twist bits have a slightly blunted nose that will not deform under pressure, but they also have a tendency to spin away from the intended hole center. The cheaper, high-carbon steel twist bits are designed specifically for drilling wood. They are not suitable for drilling hard metals such as iron and steel.

Spade bits—Spade bits are relatively inexpensive. As their name suggests, these bits have a spade-shaped blade with a sharp triangular tip. The spade is flat, and the sharp tip acts as a guide, centering the hole. Most of the drilling is done by the honed cutting edge at the shoulders of the spade. Spade bits cut fast and are long enough to drill through fairly thick workpieces. These bits are designed to cut wood and some plastics. Do not use them to drill metal. Spade bits and the holes they produce range from $\frac{1}{4}$ inch to 2 inches in diameter.

Hole-saw bits—The hole-saw bit has saw-like teeth arranged around a hollow cylinder at the end of the bit. This blade is attached to a shaft, which is inserted into the drill chuck like the shaft of any other drill bit. As the drill turns the bit, the teeth on the blade cut through the material. Use these bits to drill holes between 1 inch and $3\frac{1}{2}$ inches in diameter. Hole-saw bits are appropriate for drilling wood, plastics, and a variety of metals, such as iron, steel, and aluminum. Hole-saw bits are very useful for making holes to install piping, tubing, and conduit.

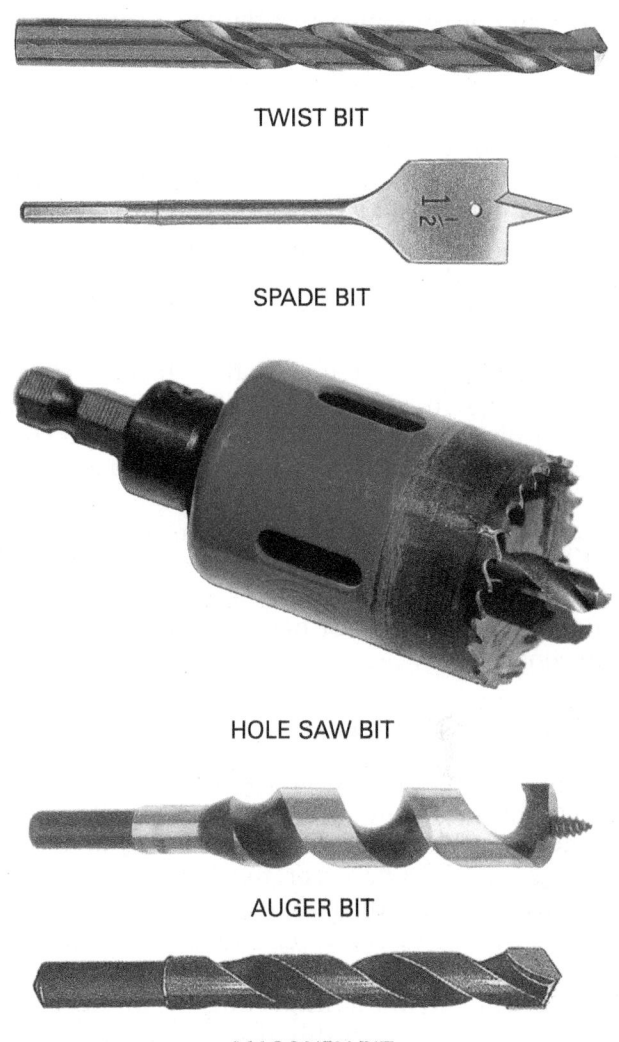

TWIST BIT

SPADE BIT

HOLE SAW BIT

AUGER BIT

MASONRY BIT

Figure 74 Types of drill bits.

Self-feeding bits—Self-feeding bits, or multispur bits, bore holes in wood quickly. These bits come in diameters ranging from $\frac{1}{2}$ inch to $4\frac{5}{8}$ inches. They are good for drilling through floor joists and wall studs and for drilling in hard-to-reach areas because you do not need to move the drill to bore the hole.

Auger bits—Auger bits come in diameters ranging from $\frac{3}{4}$ inch to 1-$\frac{1}{2}$ inches. They are designed to cut only wood. Do not use them to cut metal; it will damage the bit.

Masonry bits—A masonry bit has a carbide tip that allows for easy drilling through masonry materials, such as concrete, brick, stone, and plaster, that would quickly dull or fracture most other drill bits. Do not use these specially hardened bits on wood—they are designed only for masonry. Masonry bits should be used at relatively low drill speeds to prevent them from overheating. These bits cut relatively small holes (generally $\frac{3}{4}$ inch in diameter or less) in masonry through which to run piping.

Three-toothed bits—Three-toothed bits are used to drill large holes in wood. They are widely used in plumbing, HVAC, and other trades. The Milwaukee Big Hawg® is an example of a 3-tooth hole cutter.

Care of bits—Protect drill bits from moisture and keep them sharp. A thin coat of all-purpose oil can help prevent rust. If you notice that a bit is not sharp, replace it or have a qualified person sharpen it.

2.3.4 Impact Wrenches

An impact wrench, also called an impact drill or impact driver (*Figure 75*), is a high-torque socket wrench. An electric motor accelerates a rotating mass, called a hammer, to cause it to repeatedly strike an output shaft, called an anvil, with great force. The operator feels little torque while the tool operates. Below minimum torque, the hammer spins without impacting the anvil. This allows the tool to smoothly drive fasteners without impact. Impact wrenches are available in all common socket wrench sizes. Cordless multitools (*Figure 76*) include a variety of tool attachments that can be easily removed and swapped out for different uses. Attachments include saws, sanders, and wrenches. The rotation speed is controlled by a variable-speed trigger on the motor housing.

2.4.0 Electric Pipe-Threading Machines

The electric pipe-threading machine rotates, threads, cuts, and reams pipe (*Figure 77*). The pipe is inserted into the chuck of the machine, and the chuck is tightened. A foot pedal controls the machine. Use only your foot to operate the pedal; never place a piece of wood or other object on the pedal to depress it. For trouble-free operation, be sure to follow the preventive maintenance schedule provided by the machine's manufacturer. Electric pipe threaders are also available in handheld versions (*Figure 78*).

Figure 75 Impact wrench.

WARNING!

When you are working with a power-threading machine, do not wear loose clothing, jewelry, or gloves that could easily be caught in the rotating parts of the machine. Be sure to wear appropriate PPE. Be sure to tie back long hair to prevent it from being entangled in the machine during operation. Use a piece of cloth or cardboard to absorb any oil that spills during the machine's operation, so the oil does not get on the bottom of your work shoes and you do not track the slippery oil around the job site.

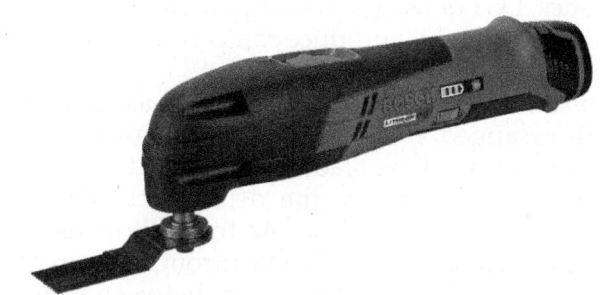

Figure 76 Cordless multitool.

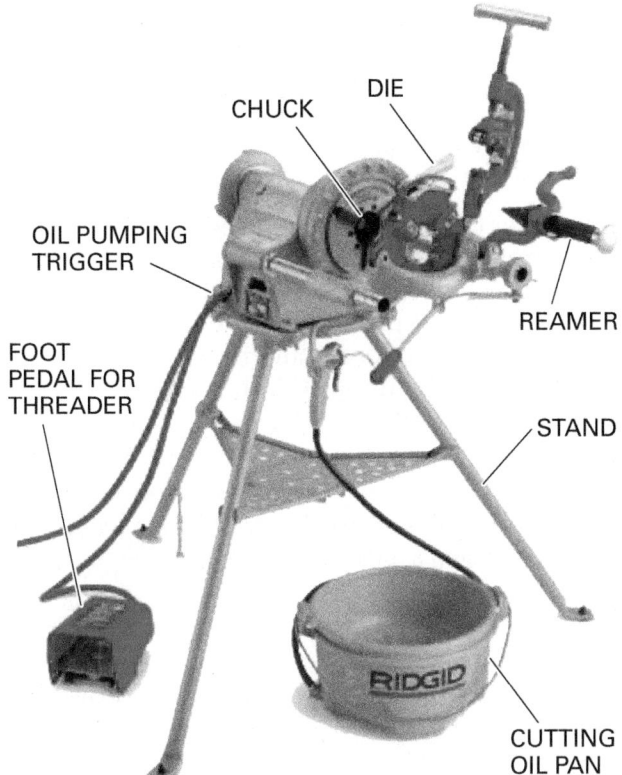

Figure 77 Electric pipe-threading machine.

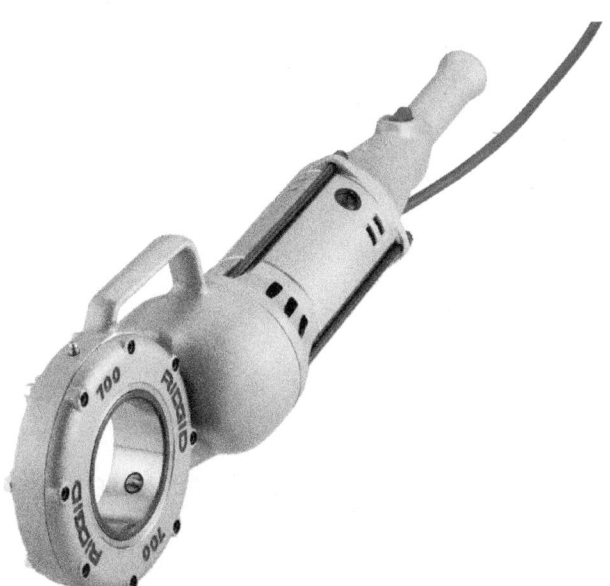

Figure 78 Handheld pipe threader.

2.5.0 Soldering Tools

Soldering is used to join rigid copper pipe and fittings. The filler material, called solder, is distributed evenly between the close-fitting surfaces of the pipes when the pipes/fittings are heated to the necessary temperature. You use a small, portable gas torch for soldering (*Figure 79*). Propane and acetylene are the two most commonly used gases. The torch heats the metal and the water-soluble flux (also known as soldering paste) to join the pipe. The droplets of solder are molten metal. They can severely burn you if they touch you. Always wear goggles, gloves, and other protective clothing when you are using a welding torch. Weld only under proper supervision and only after proper training.

> **WARNING!**
>
> Acetylene has extreme ignition characteristics. Always refer to the manufacturer's specifications for appropriate tank pressure. Avoid contact with copper, silver, and mercury. Acetylene cylinders should remain upright at all times.

> **CAUTION**
>
> Acetylene and propane gases are under extreme pressure, can cause explosions, and can release toxic gas into the atmosphere. Follow these general precautions when working with compressed gases:
> - Do not drop cylinders or strike them with force.
> - Leave the cylinder cap on until the cylinder is ready for use.
> - Transport cylinders using an appropriate cart. Do not roll, drag, or slide cylinders.
> - Do not tamper with safety devices on cylinders, regulators, or valves.
> - Do not subject compressed gas cylinders to temperatures exceeding 125°F.
> - Do not allow a flame to contact a compressed gas cylinder.
> - Use compressed gases in well-ventilated areas.
> - Use the smallest cylinder size available that is adequate for your application.
> - Use a trap or check valve to prevent liquid from backing into the regulator or cylinder when discharging gas into a liquid.
> - Refer to the label information and MSDS before handling any compressed gas.
> - Wear appropriate PPE.
> - Use caution when removing a cap from a cylinder that has been stored outdoors. Cylinders stored outside can invite wasps and other insects.
> - Turn off the cylinder valve when the cylinder is not in use.

Figure 79 Gas torch for soldering.

The torch flame produced by a soldering tool has different temperatures, all of them hot, naturally, but some hotter than others. The hottest part of any flame is at the tip of the inner cone, where the pale flame meets the deeper-colored outer flame.

Propane torches are used for soldering copper pipes. Temperatures between 400°F and 800°F can solder plumbing connections. Temperatures of 3,000°F can braze metals and cut or weld iron and mild steel. Be sure to match the torch's heat output to the task. While propane burns on its own by drawing oxygen from the air, supplementing the propane gas from a handheld torch with oxygen from a separate tank can generate even more heat.

> **WARNING!**
> Working with soldering tools can be dangerous. Soldering irons can easily burn you and objects around you. Never touch the tip and heater while the unit is on. Allow plenty of time for the iron to cool before changing the tip. Use a stand for your soldering iron and keep it a safe distance from flammable materials.

> **CAUTION**
> If you suffer from any respiratory problems, be sure to work in a well-ventilated area. Soldering irons contain rosin-based fluxes, which can irritate the respiratory system. Rosin-free solder is also available.

A striker (*Figure 80*) can be used to light torches that do not have self-igniting capability. When using a striker, ensure that the cup lip is parallel to the flame. Do not place the striker perpendicular to the front of the torch outlet, or allow the cup to cover the tip in any way. Otherwise, you risk a dangerous blowback.

> **WARNING!**
> Never use a cigarette or cigar lighter to ignite a torch. Never carry a lighter on your person when using a torch, as this poses a serious fire risk.

> **NOTE**
> Dispose of propane fuel tanks property to avoid harming the environment. Many communities have hazardous waste disposal areas set up to handle the disposal of such items.

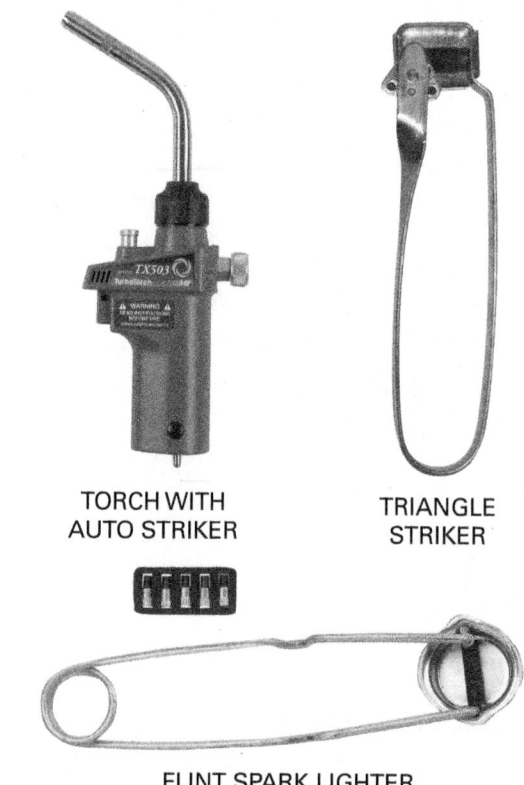

TORCH WITH
AUTO STRIKER

TRIANGLE
STRIKER

FLINT SPARK LIGHTER
WITH FLINTS

Figure 80 Strikers.

Brazing

In plumbing, a process called brazing is used to join pipe and fittings for pipelines that carry saltwater and oil and for refrigeration and chemical-handling systems, vacuum and air lines, and low-pressure steam lines. Brazing requires much hotter temperatures than those needed for soldering copper pipe.

To achieve a sufficiently hot flame for brazing, use an oxyacetylene torch unit capable of generating temperatures of 1,400°F (or 760°C) or higher. Brazing requires a torch with the correct tip, regulator valves, and both oxygen and acetylene tanks. These tanks come in two sizes. Plumbers commonly use the smaller B-tank, which has two compartments to hold the pressurized acetylene and oxygen.

Note that the torch unit shown is intended for use with oxyacetylene gas.

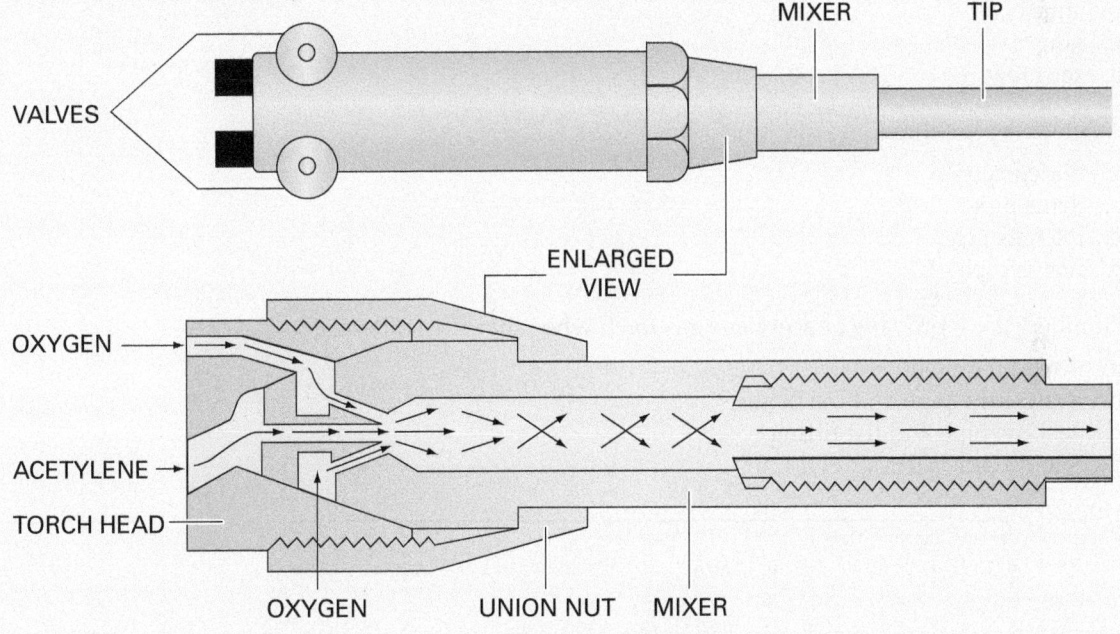

Additional Resources

Core Curriculum, Latest Edition. NCCER. New York, NY: Pearson.

2.0.0 Section Review

1. An excessively long extension cord can result in _____.

   a. a voltage drop
   b. reduced resistance
   c. decreased current
   d. harmonics

2. The saw that has a continuous blade that revolves around guides located at both ends of the saw is called a _____.

   a. portable band saw
   b. reciprocating saw
   c. chop saw
   d. demolition saw

3. The mechanism at the end of a drill motor that is used to grip or release a drill bit is called the _____.

   a. vise
   b. chuck
   c. key
   d. setscrew

4. An electric pipe-threading machine rotates, cuts, threads, and _____.

   a. widens pipe
   b. drills pipe
   c. ratchets pipe
   d. reams pipe

5. Plumbers use a propane or acetylene gas torch when they are soldering to _____.

   a. thread the pipe
   b. oxidize the pipe and the fitting
   c. heat the pipe and the fitting
   d. remove any burrs

SUMMARY

Plumbing is a sophisticated and complex profession, full of challenges. It includes designing, installing, and maintaining new systems, renovating existing systems, and repairing broken systems. To do all these jobs, you need a variety of tools, and you need to be adept in their use. Your expertise with your tools will directly affect the quality of your work.

As you begin to learn your profession, you will become familiar with the tools used in plumbing. The hand tools you will use every day include many types of measuring, marking, and cutting tools. Each tool has a specific use, safety requirements, and care requirements. Learn these until they are second nature to you. Maintenance is not just fixing or replacing a tool after it breaks but servicing and maintaining your tools to prevent them from breaking or wearing out prematurely. Take advantage of the productivity enhancements multifunction tools offer by becoming expert in their use.

You will spend a lot of your time measuring, marking, cutting, and installing different types of piping and tubing. To lay pipe that is straight and even, you may use simple tools like a plumb bob and a spirit level. You will use special tools designed to cut specific materials, like cast iron, various types of plastic, and copper. Each type of tool is designed for a specific material and a specific task. You must learn which tool to choose for which task and for which material.

Plumbing is closely related to a number of other construction trades. Your work will affect how the carpenters and electricians do their jobs. You will work with heating, ventilating, and air conditioning technicians to install systems in new buildings. To be a successful plumber, you will develop some of the same skills, including the use of tools, as these other craftworkers have. After a few years, you will find that you know a bit about each of these crafts. When other craftworkers find out that you know your own craft as well as something about theirs, it will make the job easier, earn you their respect and gratitude, and enhance your reputation as a professional.

You may find yourself using sophisticated electronic and laser equipment. As technology advances, new tools are developed, and learning to use them will keep you up-to-date and competitive. The successful professional makes a habit of continuing to learn.

1. A square that includes a try miter, protractor, and framing square in one tool is called a _____.
   a. framing square
   b. carpenter's square
   c. plumber's square
   d. speed square

2. The proper function of a spirit level depends on _____.
   a. the type of fluid in the bubble
   b. precise calibration
   c. its ability to withstand rust
   d. its length

3. A small level that is useful in confined areas or on short runs of pipe is called a _____.
   a. spirit level
   b. laser level
   c. torpedo level
   d. plumber's level

4. The level with a movable telescope to allow measurements in both the horizontal and vertical planes is called a _____.
   a. carpenter's level
   b. builder's level
   c. plumb bob
   d. transit level

5. On a hacksaw, the more teeth per inch on a blade, _____.
   a. the coarser the cut and the more debris
   b. the finer the cut and the more debris
   c. the coarser the cut with less debris
   d. the finer the cut with less debris

6. Do not use a hacksaw when you need a cut that is _____.
   a. fast
   b. square
   c. rough
   d. flexible

7. To cut into material using a chisel, you _____.
   a. strike the flat part of the chisel blade with a hammer
   b. strike the top of the handle with a hammer
   c. first cut a starter channel and then strike the upper side of the chisel blade with a hammer
   d. first clamp the chisel to the material and then strike the handle with a hammer

8. The tool on which the blade should be beveled at a precise 25-degree angle is the _____.
   a. wood chisel
   b. mushroom chisel
   c. cold chisel
   d. snap chisel

9. A common result of hammering a cold chisel is _____.
   a. smashing
   b. mushrooming
   c. spreading
   d. splaying

10. A type of cutter used for cutting small-diameter, soft materials such as copper, brass, and aluminum tubing or thin-wall conduit is a _____.
    a. pipe cutter
    b. snap cutter
    c. ratchet cutter
    d. tube cutter

11. The number of jaws that a bench vise has is _____.
    a. 2
    b. 3
    c. 4
    d. 5

12. To cut threads into galvanized and black-iron (cast-iron) pipes, use _____.

    a. augers
    b. bits
    c. dies
    d. tube cutters

13. To help turn a die tool by increasing the leverage on it, use a _____.

    a. torque wrench
    b. ratchet stock
    c. threader
    d. chuck key

14. The hammer has a flat face for striking and a round face for aligning brackets and driving out bolts is called the _____.

    a. claw hammer
    b. drywall hammer
    c. ball peen hammer
    d. sledge hammer

15. Check extension cords for a(n) _____.

    a. UL label indicating wattage
    b. NFPA label indicating fire rating
    c. OSHA label indicating appropriate uses
    d. IEEE label indicating voltage

16. Because of its brute power, a tool that is useful for demolition work is a _____.

    a. hacksaw
    b. portable band saw
    c. reciprocating saw
    d. chop saw

17. A type of drill that is useful because it works on some materials that are drilled best at high speeds, and others at lower speeds is the _____.

    a. alternating-speed drill
    b. electric-speed drill
    c. variable-speed drill
    d. reversible-speed drill

18. The type of bit that is best for cutting wood and some plastics, but which should not be used to cut metal is the _____.

    a. self-feeding bit
    b. hole-saw bit
    c. auger bit
    d. spade bit

19. A type of bit that is used to bore large holes in wood is the _____.

    a. three-toothed bit
    b. hole-saw bit
    c. twist bit
    d. self-feeding bit

20. A type of bit that will bore holes in wood quickly and is good for drilling through floor joists and wall studs, and for drilling in hard-to-reach areas is the _____.

    a. masonry bit
    b. hole-saw bit
    c. twist bit
    d. self-feeding bit

Trade Terms Quiz

Fill in the blank with the correct term that you learned from your study of this module.

1. _____ are used to mark metal.

2. _____ is the substance that helps join heated metal pipe.

3. A right-angle portable electric drill is useful when drilling between _____.

4. Pipe that does not contain iron is called _____ pipe.

5. _____ is uncut tubing material on hand for use on a project.

6. A surface that is straight vertically is considered _____.

7. _____, or _____, are used to mark a cutting line on the surface of a tube.

8. A(n) _____ can be found in the handle of some combination squares for leveling.

9. A(n) _____ is a combination of elbows or bends that brings one section of the pipe out of line but into a line parallel with the other section.

10. The process of removing burrs in a pipe is pipe _____.

11. Bar material that is uncut on a work site is _____.

12. A surface that is straight horizontally is considered _____.

13. Heads of striking tools that have _____, or are mushroomed, must be replaced or restored.

14. _____ is the twisting or turning force applied in a rotating motion.

15. _____ is the filler material used to connect joints during soldering.

16. Speed squares combine a(n) _____, protractor, and framing square in one tool.

17. The _____ and length of a pipe are determined by using measuring tools.

18. A speed square can be used to mark cutoff lines or as a(n) _____.

19. _____ describes the strength of the electrical current used by electrically-powered tools.

20. A connection that provides a path for electrical current from a device to the earth is a(n) _____.

21. An allowable variation from a measurement is a(n) _____.

22. The groove made by a saw blade is called a(n) _____.

23. Portable band saws are used to cut both _____ and nonferrous pipes.

24. The loss of voltage is called _____.

25. When drilling through thick steel, periodically withdraw the drill bit to avoid a loss of _____.

26. _____ is the use of heat to form joints.

Trade Terms

Amperage	Joist	Plumb	Straightedge
Bar stock	Keel crayon	Reaming	Temper
Burr	Kerf	Scribe	Tolerance
Diameter	Level	Soapstone	Torque
Electrical ground	Miter	Solder	Tube stock
Ferrous	Nonferrous	Soldering	Voltage drop
Flux	Offset	Spirit level	

Kenneth Allen

Plumbing Instructor, Maryland State
Department of Labor, License, and
Regulations (DLLR)

Describe how you got started in the construction industry.

My family has a long history of working in HVAC and plumbing. Growing up, I found that the occupational trades and industrial sciences interested me the most, so in high school I took welding and drafting classes. Throughout high school, I worked for my father installing water source heat pump systems, refrigeration units, and ventilation systems. After high school, I worked as an electrician for one year.

Who inspired you to enter the industry? Why?

My father inspired me, especially as I was working for him installing commercial water source heat pump systems in multi-unit condo buildings at the base of a ski area in Bartlett, New Hampshire. The plumbing foreman at that construction site took an interest in my development, and when he formed his own business (Scott Perkins Plumbing & Heating), he offered me a job working with him as a plumber. When I was laid off as an electrician, I applied for a plumbing apprenticeship with Mr. Perkins, who taught me craftsmanship, to take pride in the job, and to be proud of a job well done—traits which, as he would say, are invaluable.

After achieving journeyman status, I felt I needed to move on, so I moved to the Mid-Atlantic to find work. When I arrived, I noticed there were many new people coming into the field that needed the type of training I had received in the New England region.

After working several years as a plumber, my wife Patti suggested that I apply for an instructor job with ABC Cumberland Valley Chapter. Invigorated by all I had learned thus far in the trade, I decided to teach. I applied for a full-time teaching position. I could not anticipate how much I would come to enjoy teaching, especially in a Maryland prison teaching plumbing to inmates! I've enjoyed it so much that I have been a plumbing instructor for the last 13 years. I recently began instructing a local college in the same Maryland county where the prison is located.

What do you enjoy most about your job?

When an inmate, college student, or even a colleague requests assistance, asks for help understanding something, or comes to me seeking a specific method to accomplish a task, I give clear instructions. Moreover, when the student follows those specific directions, you can measure how well they listened and understood by the way they carry out a task. I cannot explain anything more enjoyable and life fulfilling than the student saying, "Ah-ha!"

Do you think training and education are important in construction? If so, why?

Absolutely. Without proper training and the tried, tested, and true methods of disseminating information, nothing is accomplished in life. The most important issue concerning the education of construction workers is construction safety. National statistics have proven that accidents are reduced considerably on the job site when workers are trained in the proper use of hand and power tools, and trained to handle hazardous materials, use personal protective equipment, and apply the processes they have learned.

How important are NCCER credentials to your career?

These credentials give the student a more portable credibility to move geographically from location to location because of the national recognition. For the instructor, these credentials avail the affordability to offer instruction to students looking for economic mobility in hope of achieving the American dream.

How has training/construction impacted your life and your career?

My family is represented by a long line of tradesmen; my father, my fathers' father, and my uncles all chose careers in HVAC. Since childhood, I had always been fascinated with figuring things out cognitively and applying this discovered knowledge in tangible ways. I always took great pleasure in perfecting my newly learned abilities. I continued to push myself and set my goals—especially to obtain a state certificate, which permitted me to teach the sciences associated with the occupational trades and related industries —even though I struggled with the heavy workload of working fulltime and attending classes in the evenings. I met and achieved my intended goal and obtained an APC State Department of Education credential. Moreover, every day I endeavor to honor the accolades bestowed upon me by my colleagues and I give thanks and praise for the once in a lifetime opportunity to teach. Immediately, I have come to see the outcome of my efforts, and a product of instruction put to immediate use.

Would you suggest construction as a career to other? If so, why?

The construction industries have suffered quality-wise because many people are looking for jobs as opposed to careers. Even though there has been sufficient technological advancement in the field to take some of the "back-work" out of the construction trade, there is a very real physical component that frankly the video-game generation isn't well prepared for.

I would suggest this field as a career for people who like working with their hands, who are enthusiastic about formulating solutions under a variety of conditions, and who have the courage and initiative to make their ideas a reality for which they are more than willing to stand behind. Craftworkers built America, and today it's those same grunt workers who are capable of repairing the national infrastructure. The only question anyone who wishes to pursue the construction industry as a career choice should ask is; "Where do I sign up for my apprenticeship training?" That is why I continue to teach: I instruct those who ask the right question. When people ask me what is the best program of construction education available today, my response is always NCCER.

How do you define "craftsmanship"?

I define "craftsmanship" as an effect that has a cause. When one chooses quality over quantity, the proof of pride is shown in the finished product—not only is it functional, but it is also an object of art.

I believe that some people seek to become artisans of the trade, like the stone masons and carpenters of the past. This level of craftsmanship can be achieved without a loss of productivity, as long as you stay vigilant to the task at hand.

Of course, those who aspire to ascend above the status quo usually ascend the ladder of success and achieve the level of economic satisfaction and trade leadership consistent with the richness of their efforts.

LIST OF BASIC TOOLS FOR APPRENTICE PLUMBERS

- Appropriate personal protective equipment
- Measuring and layout tools
- Measuring squares, including:
  - Speed
  - Combination
  - Framing

- Levels and precision measuring tools
- Plumb bob
- Tooth-edged cutting tools, including:
  - Hacksaw
  - Reciprocating saw
  - Portable band saw
  - Abrasive saw

- Saw blades
- Wood and cold chisels
- Smooth-edged cutting tools
- Keel crayon
- Soapstone
- Drills, including:
  - Portable electric
  - Offset
  - Cordless

- Drill bits
- Die tool set
- Soldering tool
- Wrenches, including:
  - Pipe
  - Pipe tongs
  - Strap
  - Spud
  - Open-end
  - Adjustable
  - Basin
  - Monkey
  - Torque

- Pliers
- Hammer
- Wood-splitting wedge
- Hollow-shank screwdriver
- Bits
- Vise

Trade Terms Introduced in This Module

Amperage: A measure of electrical current.

Bar stock: Uncut bar material on hand for use on a project.

Burr: Uneven or jagged edge left on metal by certain cutting tools.

Diameter: The distance across the center of a circle.

Electrical ground: A conductive connection that provides a path for electrical current to pass from an electrical component into the earth.

Ferrous: Containing iron.

Flux: A water-soluble substance that facilitates the fusion (joining) of metals and helps prevent surface oxidation (rusting, tarnishing) during welding, brazing, and soldering. Also called soldering paste.

Joist: A piece of lumber used horizontally as a support for a ceiling or a floor.

Keel crayon: A waxy crayon used to mark a cutting line on the surface of a tube. Also called soapstone.

Kerf: The cut or groove made by a saw blade, determined by the way the teeth are set on the blade.

Level: Straight on a horizontal plane.

Miter: A surface forming the beveled end or edge of a piece where a joint is made by cutting two pieces at an angle and then fitting these pieces together.

Nonferrous: Not containing iron and therefore not magnetic.

Offset: A combination of elbows or bends that brings one section of the pipe out of line but into a line parallel with the other section.

Plumb: Straight on a vertical plane.

Reaming: A process that removes burrs from pipe after it has been cut.

Scribe: A sharply pointed and hardened steel tool used for marking a surface to be cut by etching a line or a point into the surface.

Soapstone: Another term for keel crayon.

Solder: An alloy (tin plus antimony, copper, and silver) with a low melting point used to join metals or seal joints.

Soldering: A method of joining metals or sealing joints using solder and heat.

Spirit level: A level in which the adjustment to the horizon is shown by the position of a bubble in liquid contained in a nearly horizontal glass tube or a circular box with a glass cover.

Straightedge: A length of wood or metal that does not bow or twist along its length.

Temper: The strength and resilience of a metal.

Tolerance: Allowable variation in a given measurement or quantity.

Torque: Twisting or turning force applied in a rotating motion. Measurements are given in either inch-pounds or foot-pounds.

Tube stock: Uncut tubing material on hand for use on a project.

Voltage drop: The tendency of electricity traveling through the length of an extension cord to lose voltage.

Additional Resources

This module presents thorough resources for task training. The following resource material is suggested for further study.

Core Curriculum, Latest Edition. NCCER. New York, NY: Pearson.
Council Tool website. *www.counciltool.com.*
The Stanley Tools website. *www.stanleytools.com.*

Figure Credits

© Lloyd Wolf for SkillsUSA, Module Opener
Cooper Hand Tools, Figures 1, 2, 41, 46
The Stanley Works, Figures 3–8, 10, 11, 14, 16, 22, 54, 56–58, 61
RIDGID®, Figures 9, 17, 24, 27–31, 36, 39, 43, 44, 47–51 (die tool set), 52, 60, 62, 66, 77, 78
David White, Inc., Figure 12 (builder 's level)
Sokkia Corporation, Figure 12 (transit level)
Robert Bosch Tool Corporation, Figures 12 (leveling rod, three-way laser levels), 76
Laser Reference, Inc., Figure 13
Seymour Manufacturing Co., Inc., Figure 21
Wheeler Rex Pipe Tools, Figure 33
Reed Manufacturing Company, Figures 34, 37, 38, 42, 45
DeWalt Power Tools, Figures 63, 64, 68–70, 73
Mueller Industries, Figure 51 (threaded pipe)
Milwaukee Electric Tool Corporation, Figures 65 (electric), 74 (twist bit, spade bit, auger bit, and masonry bit), 75
Multiquip Inc., Figure 65 (gas powered)
Bosch Power Tools and Accessories, Figure 71
Thermadyne, Figure 79
BernzOMatic, Figure 80 (flint lighter and flints)
Goss Inc., Figure 80 (torch with auto striker and triangle striker)
Special thanks to Cemex, for their help with *Plumbing Level One*

SECTION 1.0.0

Answer	Section Reference	Objective
1. c	1.1.0	1a
2. b	1.2.2	1b
3. c	1.3.0	1c
4. c	1.4.1	1d
5. c	1.5.2	1e
6. a	1.6.2	1f
7. b	1.7.0; Figure 60	1g

SECTION 2.0.0

Answer	Section Reference	Objective
1. a	2.1.2	2a
2. a	2.2.2	2b
3. b	2.3.1	2c
4. d	2.4.0	2d
5. c	2.5.0	2e

NCCER CURRICULA — USER UPDATE

NCCER makes every effort to keep its textbooks up-to-date and free of technical errors. We appreciate your help in this process. If you find an error, a typographical mistake, or an inaccuracy in NCCER's curricula, please fill out this form (or a photocopy), or complete the online form at **www.nccer.org/olf**. Be sure to include the exact module ID number, page number, a detailed description, and your recommended correction. Your input will be brought to the attention of the Authoring Team. Thank you for your assistance.

Instructors – If you have an idea for improving this textbook, or have found that additional materials were necessary to teach this module effectively, please let us know so that we may present your suggestions to the Authoring Team.

NCCER Product Development and Revision

13614 Progress Blvd., Alachua, FL 32615

Email: curriculum@nccer.org
Online: www.nccer.org/olf

❏ Trainee Guide   ❏ Lesson Plans   ❏ Exam   ❏ PowerPoints   Other _____

Craft / Level: _____   Copyright Date: _____

Module ID Number / Title: _____

Section Number(s): _____

Description: _____

_____

_____

_____

Recommended Correction: _____

_____

_____

_____

Your Name: _____

Address: _____

_____

Email: _____   Phone: _____

This page is intentionally left blank.

Introduction to Plumbing Math

OVERVIEW

Math is used in all phases of construction. Plumbers use math to read plans, calculate pipe length, lay out fixtures, and much more. Developing good math skills will help you advance in the plumbing profession.

Module 02104

Trainees with successful module completions may be eligible for credentialing through NCCER's National Registry. To learn more, go to *www.nccer.org* or contact us at 1.888.622.3720. Our website, www.nccer.org, has information on the latest product releases and training.

Your feedback is welcome. You may email your comments to *curriculum@nccer.org*, send general comments and inquiries to *info@nccer.org*, or fill in the User Update form at the back of this module.

02104 V4.5

Objectives

Successful completion of this module prepares trainees to:

1. Perform basic mathematical calculations.
   a. Demonstrate mathematical operations using whole numbers.
   b. Demonstrate mathematical operations using fractions.
   c. Demonstrate mathematical operations using decimals.
   d. Demonstrate mathematical conversions.
   e. Describe the metric system.
   f. Demonstrate the use of squares and square roots.
2. Explain how pipe is measured.
   a. Identify the parts of a fitting.
   b. Define makeup.
   c. Define fitting allowance.
   d. Use manufacturer's tables to select pipe.
   e. Calculate pipe lengths using various methods.

Performance Tasks

Under the supervision of your instructor, you should be able to do the following:

1. Measure pipe using the following methods:
   - End-to-end
   - End-to-center
   - Center-to-center
   - End-to-face
   - Face-to-face
   - Face-to-throat
2. Determine end-to-end dimensions by figuring fitting allowances and makeup.

Trade Terms

Back
Center
Center line
Center point
Face

Fitting allowance
Remainder
Thread makeup
Throat

Industry Recognized Credentials

If you're training through an NCCER-accredited sponsor you may be eligible for credentials from NCCER's Registry. The ID number for this module is 02104. Note that this module may have been used in other NCCER curricula and may apply to other level completions. Contact NCCER's Registry at 888.622.3720 or go to *www.nccer.org* for more information.

Contents

Figures and Tables

1.0.0 BASIC MATHEMATICAL CALCULATIONS

Objective

Perform basic mathematical calculations.

  a. Demonstrate mathematical operations using whole numbers.
  b. Demonstrate mathematical operations using fractions.
  c. Demonstrate mathematical operations using decimals.
  d. Demonstrate mathematical conversions.
  e. Describe the metric system.
  f. Demonstrate the use of squares and square roots.

Trade Terms

Fitting allowance: The distance from the end of the pipe that goes into a fitting to the center of the fitting.

Remainder: The leftover amount in a division problem. For example, in the problem 34 ÷ 8, 8 goes into 34 four times (8 × 4 = 32) and 2 is left as the remainder.

Plumbers use math to read plans, calculate pipe length, and lay out fixtures. Consequently, the ability to develop accurate math skills is essential to advance in the plumbing profession. Understanding whole numbers, fractions, and decimals—and how to convert them from one form to another—is necessary for efficiency and productivity.

Plumbing fixtures are often available in English and metric sizes. Plumbers may encounter situations when they need to convert metric measurements for weight, length, or volume to the English system. Plumbing calculations often require square numbers or the square roots of numbers. Plumbers must learn to determine these values with and without the use of a calculator.

Plumbers must be able to measure pipe quickly and accurately. Being able to identify the basic parts of a fitting allows a plumber to define the beginning and ending points of the measurement. Plumbers must also be able to determine fitting allowance—the distance from the end of the pipe that goes into the fitting. Plumbers can calculate this distance or refer to manufacturers' tables. Three pipe-measuring techniques—center-to-center, end-to-center, and face-to-face—are used to calculate the length of pipe to cut. Skilled plumbers understand which method to use based on the available information and the type of installation.

Mathematics is one of the most important tools you'll ever use. Like your plumbing skills, your math skills actually get better the more you use them. This module introduces some of the basic math used by plumbers in the field and explains how to use it to calculate pipe length.

You learned basic math in *Core Curriculum*. The following sections review whole numbers, fractions, decimals, and conversion processes.

1.1.0 Review of Whole Numbers

Whole numbers are complete units without any fractions or decimals. The following are examples of whole numbers:

$$15 \quad 32 \quad 60 \quad 144 \quad 2,436$$

1.1.1 Place Value

Each digit in a whole number represents a place value. Each digit has a value that depends on its place, or location, in the whole number. In the whole number 5,679, for example, the place value of the 5 is five thousand, and the place value of the 6 is six hundred.

Numbers larger than zero are positive (+) numbers (such as 1, 2, 3). Numbers less than zero are negative (–) numbers (such as –1, –2, –3). Zero is neither positive nor negative. Except for zero, numbers without a minus sign in front of them are positive.

1.1.2 Addition

To add means to combine the value of two or more numbers. The total when you add two or more numbers is called the sum. The sign for addition is the plus sign (+).

$$56 + 32 = 88$$

1.1.3 Subtraction

Subtraction means finding the difference between two numbers or taking one number away from another. The subtraction sign (–) is also called the minus sign. The result (answer) of a subtraction problem is called the difference.

$$56 - 32 = 24$$

1.1.4 Multiplication

Multiplication is the quick way to add the same number together many times. The symbol for multiplication is the × sign. For example, you must deliver five boxes to each of eight job sites. How many boxes do you deliver in all?

$$5 \times 8 = 40 \text{ boxes}$$
$$5 + 5 + 5 + 5 + 5 + 5 + 5 + 5 = 40 \text{ boxes}$$

1.1.5 Division

Division is the opposite of multiplication. The symbol for division is the ÷ sign. Instead of adding a number several times as in multiplication, when you are dividing you subtract it several times. You can solve a problem faster by using division instead of subtracting the same number over and over. For example, you have 40 boxes to deliver to 8 different job sites, and you need to find out how many boxes go to each site.

$$40 \div 8 = 5$$
$$40 - 8 - 8 - 8 - 8 - 8 = 0$$
(You had to subtract 8 five times,
so the answer is 5.)

1.2.0 Review of Fractions

A fraction divides whole units into parts. Fractions are written as two numbers separated by a slash or by a horizontal line, like this:

$$\frac{1}{2} \text{ or } \frac{1}{2}$$

Think of a fraction as a division problem. The lower number (denominator) of the fraction tells you the number of parts by which the upper number (numerator) is being divided. The slash or horizontal line means the same thing as the ÷ sign. The fraction ½ means 1 divided by 2, or one divided into two equal parts. Read this fraction as "one-half."

1.2.1 Equivalent Fractions

Equivalent fractions have the same value, or are equal. For example, ½, ²⁄₄, ⁴⁄₈, and ⁸⁄₁₆ are equivalent fractions. If you cut off a piece of wood ⁸⁄₁₆" long and a co-worker cuts off a piece ½" long, the two pieces would be the same length.

You need to know how to find equivalent fractions so that you can compare, add, and subtract fractional measurements. For example, to find out how many eighths of an inch are equal to ½", multiply the numerator and denominator by the same number. Ask yourself what number you

would multiply by 2 to get 8. The answer is 4, so you multiply the numerator and denominator by 4.

$$\frac{1}{2} \times \frac{4}{4} = \frac{4}{8}$$

The answer is that ⁴⁄₈" is equivalent to ½".

1.2.2 Lowest Terms

When you are working with fractions, it is best to reduce them to lowest terms. To reduce a fraction, ask yourself, "What is the largest number that can be divided evenly into both the numerator and the denominator?" If you're unsure about the largest number that divides evenly into both the numerator and the denominator, try using the number 2 or 3 to start, and then continue until the fraction is in its lowest form. If there is not a number (other than 1) that divides evenly into both numbers, the fraction is already in its lowest terms. Look at the following examples:

$$²⁄₄ = ½$$
$$⁴⁄₁₆ = ¼$$
$$⁸⁄₃₂ = ¼$$
$$⅜ = ⅜ \text{ (already in lowest terms)}$$
$$⁷⁄₁₆ = ⁷⁄₁₆ \text{ (already in lowest terms)}$$

1.2.3 Common Denominator

A common denominator means that the fraction's bottom number, or denominator, in a group of fractions is the same. For example, ¼, ²⁄₄, and ¾ have common denominators. Before you can compare fractions, they must have the same denominators. Which of these fractions is larger?

$$\frac{3}{4} \text{ or } \frac{2}{3}$$

To compare, you need to find a common denominator for the fractions. The common denominator is a number that both denominators can go into evenly. Here's one way to find a common denominator.

Step 1 Multiply the two denominators (4 × 3 = 12). The result is a common denominator for ¾ and ⅔. You found a common denominator so that you can easily compare the fractions.

Step 2 Convert the two fractions so they will have the same denominator by multiplying the numerator and denominator of each fraction by the denominator of the other fraction. Convert ¾ and ⅔ to fractions having the common denominator of 12.

$$\frac{3}{4} \times \frac{3}{3} = \frac{9}{12}$$

$$\frac{2}{3} \times \frac{4}{4} = \frac{8}{12}$$

Now it's easy to compare the two fractions to see which is larger: $9/12$, or $3/4$, is larger than $8/12$, or $2/3$.

In construction, many fractions have denominators of 2, 4, 8, or 16 due to common units of measurement. If one of the denominators is a multiple of the other, the larger number is the common denominator. For example, the common denominator for $3/4$ and $5/8$ is 8. Since the lowest common denominator is 8, you would then multiply $3/4$ by a number that will convert the denominator to an 8. That number is 2.

$$\frac{3}{4} \times \frac{2}{2} = \frac{6}{8}$$

Now you can see that $3/4$, which is equivalent to $6/8$, is more than $5/8$.

1.2.4 Improper Fractions

An improper fraction has a numerator larger than the denominator. For example, $11/8$ is an improper fraction.

To reduce the improper fraction $11/8$ to its lowest terms, convert it to a mixed number. A mixed number is a combination of a whole number and a fraction.

$$11/8 = 8/8 + 3/8 = 1\,3/8$$

To change a whole number into an improper fraction, simply place the number over 1. Remember, fractions express division. When you divide any number by 1, the result is the same number. Here is an example:

$$4 = \frac{4}{1}$$

To change a mixed number (for example, $2\,1/8$) to an improper fraction, follow these steps:

Step 1 Multiply the whole number by the denominator.

$$2 \times 8 = 16$$

Step 2 Add the result to the numerator. Put this total over the denominator.

$$16 + 1 = 17$$
$$17/8$$

1.2.5 Addition

How many total inches will you have if you add $3/4$" to $5/8$"? To answer this question, you will have to add the two fractions using the following steps.

Step 1 Find the common denominator of the fractions you wish to add. Since 8 is a multiple of 4, it is the common denominator.

Step 2 Convert the fractions to equivalent fractions with a common denominator.

$$\frac{3}{4} \times \frac{2}{2} = \frac{6}{8}$$

Step 3 Add the numerators of the fractions. Place this sum over the common denominator.

$$\frac{6}{8} + \frac{5}{8} = \frac{11}{8}$$

Step 4 Reduce the fraction to its lowest terms. For this example, it is $1\,3/8$.

1.2.6 Subtraction

Subtracting fractions is very much like adding fractions. You must find a common denominator before you subtract. For example, subtract the following fractions:

$$\frac{7}{8} - \frac{1}{4}$$

Step 1 Find the common denominator. Convert the fractions to equivalent fractions with the same denominator. Multiply the numerator and denominator by 2 to get a new fraction with the common denominator of 8.

$$\frac{1}{4} \times \frac{2}{2} = \frac{2}{8}$$

Step 2 Rewrite the equation and subtract the numerators. The difference is $5/8$.

$$\frac{7}{8} - \frac{2}{8} = \frac{5}{8}$$

1.2.7 Multiplication and Division

Multiplying and dividing fractions is different from adding and subtracting them. You do not have to find a common denominator when you multiply or divide fractions. Using $3/4 \times 5/6$ as an example, follow these steps:

Step 1 Multiply the numerators to get a new numerator. Multiply the denominators to get a new denominator.

$$\frac{3}{4} \times \frac{5}{6} = \frac{15}{24}$$

Step 2 Reduce the fraction to its lowest terms.

$$^{15}/_{24} = {}^5/_8$$

Dividing fractions is very much like multiplying fractions, with one difference. You must invert, or flip, the fraction you are dividing by. Using $^3/_8 \div {}^3/_4$ as an example, follow these steps:

Step 1 Invert the fraction you are dividing by.

$$\frac{3}{4} \text{ becomes } \frac{4}{3}$$

Step 2 Change the division sign to a multiplication sign.

$$\frac{3}{8} \div \frac{3}{4} = \frac{3}{8} \times \frac{4}{3}$$

Step 3 Multiply the fractions.

$$\frac{3}{8} \times \frac{4}{3} = \frac{12}{24}$$

Step 4 Reduce the fraction to its lowest terms.

$$^{12}/_{24} = {}^1/_2$$

If you are working with a whole number or a mixed number, you must convert it to an improper fraction before you invert it. Using $5^3/_4 \div 4$ as an example, follow these steps:

Step 1 Convert $5^3/_4$ to an improper fraction. Multiply the whole number by the denominator. Add the result to the numerator. Put this total over the denominator.

$$5 \times 4 = 20$$
$$20 + 3 = 23$$
$$5^3/_4 = {}^{23}/_4$$

Step 2 Convert 4 to an improper fraction.

$$4 = 4/1$$

Step 3 Now you have the following equation. Invert $^4/_1$ and multiply the fractions.

$$\frac{23}{4} \div \frac{4}{1} = \frac{23}{4} \times \frac{1}{4} = \frac{23}{16}$$

Step 4 Reduce the fraction to its lowest terms.

$$^{23}/_{16} = 1^7/_{16}$$

1.2.8 Whole Numbers and Fractions: Practice Exercises

Solve the following problems. Remember to show all your work.

1.
$$\begin{array}{r} 178 \\ 568 \\ + 10 \end{array}$$

2.
$$\begin{array}{r} 923 \\ -598 \end{array}$$

3.
$$\begin{array}{r} 9 \\ \times 6 \end{array}$$

4.
$$\begin{array}{r} 7 \\ \times 45 \end{array}$$

5. $20 \div 5 =$

6. Find the equivalent of the following measurement: $^5/_8" =$

7. Reduce $^{108}/_{288}$ to its lowest terms.

8. The lowest form of the fraction $^5/_{16}$ is

9. Of $^{15}/_{16}$ and $^5/_8$, which fraction is larger?

10. Reduce $^{57}/_{16}$ to its lowest terms.

11. $^1/_2 + {}^4/_{16} =$

12. $^1/_2 - {}^4/_{16} =$

13. $^5/_{32} \times {}^1/_4 =$

14. $^{17}/_{32} \div {}^3/_4 =$

15. $^6/_8 \div {}^1/_2 =$

1.3.0 Review of Decimals

Decimals represent values less than one whole unit. They are fractions expressed in a different form. You are already familiar with decimals in the form of money.

$$25¢ = 0.25 \text{ or } {}^{25}/_{100}$$
$$10¢ = 0.10 \text{ or } {}^{10}/_{100}$$
$$50¢ = 0.50 \text{ or } {}^{50}/_{100}$$

On the job, you may need to use decimals to read instruments or calculate flow rates. You will also encounter them when reading civil drawings and calculating water pressure levels.

The following chart compares whole number place values with decimal place values:

To read a decimal, say the number as it is written and then the name of its place value. For example, read 0.05 as "five hundredths." Mixed

Whole Numbers	Decimals
1 ones	
10 tens	0.1 tenths
100 hundreds	0.01 hundredths
1,000 thousands	0.001 thousandths

numbers also appear in decimals. You would read 1.05 as "one and five hundredths," for example.

1.3.1 Addition and Subtraction

The one major rule to remember when adding and subtracting decimals is to keep your decimal points lined up.

Suppose you want to add 4.76 and 0.834. Line up the problem like this:

$$\begin{array}{r} 4.760 \\ +0.834 \\ \hline 5.594 \end{array}$$

You can add a 0 as a place holder to help keep the numbers lined up.

The same is true for subtraction. Suppose you want to subtract 2.724 from 5.6. Line up the decimal points and add two zeros to the end of the first number to make it easier to subtract.

$$\begin{array}{r} 5.600 \\ -2.724 \\ \hline 2.876 \end{array}$$

1.3.2 Multiplication

To multiply decimals, set up the problem as you would with whole numbers.

$$\begin{array}{r} 3.6 \\ \times 9 \end{array}$$

Step 1 Multiply.

$$\begin{array}{r} 3.6 \\ \times 9 \\ \hline 324 \end{array}$$

Step 2 Once you have the answer, count the number of digits to the right of the decimal point in both numbers being multiplied. In this example, there is only one decimal point and only one number to the right of it.

Step 3 In the answer, count over the same number of digits (from right to left) and place the decimal point there.
- Count one total digit to the right of the decimal point in the two numbers.

$$\begin{array}{r} 3.6 \\ \times 9 \\ \hline 32.4 \end{array}$$

- Count one digit from right to left in the answer and place the decimal point there.

1.3.3 Division

Three types of division problems involve decimals:

- Those that have a decimal point in the number being divided (the dividend)

$$22\overline{)44.5}$$

- Those that have a decimal point in the number you are dividing by (the divisor)

$$0.22\overline{)4450}$$

- Those that have decimal points in both numbers (the dividend and the divisor)

$$0.22\overline{)44.5}$$

Solve the first problem: How many 22" lengths of pipe can you cut from a piece of pipe measuring 44.5"?

Step 1 Place a decimal point directly above the decimal point in the dividend.

$$22\overline{)44.5}$$

Step 2 Divide as usual.

$$\begin{array}{r} 2.0 \\ 22\overline{)44.5} \\ -44 \\ \hline 00.5 \\ -0 \\ \hline 00.5r \end{array}$$

How many 22" pieces of pipe will you have? The answer: two, with a little left over. This left-over portion is called the **remainder** and is expressed as *r*. In this case, $r = 0.5$.

Now solve the second problem: If couplings cost $0.22 each, how many can you buy with $4,450?

Step 1 Move the decimal point in the divisor to the right until you have a whole number.

$$0.22\overline{)4450}$$

- When you move the decimal point in the divisor two places to the right, 0.22 becomes 22.

Step 2 Move the decimal point in the dividend the same number of places to the right. You may have to add zeros first. Then divide as usual.

```
            20227.2
     22)4450.00,0
        −44
          0500
        −0044
          00060
         −00044
           000160
          −000154
             0000060
            −0000044
               0000016r
```

- After you have added the zeros and moved the decimal in the dividend two places to the right, the number becomes 445,000.

Now solve the third problem: You have $44.50 to buy 200 nuts costing $0.22 each. Can you stay within budget?

Step 1 Move the decimal point in the divisor to the right until you have a whole number.

$$0.22)\overline{4450}$$

- After you have moved the decimal point in the divisor to the right, 0.22 becomes 22.

Step 2 Move the decimal point in the dividend the same number of places to the right (so 44.5 becomes 4,450). Then divide as usual.

```
          202
     22)4450,
        −44
          050
         −44
           06r
```

- To find the remainder in terms of money, you have to undo the changes you made to the decimals in the original problem. In this case, move the decimal point back two places to the left: 0.06.

You will be able to buy 202 nuts with the $44.50, with $0.06 left over. Therefore, you will stay within budget.

1.3.4 Rounding Decimals

Often, calculations with decimals produce very precise answers, such as 25.29. But in most practical measurements, you probably need an answer only to the nearest tenth (0.1). For this exercise, you will round 25.29 to the nearest tenth:

Step 1 Underline the place to which you are rounding.

25.2̲9

Step 2 Look at the digit one place to its right.

25.2̲**9**

Step 3 If the digit to the right is 5 or more, you will round up by adding 1 to the underlined digit. If the digit is less than 5, leave the underlined digit the same. In this example, the digit to the right is 9, so you round up by adding 1 to the underlined digit.

25.3̲9

Step 4 Drop all other digits to the right.

25.3

1.3.5 Decimals: Practice Exercises

Solve the following problems. Remember to show all your work and round to the nearest tenth if necessary:

1.

```
    7.8
   13.4
  +0.8
  ─────
```

2.

```
  129.6
  −54.9
  ─────
```

3.

$$\begin{array}{r} 8.9 \\ \times 3.5 \\ \hline \end{array}$$

4. $785 \div 6.7 =$

5. $23000.50 \div 7.5 =$

1.4.0 Review of Conversion Processes

In some situations, you need to convert the numbers you want to work with so that all your numbers are in the same form. For example, you may have some numbers that appear as decimals, some that appear as percentages, and some that appear as fractions. Decimals, percentages, and fractions are all just different ways of expressing similar things. The decimal 0.25, the percentage 25%, and the fraction ¼ all mean the same thing. To work successfully with the different forms of numbers like these, you need to know how to convert them from one form into another.

1.4.1 Decimals to Percentages and Percentages to Decimals

What are percentages? Think of a whole number divided into 100 parts. You can express any part of the whole as a percentage. Percentages are an easy way to express parts of a whole. Decimals and fractions also express parts of a whole. Let's look at the relationship among percentages, decimals, and fractions.

To convert a decimal to a percentage, simply multiply the decimal by 100 and add a percent sign. Try converting 7.35 to a percentage.

Step 1 Multiply the decimal by 100.

$$7.35 \times 100 = 735$$

Step 2 Add a percent sign.

735%

Suppose you are preparing a gallon of cleaning solution. The mixture should contain 10 to 15 percent of the cleaning agent, the rest should be water. You have 0.12 gallon of cleaning agent. Will you have enough to prepare a gallon of the solution? To answer the question, you must convert a decimal (0.12) to a percentage.

Step 1 Multiply the decimal by 100. (Move the decimal point two places to the right.)

$$0.12 \times 100 = 12$$

Step 2 Add a percent sign.

12%

- You have enough cleaning agent to make the solution. Recall that the mixture should be 10 to 15 percent cleaning agent. You have 12 percent.

You may also need to convert percentages to decimals, as illustrated in the following example. Let's say that another mixture should contain 22 percent of a certain chemical by weight. You're making 1 pound of the mixture. You weigh the ingredients on a digital scale. How much of the chemical should you add? To answer this, you must convert a percentage (22%) to a decimal:

Step 1 Drop the percent sign.

22

Step 2 Divide the number by 100. (Move the decimal point two places to the left.)

$$22 \div 100 = .22$$

- So you would add 0.22 pounds of the chemical to make a 22 percent mixture.

1.4.2 Fractions to Decimals

To convert a fraction to a decimal, do what the fraction tells you to do; that is, divide the numerator by the denominator. Convert ⁷⁄₉ to its equivalent decimal.

Step 1 Set up the division problem. Divide the numerator by the denominator.

$$9\overline{)7}$$

Step 2 In this example, you need to put the decimal point and a zero after the number 7 because you need a number large enough to divide by 9. Put the decimal point directly above its location in the number within the division symbol.

$$9\overline{)7.0}^{.?}$$

Step 3 Once the decimal point is in its proper place above the line, you can divide as you normally would. The decimal point holds everything in place.

$$\begin{array}{r} .777 \\ 9\overline{)7.00} \\ \underline{-6.3} \\ 7.0 \\ \underline{-6.3} \\ 0.70 \\ \underline{-0.63} \\ 0.7r \end{array}$$

- Dividing to three places is satisfactory for plumbing calculations.

Suppose you need 1¾ of a pound of material and need to know its decimal equivalent.

Step 1 Convert the proper fraction into an improper fraction. Multiply the whole number by the denominator and add the numerator. Place this number over the denominator.

$$1\tfrac{3}{4} = \tfrac{7}{4}$$

Step 2 Set up the division problem. Divide the numerator by the denominator.

$$4\overline{)7.0}$$

Step 3 Put the decimal point directly above its location within the division symbol.

$$\overset{.?}{4\overline{)7.0}}$$

Step 4 Divide.

$$\begin{array}{r} 1.75 \\ 4\overline{)7.00} \\ \underline{-4} \\ 30 \\ \underline{-28} \\ 20 \\ \underline{-20} \\ 0 \end{array}$$

The fraction ⁷⁄₄ converted to a decimal is 1.75, so 1¾ pounds is the same as 1.75 pounds.

1.4.3 Decimals to Fractions

Not only will you need to convert fractions to decimals, you must know how to go the other way and convert decimals into fractions. For example, what fraction of a pound is 0.25?

Step 1 Say the decimal in words.

- 0.25 is expressed as "twenty-five hundredths."

Step 2 Write the decimal as a fraction.

- 0.25 written as a fraction is ²⁵⁄₁₀₀.

Step 3 Reduce it to its lowest terms.

$$\frac{25}{100} = \frac{25 \div 25}{100 \div 25} = \frac{1}{4}$$

You can see that 0.25 converted to a fraction is ¼.

Let's say you have 17.35 pounds of nails. What fraction of a pound is that?

Step 1 Say the decimal in words.
17.35 is expressed as "17 and 35 hundredths."

Step 2 Write the decimal as a fraction.

$$17.35 = 17\tfrac{35}{100}$$

Step 3 Reduce the fraction portion to its lowest terms.

$$\frac{35}{100} = \frac{35 \div 5}{100 \div 5} = \frac{7}{20}$$

So 17.35 pounds of nails is the same as 17⁷⁄₂₀ pounds of nails.

When calculating the cut length of pipe, plumbers need to be able to convert a decimal to the nearest ¹⁄₁₆ of an inch. Here are the steps for converting to sixteenths of an inch, using 0.91' as an example.

Step 1 Multiply the decimal part of a foot by 12, which is the number of inches in a foot. Thus, 0.91 equals 10" with a remainder of 0.92.

$$0.91 \times 12 = 10.92$$

Step 2 Now convert the remainder to sixteenths of an inch by multiplying it by 16.

$$0.92 \times 16 = 14.72$$

Because this answer is not a whole number, round it. So, 14.72 rounds up to 15. Hence, 0.91' = 10¹⁵⁄₁₆".

Convert 13.67' into the nearest sixteenth of an inch. You already know you have 13'. You need to convert 0.67' into inches.

Step 1 Multiply the decimal part of a foot by 12.

$$0.67 \times 12 = 8.04$$

Thus, 0.67 equals 8" with a remainder of 0.04.

Step 2 Now multiply the remainder by 16.

$$0.04 \times 16 = 0.64$$

Table 1 Inches Converted to Decimals of a Foot

Inches	Decimals of a Foot	Inches	Decimals of a Foot	Inches	Decimals of a Foot	Inches	Decimals of a Foot
$1/16$	0.005	$3\ 1/16$	0.255	$6\ 1/16$	0.505	$9\ 1/16$	0.755
$1/8$	0.010	$3\ 1/8$	0.260	$6\ 1/8$	0.510	$9\ 1/8$	0.760
$3/16$	0.016	$3\ 3/16$	0.266	$6\ 3/16$	0.516	$9\ 3/16$	0.766
$1/4$	0.021	$3\ 1/4$	0.271	$6\ 1/4$	0.521	$9\ 1/4$	0.771
$5/16$	0.026	$3\ 5/16$	0.276	$6\ 5/16$	0.526	$9\ 5/16$	0.776
$3/8$	0.031	$3\ 3/8$	0.281	$6\ 3/8$	0.531	$9\ 3/8$	0.781
$7/16$	0.036	$3\ 7/16$	0.286	$6\ 7/16$	0.536	$9\ 7/16$	0.786
$1/2$	0.042	$3\ 1/2$	0.292	$6\ 1/2$	0.542	$9\ 1/2$	0.792
$9/16$	0.047	$3\ 9/16$	0.297	$6\ 9/16$	0.547	$9\ 9/16$	0.797
$5/8$	0.052	$3\ 5/8$	0.302	$6\ 5/8$	0.552	$9\ 5/8$	0.802
$11/16$	0.057	$3\ 11/16$	0.307	$6\ 11/16$	0.557	$9\ 11/16$	0.807
$3/4$	0.063	$3\ 3/4$	0.313	$6\ 3/4$	0.563	$9\ 3/4$	0.813
$13/16$	0.068	$3\ 13/16$	0.318	$6\ 13/16$	0.568	$9\ 13/16$	0.818
$7/8$	0.073	$3\ 7/8$	0.323	$6\ 7/8$	0.573	$9\ 7/8$	0.823
$15/16$	0.078	$3\ 15/16$	0.328	$6\ 15/16$	0.578	$9\ 15/16$	0.828
1	0.083	4	0.333	7	0.583	10	0.833
$1\ 1/16$	0.089	$4\ 1/16$	0.339	$7\ 1/16$	0.589	$10\ 1/16$	0.839
$1\ 1/8$	0.094	$4\ 1/8$	0.344	$7\ 1/8$	0.594	$10\ 1/8$	0.844
$1\ 3/16$	0.099	$4\ 3/16$	0.349	$7\ 3/16$	0.599	$10\ 3/16$	0.849
$1\ 1/4$	0.104	$4\ 1/4$	0.354	$7\ 1/4$	0.604	$10\ 1/4$	0.854
$1\ 5/16$	0.109	$4\ 5/16$	0.359	$7\ 5/16$	0.609	$10\ 5/16$	0.859
$1\ 3/8$	0.115	$4\ 3/8$	0.365	$7\ 3/8$	0.615	$10\ 3/8$	0.865
$1\ 7/16$	0.120	$4\ 7/16$	0.370	$7\ 7/16$	0.620	$10\ 7/16$	0.870
$1\ 1/2$	0.125	$4\ 1/2$	0.374	$7\ 1/2$	0.625	$10\ 1/2$	0.875
$1\ 9/16$	0.130	$4\ 9/16$	0.380	$7\ 9/16$	0.630	$10\ 9/16$	0.880
$1\ 5/8$	0.135	$4\ 5/8$	0.385	$7\ 5/8$	0.635	$10\ 5/8$	0.885
$1\ 11/16$	0.141	$4\ 11/16$	0.391	$7\ 11/16$	0.641	$10\ 11/16$	0.891
$1\ 3/4$	0.146	$4\ 3/4$	0.396	$7\ 3/4$	0.646	$10\ 3/4$	0.896
$1\ 13/16$	0.151	$4\ 13/16$	0.401	$7\ 13/16$	0.651	$10\ 13/16$	0.901
$1\ 7/8$	0.156	$4\ 7/8$	0.406	$7\ 7/8$	0.656	$10\ 7/8$	0.906
$1\ 15/16$	0.161	$4\ 15/16$	0.411	$7\ 15/16$	0.661	$10\ 15/16$	0.911
2	0.167	5	0.417	8	0.667	11	0.917
$2\ 1/16$	0.172	$5\ 1/16$	0.422	$8\ 1/16$	0.672	$11\ 1/16$	0.922
$2\ 1/8$	0.177	$5\ 1/8$	0.427	$8\ 1/8$	0.677	$11\ 1/8$	0.927
$2\ 3/16$	0.182	$5\ 3/16$	0.432	$8\ 3/16$	0.682	$11\ 3/16$	0.932

Table 1 Inches Converted to Decimals of a Foot (continued)

Inches	Decimals of a Foot	Inches	Decimals of a Foot	Inches	Decimals of a Foot	Inches	Decimals of a Foot
2¼	0.188	5¼	0.438	8¼	0.688	11¼	0.938
2⁵⁄₁₆	0.193	5⁵⁄₁₆	0.443	8⁵⁄₁₆	0.693	11⁵⁄₁₆	0.943
2⅜	0.198	5⅜	0.448	8⅜	0.698	11⅜	0.948
2⁷⁄₁₆	0.203	5⁷⁄₁₆	0.453	8⁷⁄₁₆	0.703	11⁷⁄₁₆	0.953
2½	0.208	5½	0.458	8½	0.708	11½	0.958
2⁹⁄₁₆	0.214	5⁹⁄₁₆	0.464	8⁹⁄₁₆	0.714	11⁹⁄₁₆	0.964
2⅝	0.219	5⅝	0.469	8⅝	0.719	11⅝	0.969
2¹¹⁄₁₆	0.224	5¹¹⁄₁₆	0.474	8¹¹⁄₁₆	0.724	11¹¹⁄₁₆	0.974
2¾	0.229	5¾	0.479	8¾	0.729	11¾	0.979
2¹³⁄₁₆	0.234	5¹³⁄₁₆	0.484	8¹³⁄₁₆	0.734	11¹³⁄₁₆	0.984
2⅞	0.240	5⅞	0.490	8⅞	0.740	11⅞	0.990
2¹⁵⁄₁₆	0.245	5¹⁵⁄₁₆	0.495	8¹⁵⁄₁₆	0.745	11¹⁵⁄₁₆	0.995
3	0.250	6	0.500	9	0.750	12	1.000

Because this answer is not a whole number, round it. 0.64 rounds up to 1. The answer is 13' 8¹⁄₁₆".

An easier way to convert a decimal into sixteenths (or other fractions) of an inch is to use a conversion table (see *Table 1*). For instance, if you have 0.86', find the number closest to that number in the Decimals of a Foot column (0.859) and look in the same row of the corresponding Inches column. So 0.86' = 10⁵⁄₁₆".

1.4.4 Inches to Decimal Equivalents in Feet

Sometimes you need to convert inches to their decimal equivalents in feet. To do so, first express the inches as a fraction that has 12 as the denominator. You use 12 because there are 12" in a foot. Then reduce the fraction and convert it to a decimal.

For example, 3" equals what decimal equivalent in feet?

Step 1 First, place the number of inches over 12.

$$^3/_{12}$$

Step 2 Reduce the fraction to its lowest terms.

$$^3/_{12} = ^1/_4$$

Step 3 Convert the fraction ¼ to a decimal by dividing the 4 into 1.00:

```
      0.25
  4)1.00
   -0.8
    0.20
   -0.20
      0
```

Thus, 3" converts to 0.25'.

Now try a problem that includes decimals: 14.4" equals what decimal equivalent in feet?

Step 1 Convert the decimal to a fraction. If this fraction is in proper form, change it to improper form.

$$14.4" = 14\ ^4/_{10}" = \ ^{144}/_{10}$$

Step 2 Multiply by ¹⁄₁₂.

$$\frac{144}{10} \times \frac{1}{12} = \frac{144}{120}$$

Step 3 Convert the fraction ¹⁴⁴⁄₁₂₀ to a decimal by dividing the denominator (120) into the numerator (144) and rounding the answer if necessary. In this case, the quotient stops at the tenths.

```
        1.2
  120)144.0
     -120
      240
     -240
        0
```

Thus, 14.4″ converts to 1.2′. Do you see that this is the same as saying "14.4 divided by 12?"

1.4.5 Conversions: Practice Exercises

Round answers to the nearest thousandth where necessary.

1. Convert 0.85 to its equivalent percentage.
2. Convert 4.35 to its equivalent percentage.
3. Convert 20 percent to its equivalent decimal.
4. Convert 150 percent to its equivalent decimal.
5. Convert ⅜ to its equivalent decimal.
6. Convert 1½ to its equivalent decimal
7. Convert 6/4 to its equivalent decimal.
8. Convert 0.75 to its equivalent fraction.
9. Convert 13.4 to its equivalent fraction.
10. Convert 133.365 to its equivalent fraction.
11. Convert 0.3′ into the nearest sixteenth of an inch.
12. Convert 11.86′ into the nearest sixteenth of an inch.
13. Convert 75.8′ into the nearest sixteenth of an inch.
14. Convert 16″ to its decimal equivalent in feet.
15. Convert 115.6″ to its decimal equivalent in feet.

1.5.0 Metric System

The metric system is a system of measurement that uses a base-ten method of determining weight, length, volume, and temperature. In other words, all measurements are counted in tens.

There are only seven basic units of measurement in the modern metric system, including the meter, liter, and gram. Multiples or fractions of the basic units are expressed as powers of 10. A standard set of prefixes is used to denote these larger or smaller numbers (*Table 2*). That way, you will always know that a kilometer is 1,000 meters just by looking at the prefix.

Table 2 Metric Prefixes

deka = 10	deci = 0.1
hecto = 100	centi = 0.01
kilo = 1,000	milli = 0.001
mega = 1,000,000	micro = 0.000001

1.5.1 Converting within the Metric System

To convert from one metric unit to another, simply move the decimal point the required number of spaces. For instance, to find how many millimeters are in 5 meters, move the decimal point three places to the right (multiply by 1,000). If you want to know how many meters equals 1,500 centimeters, you would move the decimal point two places to the left (divide by 100).

Figure 1 shows an easy way to do these calculations. For every jump on the number line, move the decimal one place in the same direction. Add zeros if necessary. For example, if you have 100 centiliters and want to convert to hectoliters, you move the decimal four places to the left, adding a zero (or divide by 10,000). So, 100 centiliters equals 0.01 hectoliter.

1.5.2 Converting Measurements

Some countries use the metric system, and others use the standard (or imperial) system. The standard system uses measurements such as feet and pounds. Plumbing fixtures are often available in both standard and metric sizes. You may encounter situations where pipe measurements need to be converted. For example, imported materials may come in metric measurements, and you may need to convert them to the standard system. Also, federal agencies use the metric system. To convert between the standard and metric systems, you can use a conversion table such as *Table 3*.

> **CAUTION**
>
> Be sure to express measurements using the correct system. Errors caused by using the wrong system of measurement can cost time and money. Never use a metric tool on standard-system pipe and fittings, or vice versa. You could damage both the tool and the fitting.

To convert a measurement from one system to the other, multiply the measurement by the number in the far-right column of *Table 3*. For example, you may want to know how many kilograms equal 250 pounds. Looking at the table, you find pounds in the *To convert* column and kilograms in the *Into* column, and you see that the number must be multiplied by 0.45 (250 × 0.45). The answer is 112.5 kilograms.

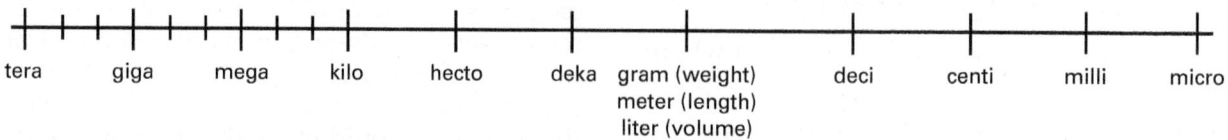

tera	giga	mega	kilo	hecto	deka	gram (weight)	deci	centi	milli	micro

meter (length)
liter (volume)

$10^0 = 1$	BASE UNIT:	GRAM (WEIGHT), METER (LENGTH), LITER (VOLUME)		$10^0 = 1$	
MULTIPLIER	**PREFIX**	**MEANING**	**PREFIX**	**MEANING**	**MULTIPLIER**
$10^1 = 10$	deka-	ten	deci-	tenth	$10^{-1} = 0.1$
$10^2 = 100$	hecto-	hundred	centi-	hundredth	$10^{-2} = 0.01$
$10^3 = 1,000$	kilo-	thousand	milli-	thousandth	$10^{-3} = 0.001$
$10^6 = 1,000,000$	mega-	million	micro-	millionth	$10^{-6} = 0.000001$
$10^9 = 1,000,000,000$	giga-	billion	nano-	billionth	$10^{-9} = 0.000000001$
$10^{12} = 1,000,000,000,000$	tera-	trillion	pico-	trillionth	$10^{-12} = 0.000000000001$

Figure 1 The decimal scale.

Table 3 Conversion Table

A. English to Metric			
	To convert...	**Into...**	**Perform this...**
LENGTH	Inches	Millimeters	multiply the English unit by 25.40
	Feet	Centimeters	multiply the English unit by 30.00
	Yards	Meters	multiply the English unit by 0.90
	Miles	Kilometers	multiply the English unit by 1.60
AREA	Square inches	Square centimeters	multiply the English unit by 6.50
	Square feet	Square meters	multiply the English unit by 0.09
	Square yards	Square meters	multiply the English unit by 0.80
	Square miles	Square kilometers	multiply the English unit by 2.60
	Acres	Hectares	multiply the English unit by 0.40
MASS and WEIGHT	Fluid ounces	Grams	multiply the English unit by 28.00
	Pounds	Kilograms	multiply the English unit by 0.45
	Short tons	Megagrams	multiply the English unit by 0.90
LIQUID MEASURE	Ounces	Milliliters	multiply the English unit by 30.00
	Pints	Liters	multiply the English unit by 0.47
	Quarts	Liters	multiply the English unit by 0.95
	Gallons	Liters	multiply the English unit by 3.80
TEMPERATURE	Fahrenheit	Celsius	subtract 32 from the English unit, then multiply the result by 5, then divide that result by 9

Table 3 Conversion Table (continued)

B. Metric to English			
	To convert...	**Into...**	**Perform this...**
LENGTH	Millimeters	Inches	multiply the Metric unit by 0.040
	Centimeters	Feet	multiply the Metric unit by 0.033
	Meters	Yards	multiply the Metric unit by 1.100
	Kilometers	Miles	multiply the Metric unit by 0.620
AREA	Square centimeters	Square inches	multiply the Metric unit by 0.160
	Square meters	Square yards	multiply the Metric unit by 1.200
	Square kilometers	Square miles	multiply the Metric unit by 0.400
	Hectares	Acres	multiply the Metric unit by 2.500
MASS and WEIGHT	Grams	Fluid ounces	multiply the Metric unit by 0.035
	Kilograms	Pounds	multiply the Metric unit by 2.200
	Megagrams	Short tons	multiply the Metric unit by 1.100
LIQUID MEASURE	Milliliters	Ounces	multiply the Metric unit by 0.034
	Liters	Pints	multiply the Metric unit by 2.100
	Liters	Quarts	multiply the Metric unit by 1.060
	Liters	Gallons	multiply the Metric unit by 0.260
TEMPERATURE	Celsius	Fahrenheit	multiply the Metric unit by 9, then divide the result by 5, then add 32 to that result

1.5.3 Metric Conversion: Practice Exercises

Convert the following measurements:

1. 75 meters = _____ centimeters
2. 0.25 milliliters = _____ dekaliters
3. 675,000 micrometers = _____ hectometers
4. 8.9 kilograms = _____ dekagrams
5. 1,011 millimeters = _____ centimeters
6. 68 inches = _____ millimeters
7. 4.5 fluid ounces = _____ grams
8. 35 centimeters = _____ feet
9. 875 liters = _____ gallons
10. 1,568 millimeters = _____ inches
11. 62°C = _____ °F

1.6.0 Squares and Square Roots

Plumbing calculations may require you to square numbers as well as take square roots of numbers.

To square a number, multiply the number by itself. To show that a number is being squared, write a superscript 2 to the right of the number.

$$3 \text{ squared} = 3^2 = 3 \times 3 = 9$$

To take the square root of a number means to find the number that, when multiplied by itself, results in the number under the square root sign. The radical sign ($\sqrt{}$) indicates a root of a number.

$$\text{The square root of } 9 = \sqrt{9} = 3$$

The easiest way to square numbers and take square roots of numbers is to use a calculator (*Figure 2*). To square a number using a calculator,

Figure 2 Construction calculators.

simply enter the number, press the × key, and then enter the number again and press =. With more complicated numbers, you can save time by using the special square function. Enter the number and then press the x2 key on the calculator. The answer will appear in the display. You do not need to press the = sign.

To find the square root of a number, enter the number. Then press the √ key. The answer will appear in the display. You do not need to press the = sign.

Did You Know?

On some calculators, the x2 or √ key may be written above another key. In this case you will have to press a shift or 2nd key and then press the key underneath the x2 or √. For more guidance, refer to the manufacturer's instructions.

1.6.1 Squares and Square Roots: Practice Exercises

Use a calculator when necessary to solve the following problems. Round your answers to the nearest tenth.

1. 8 squared is
2. $12^2 =$
3. The square root of 156 is
4. $\sqrt{3,136}$ is
5. $\sqrt{54}$ is

1.0.0 Section Review

1. Reduce all fractions to lowest terms, use a calculator when needed, and round answers to the nearest tenth when necessary. You have lengths of tube that are 2.8', 4.2', and 6' long. The total length of tubing is _____.

   a. 12.8'
   b. 13.0'
   c. 113'
   d. 130'

2. Reduce all fractions to lowest terms, use a calculator when needed, and round answers to the nearest tenth when necessary. What is the common denominator for ⅔ and ⅜?

   a. 6
   b. 12
   c. 16
   d. 24

3. Reduce all fractions to lowest terms, use a calculator when needed, and round answers to the nearest tenth when necessary. Rounding 118.76 to the nearest tenth gives _____.

   a. 118.5
   b. 118.7
   c. 118.8
   d. 119

4. Reduce all fractions to lowest terms, use a calculator when needed, and round answers to the nearest tenth when necessary. 6.85" converted to its decimal equivalent in feet is _____.

   a. 0.50'
   b. 0.57'
   c. 0.60'
   d. 0.75'

5. Reduce all fractions to lowest terms, use a calculator when needed, and round answers to the nearest tenth when necessary. The metric prefix tera represents one _____.

   a. hundred
   b. thousand
   c. million
   d. trillion

6. Reduce all fractions to lowest terms, use a calculator when needed, and round answers to the nearest tenth when necessary. $13^2 = $ _____.

   a. 3.6
   b. 26
   c. 169
   d. 260

2.0.0 MEASURING PIPE

Objective

Explain how pipe is measured.
a. Identify the parts of a fitting.
b. Define makeup.
c. Define fitting allowance.
d. Use manufacturer's tables to select pipe.
e. Calculate pipe lengths using various methods.

Performance Tasks

1. Measure pipe using the following methods:
   • End-to-end
   • End-to-center
   • Center-to-center
   • End-to-face
   • Face-to-face
   • Face-to-throat
2. Determine end-to-end dimensions by figuring fitting allowances and makeup.

Trade Terms

Back: Part of the fitting that is opposite to the side with an opening or face.

Center: A point exactly halfway between two other points or surfaces.

Center line: On a drawing, a line that shows the center of an object.

Center point: Point created where the center line of the pipe and the center line of the fitting meet within the fitting. The center point is used to determine the correct length of a pipe.

Face: The open end of a fitting where a pipe is joined to the fitting, such as the opening of the inlet or either end.

Thread makeup: The distance that a pipe screws into a fitting. Also called thread engagement or thread-in.

Throat: The part of the fitting where you thread in another pipe or fitting.

A plumber must be able to measure pipe accurately and quickly. Measuring is basic to the plumbing profession and lies at the heart of many other trade skills as well. Pipe, like other construction materials such as 2 × 4 lumber, comes in nominal sizes. Nominal sizes are used for the purpose of general identification. The actual size of the piece is approximately the same as the nominal size, but won't be exactly the same.

In the fabrication and installation of piping systems, plumbers often use the center-to-center (C–C) measurements between two fittings. The lengths of pipe and fittings are measured along center lines. The extensions of the center line of the pipe and the center line of the fitting meet inside the fitting to create a center point.

2.1.0 Parts of a Fitting

The basic parts of a fitting are shown in *Figure 3*. The terms face, center, and back are used to describe these parts. These elements are important when you measure pipe length because they define the beginning and ending points of the measurement. You will often use a face-of-fitting to center-of-fitting measurement. The throat of the fitting is not used as often in measuring pipe.

Figure 4 shows the various methods of measuring pipe lengths. You will have a chance to practice some of these methods later in the module.

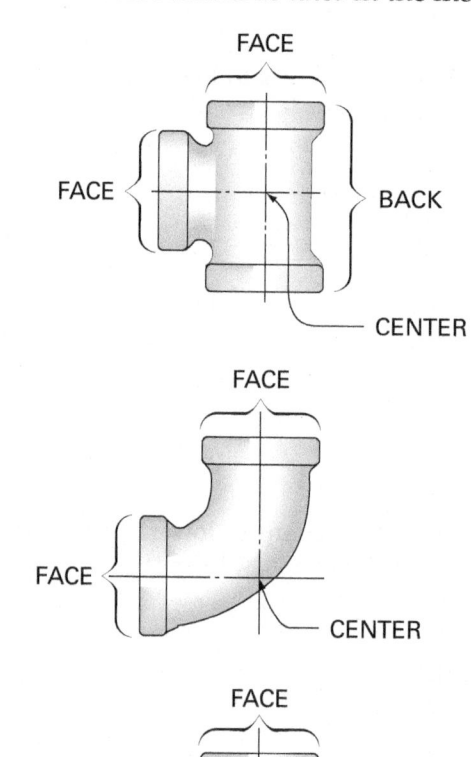

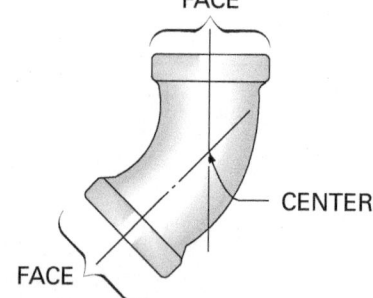

Figure 3 Basic fitting parts.

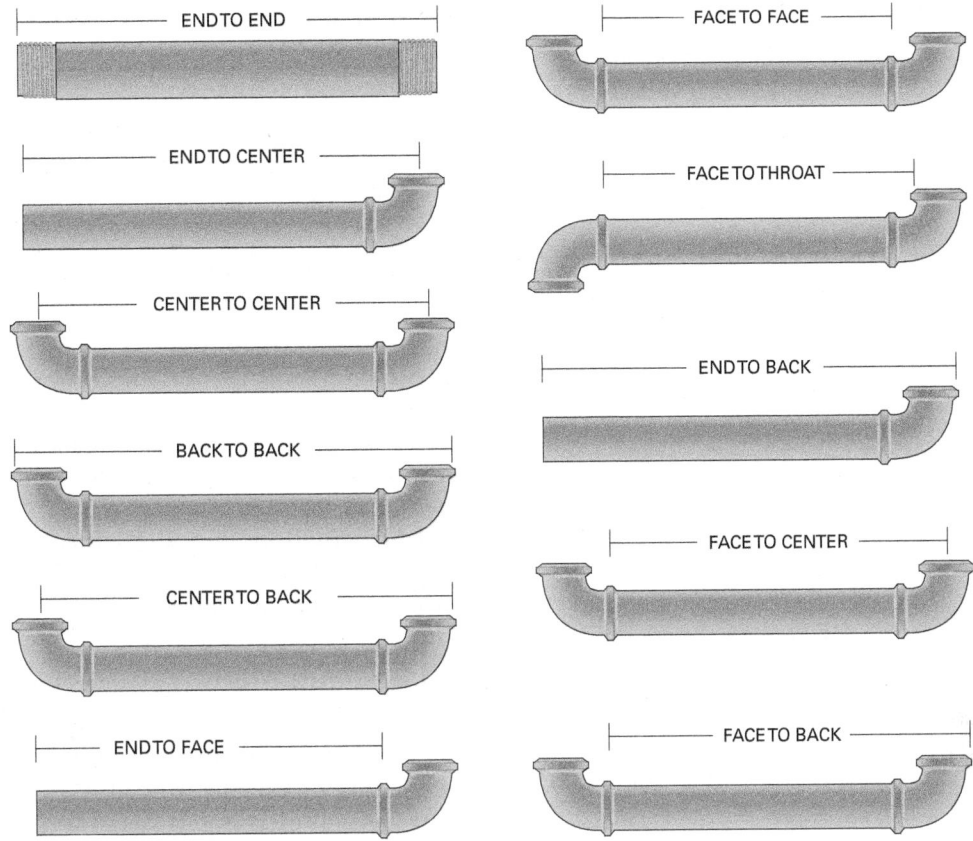

Figure 4 Measuring techniques.

Figure 5 shows the same information in single line drawing form. It is more convenient to draw this way than to draw the whole pipe or fitting. Compare *Figure 4* and *Figure 5* and note the differences.

The dimensions of the pipe assemblies shown in *Figure 4* and *Figure 5* can also be expressed verbally, as *Table 4* shows. Speaking this way may seem strange at first, but this is part of the language of plumbing. To be a plumber, you must be able to understand, speak, and write this language. To get accustomed to this language, read each statement out loud and relate each to one of the sketches in *Figure 5*, where possible. For instance, "8" end to end" corresponds to the first drawing in *Figure 5*.

Table 4 Verbal Expression of Pipe

Pipe Size	Statement	Abbreviation
½"	8 inches end to end	½" 8" E–E
¾"	10 inches end to center of ell	¾" 10" E–C ell
1"	8 inches end to face of ell	1" 8" E–F ell
3"	17 inches end to back of ell	3" 17" E–B ell
1"	11 inches center of ell to center of tee	1" 11" C ell–C tee
2"	8½ inches face of ell to throat of ell	2" 8½" F ell–TH ell
4"	8 inches face of ell to face of wye	4" 8" F ell–F wye
¾"	42 inches center of wye to face of ell	¾" 42" C wye–F ell
6"	14 inches center of tee to back of ell	6" 14" C tee–B ell
1½"	69 inches face of ell to back of ell	1½" 69" F ell–B ell

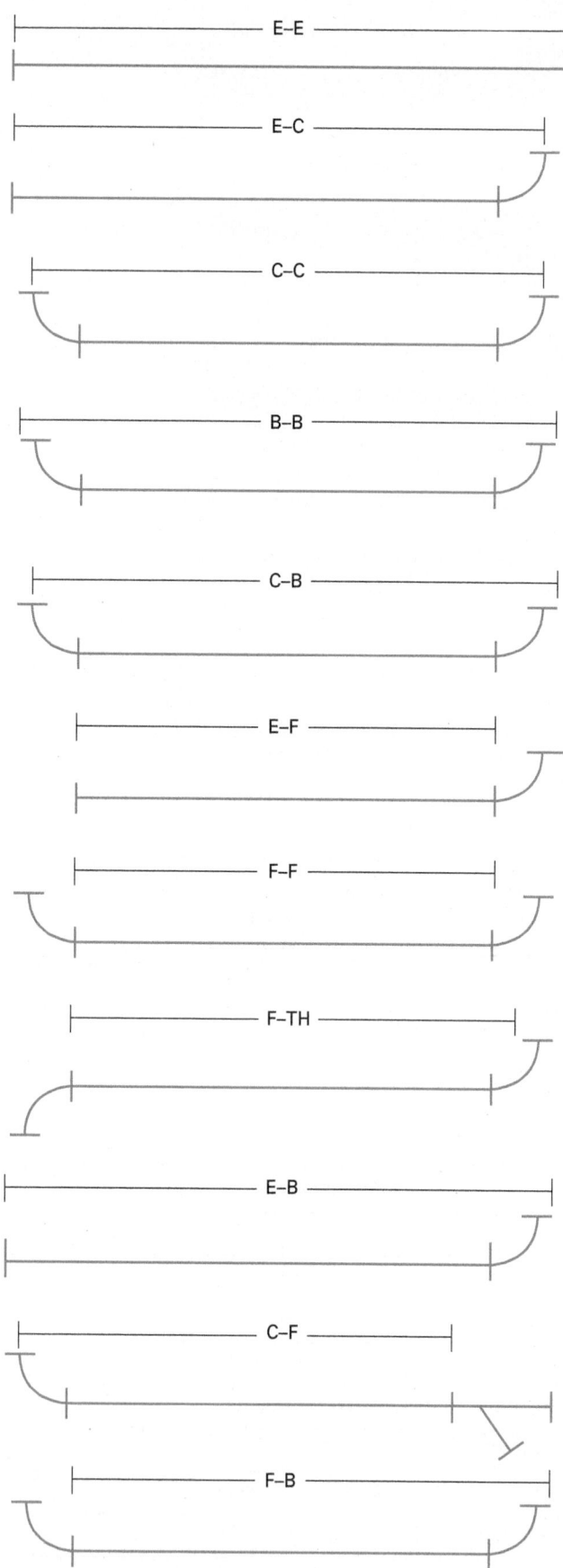

Figure 5 Schematic views.

2.2.0 Makeup

Thread makeup, also known as thread-in or thread engagement, is the distance that the pipe screws into the threaded fitting. The most convenient way to determine thread makeup is simply to measure the depth of the fitting's socket or thread.

2.3.0 Fitting Allowance

Fitting allowance is the distance from the end of the pipe that goes into the fitting (face of fitting) to the center of the fitting. Some joints, such as plastic glued or threaded joints, have fitting allowances. Others, including cast-iron no-hub and face-to-face flanged joints, do not have fitting allowances.

To calculate fitting allowance, measure the fitting from its center to its face. Then subtract the thread-in measurement.

2.4.0 Using Manufacturer Tables

You can also determine fitting allowances by referring to tables published by fitting manufacturers. *Figure 6* shows fitting dimensions for copper pipe, which also is not threaded. Fitting dimensions vary for different types of fittings made by different manufacturers. For instance, *Figure 7* shows fitting dimensions for the following plastic fittings: elbows, tees, and wyes. Cast-iron (soil) pipe, which is not threaded, comes in two types: cast-iron pipe with hub-and-spigot fittings and cast-iron pipe with no-hub fittings. *Figure 8* shows fitting dimensions for both types of cast-iron pipe.

Figure 9 shows fitting dimensions for threaded steel tees and elbows. A capital letter on the drawing of the fitting is keyed to a column on the table. To use the tables, find the pipe size in the first column and then find the dimensions in the columns to the right. For example, the fitting allowance for a threaded, steel, 1-inch, 90-degree elbow fitting is 1 inch because $C = 1\frac{1}{2}$" and $T = \frac{1}{2}$", so $1\frac{1}{2}" - \frac{1}{2}" = 1$".

2.5.0 Calculating Pipe Length

The measuring technique you use determines how you calculate the length of pipe to cut. You will use fitting dimensions as part of your calculations in most pipe-measuring techniques.

2.5.1 Center-to-Center Method

To determine the length of pipe between two fittings using the center-to-center method, subtract the fitting allowances from the center-to-center measurement (see *Figure 10*).

ELLS

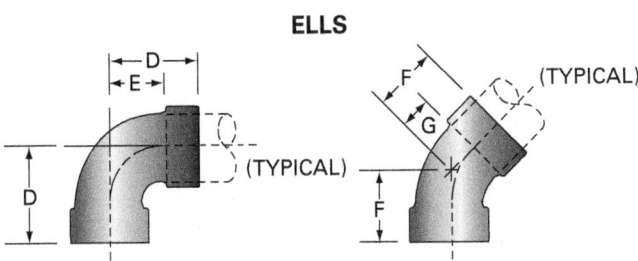

TEE

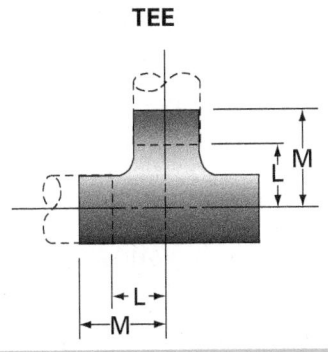

NOTES:
1. DIMENSIONS D AND F ARE CENTER TO END OF FITTING.
2. DIMENSIONS E AND G ARE CENTER OF FITTING TO END OF PIPE.
3. DIMENSIONS ARE IN INCHES.

NOTES:
1. DIMENSION M IS THE CENTER TO END OF THE FITTING.
2. DIMENSION L IS THE CENTER OF FITTING TO END OF PIPE.
3. DIMENSIONS ARE IN INCHES.

NOMINAL PIPE SIZE (IN)	90° ELBOWS			45° ELBOWS		
	D	E	WT. 90 (LB)	F	G	WT. 45 (LB)
¼	1.88	1.56	0.13	1.02	0.70	0.09
⅜	1.94	1.62	0.17	1.20	0.89	0.11
½	2.25	1.94	0.25	1.38	1.06	0.17
¾	2.25	1.94	0.34	1.38	0.94	0.24
1	2.38	2.00	0.53	1.50	1.12	0.35
1¼	2.88	2.44	0.76	1.78	1.34	0.51
1½	3.50	2.94	0.96	2.19	1.62	0.67
2	4.28	3.72	1.81	2.53	1.97	1.21
2½	5.38	4.62	2.65	3.19	2.44	1.77
3	6.17	5.41	4.54	3.52	2.76	2.68
4	7.78	6.97	8.55	4.27	3.45	4.88
5	9.53	8.59	13.32	5.14	4.20	6.87
6	11.17	10.17	19.90	5.91	4.91	11.61
8	14.62	13.38	38.50	7.59	6.34	28.08
10	18.00	16.56		9.22	7.78	
12	21.25	19.69		10.70	9.14	

NOMINAL PIPE SIZE (IN)	STRAIGHT TEES		
	M	L	WT. (LB)
¼	0.69	0.38	0.13
⅜	0.91	0.59	0.17
½	1.09	0.78	0.23
¾	1.25	0.94	0.34
1	1.50	1.12	0.53
1¼	1.88	1.44	0.87
1½	2.25	1.69	1.21
2	2.50	1.94	1.87
2½	3.00	2.25	2.80
3	3.38	2.62	4.33
4	4.12	3.31	8.09
5	4.88	3.94	11.17
6	5.62	4.62	16.58
8	7.00	5.75	27.27
10	8.50	7.06	
12	10.00	8.43	

Figure 6 Fitting dimensions for copper pipe.

For example, say you want to know what length of a threaded steel pipe to cut to fill a space between two threaded tees. The center-to-center measurement between the two fittings is 50". You are using 1¼" threaded steel pipe, which, as you can see by referring to *Figure 9*, has a ½" thread-in (T) and a 1¾" center-to-face measurement (C).

Step 1 Determine the fitting allowance by subtracting T from C.

C – T = fitting allowance
1¾" – ½" = 1¼"

Step 2 Because you have two fittings, you must add the allowances together to get the total fitting allowance.

Total fitting allowance
= 1¼" + 1¼" = 2½"

Step 3 Subtract the total fitting allowance from the 50" center-to-center measurement.

50" – 2½" = 47½"

- You need to cut a pipe that is 47½" long.

2.5.2 End-to-Center Method

To calculate the end-to-end dimension when you know the end-to-center measurement, subtract one fitting allowance, because there is just one fitting (see *Figure 11*).

Suppose you need to connect two $\frac{3}{4}$" threaded steel tees. The end-to-center measurement is $17\frac{1}{2}$". Calculate the length of pipe you need to cut.

Step 1 Determine the fitting allowance by subtracting the thread-in from the center-to-face measurement (see *Figure 11*):

Fitting Allowance
$= 1\frac{3}{8}" - \frac{1}{2}" = \frac{11}{8}" - \frac{4}{8}" = \frac{7}{8}"$

Step 2 Subtract the fitting allowance from the end-to-center measurement.

$$17\frac{1}{2}" - \frac{7}{8}" = 16\frac{12}{8}" - \frac{7}{8}" = 16\frac{5}{8}"$$

You need $16\frac{5}{8}$" of $\frac{3}{4}$" steel pipe.

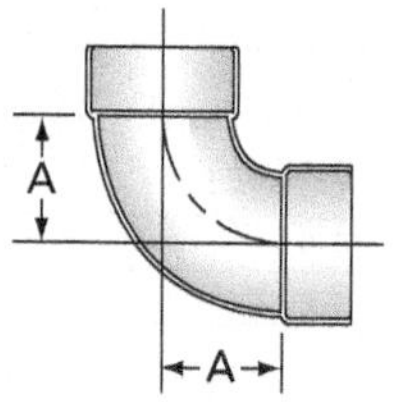

BEND (SANITARY 90 ELL) ALL HUB			
SIZE	A	B	C
1½"	1¾"		
2"	2 5/16"		
3"	3 1/16"		
4"	3 7/8"		
6"	5"		
8"	6"		

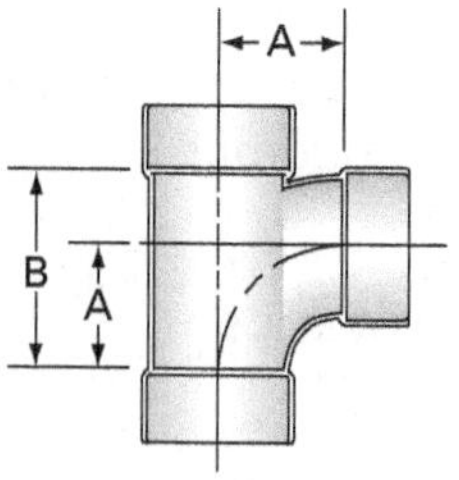

SANITARY TEE ALL HUB			
SIZE	A	B	C
1½"	1¾"	2¾"	
2"	2 5/16"	3 11/16"	
3"	3 1/16"	4 7/8"	
4"	3 7/8"	6 1/8"	
6"	5"	8½"	
8"	6"	10½"	

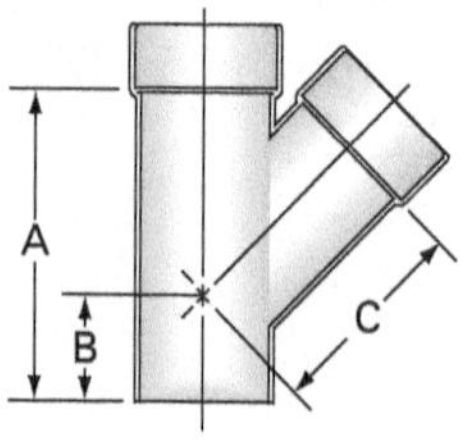

WYE, STREET (45 WYE) SPIGOT × HUB × HUB			
SIZE	A	B	C
1½"	4¾"	1 7/8"	2 7/8"
2"	5 7/8"	2¼"	3 5/8"
3"	8 1/8"	3 1/8"	5"
4"	10"	3 5/8"	6 3/8"

AVAILABLE IN PVC ONLY

Figure 7 Fitting dimensions for plastic pipe.

2.5.3 Face-to-Face Method

To determine the required end-to-end dimension given a face-to-face measurement, you need to add the thread makeup for each fitting.

Example: Calculate the length of steel pipe required to connect $\frac{1}{2}$" threaded steel tees if the face-to-face measurement is $6\frac{3}{4}$".

$$6\frac{3}{4}" + \frac{1}{2}" + \frac{1}{2}" = 7\frac{3}{4}"$$

You need $7\frac{3}{4}$" of $\frac{1}{2}$" steel pipe.

2.5.4 General Principles of Measurement

As shown in *Figure 4*, there are many measuring techniques, depending on the information you have and the type of installation. You may have noticed the following general principles underlying all of these methods of measurement:

- When you have a center dimension, subtract the fitting allowance for that end.
- When you have a face dimension, add the thread makeup for that end.
- When you have an end dimension, you do not need to add or subtract anything.

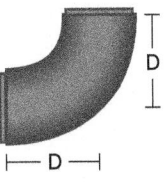

QUARTER BEND

SIZE (IN)	D	WEIGHT (LB)
1½	4¼	1.9
2	4½	2.4
3	5	3.9
4	5½	6.0
5	6½	10.0
6	7	12.0
8	8½	25.0
4 × 3	5½	6.0

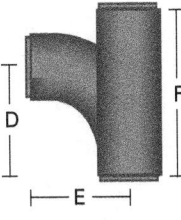

SANITARY TEE

SIZE (IN)	E	F	D	WEIGHT (LB)
1½ × 1½	4¼	6½	4¼	2.3
2 × 1½	4½	6⅝	4¼	3.0
2 × 2	4½	6⅞	4¼	2.9
3 × 1½	5	6	4¼	5.0
3 × 2	5	6⅞	4½	4.0
3	5	8	5	4.7
4 × 2	5½	6⅞	4½	6.0
4 × 3	5½	8	5	7.2
4	5½	9⅛	5½	8.0
5 × 2	6½	8½	5	8.7
5 × 3	6	9⁵⁄₁₆	5½	10.5
5 × 4	6	10¹³⁄₃₂	6	11.5
5	6½	11⁷⁄₁₆	6½	15.0
6 × 2	6½	8³⁄₁₆	5	12.0
6 × 3	6½	9³⁄₁₆	5½	23.0
6 × 4	6½	10¹⁄₁₆	6	11.5
6 × 5	7	11½	6½	27.0
6	7	12½	7	15.0
8 × 4	7½	11½	6½	21.0
8 × 5	8	12½	7	24.0
8 × 6	8	13½	7½	24.0
8	8	15½	9½	28.0

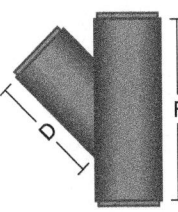

WYE

SIZE (IN)	D	F	WEIGHT (LB)
1½ × 1½	4	6	2.2
2 × 2	4⅝	6⅝	2.6
3 × 1½	4⅝	6⅝	3.0
3 × 2	5⁵⁄₁₆	6⅝	3.7
3	5¾	8	4.9
4 × 2	6	6⅝	5.5
4 × 3	6½	8	7.2
4	7¹⁄₁₆	9½	8.8
5 × 2	7½	8¹⁄₁₆	7.2
5 × 3	8	9¹¹⁄₁₆	10.5
5 × 4	8½	11³⁄₁₆	12.0
5	9½	12⅝	14.5
6 × 2	8¼	8⁵⁄₁₆	9.5
6 × 3	8¾	9¾	11.0
6 × 4	9¼	11³⁄₁₆	13.0
6 × 5	10¼	12½	16.0
6	10¾	14¹⁄₁₆	18.0
8 × 3	9⅞	10	19.0
8 × 4	10⅜	11⁷⁄₁₆	22.0
8 × 5	11⅜	12⅞	25.0
8 × 6	11¹³⁄₁₆	14¹³⁄₁₆	28.0
8	13⅜	17⅛	38.5
10 × 4	11¹¹⁄₁₆	12⅝	37.0
10 × 6	13⅛	15⁷⁄₁₆	46.0
10 × 8	14¹¹⁄₁₆	18⅜	52.0
10	16½	21½	72.0

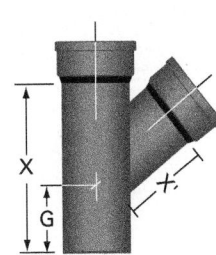

WYE

SIZE (IN)	G	X	X'	WEIGHT (LB) SV	WEIGHT (LB) XH
2 × 2	4	8	4	7	8
3 × 2	4³⁄₁₆	9	5	10	14
3 × 3	5	10½	5½	13	17
4 × 2	3⅝	9	5¾	12	17
4 × 3	4⁷⁄₁₆	10½	6¼	14	20
4 × 4	5¼	12	6¾	17	24
5 × 2	3⅛	9	6½	14	–
5 × 3	3⅞	10½	7	17	24
5 × 4	4¹¹⁄₁₆	12	7½	19	27
5 × 5	5½	13½	8	22	32
6 × 2	2⁹⁄₁₆	9	7¼	17	–
6 × 3	3⅜	10½	7¾	19	27
6 × 4	4³⁄₁₆	12	8¼	22	31
6 × 5	4¹⁵⁄₁₆	13½	8¾	24	35
6 × 6	5¾	15	9¼	28	40
8 × 2	3⅛	10½	8½	29	–
8 × 3	3¹⁵⁄₁₆	12	9	32	–
8 × 4	4¾	13½	9½	36	52
8 × 5	5½	15	10	39	–
8 × 6	6⁵⁄₁₆	16½	10½	44	63
8 × 8	7¹¹⁄₁₆	19½	11¹³⁄₁₆	55	82
10 × 3	2⁹⁄₁₆	12	11	50	–
10 × 4	3⁹⁄₁₆	13½	11⅛	53	74
10 × 5	4⁵⁄₁₆	15	11⅝	57	–
10 × 6	5⅛	16½	12⅛	61	86
10 × 8	6½	19½	13⁷⁄₁₆	77	110
10 × 10	8	22½	14½	94	133
12 × 4	4⅛	15	12⁷⁄₁₆	70	–
12 × 5	4⅞	16½	12¹⁵⁄₁₆	74	–
12 × 6	5¹¹⁄₁₆	18	13⁷⁄₁₆	80	111
12 × 8	7¹⁄₁₆	21	14¾	96	136
12 × 10	8⁹⁄₁₆	24	15¹³⁄₁₆	115	160
12 × 12	10⅛	27	16⅞	135	186
15 × 4	2¼	15	15	130	130
15 × 6	4	18	15¾	109	–
15 × 8	5⅜	21	17¹⁄₁₆	127	182
15 × 10	6⅞	24	18⅛	152	213
15 × 12	8⁷⁄₁₆	27	19³⁄₁₆	242	242
15 × 15	10¾	31½	20¾	338	338

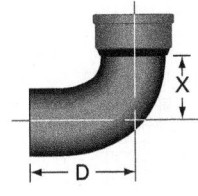

QUARTER BEND

SIZE (IN)	D	X	WEIGHT (LB) SV	WEIGHT (LB) XH
2	6	3¼	4	5
3	7	4	7	10
4	8	4½	11	15
5	8½	5	13	19
6	9	5½	17	24
8	11½	6⅝	34	51
10	12½	7⅝	55	78
12	15	8¾	80	111
15	16½	10¼	169	169

Figure 8 Fitting dimensions for cast-iron pipe.

4¾" 3⁵⁄₁₆"
5⅞" 4¹⁄₁₆"
5⅞" 4¼"
6¼" 4⅝"

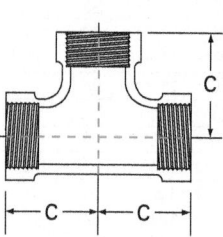

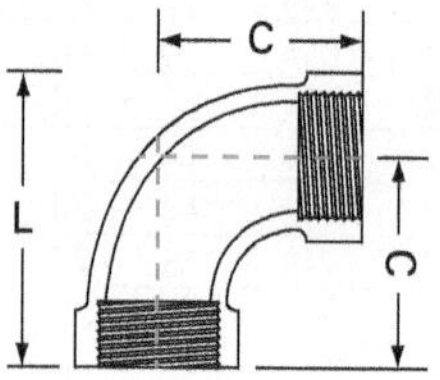

THREADED 90-DEGREE ELBOW

A IS CENTER TO FACE
B IS FACE

NOMINAL PIPE SIZE	A	B
⅛"	⅞"	⅞"
¼"	⅞"	⅞"
⅜"	³¹⁄₃₂"	1¹⁄₁₆"
½"	1⅛"	1⁵⁄₁₆"
¾"	1⁵⁄₁₆"	1½"
1"	1½"	1¹³⁄₁₆"
1¼"	1¾"	2³⁄₁₆"
1½"	2"	2½"
2"	2⅜"	3"
2½"	3"	3⅝"
3"	3⅜"	4⁵⁄₁₆"
4"	4³⁄₁₆"	5⅞"

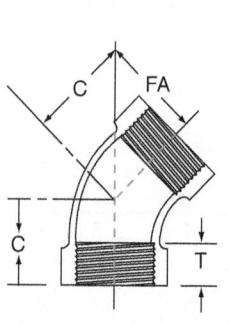

THREADED TEES

C IS CENTER TO FACE
T IS THREAD-IN OR MAKEUP

NOMINAL PIPE SIZE	C	T
⅜"	1"	⅜"
½"	1⅛"	½"
¾"	1⅜"	½"
1"	1½"	½"
1¼"	1¾"	½"
1½"	1⅞"	½"
2"	2¼"	½"
2" × 1½"	2¼"	½"
2½"	2¾"	¾"

THREADED ELBOWS

FA IS FITTING ANGLE
C IS CENTER TO FACE
T IS THREAD-IN OR MAKEUP

NOMINAL PIPE SIZE	C					T
	FITTING ANGLES					
	90°	60°	45°	22½°	11¼°	
⅜"	1"		¾"			⅜"
½"	1⅛"		¾"			½"
¾"	1⅜"		1"			½"
1"	1½"		1⅛"			½"
1¼"	1¾"	1¼"	1¼"	1⅛"	1"	½"
1½"	1⅞"	1¾"	1½"	1¼"	1⅛"	½"
2"	2¼"	2¼"	1⅝"	1⅜"	1¼"	½"
2½"	2¾"	2½"	1¾"	1½"	1⅜"	¾"

Figure 9 Fitting dimensions for threaded steel tees and elbows.

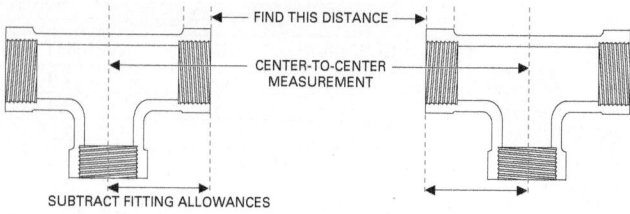

Figure 10 Center-to-center method.

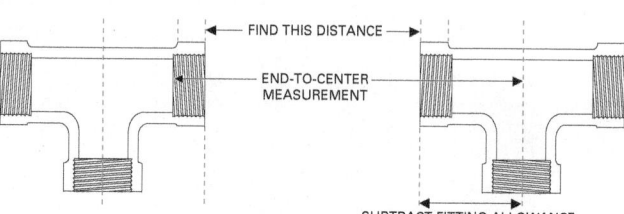

Figure 11 End-to-center method.

2.0.0 Section Review

1. A nominal ½" pipe that is 10" long from the end to the center of an ell is abbreviated as _____.

   a. ½"–10E–C
   b. ½" C–Ell
   c. ½" 10" E–C ell
   d. 10" E–Ell C

2. The most convenient way to determine thread makeup is to measure the _____.

   a. depth of the fitting's socket or thread
   b. fitting from its center to its face
   c. fitting from center to face plus the fitting socket or thread
   d. fitting from center to face minus the fitting socket or thread

3. The distance from the end of the pipe that goes into the fitting to the center of the fitting is _____.

   a. center-to-center
   b. thread-in
   c. makeup
   d. fitting allowance

4. Referring to *Figure 9*, the fitting allowance for a threaded steel 2-inch 45-degree elbow is _____.

   a. 1⅛"
   b. 1⅜"
   c. 1½"
   d. 1⅝"

5. You need to cut a threaded steel pipe to connect two ¾" threaded tees. The center-to-center measurement is 41⅞". Referring to *Figure 9*, the end-to-end dimension required for this pipe is _____.

   a. ⅞"
   b. 40⅛"
   c. 41⅞"
   d. 41"

SUMMARY

Skilled, productive plumbers know mathematics. Being able to use math easily and accurately in the field leads to more efficiency and greater productivity. This helps make you a more valued employee, and your career will benefit as a result.

You have been introduced to methods of measuring pipe and have learned how to find the cut length of pipe that runs between fittings. With this groundwork, you can progress in your math skills and your trade skills. The more you know, the more you can take pride in your abilities.

1. You already have 2,032.25 pounds of bricks. If you buy 7,525.5 more pounds, the total pounds of bricks you will have is _____.

   a. 2,758
   b. 2,775.80
   c. 9,557.75
   d. 95,577.5

2. If cast-iron pipe weighs 4.75 pounds per foot, 129.4' weighs _____.

   a. 614.65 pounds
   b. 1,770.40 pounds
   c. 6,146.50 pounds
   d. 17,704 pounds

3. If you have a length of pipe measuring 1,780", you can cut _____.

   a. 140 12" pieces
   b. 148 12" pieces
   c. 151 12" pieces
   d. 156 12" pieces

4. If tubing costs $4.20 per foot and you pay a total of $120.95, the total feet of tubing you have purchased is _____.

   a. 2.8
   b. 28.6
   c. 28.8
   d. 287.9

5. Converted to a decimal, $\frac{3}{16}$ equals _____.

   a. 0.1875
   b. 0.316
   c. 1.875
   d. 3.187

6. 88.47' converted into the nearest sixteenth of an inch is _____.

   a. $5\frac{10}{16}$"
   b. $88'5\frac{17}{25}$"
   c. $88'5\frac{5}{8}$"
   d. 88'10"

7. $1\frac{5}{16}$" is equal to _____.

   a. 0.047'
   b. 0.063'
   c. 0.109'
   d. 0.172'

8. 25^2 = _____.

   a. 3.87
   b. 30
   c. 125
   d. 625

9. The extensions of the center line of the pipe and the center line of the fitting intersect inside the fitting to create a(n) _____.

   a. center line
   b. single line
   c. center point
   d. end-to-end dimension

10. A nominal 4" pipe that is 12" long from end to end is abbreviated as _____.

    a. 4–12E
    b. 4" C–Ell
    c. 4" 12" E–E
    d. 12" E

11. The distance that a pipe screws into a threaded fitting is called the _____.

    a. fitting allowance
    b. thread makeup
    c. center point
    d. center-to-face dimension

12. Referring to *Figure RQ01*, the fitting allowance for a threaded steel ¾" 45-degree elbow is _____.

4¾" 3⁵⁄₁₆"
5⅞" 4¹⁄₁₆"
5⅞" 4¼"
6¼" 4⅝"

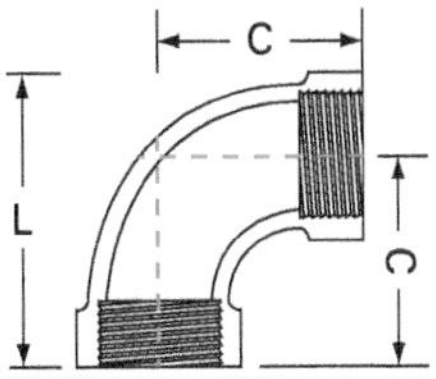

THREADED 90-DEGREE ELBOW

A IS CENTER TO FACE
B IS FACE

NOMINAL PIPE SIZE	A	B
⅛"	⅞"	⅞"
¼"	⅞"	⅞"
⅜"	3¹⁄₃₂"	1¹⁄₁₆"
½"	1⅛"	1⁵⁄₁₆"
¾"	1⁵⁄₁₆"	1½"
1"	1½"	1¹³⁄₁₆"
1¼"	1¾"	2³⁄₁₆"
1½"	2"	2½"
2"	2⅜"	3"
2½"	3"	3⅝"
3"	3⅜"	4⁵⁄₁₆"
4"	4³⁄₁₆"	5⅞"

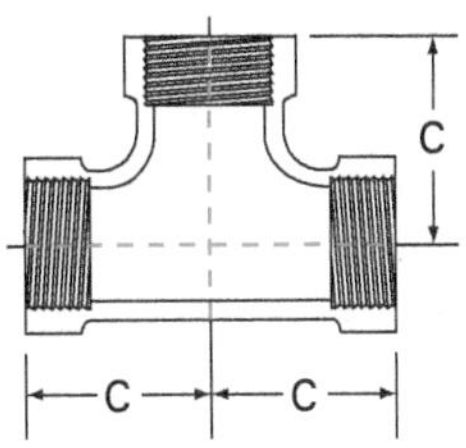

THREADED TEES

C IS CENTER TO FACE
T IS THREAD-IN OR MAKEUP

NOMINAL PIPE SIZE	C	T
⅜"	1"	⅜"
½"	1⅛"	½"
¾"	1⅜"	½"
1"	1½"	½"
1¼"	1¾"	½"
1½"	1⅞"	½"
2"	2¼"	½"
2" × 1½"	2¼"	½"
2½"	2¾"	¾"

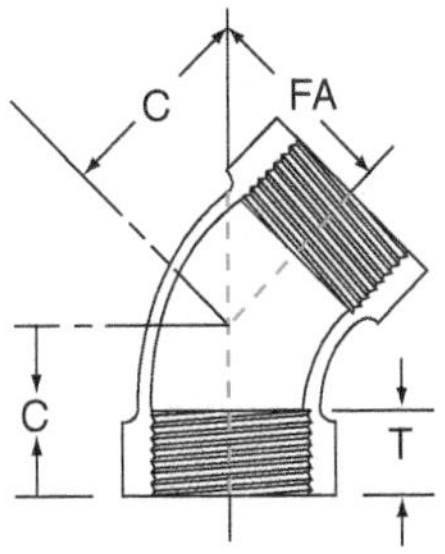

THREADED ELBOWS

FA IS FITTING ANGLE
C IS CENTER TO FACE
T IS THREAD-IN OR MAKEUP

NOMINAL PIPE SIZE	C					T
	FITTING ANGLES					
	90°	60°	45°	22½°	11¼°	
⅜"	1"		¾"			⅜"
½"	1⅛"		¾"			½"
¾"	1⅜"		1"			½"
1"	1½"		1⅛"			½"
1¼"	1¾"	1¼"	1¼"	1⅛"	1"	½"
1½"	1⅞"	1¾"	1½"	1¼"	1⅛"	½"
2"	2¼"	2¼"	1⅝"	1⅜"	1¼"	½"
2½"	2¾"	2½"	1¾"	1½"	1⅜"	¾"

Figure RQ01

a. ⅛"
b. ¼"
c. ½"
d. ⅝"

13. In an installation, two 1" threaded tees need to be connected. The center-to-center measurement is 42". The end-to-end dimension required for this pipe is _____.

a. 1½"
b. 40"
c. 42"
d. 44"

14. The end-to-end dimension required for a ½" threaded pipe between two tees with an end-to-center length of 29" is _____.

a. ⅝"
b. ¹¹⁄₈"
c. 28⅜"
d. 29"

15. The end-to-end dimension required for a 1½" threaded pipe between two tees with a face-to-face length of 18⅞" is _____.

a. ½"
b. 18⅞"
c. 19⅞"
d. 21⅞"

Fill in the blank with the correct term that you learned from your study of this module.

1. The _____, _____, and _____ describe the basic parts of a fitting; they define the beginning and ending points of the measurement.

2. The lengths of pipe and fittings are measured along the _____.

3. The _____ is where the extension of the center line of the pipe and the center line of the fitting meet inside the fitting.

4. To calculate _____, subtract the thread-in measurement from the center-to-face measurement.

5. $89 \div 3 = 29$ with a(n) _____ of 2.

6. For threaded fittings, the _____ is the distance that the pipe screws into the fitting.

7. The _____ of a fitting is not used often in measuring pipe.

Trade Terms

Back
Center
Center line

Center point
Face
Fitting allowance

Remainder
Thread makeup
Throat

Trade Terms Introduced in This Module

Back: Part of the fitting that is opposite to the side with an opening or face.

Center: A point exactly halfway between two other points or surfaces.

Center line: On a drawing, a line that shows the center of an object.

Center point: Point created where the center line of the pipe and the center line of the fitting meet within the fitting. The center point is used to determine the correct length of a pipe.

Face: The open end of a fitting where a pipe is joined to the fitting, such as the opening of the inlet or either end.

Fitting allowance: The distance from the end of the pipe that goes into a fitting to the center of the fitting.

Remainder: The leftover amount in a division problem. For example, in the problem $34 \div 8$, 8 goes into 34 four times ($8 \times 4 = 32$) and 2 is left as the remainder.

Thread makeup: The distance that a pipe screws into a fitting. Also called thread engagement or thread-in.

Throat: The part of the fitting where you thread in another pipe or fitting.

Additional Resources

This module presents thorough resources for task training. The following resource material is suggested for further study.

Plumber's and Pipefitter's Calculations Manual, 1999. R. Dodge Woodson. McGraw-Hill Professional.

Figure Credits

© Lloyd Wolf for SkillsUSA, Module Opener
Calculated Industries, Figure 2
Mueller, Figure 9 (threaded 90-degree fitting table)

SECTION 1.0.0

Answer	Section Reference
1. c	1.1.2
2. a	1.1.4
3. b	1.1.5
4. c	1.1.5
5. a	1.4.2
6. c	1.4.3
7. c	1.4.3; Table 1

SECTION 2.0.0

Answer	Section Reference
8. d	1.6.0
9. c	2.0.0
10. c	2.1.0; Table 4
11. b	2.2.0
12. c	2.4.0
13. b	2.5.1
14. c	2.5.2
15. c	2.5.3

Appendix
Practice Exercise Answers

Section 1.2.8

1. 756
2. 325
3. 54
4. 315
5. 4
6. $^{20}/_{32}$"
7. $^{3}/_{8}$
8. $^{5}/_{16}$ is already in lowest form
9. $^{5}/_{8}$
10. $3^{9}/_{16}$
11. $^{3}/_{4}$
12. $^{1}/_{4}$
13. $^{5}/_{128}$
14. $^{17}/_{24}$
15. $^{3}/_{2}$ or $1^{1}/_{2}$

Section 1.3.5

1. 22.0
2. 74.7
3. 31.15
4. 117.16 rounded to the nearest tenth 117.2
5. .3066.73 rounded to the nearest tenth 3066.7

Section 1.4.5

1. 85%
2. 435%
3. 0.20
4. 1.5
5. 0.375
6. 1.5
7. 1.5
8. $^{3}/_{4}$
9. 13
10. $13^{2}/_{5}$
11. $133^{73}/_{200}$
12. $3^{10}/_{6}$"

13. $11'10^{5}/_{16}$"
14. 1.33'
15. 9.63'

Section 1.5.3

1. 7,500
2. 0.000025
3. 0.675
4. 890
5. 101.1
6. 1,727.2
7. 126
8. 1.155
9. 227.5
10. 62.72
11. 143.6

Section 1.6.1

1. 64
2. 144
3. 12.49
4. 56
5. 7.35

NCCER CURRICULA — USER UPDATE

NCCER makes every effort to keep its textbooks up-to-date and free of technical errors. We appreciate your help in this process. If you find an error, a typographical mistake, or an inaccuracy in NCCER's curricula, please fill out this form (or a photocopy), or complete the online form at **www.nccer.org/olf**. Be sure to include the exact module ID number, page number, a detailed description, and your recommended correction. Your input will be brought to the attention of the Authoring Team. Thank you for your assistance.

Instructors – If you have an idea for improving this textbook, or have found that additional materials were necessary to teach this module effectively, please let us know so that we may present your suggestions to the Authoring Team.

NCCER Product Development and Revision

13614 Progress Blvd., Alachua, FL 32615

Email: curriculum@nccer.org
Online: www.nccer.org/olf

❏ Trainee Guide   ❏ Lesson Plans   ❏ Exam   ❏ PowerPoints   Other _____

Craft / Level: _____ Copyright Date: _____

Module ID Number / Title: _____

Section Number(s): _____

Description: _____

_____

_____

_____

Recommended Correction: _____

_____

_____

_____

_____

Your Name: _____

Address: _____

_____

Email: _____ Phone: _____

This page is intentionally left blank.

Introduction to Plumbing Drawings

OVERVIEW

Every building's design begins on paper. Construction drawings (or blueprints) illustrate design, location, and dimensions. Plumbers must learn to work with and interpret the complete set of construction drawings. Being able to do so enables plumbers to direct work and follow written instructions included on the drawings.

Module 02105

02105 V4.5

Objectives

Successful completion of this module prepares trainees to:

1. Understand how to read drawings.
   a. Explain how to scale and dimension a drawing.
   b. Identify symbols used on construction drawings.
   c. Explain the role of specifications and codes in plumbing.
   d. Identify the elements of a drawing set.
2. Identify the different types of drawings used to install plumbing systems.
   a. Identify the types of pictorial drawings used by plumbers.
   b. Identify schematic diagrams.
   c. Describe the purpose of orthographic drawings.
   d. Identify plumbing-specific drawings, including submittals, fixture drawings, exploded views, and cutaways.

Performance Task

Under the supervision of your instructor, you should be able to do the following:

1. Sketch orthographic and isometric drawings.

Trade Terms

Approved submittal data
Architect's scale
Catalog drawing
Computer-aided drafting (CAD)
Construction drawing
Coordination drawing
Cutaway drawing
Details
Dimension line
Easement
Electrical drawings
Elevation drawings
Exploded drawing
Extension line
Fixture drawing
Floor plan
Foundation plan

HVAC (Heating, Ventilating, and Air Conditioning) drawings
Isometric drawings
Oblique drawing
Orthographic drawing
Pictorial drawings
Plot plan
Plumbing drawing
Riser diagram
Scale
Sepia
Setback
Side yards
Single-line drawings
Specifications
Symbols
Takeoffs

If you're training through an NCCER-accredited sponsor you may be eligible for credentials from NCCER's Registry. The ID number for this module is 02105. Note that this module may have been used in other NCCER curricula and may apply to other level completions. Contact NCCER's Registry at 888.622.3720 or go to *www.nccer.org* for more information.

CODE NOTE

Codes vary among jurisdictions. Because of the variations in code, consult the applicable code whenever regulations are in question. Referencing an incorrect set of codes can cause as much trouble as failing to reference codes altogether. Obtain, review, and familiarize yourself with your local adopted code. Safety codes are developed by the US Occupational Safety and Health Administration (OSHA).

Contents

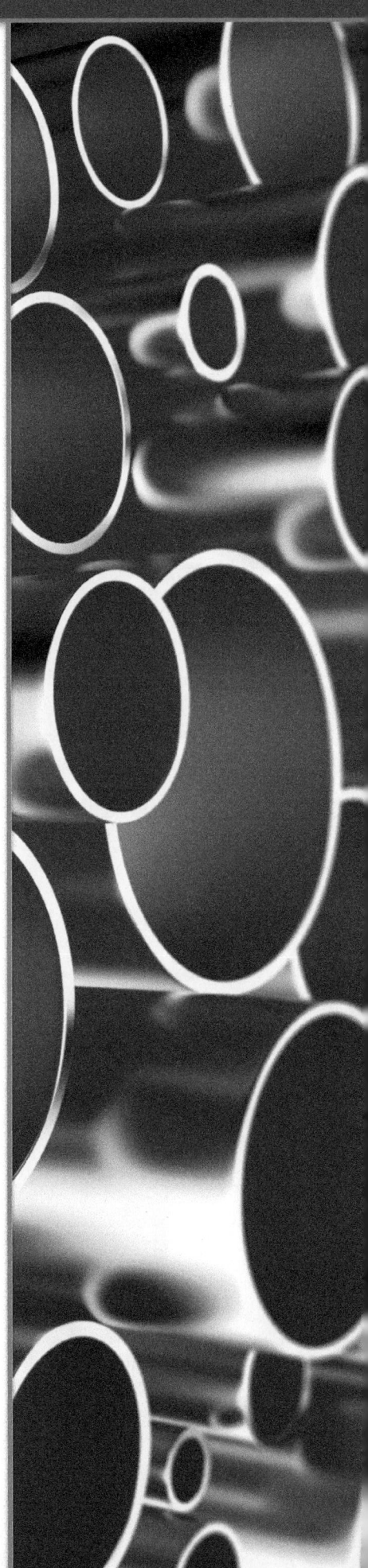

Figures and Tables

1.0.0 READING DRAWINGS

Objective

Understand how to read drawings.
  a. Explain how to scale and dimension a drawing.
  b. Identify symbols used on construction drawings.
  c. Explain the role of specifications and codes in plumbing.
  d. Identify the elements of a drawing set.

Trade Terms

Approved submittal data: The fixtures and fittings in catalog drawings that have been approved by the architect or engineer.

Architect's scale: A measuring device that uses smaller units, such as ½" or ¼", to represent 1'. Architect's scales are also issued in metric units. The units are used so that all building measurements are in proportion to their actual measurements but able to fit in the drawings.

Computer-aided drafting (CAD): A sophisticated design program used on computers. It allows designers to create drawings in two or three dimensions.

Construction drawing: A drawing that shows the design, location, and dimensions of a building and its various components.

Coordination drawing: A dimensioned drawing with elevations and sections that indicate the proposed routing of system components. The dimensioned coordination drawing includes the actual dimensioned equipment as well as the access space this equipment will need.

Details: Sections of construction drawings that are enlarged to make them clearer.

Dimension line: A line on a drawing with a measurement indicating actual length.

Easement: A designated right-of-way, such as the access guaranteed to utility companies for repair of utilities that are located on, or cross over, private land, or for vehicles to cross private property.

Electrical drawings: Drawings that show the location of outlets, switches, and electrical fixtures. An electrical drawing may be superimposed on the floor plan.

Elevation drawings: Drawings of structures showing a side, front, or back view.

Extension line: A line used on a drawing to locate a dimension away from the actual points of the dimension. This method is used when a drawing would be too crowded or cluttered if the dimension were shown within the two points.

Floor plan: A construction drawing of a building looking down at the floor from above (bird's-eye view). The plan shows at least the outline of the wall locations and lengths to scale. Normally, this drawing is oriented such that the top of the page represents north.

Foundation plan: A construction drawing showing the placement and dimensions of a building foundation.

HVAC (heating, ventilating, and air conditioning) drawings: Construction drawings that show the placement of the furnace and air conditioning equipment and the location of ducts and registers or pipes and radiators.

Isometric drawings: Pictorial drawings that create the illusion of a three-dimensional object. All horizontal lines are projected at a 30-degree angle.

Plot plan: A drawing of a structure that includes the dimensions of the building site, location of the structure in relation to the property boundaries, elevation of key points, existing and finish contour lines, utility services, and compass directions. Also referred to as a site plan.

Plumbing drawing: A construction drawing that shows the location of fixtures and pipe runs and gives the size and type of pipe to be installed.

Scale: The relationship of the dimensions on a drawing to the actual dimensions of the structure. For example, in a ¼ scale, 0.25" represent 1'. Scale is often provided in both English and metric units.

Sepia: A print or construction drawing with dark reddish-brown lines on a light background.

Setback: The distance a code requires between a building and a property line, such as the street.

Side yards: The spaces along the sides of a structure that provide access to rear yards, reduce the possibility of fire jumping from one building to the next, and promote ventilation around the structure.

Specifications: Written requirements included with the drawings or blueprints of a construction project. They provide more details or descriptions of the technical standards that must be met during construction. Specifications usually override drawings but are overridden by the contract. Also referred to as specs.

Symbols: Marks or drawings used to indicate a specific object, material, class, or entity. A legend shows the symbols used on a drawing and their meanings.

Before the first shovel of earth is moved, a building has already had a long life on paper. Every building, regardless of its size, begins life on a designer's drawing board or computer terminal. These drawings illustrate the elements of a building, including its design, location, and dimensions. The ability to read drawings is an essential skill. Drawings allow plumbers to visualize a finished project. By visualizing, plumbers can plan ahead and anticipate potential problems, which saves time and money on the job. Developing this skill takes time and experience.

The term construction drawing is often used interchangeably with the term *blueprint*. Blueprints originally referred to architects' drawings that appeared as white line drawings on a solid blue background. Today's construction drawings are likely to be reproduced on white paper, usually with blue lines but sometimes with black or dark reddish lines. A drawing with dark reddish lines is called a sepia. Most people in the building professions today produce drawings by using computer-aided drafting (CAD), which generate drawings from computer programs. CAD systems significantly increase productivity over drafting by hand because they automate much of the repetitive work of drafting. CAD also makes it relatively easy to make changes to drawings.

Understanding the components of a drawing is essential to being able to interpret it accurately. Basic information about each drawing is included in the title block. The title block usually appears in the lower, right-hand corner of the drawing, but the location can vary depending on the company's placement system. The title block includes information such as the title of the project, the date of the drawing, the logo of the company that created the drawing, the name or initials of the person who drafted the drawing, and the owners and address of the property. The title block of a drawing also shows the drawing scale. When plumbers make any construction drawings, they always put a title block on that drawing. The title block is the first thing plumbers should read on construction drawings.

In addition to interpreting existing drawings, plumbers also must be able to create simple sketches and isometric drawings. These drawings enable plumbers to communicate with other workers and convey specific information about an installation. Manufacturers also produce drawings that contain detailed product information. The ability to read these drawings to determine specifications and interpret details about piping assemblies is essential to the plumbing trade.

1.1.0 Scaling and Dimensioning

It is essential that plumbers be able to correctly interpret a drawing's scale and dimensions. These measurements indicate actual sizes and distances, and are critical to the correct installation of piping, fittings, and fixtures. The drawing scale can typically be found in the drawing's title block, while dimensions are listed on the drawing itself.

1.1.1 Scale

Construction drawings are usually drawn to scale. The scale indicates the size relationship between an object in the drawing and the object's actual size. The type of scale used on a drawing depends on the size of the objects being shown, the space available on the paper, and the type of plan. Scale drawings show objects such as buildings, rooms, doors and windows, or piping assemblies reduced to a smaller size.

For example, a wall that measures 8' high and 14' wide obviously cannot be drawn full size. By using scale, an architect, designer, or plumber can represent the wall in a smaller size while keeping the same proportions of height and width as the full-size wall. The proportional reduction allows plumbers to determine the actual size of the object from the drawing.

Typical scales for a plumbing drawing are $\frac{1}{8}$" = 1' or $\frac{1}{4}$" = 1'. If the scale is $\frac{1}{4}$" = 1', this means that for every $\frac{1}{4}$" on the drawing, the real object takes up 1'. In other words, the drawing is $\frac{1}{48}$ the size of the real object. Scale makes drawings convenient to handle. The scale of the drawing usually appears in the title block of the drawing. Be aware that some drawings are not drawn to scale. A note on such drawings will read Not To Scale or NTS.

> **CAUTION**
>
> When a plan is marked NTS, plumbers cannot measure dimensions on the drawing and use them to build the project. Not-to-scale drawings give relative positions and sizes. These measurements are approximate and are not accurate enough for construction.

The term scale is also used to describe the three measuring tools that are used to draw or measure the lines of a construction drawing: architect's scale, (which uses small units to represent 1'), engineer's scale, and metric scale.

To draw and measure scaled drawings, plumbers use an architect's scale (*Figure 1*). The architect's scale is used on all plans other than site plans. The architect's scale is a ruler that is either flat or triangular in shape. If it is flat, it contains

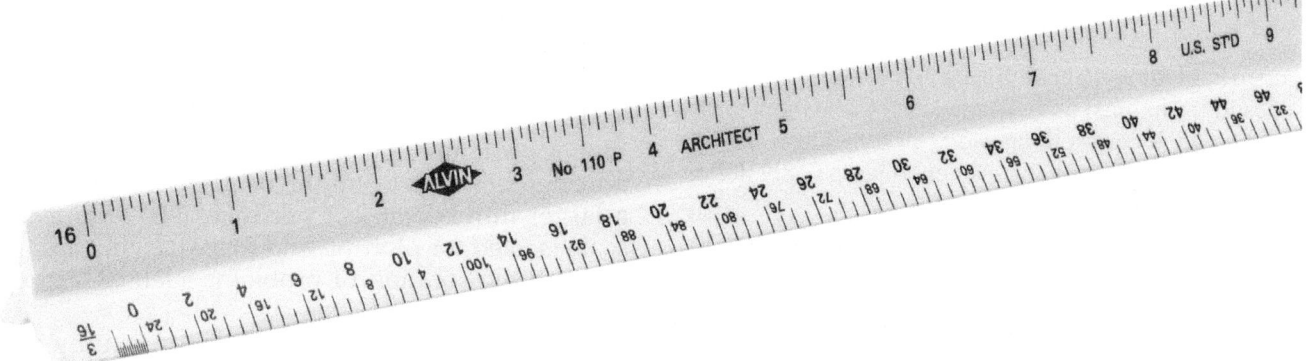

Figure 1 Architect's scale.

four different scales; if it is triangular, it may contain 10, 11, or 12 scales depending on the manufacturer. The triangular form is commonly used because it contains a variety of scales on a single tool. Plumbers can read an architect's scale from left to right or from right to left, depending on which scale a plumber is reading. For example, a plumber would read the ¼ scale from right to left and the ⅛ scale from left to right. *Table 1* lists the scales on a triangular architect's scale. The full size appears in the first row of the table. Notice that all scales are given in reference to 1'.

Each scale is set up the same way. First, plumbers see the designation of the scale itself, such as ¼ or ³⁄₃₂. The designation is followed by a series of lines, called graduations. These graduations represent exactly 1' as it is drawn with that particular scale.

Notice that the number 0 appears at one end of the 1' representation on every scale. This is where a plumber starts measuring. On the ¼ scale, for example, graduations to the left of the zero signify feet, and graduations to the right signify inches (*Figure 2*).

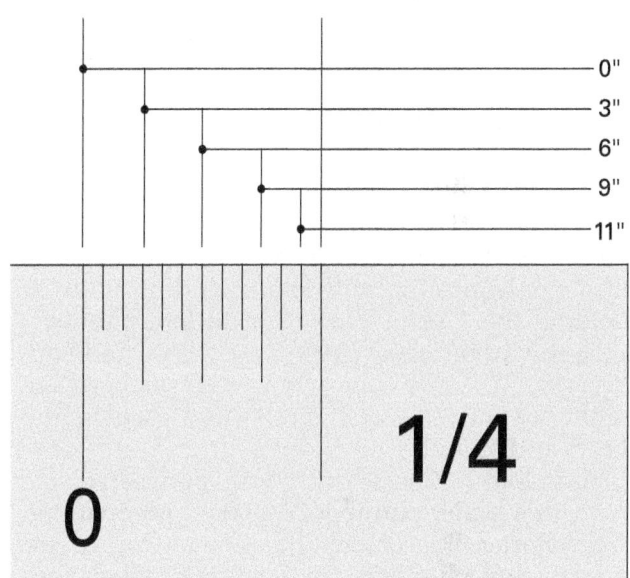

Figure 2 The ¼ scale.

Table 1 The Scales on a Triangular Architect's Scale and Corresponding Reduction

Designation	Scale	Reduction
12	12 = 1'	Full size
3	3 = 1'	One-quarter size
1½	1½ = 1'	One-eighth size
1	1 = 1'	One-twelfth size
¾	¾ = 1'	One-sixteenth size
½	½ = 1'	One twenty-fourth size
3⁄8	3⁄8 = 1'	One thirty-second size
¼	¼ = 1'	One forty-eighth size
3⁄16	3⁄16 = 1'	One sixty-fourth size
1⁄8	1⁄8 = 1'	One ninety-sixth size
3⁄32	3⁄32 = 1'	One one-hundred twenty-eighth size

Each scale has a different number of graduations to the right of zero. Not all of them signify inches. As a plumber looks at the architect's scale, the numbers may confuse a plumber. Refer back to the ¼ scale in *Figure 2*. As a plumber move to the left, they will see a line of numbers beginning with 0 and ending with 46, as well as a line of numbers beginning with 92 and ending with 0. Remember, a plumber reads the ¼ scale from right to left. Only the numbers from 0 to 46 are used on the ¼ scale. The other numbers are for the ⅛ scale, which is read from left to right.

The system is the same for every face of the architect's scale. Plumbers should rotate their scale until they see the 1 and ½ scales. Reading the ½ scale from left to right, plukmbers see a row of numbers beginning with 0 and ending with 20. Reading the 1 scale from right to left, plumbers see a row of numbers beginning with 0 and ending with 10. A plumber will see the same system on the ³⁄₃₂ and ³⁄₁₆ scales, the 1½ and 3 scales, and the ⅜ and ¾ scales.

Refer to the ¼ scale in *Figure 3*. Between 0' and 2' on this scale, plumbers see four graduations. (The shorter lines along the top belong to the ⅛ scale, but plumbers also use them when they read the ¼ scale.) Each graduation along the ¼ scale represents 6" in the length of an actual object. Thus, a line drawn from 0 to the graduation before the 2 would be 18", or 1'6" long. Plumbers can see how 1'10" is represented on the ¼ scale in *Figure 4*.

The engineer's scale is divided into decimal graduations (10, 20, 30, 40, 50, and 60 divisions to the inch). The engineer's scale is used when an area is too large to be represented by the usual scale. For example, the usual scale has a single foot represented by a portion of an inch, and an engineer's scale might be used to represent a larger number of feet per inch. The engineer's scale is used for plotting and map drawing and in the graphic solution of problems such as survey and site plans.

The metric scale is divided into centimeters (cm), with each centimeter divided into 10 millimeters (mm) or 20 half-millimeters. Some scales are made with metric divisions on one edge and inch divisions on the opposite edge. Many companies express measurements in both metric and English units.

1.1.2 Dimensions

Dimensions that appear on a drawing show actual distances. To indicate the dimension of a distance on a drawing, plumbers use a dimension line. Plumbers can create these lines with arrows or slashes at a termination line drawn perpendicular to the dimension line. Because of space constraints on the drawing, plumbers may need to use an extension line to indicate the limits of the dimension away from the actual points of the dimension (*Figure 5*). Be sure to note the point from which the dimension is taken. Some dimensions are shown from end to end, while others are shown from center lines.

1.2.0 Symbols and Notes

Drawings include symbols and notes to provide information about the installation and the piping and fixtures to be installed. Notes typically provide additional installation or product details that correspond to numbers on a drawing.

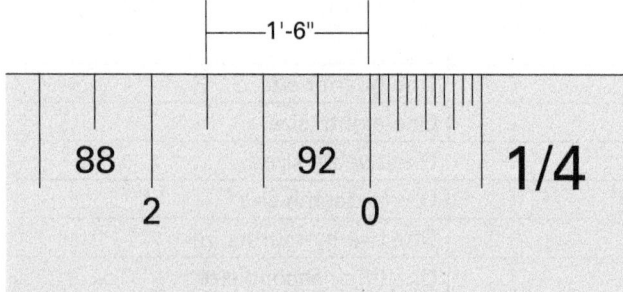

Figure 3 1', 6" on the ¼ scale.

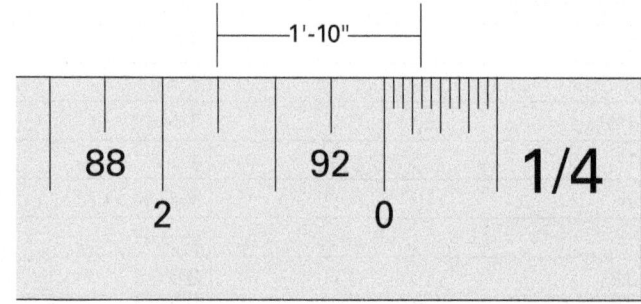

Figure 4 1', 10" on the ¼ scale.

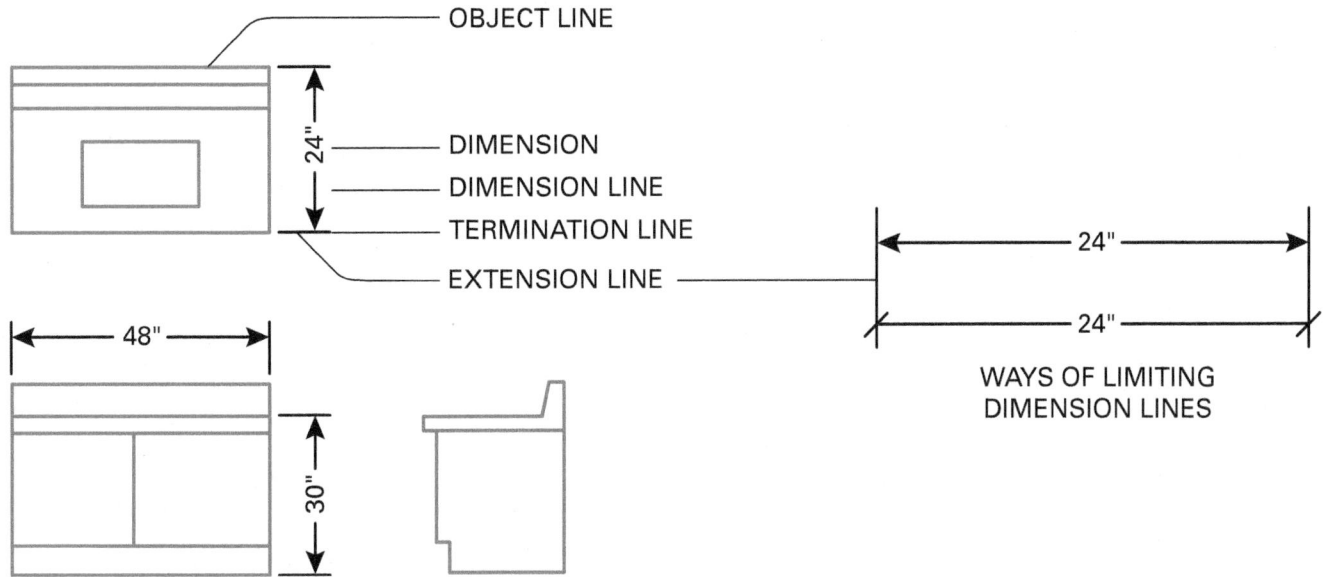

Figure 5 Extension lines and dimension lines.

1.2.1 Symbols

Plumbers often use symbols when they are drawing plans for pipe and fittings. A symbol is a mark or drawing that represents a specific object or material. Using symbols allows plumbers to produce drawings quickly and make plans easy for others to understand.

> **NOTE**
> The fitting, pipe, and fixture symbols used on drawings in your jurisdiction may vary from the ones depicted in this module.

Be aware that people do not always use the same symbols to represent the same objects. Most drawings include a legend that explains the meaning of the various symbols. Always consult the legend before reading the drawing. Seemingly small differences among symbols (*Figure 6*) actually make a huge difference in identifying various fittings.

See the typical symbols for pipe in *Figure 7*. Note that, generally, drain lines are represented by solid lines and vent lines are represented by broken lines. If the pipe symbols are combined with the fitting symbols, almost any piping system can be illustrated. Plumbers can also use symbols for plumbing fixtures to show what fixtures are connected to various runs of pipe (*Figure 8A* and *Figure 8B*).

1.2.2 Notes

Some of the information on a set of drawings is found under the heading of Notes. Notes may correspond to numbers on the drawing, and they provide more detail about the materials plumbers need and special installation requirements (*Figure 9*). Always look for the Notes section on drawings.

> **NOTE**
> Remember that people use different symbols to represent the same objects. To explain what the various symbols mean, most drawings include a legend. Always consult the legend before reading the drawing.

1.3.0 Codes and Specifications

Codes and specifications include information about the quality of the work, the materials to be used, and the method of installation. If specifications are not included in a set of drawings, plumbers follow applicable codes. Every jurisdiction adopts a specific plumbing code to ensure 432plumbing systems are installed properly.

1.3.1 Codes

The basic purpose of codes is to set the minimum standards that protect the health and safety of the people who use the system. Codes are the law in a community and are enforceable through

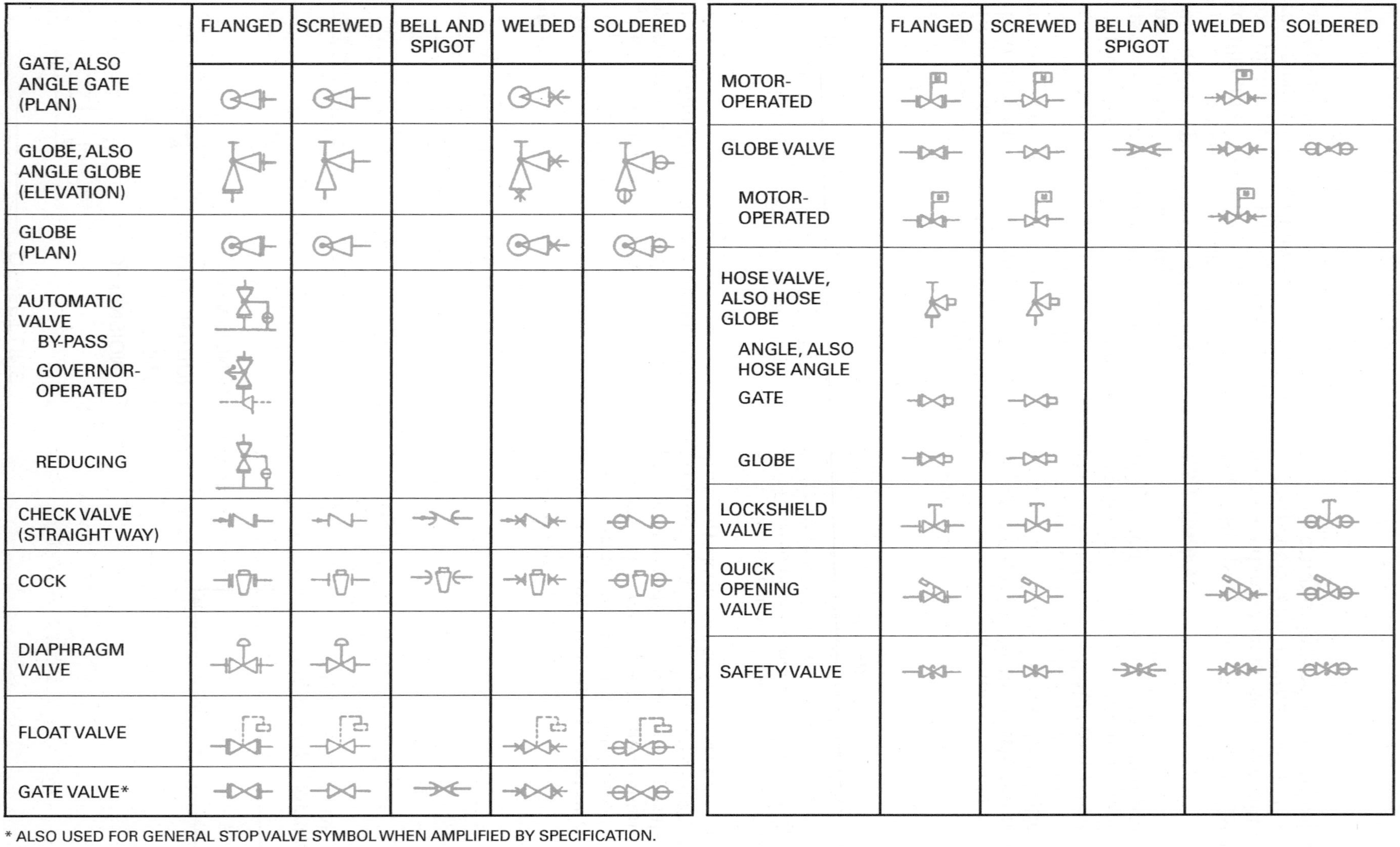

*ALSO USED FOR GENERAL STOP VALVE SYMBOL WHEN AMPLIFIED BY SPECIFICATION.

Top table

	FLANGED	SCREWED	BELL AND SPIGOT	WELDED	SOLDERED
GATE, ALSO ANGLE GATE (PLAN)					
GLOBE, ALSO ANGLE GLOBE (ELEVATION)					
GLOBE (PLAN)					
AUTOMATIC VALVE BY-PASS					
GOVERNOR-OPERATED					
REDUCING					
CHECK VALVE (STRAIGHT WAY)					
COCK					
DIAPHRAGM VALVE					
FLOAT VALVE					
GATE VALVE*					

Bottom table

	FLANGED	SCREWED	BELL AND SPIGOT	WELDED	SOLDERED
MOTOR-OPERATED GLOBE VALVE					
MOTOR-OPERATED					
HOSE VALVE, ALSO HOSE GLOBE					
ANGLE, ALSO HOSE ANGLE					
GATE					
GLOBE					
LOCKSHIELD VALVE					
QUICK OPENING VALVE					
SAFETY VALVE					

Figure 6 Fitting symbols.

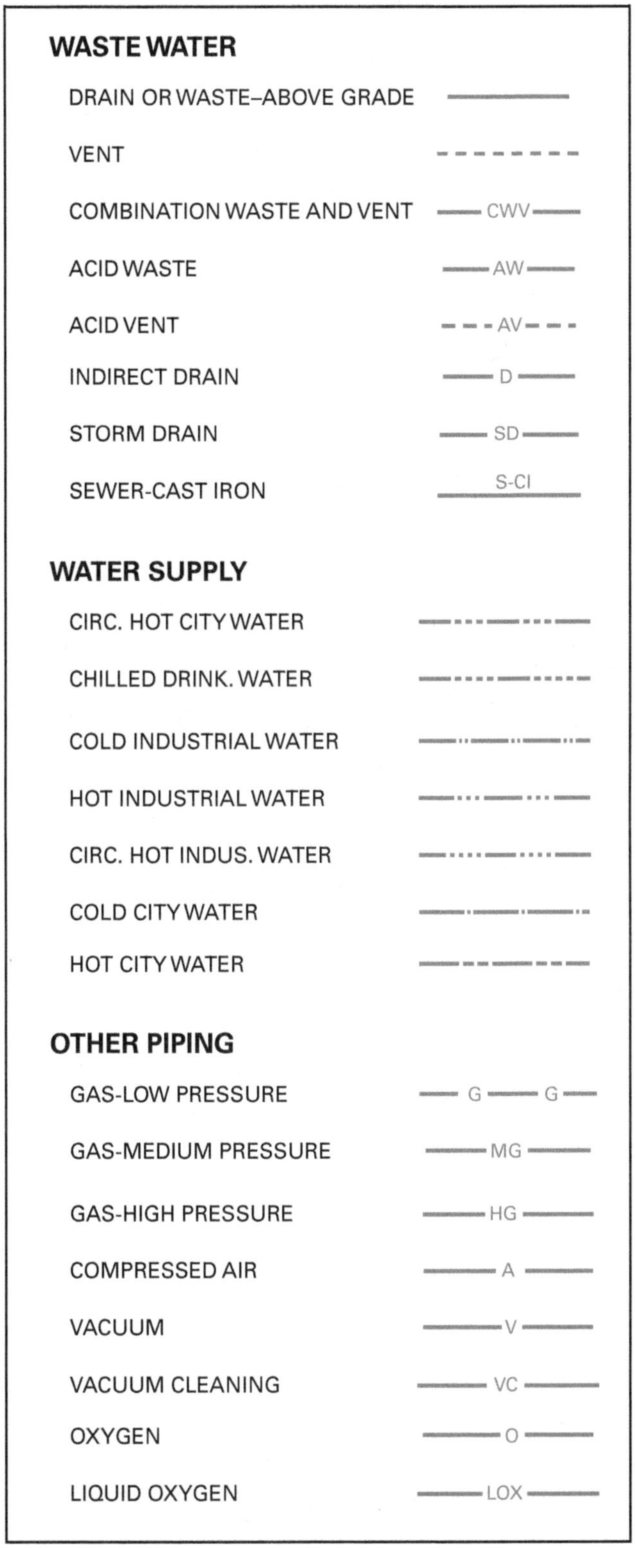

WASTE WATER

DRAIN OR WASTE–ABOVE GRADE	————————
VENT	- - - - - - - -
COMBINATION WASTE AND VENT	—— CWV ——
ACID WASTE	—— AW ——
ACID VENT	- - - AV - - -
INDIRECT DRAIN	—— D ——
STORM DRAIN	—— SD ——
SEWER-CAST IRON	S-CI

WATER SUPPLY

CIRC. HOT CITY WATER	—··—··—··—··
CHILLED DRINK. WATER	—·—·—·—·—
COLD INDUSTRIAL WATER	—··—··—··
HOT INDUSTRIAL WATER	—···—···—···
CIRC. HOT INDUS. WATER	—····—····—
COLD CITY WATER	—·—··—·—··
HOT CITY WATER	—— — —— —

OTHER PIPING

GAS-LOW PRESSURE	— G — G —
GAS-MEDIUM PRESSURE	—— MG ——
GAS-HIGH PRESSURE	—— HG ——
COMPRESSED AIR	—— A ——
VACUUM	—— V ——
VACUUM CLEANING	—— VC ——
OXYGEN	—— O ——
LIQUID OXYGEN	—— LOX ——

Figure 7 Typical pipe symbols.

Symbols for Other Trades

As plumbers work with construction drawings, they will see many types of building symbols that different building professions use. In addition to plumbing symbols, they may see architectural symbols, engineering symbols, mechanical symbols, and electrical symbols. Plumbers need to become familiar with the symbols that electricians and carpenters use, for example, so that when they see them on drawings, they will understand how they affect your work. Knowing that an electrical conduit is located close to your plumbing pipes, for example, may allow them to prevent costly errors during installation.

the courts. Each state, county, city, or other political entity establishes an Authority Having Jurisdiction (AHJ) that is authorized to enforce the plumbing codes adopted by that community. The AHJ may be an agency, department, or board.

The plans and specifications for a particular building may call for installation that is of higher quality (that is, above the minimum standards) than that required by applicable codes, which is perfectly acceptable. But any plan or specification that calls for a plumbing installation of lower quality than that required by the applicable codes is in violation and should be addressed through proper channels. Procedures are usually included in the architectural specifications, also called the request for information or RFI.

Plumbing Codes

Plumbing codes have been developed to protect the water supply and public health and safety. Most cities, counties, and states have adopted model codes that are based on suggested national standards. These model codes may be subject to local interpretation and usually reflect local conditions. For example, an area that often has earthquakes or floods usually has special requirements to deal with these conditions. The specifications for each job should detail these requirements. Installers take responsibility for correcting code violations, which can be expensive. While standards are guidelines to improve performance or reliability, codes and ordinances are mandatory, and must be followed.

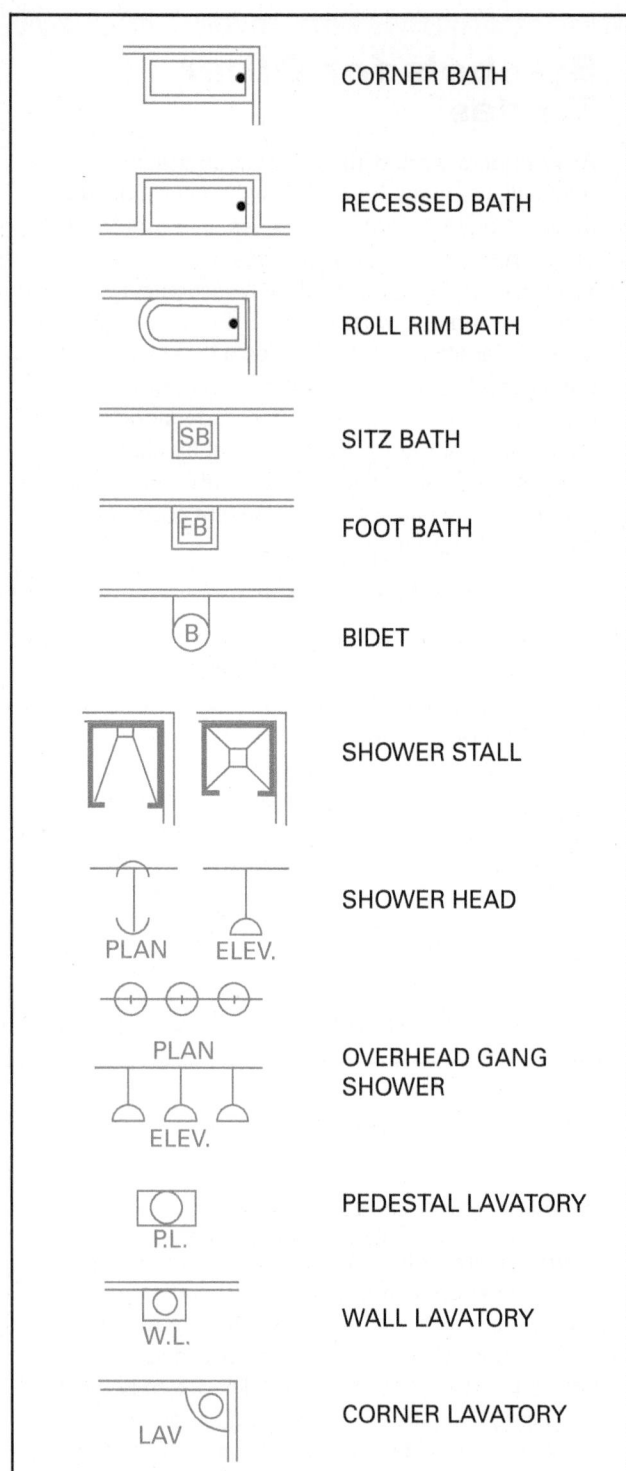

	CORNER BATH
	RECESSED BATH
	ROLL RIM BATH
SB	SITZ BATH
FB	FOOT BATH
B	BIDET
	SHOWER STALL
PLAN ELEV.	SHOWER HEAD
PLAN ELEV.	OVERHEAD GANG SHOWER
P.L.	PEDESTAL LAVATORY
W.L.	WALL LAVATORY
LAV	CORNER LAVATORY

ML	MANICURE LAVATORY
DENTAL LAV.	DENTAL LAVATORY
S	PLAIN KITCHEN SINK
	KITCHEN SINK, R & L DRAIN BOARD
	KITCHEN SINK, LH DRAIN BOARD
	COMBINATION SINK AND DISH WATER
S & T	COMBINATION SINK AND LAUNDRY TRAY
SS	SERVICE SINK
	WASH SINK, WALL TYPE
	WASH SINK

Figure 8A Fixture symbols.

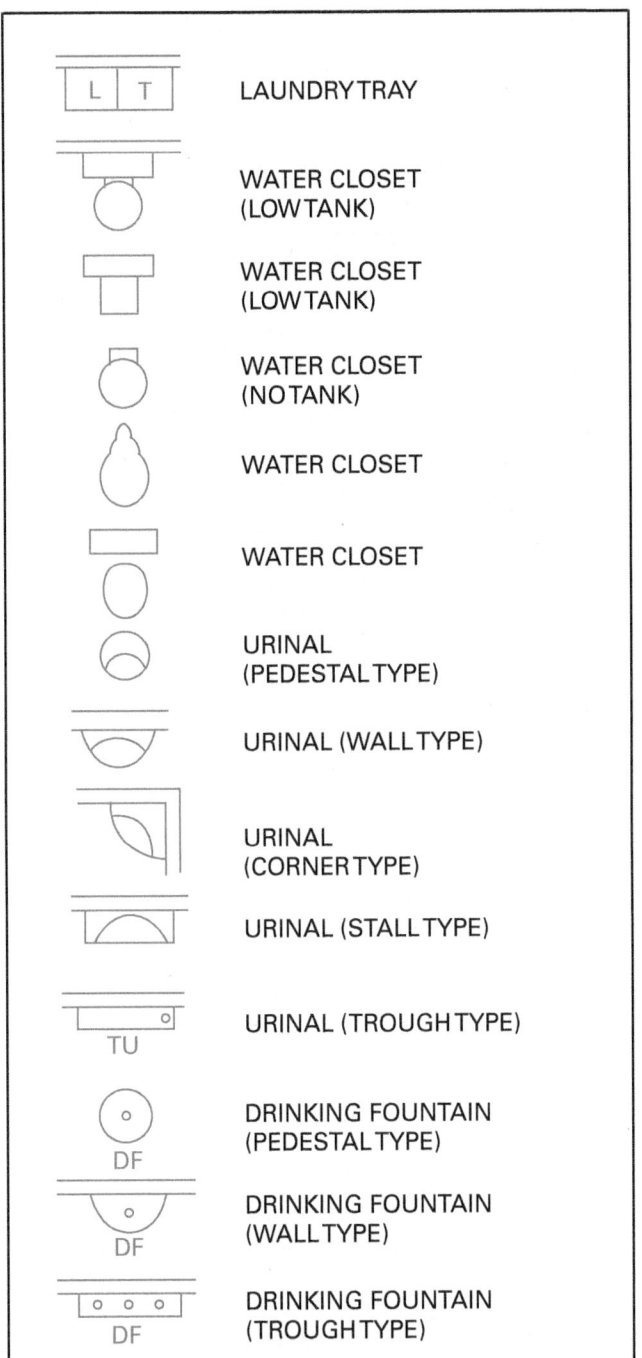

	LAUNDRY TRAY
	WATER CLOSET (LOW TANK)
	WATER CLOSET (LOW TANK)
	WATER CLOSET (NO TANK)
	WATER CLOSET
	WATER CLOSET
	URINAL (PEDESTAL TYPE)
	URINAL (WALL TYPE)
	URINAL (CORNER TYPE)
	URINAL (STALL TYPE)
	URINAL (TROUGH TYPE)
	DRINKING FOUNTAIN (PEDESTAL TYPE)
	DRINKING FOUNTAIN (WALL TYPE)
	DRINKING FOUNTAIN (TROUGH TYPE)
HWT	HOT WATER TANK
WH	WATER HEATER
M	METER
HR	HOSE RACK
HB	HOSE BIBB
G	GAS OUTLET
	VACUUM OUTLET
D	DRAIN
G	GREASE SEPARATOR
O	OIL SEPARATOR
C/O	CLEANOUT
	GARAGE DRAIN
	FLOOR DRAIN WITH BACKWATER VALVE
	ROOF SUMP

Figure 8B Fixture symbols.

```
NOTES BY SYMBOL "○"
─────────────────────────────────

①  CAP EXISTING INLETS AND OUTLETS
    TO EXISTING ACID DILUTION BASIN
    BELOW FLOOR. REWORK EXISTING
    ACID WASTE LINE TO CONNECT
    TO NEW 4" ACID WASTE FROM
    ABOVE.

②  REWORK EXISTING 3" AW TO
    CONNECT TO NEW 4" AW
    DOWN.

③  REFER TO ARCH. DRAWINGS FOR
    FURRINGS

④  2½ " CW, 1½ " HW–UP TO 1ST FLOOR CLG.
    ¾ " HWR DN. TO BELOW FLOOR
    RE: ARCH. DRAWINGS FOR FURRING.

⑤  SLEEVE STRUCTURAL WALL IN CRAWL SPACE.
    REFER TO STRUCTURAL DRAWINGS.

⑥  2" WASTE & 4" AW DOWN IN WALL/CHASE.
    COORDINATE WITH HVAC PIPING.

⑦  REFER TO ARCHITECTURAL DRAWINGS FOR
    WALL FURRING.
```

Figure 9 Sample notes.

1.3.2 Specifications

Specifications, also called *specs*, provide detailed information about the quality of workmanship required on a project as well as the specific materials and method of installation plumbers must use. The person on the project who has been designated responsible for overseeing the project will have a set of specs. Specs may also indicate that the worker must follow certain practices in testing the plumbing system. Not all drawings contain a separate set of specs. In that case, workers should follow the applicable plumbing codes for the type and quality of materials and installation practices that must be used. Often, workers will find that the plans and specifications for a building are not complete. For example, plumbing drawings may not be included beyond the location of fixtures on the floor plan. When this happens, it is the plumber's responsibility to design the piping system so that it meets all applicable requirements. Plumbers need much more knowledge about plumbing codes and sizing of pipe before they can undertake this responsibility.

1.4.0 Components of Construction Drawings

A complete set of construction drawings typically includes a plot plan, a foundation plan, floor plans, elevation drawings, details, electrical drawings, HVAC (heating, ventilating, and air conditioning) drawings, and plumbing drawings. Construction projects involve professionals from many different building professions, from architects to engineers to plumbers. Construction drawings form the basis of agreement for how a building will be built. To complete a plumbing installation, a worker will need to understand how to read all types of drawings.

1.4.1 Plot Plan

The plot plan, also known as a *site plan*, shows the location of the building on the property and shows where utilities, such as the water, storm sewer, sanitary sewer, and gas mains, are installed (see the large-scale example in the first drawing in the Appendix). The plot plan also indicates where workers cannot build on certain parts of the property. For example, setback requirements establish the minimum distance that must be maintained between the property line and the structure. An easement specifies places on the property where utilities can be installed (see *Figure 10*). Codes may set a minimum width for side yards to provide access to rear yards, to reduce the possibility of a fire jumping from one building to another, or to promote ventilation. In a set of construction drawings, plot plans are labeled A in the title block.

1.4.2 Foundation Plan

The foundation plan shows the lowest level of a building, including concrete footings, slabs, and foundation walls. It may also show the drainage, waste, and vent (DWV) piping and the water supply piping that must be installed below the first floor. A foundation plan may also indicate the positions of floor drains and other piping that must be installed beneath the basement floor. In a set of construction drawings, foundation plans may be labeled S in the title block.

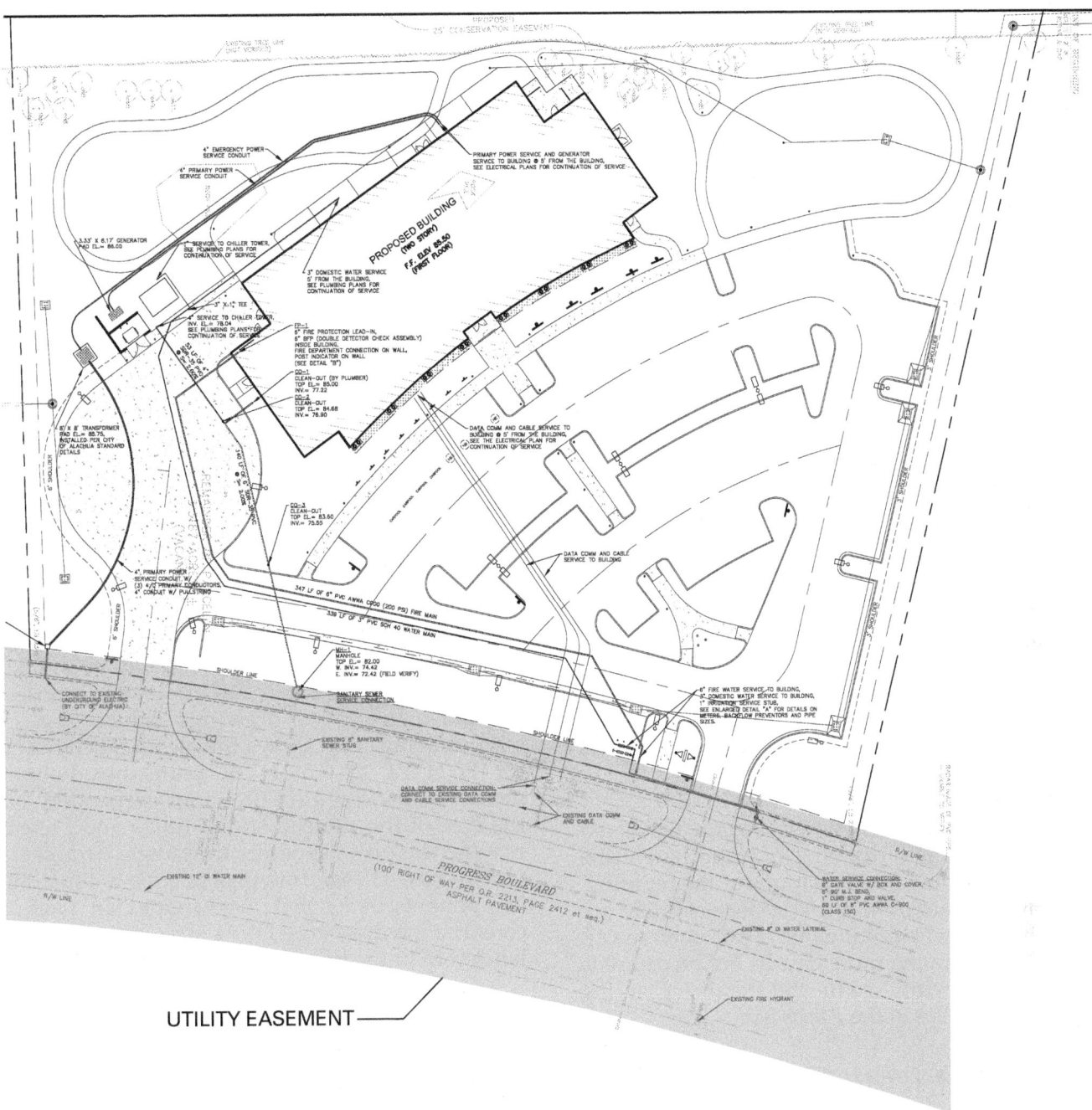

Figure 10 Utility easement.

1.4.3 Floor Plan

The floor plan shows an aerial view of the layout of rooms on each level of a building and the shape and size of each room (see the large-scale example in *Figure A02* in the *Appendix*). Think of the floor plans as showing a view of a building from above if the roof were removed. Plumbers will use floor plan drawings to locate the position of water supply and waste piping. In a set of construction drawings, floor plans may also be labeled A in the title block.

1.4.4 Elevation Drawings

Elevation drawings show the vertical elements of a building. Elevation drawings and approved submittal data (list of approved fixtures and fittings) should relate to the dimension sheets and match up to the architectural elevations and details. This ensures a proper fit and accurate installation. They show the height and width of interior or exterior walls. Plumbers will find the elevation drawings in the overall set of construction drawings for a project. See an example of an interior elevation drawing in *Figure 11*.

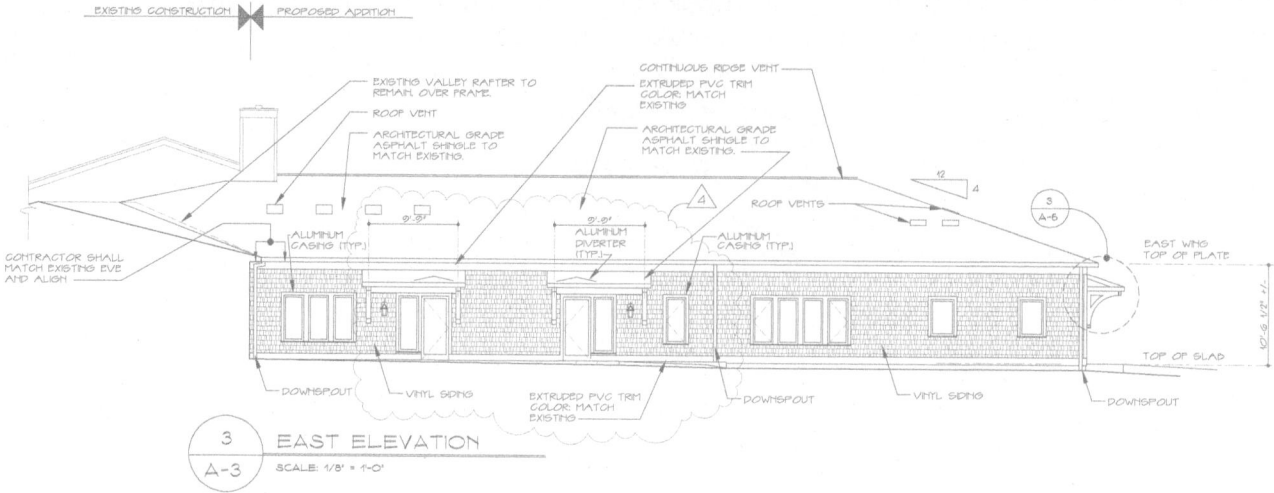

Figure 11 Detail of elevation drawing.

For another example, find the interior elevations in *Figure A02* in the *Appendix*. Plumbers can use elevation drawings to find information about where exterior hose bibbs go, and where storm drainage should be terminated. In a set of construction drawings, external elevation drawings are often labeled S in the title block; internal elevations are often labeled I.

1.4.5 Details

Details are enlarged drawings, usually presented as part of the main drawing, of important structural elements or special features of the building, such as cabinets or staircases. Because they are enlarged, detail drawings let a plumber see an area of the drawing more clearly. For an example of a detail drawing, see the Sill Detail at Planting Bed in *Figure A04* in the *Appendix*.

1.4.6 Electrical Drawings

Electrical drawings are sometimes superimposed on the floor plan. They show how the lighting and power layout appear on the plan. They also show the location of outlets, switches, and electrical fixtures, including electrical controls and programmable controllers. Electrical symbols illustrate how a specific part of the system works. These drawings also explain how to execute the electrical plan and include electrical specifications. See the large-scale example in *Figure A05* in the *Appendix*. In a set of construction drawings, electrical drawings are often labeled E in the title block.

1.4.7 HVAC/Mechanical Drawings

HVAC drawings, also called mechanical drawings, show the location of the furnace and air conditioning equipment, as well as the location of ducts and registers or piping and radiators. Like electrical drawings, they are also superimposed on floor plans. HVAC drawings include an electrical diagram that shows the electrical circuitry for the HVAC system. HVAC plans include both mechanical and control drawings. Depending on the size of the project, HVAC ducts and piping may be shown in separate drawings. Architects may redraw HVAC drawings to scale to illustrate components using their true shapes. In a set of construction drawings, HVAC drawings are often labeled M in the title block.

1.4.8 Plumbing Drawings

Like electrical and HVAC drawings, plumbing drawings are superimposed on floor plans. Plumbing drawings show the piping system, with the location of fixtures and pipe runs and the size and type of pipe to be installed (see the large-scale example in *Figure A06* in the *Appendix*). They show the diameter of each run of pipe and the fittings required to make the installation. A plumber can estimate the length of each run of pipe by looking at the measurements on the floor plans, but they will not know the exact length of pipe until they are actually installing it. In a set of construction drawings, plumbing drawings are often labeled P in the title block.

When looking at plumbing drawings, understand that it is standard practice to install the cold-water supply at the right of each fixture. Having the cold water controlled by the right-hand faucet and the hot controlled by the left-hand faucet prevents burns, because people are familiar with this setup.

1.4.9 Coordination Drawings

A coordination drawing is a dimensioned drawing with elevations and sections that, if needed, indicate proposed routing of system components. Based as nearly as possible on contract drawings, the intent of a coordination drawing is to identify and resolve conflicts that may occur with building elements and other trade contractors. This drawing should also include the actual equipment that will be provided as well as clearances required by that equipment.

Additional Resources

Core Curriculum, Latest Edition. NCCER. New York, NY. Pearson.

Hammer's Blueprint Reading Basics, 2017. Charles A. Gillis and Warren Hammer. New York: Industrial Press.

1.0.0 Section Review

1. The instrument used to draw and measure scaled drawings is a(n) _____.

   a. dimension scale
   b. architect's scale
   c. orthographic scale
   d. plumber's scale

2. On a plumbing drawing, drain lines are usually represented by _____.

   a. solid lines
   b. broken lines
   c. dashed lines
   d. dotted lines

3. Compared to the requirements specified in applicable codes, it is acceptable for the plans and specifications for a particular building to call for an installation that is of _____.

   a. lower quality than the applicable codes
   b. higher quality than the applicable codes
   c. lower cost than the industry standard
   d. higher cost than the industry standard

4. The location of the sewer and water mains and the location of the building on the property is shown on the _____.

   a. foundation plan
   b. plot plan
   c. elevation drawing
   d. plumbing drawing

2.0.0 TYPES OF DRAWINGS

Objective

Identify the different types of drawings used to install plumbing systems.

a. Identify the types of pictorial drawings used by plumbers.
b. Identify schematic diagrams.
c. Describe the purpose of orthographic drawings.
d. Identify plumbing-specific drawings, including submittals, fixture drawings, exploded views, and cutaways.

Performance Task

1. Sketch orthographic and isometric drawings.

Trade Terms

Catalog drawing: A drawing of a plumbing fixture that is found in manufacturers' catalogs. Also referred to as submittal data.

Cutaway drawing: A section drawing that shows the construction elements of a particular part of a building or fixture.

Exploded drawing: A drawing that shows how to assemble a complex product. It also shows the relationship of the individual parts to the object as a whole. Also referred to as an assembly drawing.

Fixture drawing: A drawing that shows the components of a fixture in detail.

Oblique drawing: A pictorial drawing that shows the shape of an object. It shows the front of the object with the body of the object at a slight angle.

Orthographic drawing: A construction drawing that shows straight-on views of the different sides of an object. Orthographic drawings are used for elevation drawings.

Pictorial drawings: Drawings that show a three-dimensional view of an object.

Riser diagram: A drawing that shows vertical and horizontal piping along with sizes and a riser number that refers back to the full set of plumbing drawings.

Single-line drawings: Plumbing drawings that use a single line to represent the centerline of a pipe. Single-line drawings can be used to represent pipe of any diameter. Also referred to as schematic drawing.

Takeoffs: Detailed lists that are compiled, based on drawings and specifications, of all the material and equipment necessary to construct a project. Such a list is also called a material takeoff.

Plumbers will work with many different types of drawings. They must be able to read and create simple sketches as well as complex isometric drawings (drawings that create the illusion of a three-dimensional object). Even though each type of drawing may use a different technique to convey information, drawings have many similarities.

2.1.0 Pictorial Drawings

Pictorial drawings show a three-dimensional view of an object. Much like a photograph, pictorial drawings present a picture of the object that is very similar to the actual object. Plumbers use two types of pictorial drawings: an isometric drawing and an oblique drawing (showing an object's shape).

2.1.1 Isometric and Oblique Drawings

An isometric drawing is a pictorial drawing in which all vertical lines are shown vertically but all horizontal lines are projected at a 30-degree angle and appear to go back into the horizon. As explained earlier, isometrics create an illusion of three dimensions. Plumbers encounter isometric drawings often.

When converting floor plan drawings to isometrics, remember that any horizontal lines from the floor plan convert to 30-degree angled lines on an isometric drawing. A drawing of a simple box may be the easiest way to explain this process because it relates directly to the shape of most buildings and rooms (*Figure 12*). Note that the vertical edges on the box remain vertical in the sketch. Lines that are horizontal in the front view are drawn 30 degrees from horizontal. Lines that are horizontal in the right-side view are also drawn at 30 degrees from horizontal. The dashed lines indicate the hidden back and bottom edges of the box.

Building Information Modeling (BIM)

Traditionally, buildings and their components have been represented in drawings—first on paper, and more recently in CAD files. However, drawings have some big drawbacks: because it takes many drawings to show an entire building, errors can creep in. Moving or changing a part may not show clearly how it will affect other parts of the structure. And the properties of a part—its cost, dimensions, weight, and material—are listed separately from the part itself and might not be up to date.

All of this means wasted time and money. An August 2004 report by the US National Institute of Standards and Technology (NIST) estimated that every year, the nation's construction industry loses at least $15.8 billion because of the inefficiency of paper-based drawings.

Building Information Modeling, or BIM, uses sophisticated computer databases to pull all of those things together into a dynamic model of an entire structure. BIM allows designers to experiment with changes before they're built, to see immediately how those changes will affect the rest of the building and whether they will save on time, materials, and labor. It helps identify errors and conflicts before they happen. It keeps track of parts inventory and prices, so that cost estimates and inventory control are much easier to perform. It helps contractors schedule the construction more efficiently for all the trades involved. BIM can even show a plumber where sunlight will fall in a room in the morning, afternoon, and evening, how energy efficient the room is, and let them test how sounds will echo inside of it.

Also called *Virtual Building Environment (VBE)* and *Virtual Design to Construction Project Manager (VDC)*, BIM is much more than just 3-D rendering software. It's a whole new way of thinking about the job.

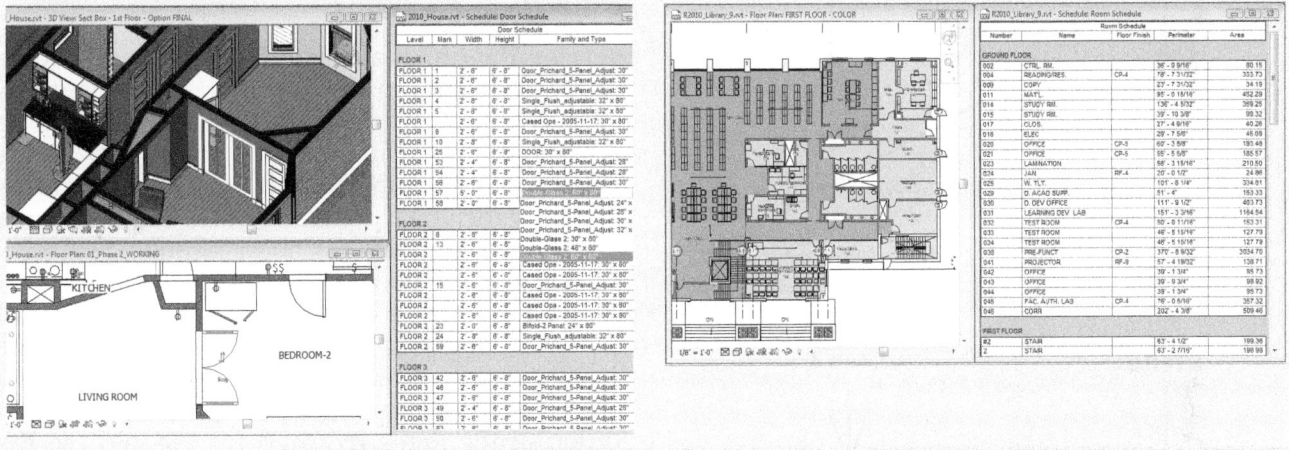

Sketches

Every plumber must be able to communicate with other workers. For example, they must be able to show another plumber how pipe is to be installed or how the building structure must be modified to install plumbing. Plumbers often make sketches on any convenient surface: scraps of paper, wood, the floor, or the wall. These simple sketches can communicate very effectively.

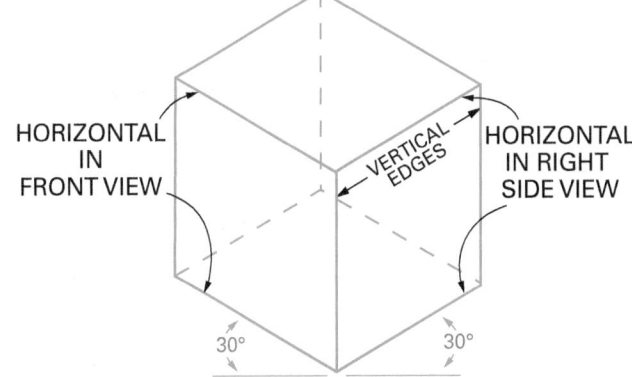

Figure 12 Isometric sketch of a single box.

Plumbers must be able to read and create isometric sketches of piping assemblies (*Figure 13*). The piping system typically needs to fit within or near the walls. Sketches can be made without including the walls as part of the sketch. However, they must know where the walls are located and position the fitting to make the required bends in the piping system.

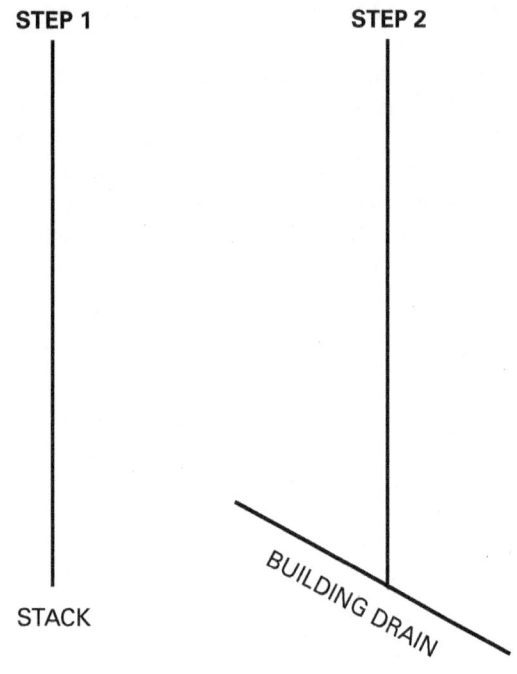

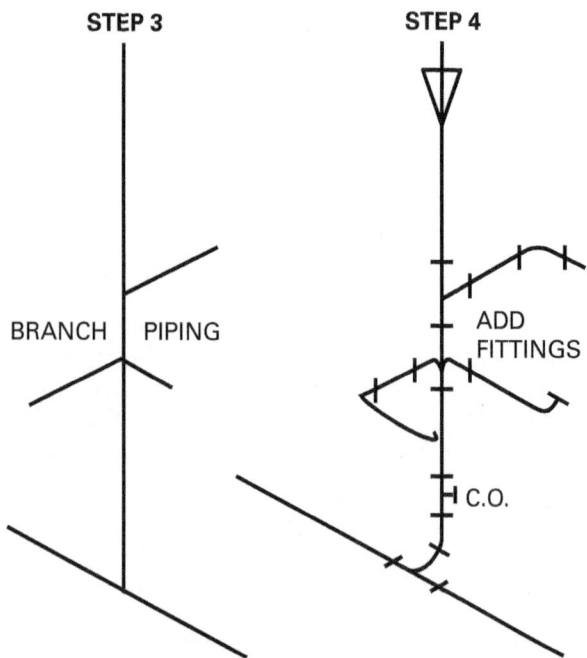

Figure 13 Steps for making sketches.

Dimensions on isometric drawings are usually not to scale. Distortion is permitted on these drawings so that they can illustrate three dimensions, and this distortion makes scaling difficult. Plumbers can use isometric paper to draft isometric drawings (*Figure 14*).

See the isometric representation of a stack vent in *Figure 15*. A full circle drawn at the end of an angled line indicates that the line continues up from that end of the horizontal line. A half-circle indicates that the pipe continues downward from the end of the horizontal line. These lines refer to vertical lines on the floor plan. To practice creating isometric drawings, draw what you see in *Figure 15* by following the steps shown in *Figure 13*.

An oblique drawing is a type of pictorial drawing that creates the illusion of depth by using lines that are drawn at 45 or 60 degrees to the horizontal (*Figure 16*).

2.1.2 Riser Diagrams

A riser diagram is a single-line isometric drawing of the plumbing in the building. Riser diagrams show vertical and horizontal piping along with sizes and a riser number that refers back to the full set of plumbing drawings. These diagrams typically include a domestic water diagram and a DWV diagram. They can show plumbing systems on one or more floors.

Riser diagrams can be used to do material takeoffs. Takeoffs are detailed lists of all the material and equipment necessary to construct a project. Plumbers use them as checklists when ordering the items required for the job (see the large-scale example in *Figure A07* in the *Appendix*).

2.2.0 Schematic Drawings

Most plumbing drawings are single-line drawings, also called *schematic drawings*. In a single-line drawing, the line represents the centerline of the pipe, so these drawings can be used to represent pipe of any diameter. See the schematic sketch of a piping stack in *Figure 17*. For comparison, refer back to the isometric representation of a stack vent in *Figure 15*. Note the similarities and differences between the two.

2.3.0 Orthographic Drawings

To communicate information about the exact size and shape of an object, designers often use an orthographic drawing, which shows dimensions that are proportional to the actual dimensions of the object. In orthographic

Figure 14 Isometric paper.

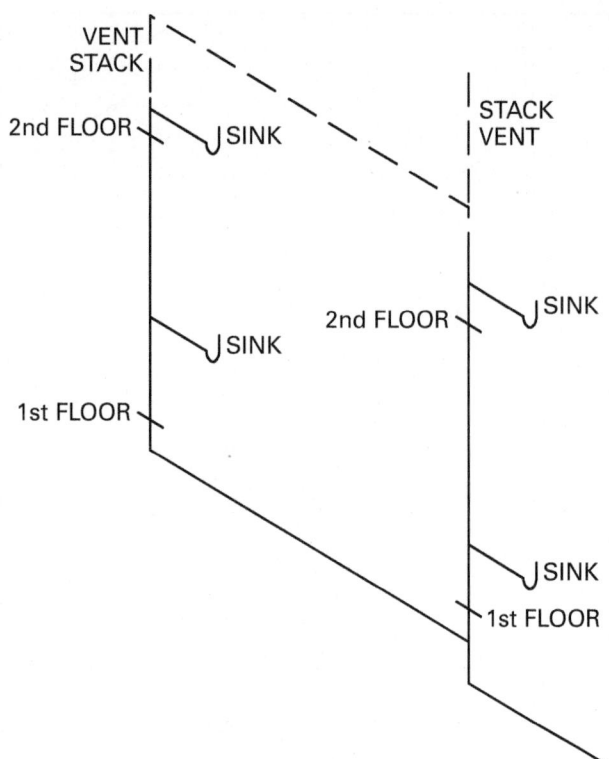

Figure 15 Isometric sketch of a stack vent.

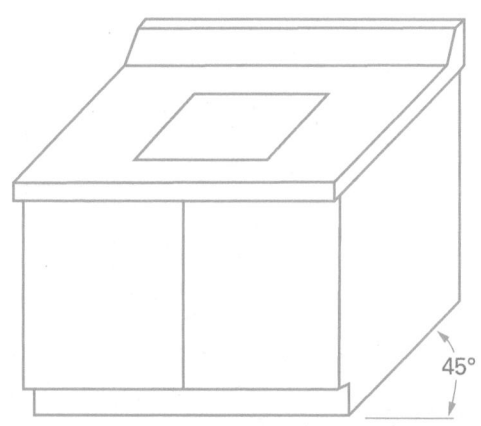

Figure 16 Oblique drawing.

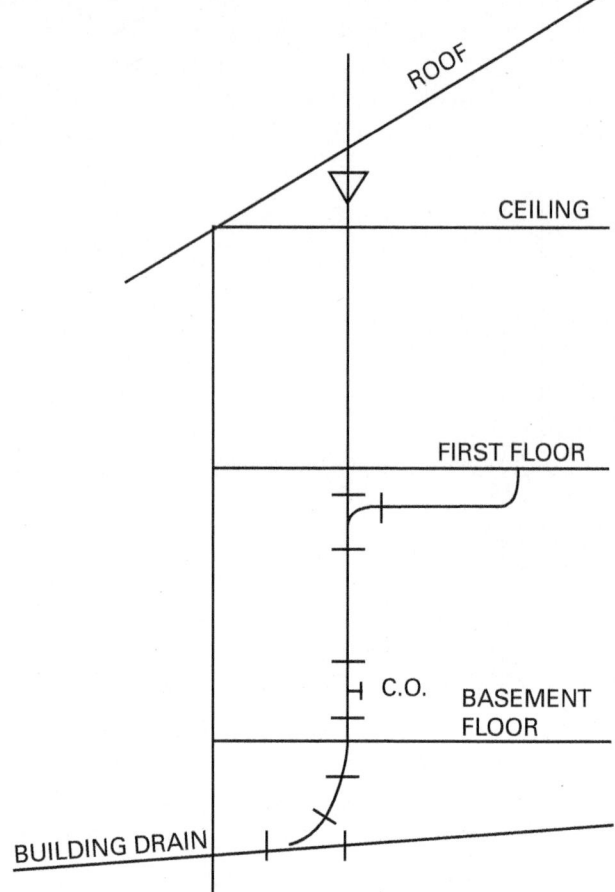

Figure 17 Schematic sketch of a piping stack.

drawings, designers draw lines that are scaled-down representations of real dimensions. Every 12", for example, may be represented by ¼" on the drawing. An orthographic drawing is created by viewing the object from one or more of six possible directions and recording how the object looks in each of the views (*Figure 18*).

Plumbers can create more views of the object by looking at it from other angles (*Figure 19*). Although plumbers can present as many as six different views, generally they will not need more than three to show the object completely.

To make orthographic drawings understandable, designers often draw the views in the proper relationship to one another. Notice in *Figure 20* how the views from different angles are related to the finished drawing.

To read an orthographic drawing, plumbers have to learn to visualize what the finished object looks like from the views provided, a skill requires a lot of practice. Examine the orthographic and isometric drawings of the same object in *Figure 21*. Study and compare these drawings carefully. Also look at the two drawings in *Figure 22*. One is a schematic representation of a piping assembly; the other is an orthographic drawing of the same thing. Note the differences between these two drawings.

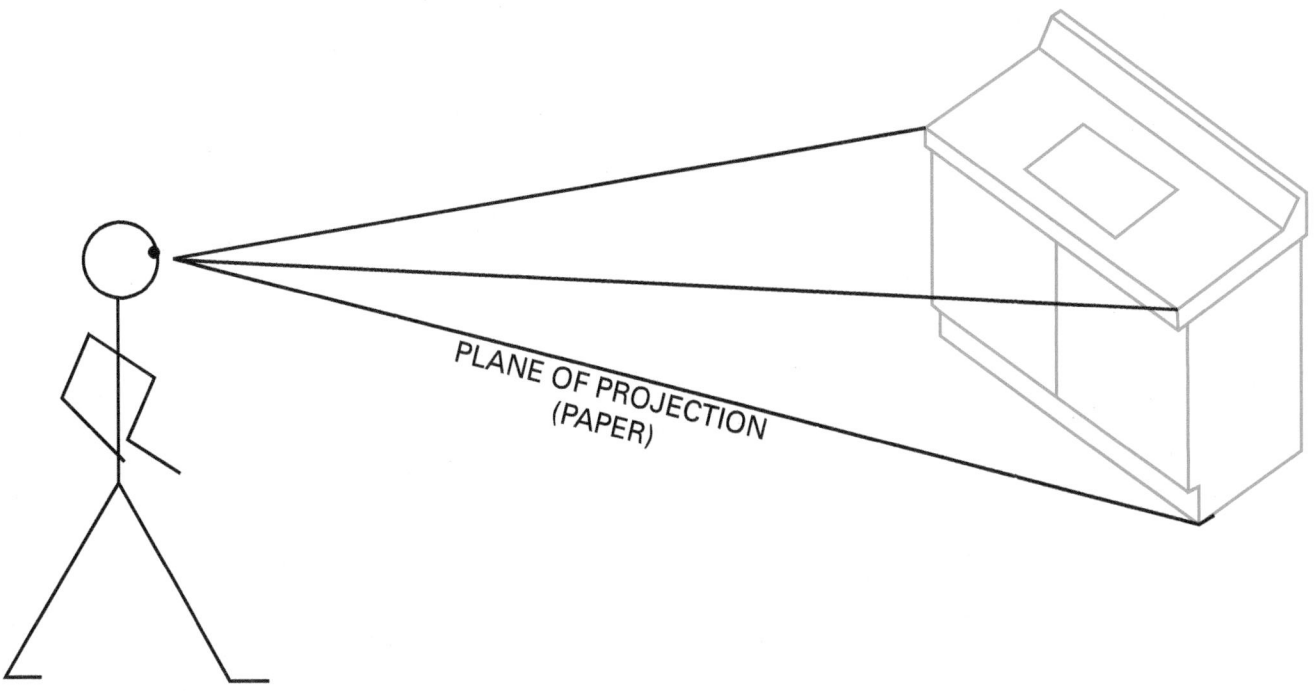

Figure 18 Orthographic projection.

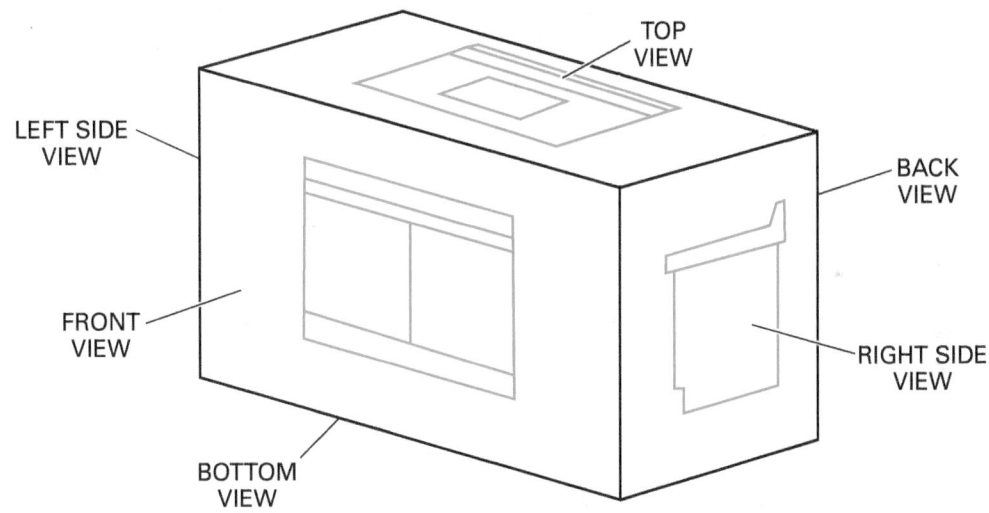

Figure 19 Additional planes for orthographic projection.

2.4.0 Types of Plumbing Drawings

Most plumbing drawings include riser diagrams, floor plans, isometric views, and details. Plumbers work with many different types of drawings. For example, plumbers must be able to find the size and dimensions of fittings from manufacturers' catalogs, and they must also be able to interpret details about assemblies. Study each drawing in this section carefully and think about what information it contains and how it shows that information.

2.4.1 Approved Submittal Data

Manufacturers of pipe and fittings publish catalogs that contain information about their products. Many of them publish dimensional catalogs that show each of their fittings along with overall sizes and, in some cases, weights. Plumbers use a catalog drawing, also called *submittal data*, to determine such factors as overall dimensions, actual inside dimensions, fitting depth, and laying length of the fittings. When the architect or engineer approves the fixture or fitting in a catalog drawing, it becomes approved submittal data.

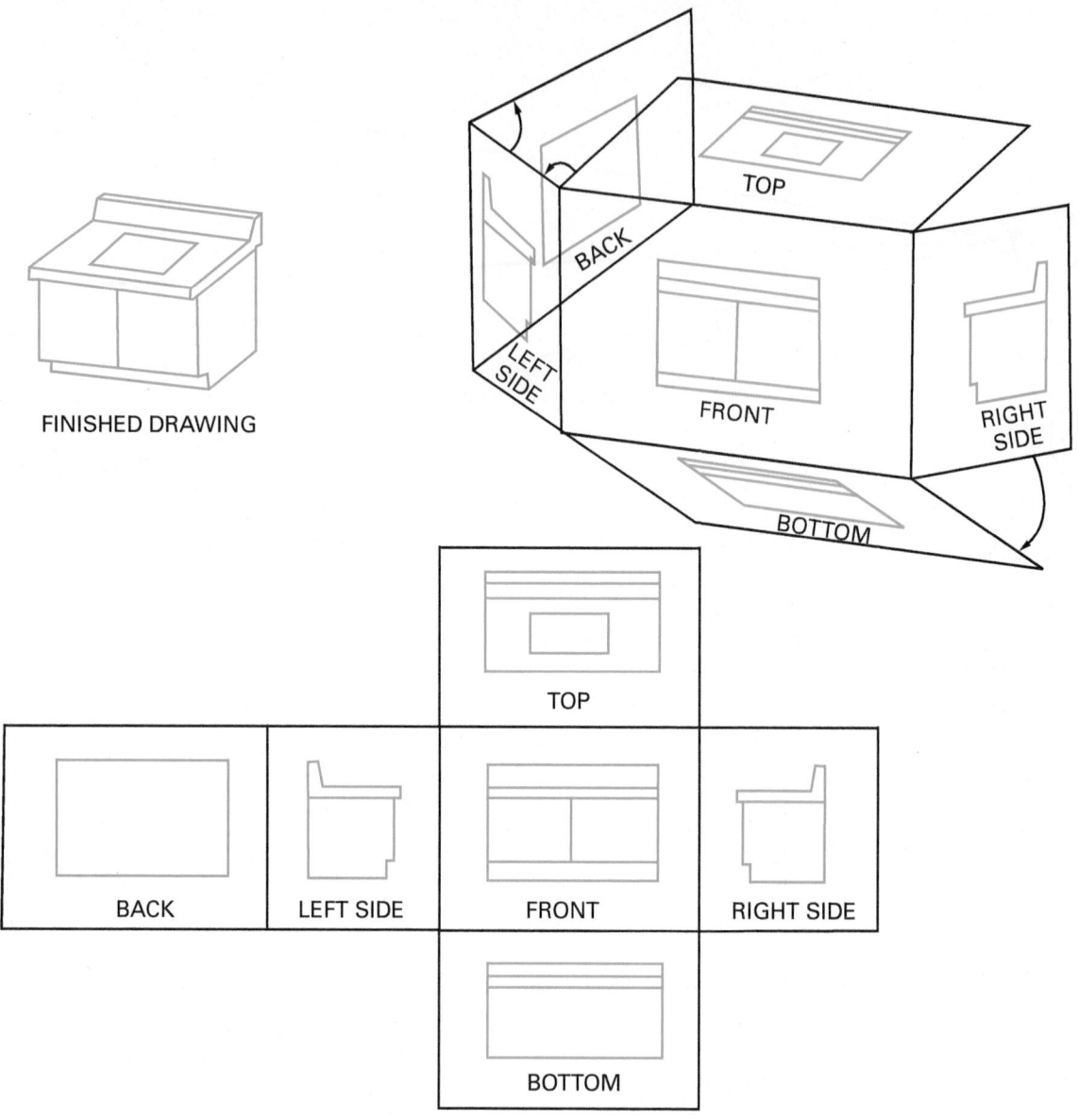

Figure 20 Relationship of orthographic views.

Figure 23 shows an example of a plastic DWV system P-trap with a solvent-welded joint from a catalog. The method used in this drawing to show the dimensions of the fitting is typical. The drawing uses capital letters (A through E) that are keyed to a table. The table provides information about available sizes. This method is efficient because manufacturers can use the same drawing to represent fittings of different sizes.

Figure 24 uses a similar technique to give information about a cast-iron wye. The table provides information for 39 separate fittings. By checking the various combinations of pipe sizes (G, X, and X'), plumbers can find a combination that will work for their installation. (Note that the SV and XH in the table stand for service weight and extra-heavy pipes, respectively.)

2.4.2 Fixture Drawings

A plumber will often use a fixture drawing, also called cut sheets, to determine whether a certain fixture will fit in the available space. See the fixture drawing of a typical water closet in *Figure 25*. The drawing shows the overall dimensions and the location of the water supply piping. The symbol CL stands for center line and shows the measured center of an object such as a pipe or plumbing fixture.

Figure 26 is a drawing of a typical lavatory. Two views are required to show the size and shape of this lavatory. The dimensions for roughing-in the water supply pipes and the drains vary depending on the type of drain installed. Study this drawing and answer the following questions:

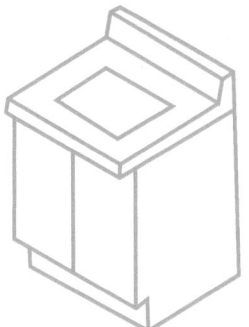

PICTORIAL

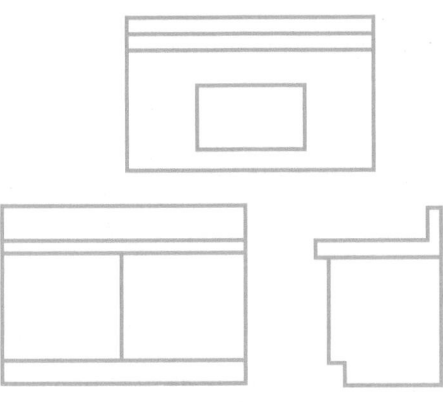

ORTHOGRAPHIC

Figure 21 Isometric (top) and orthographic views of the same object.

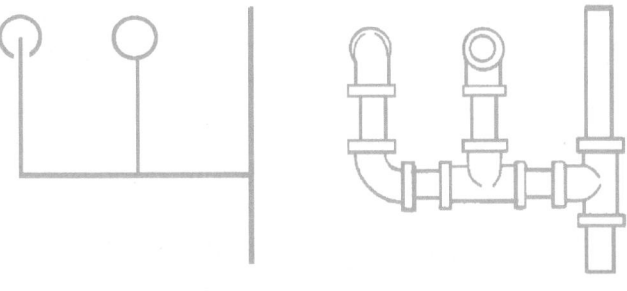

SCHEMATIC   ORTHOGRAPHIC

Figure 22 Schematic and orthographic drawings of the same piping assembly.

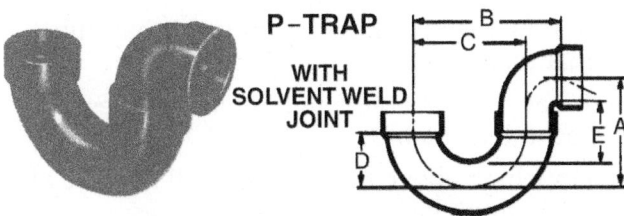

P-TRAP

WITH
SOLVENT WELD
JOINT

SIZE	DESCRIPTION H x H	A	B	C	D	E
1½		3¾	5⁷⁄₃₂	4	1¹³⁄₁₆	2⅛
2		4⁹⁄₁₆	7½	5¼	2¼	2½
3		6⁵⁄₁₆	8⁷⁄₁₆	6⁵⁄₁₆	2¹¹⁄₁₆	3¼
4		7⅞	10¹³⁄₁₆	8⅛	3½	3⅞

Figure 23 Catalog drawing of a P-trap.

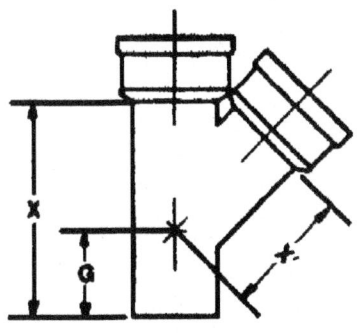

		Wye			
				Weights	
Size	**G**	**X**	**X'**	**SV**	**XH**
2 × 2	4	8	4	7	8
3 × 2	4³⁄₁₆	9	5	10	14
3 × 3	5	10½	5½	13	17
4 × 2	3⅝	9	5¾	12	17
4 × 3	4⁷⁄₁₆	10½	6¼	14	20
4 × 4	5¼	12	6¾	17	24
5 × 2	3⅛	9	6½	14	20
5 × 3	3⅞	10½	7	17	24
5 × 4	4¹¹⁄₁₆	12	7½	19	27
5 × 5	5½	13½	8	22	32
6 × 2	2⁹⁄₁₆	9	7¼	17	23
6 × 3	3⅜	10½	7¾	19	27
6 × 4	4³⁄₁₆	12	8¼	22	31
6 × 5	4¹⁵⁄₁₆	13½	8¾	24	35
6 × 6	5¾	15	9¼	28	40
8 × 2	3⅛	10½	8½	29	42
8 × 3	3¹⁵⁄₁₆	12	9	32	47
8 × 4	4¾	13½	9½	36	52
8 × 5	5½	15	10	39	57
8 × 6	6⁵⁄₁₆	16½	10½	44	63
8 × 8	7¹¹⁄₁₆	19½	11¹³⁄₁₆	55	82
10 × 3	2⁹⁄₁₆	12	11	50	71
10 × 4	3⁹⁄₁₆	13½	11⅛	53	74
10 × 5	4⁵⁄₁₆	15	11⅝	57	80
10 × 6	5⅛	16½	12⅛	61	86
10 × 8	6½	19½	13⁷⁄₁₆	77	110
10 × 10	8	22½	14½	94	133
12 × 4	4⅛	15	12⁷⁄₁₆	70	97
12 × 5	4⅞	16½	12¹⁵⁄₁₆	74	104
12 × 6	5¹¹⁄₁₆	18	13⁷⁄₁₆	80	111
12 × 8	7¹⁄₁₆	21	14¾	96	136
12 × 10	8⁹⁄₁₆	24	15¹³⁄₁₆	115	160
12 × 12	10⅛	27	16⅞	135	186
15 × 4	2¼	15	15	130	130
15 × 6	4	18	15¾	109	152
15 × 8	5⅜	21	17¹⁄₁₆	127	182
15 × 10	6⅞	24	18⅛	152	213
15 × 12	8⁷⁄₁₆	27	19³⁄₁₆	170	242
15 × 15	10¾	31½	20¾	204	290

Figure 24 Catalog drawing of a wye.

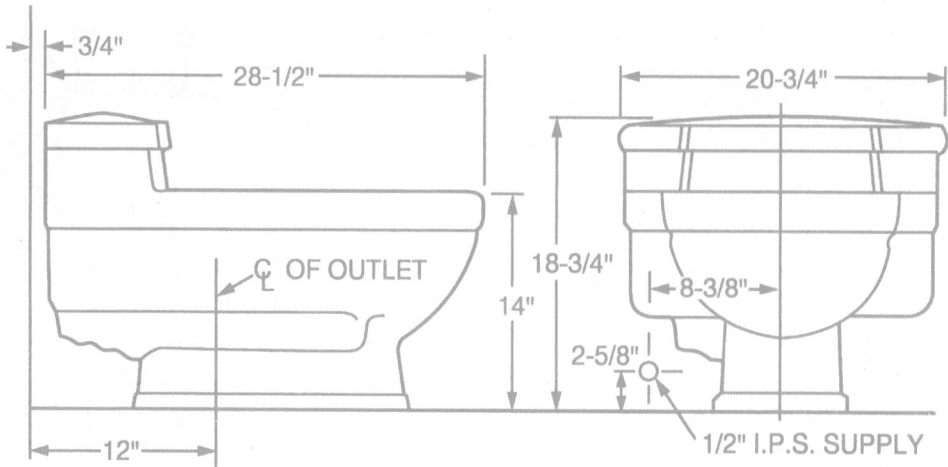

Figure 25 Fixture rough-in drawing of a water closet.

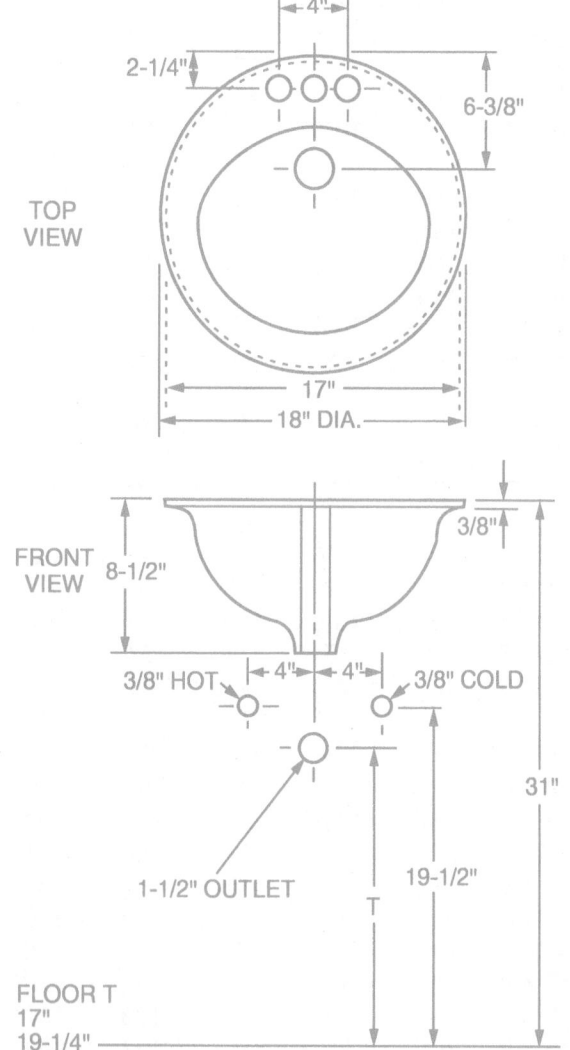

Figure 26 Fixture rough-in of a lavatory.

1. What is the diameter of the opening that must be cut out of the countertop?

2. What is the overall depth of the lavatory?

3. If a pop-up drain is used, how far from the floor should the drainpipe be located?

4. How far apart must the cold and hot water connections be on the faucet that is installed in this lavatory?

Fixture drawings such as the ones found in *Figure 25* and *Figure 26* are also called fixture rough-in sheets, manufacturers' fixture spec sheets, and cut sheets. Changes in the building structure, such as wall placement and wall covering, could affect your work. Plumbers must remeasure all fixtures before installing them to ensure proper fit.

Similar drawings are used to show other types of fixtures, such as bathtubs (*Figure 27*). The top view is typical of views shown in orthographic drawings, but the side and end views are drawn as section views. Section views illustrate the internal shape of objects. Also, part of the end view has been eliminated, and the side view has been moved to the left of its normal location. Study this drawing carefully and then answer the following questions:

1. In how many different sizes is this bathtub manufactured?

2. What is the overall length of the smallest bathtub available?

3. How high above the subfloor will the side of the bathtub be?

4. Will an opening need to be cut in the subfloor to permit installation of the bathtub?

5. How far above the subfloor will the stub-out for the showerhead be located?

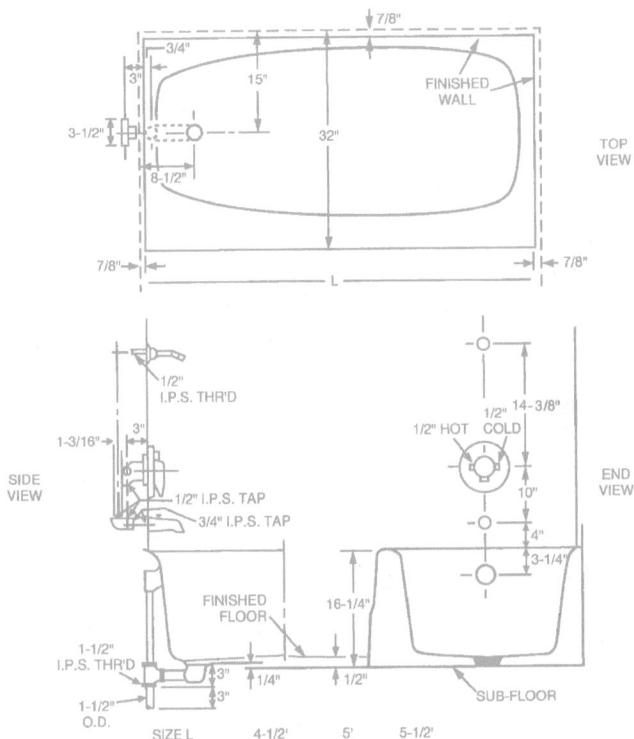

Figure 27 Fixture rough-in drawing of a bathtub.

2.4.3 Exploded Drawings

An **exploded drawing**, also called *assembly drawings*, illustrates how to assemble complex objects. They show the relationship of the individual parts to the whole object. This type of drawing may also show the illustrated parts breakdown (IPB). *Figure 28* shows an exploded view of a pop-up assembly with all its parts. Note how the parts are shown so that a reader can easily visualize how to assemble the pop-up.

Figure 29 shows more complicated exploded drawings than *Figure 28*. Each part is numbered to correspond to a number in a parts list (not shown). In this image, both a handle version and a push-button version of the flush valve are shown on the same drawing. Additionally, two

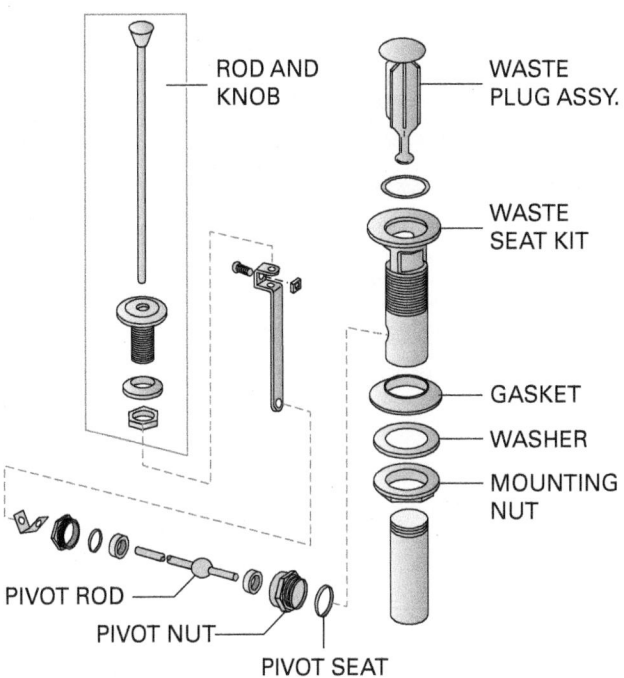

Figure 28 Exploded view of a pop-up assembly.

types of control stops may be installed. The figure also shows an electronic version of a flush valve. An exploded drawing of a kitchen faucet appears in *Figure 30*. Using these types of drawings, you should be able to disassemble and reassemble any plumbing product and identify needed repair parts.

2.4.4 Cutaway Drawings

For many products, a **cutaway drawing** provides enough detail to show how the product is constructed (*Figure 31*). Note in the drawing that each part is numbered to correspond to the parts list. Also note that the dimensions of many sizes of this valve are given in the specifications table. This method eliminates the need to make a separate drawing for each size valve.

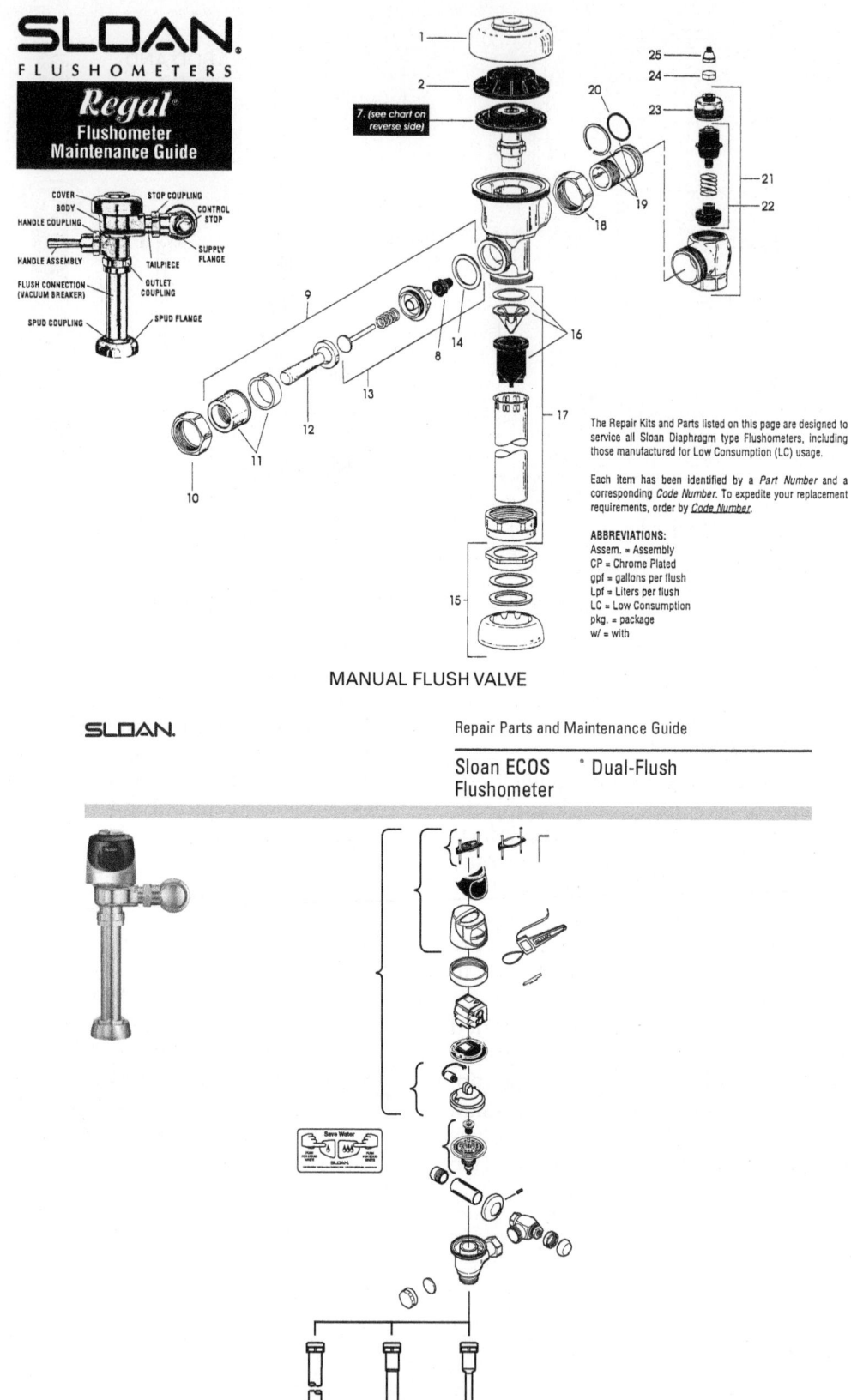

MANUAL FLUSH VALVE

ELECTRONIC FLUSH VALVE

Figure 29 Exploded drawings with IPBs of manual and electronic flush valves.

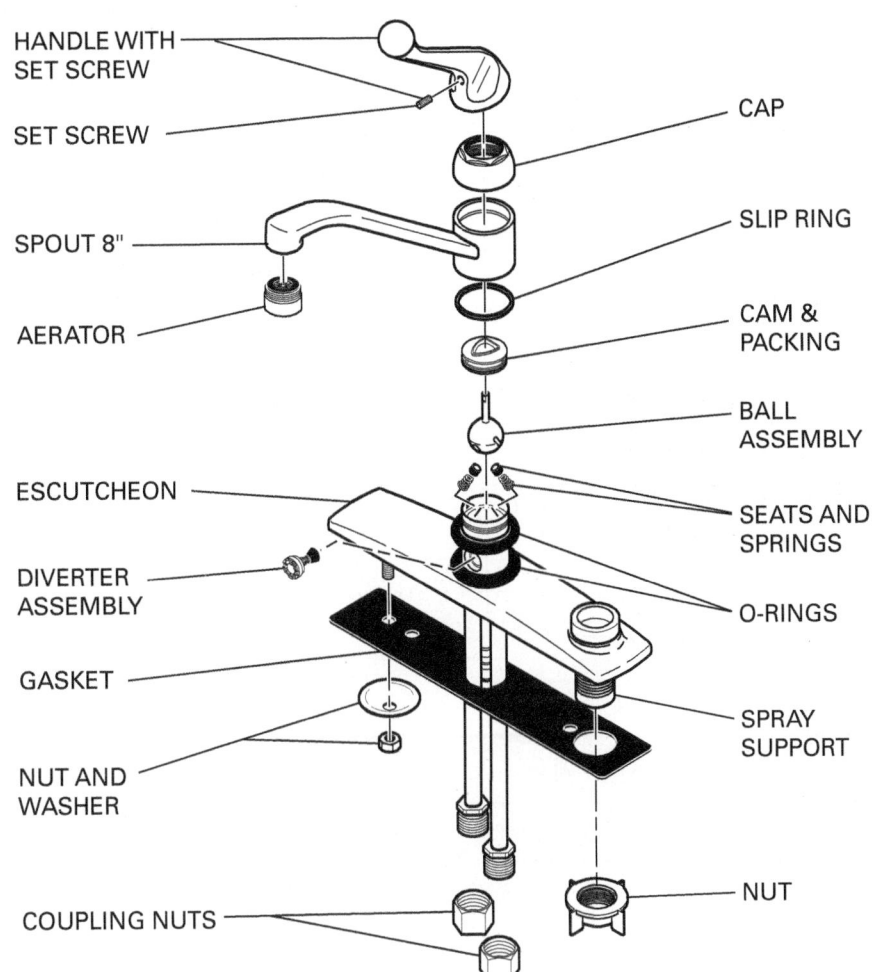

HANDLE WITH
SET SCREW

SET SCREW

SPOUT 8"

AERATOR

ESCUTCHEON

DIVERTER
ASSEMBLY

GASKET

NUT AND
WASHER

COUPLING NUTS

CAP

SLIP RING

CAM &
PACKING

BALL
ASSEMBLY

SEATS AND
SPRINGS

O-RINGS

SPRAY
SUPPORT

NUT

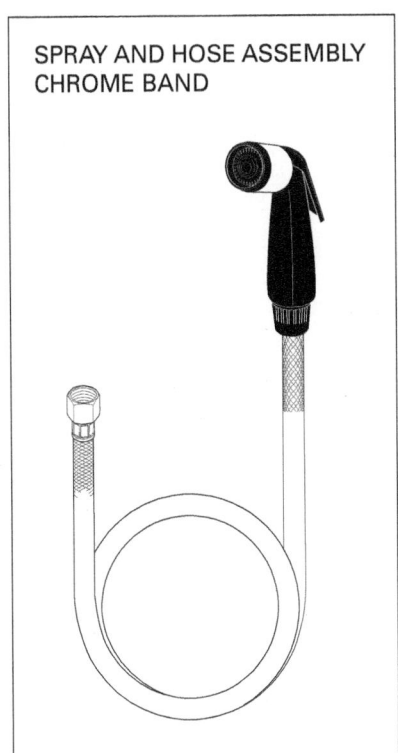

SPRAY AND HOSE ASSEMBLY
CHROME BAND

Figure 30 Exploded drawing of a kitchen faucet.

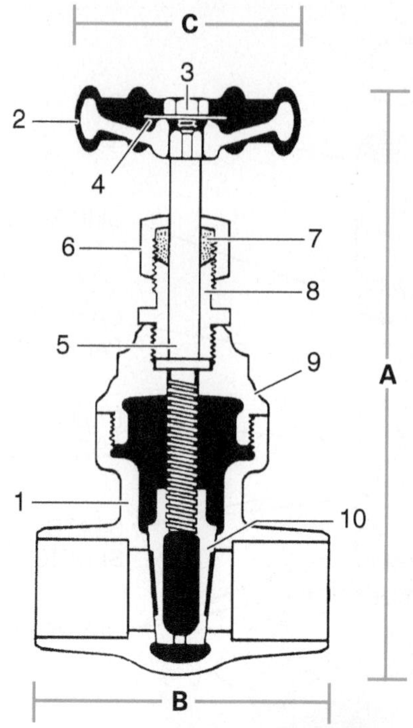

PARTS AND MATERIALS		
Item	Description	Material
1	Body	ASTM B-62
2	Handwheel	Malleable iron
3	Handwheel nut	ASTM B-16
4	Identification plate	Brass
*5	Stem	ASTM B-16, B-62
*6	Packing nut	ASTM B-16, B-62
7	Packing	Teflon®-Asbestos
*8	Stuffing box	ASTM B-16, B-62
9	Centerpiece	ASTM B-62
*10	Disc	ASTM B-62

*B-62 used in larger sizes

SPECIFICATIONS				
Valve Size	Dim. A	Dim. B	Dim. C	Weight Pounds
3/8"	4 1/8"	1 5/8"	2 1/16"	0.8
1/2"	4 1/4"	2 3/16"	2 1/16"	0.9
3/4"	4 7/16"	2 7/8"	2 1/2"	1.4
1"	5 7/16"	3 3/8"	2 1/2"	2.1
1 1/4"	6 1/4"	3 7/16"	2 1/2"	2.9
1 1/2"	7 1/16"	3 3/4"	3"	3.8
2"	8 1/4"	4 3/8"	3 1/2"	6.3
2 1/2"	9 3/16"	5 3/8"	4 1/8"	10.9
3"	11 1/8"	6 1/4"	5 1/4"	16.4

Figure 31 Cutaway drawing of a gate valve.

Additional Resources

Hammer's Blueprint Reading Basics, 2017. Charles A. Gillis and Warren Hammer. New York: Industrial Press.

Technical Drawing, 15th Edition, 2016. Frederick E. Giesecke, et al. New York, NY. Pearson.

2.0.0 Section Review

1. A representation of an object that gives the impression of three dimensions resembling the physical object is called _____.

   a. orthographic
   b. schematic
   c. isometric
   d. linear

2. Schematic drawings are also known as _____.

   a. orthographic drawings
   b. oblique drawings
   c. projection drawings
   d. single-line drawings

3. To communicate information about the exact size and shape of an object, designers use a(n) _____.

   a. projection drawing
   b. orthographic drawing
   c. oblique drawing
   d. pictorial drawing

4. To determine whether a certain fixture will fit in the available space, plumbers often use _____.

   a. fixture drawings
   b. schematic drawings
   c. orthographic drawings
   d. linear drawings

SUMMARY

Information found on plumbing drawings is critical to performing quality work. Remember that it takes many different types of drawings, along with the specifications for the project, to get a complete picture of the plumbing requirements. Pictorial drawings show depth. Orthographic drawings show objects as if you were looking at them head-on from certain angles. The ability to read and understand different kinds of drawings is an essential plumbing skill, as is the ability to make orthographic or isometric sketches and to draw plumbing assemblies. Learning the basics of reading and interpreting drawings is a building block for future study. Plumbers must practice construction-drawing reading, like all other plumbing skills, and study throughout your career.

1. The act of showing the actual building, room, or object reduced in proportion from full size to a smaller size is called _____.

   a. dimensioning the drawing
   b. laying out the drawing
   c. measuring the drawing
   d. scaling the drawing

2. To establish the limits of a measurement on a drawing, use _____.

   a. specifications
   b. scales
   c. extension lines
   d. notes

3. To check for certain practices that must be used in installing and testing a plumbing system, a plumber refers to the _____.

   a. title block
   b. scale
   c. dimensions
   d. specifications

4. Plot plans usually show _____.

   a. plumbing fixtures
   b. electrical wiring and fixtures
   c. utility easements
   d. an aerial layout of a building's rooms

5. To identify where floor drains and DWV pipes may be installed beneath the basement floor, refer to the _____.

   a. floor plan
   b. detail drawing
   c. foundation plan
   d. coordination drawing

6. The layout of rooms in a building is shown on the _____.

   a. plot plan
   b. elevation drawing
   c. foundation plan
   d. floor plan

7. To ensure that there are no conflicts with furnace and air conditioning equipment, refer to the _____.

   a. floor plan
   b. HVAC/mechanical drawing
   c. foundation plan
   d. elevation drawing

8. Vertical and horizontal piping along with sizes and a riser number that refers back to the full set of plumbing drawings are shown on _____.

   a. riser diagrams
   b. water diagrams
   c. electrical drawings
   d. HVAC/mechanical diagrams

9. Drawings that can be used to determine such factors as overall dimensions, actual inside dimensions, fitting depth, and laying length of the fittings are known as_____.

   a. schematic drawings
   b. exploded drawings
   c. riser diagrams
   d. catalog drawings

10. Drawings of products that provide enough detail to show how the product is constructed are called _____.

    a. schematic drawings
    b. cutaway drawings
    c. exploded drawings
    d. catalog drawings

Trade Terms Quiz

Fill in the blank with the correct term that you learned from your study of this module.

1. You will find the location of the furnace and air conditioning equipment, as well as the location of ducts and registers or piping and radiators, in _____.

2. A(n) _____ is created by viewing an object from one or more of six possible directions and recording how the object looks in each of the views.

3. The _____ may include information about the drainage, waste, and vent (DWV) piping and the water supply piping that must be installed below the first floor.

4. _____ allow you to determine whether a certain fixture will fit in the available space.

5. _____ are lines with arrows or slashes at a termination line used to indicate the dimension of a distance on a drawing.

6. The _____ is the minimum distance that must be maintained between a structure and the structure's property line, such as a street.

7. In _____, the line represents the centerline of the pipe, so these drawings can be used to represent pipe of any diameter.

8. _____ are marks or drawings that represent specific objects or materials.

9. The reasons that codes may set a minimum width for _____ include providing access to rear yards and reducing the possibility of a fire jumping from one building to another.

10. _____ illustrate how to assemble complex objects and show the relationship of the individual parts to the whole object.

11. A pictorial drawing that creates the illusion of depth by using lines that are drawn at 45 or 60 degrees to the horizontal is a(n) _____.

12. _____ indicates the size relationship between an object in a drawing and the completed object itself.

13. _____ specify places on the property where utilities can be installed.

14. A drawing with dark reddish lines is called a(n) _____.

15. A section drawing that shows the details of how a product is constructed is called a(n) _____.

16. _____ are detailed lists of all the material and equipment necessary to construct a project.

17. A(n) _____ provides information about where you may not build on certain parts of the property.

18. After the architect or engineer approves the fixture or fitting in a catalog drawing, it becomes _____.

19. _____ aids productivity by automating the repetitive work of drafting.

20. _____ present a three-dimensional picture of an object that is very similar to the actual object.

21. In _____, all vertical lines are shown vertically but all horizontal lines are projected at a 30-degree angle and appear to go back into the horizon.

22. Limited space may require you to use a(n) _____ to indicate the limits of a dimension on a drawing away from the actual points of the dimension.

23. Sections of the main drawing that are enlarged to show important structural elements more clearly are called _____.

24. The _____ shows an aerial view of the layout of rooms on each level of a building and the shape and size of each room.

25. A _____ shows the design, location, and dimensions of a building and its various components.

26. A(n) _____ is a dimensioned drawing with elevations and sections that indicate a proposed routing of system components.

27. _____ show the location of outlets, switches, and electrical fixtures in a structure.

28. A _____ is a complex isometric drawing that shows vertical and horizontal piping along with sizes and a riser number that refers back to the full set of plumbing drawings.

29. A drawing of plumbing fixtures that is found in manufacturers' catalogs is called a _____.

30. A _____ shows the piping system with the location of fixtures and pipe runs and the size and type of pipe to be installed.

31. The requirements that provide detailed information about the quality of workmanship required on a project, as well as the specific materials and method of installation you must use, are called _____.

32. The type of measuring device used to draw and measure scaled drawings is a(n) _____.

33. _____ show the vertical elements of a building, such as the height and width of interior or exterior walls.

Trade Terms

Approved submittal data
Architect's scale
Catalog drawing
Computer-aided drafting (CAD)
Construction drawing
Coordination drawing
Cutaway drawing
Details
Dimension line

Easement
Electrical drawings
Elevation drawings
Exploded drawing
Extension line
Fixture drawing
Floor plan
Foundation plan

HVAC (heating, ventilating, and air conditioning) drawings
Isometric drawings
Oblique drawing
Orthographic drawing
Pictorial drawings
Plot plan
Plumbing drawing
Riser diagram

Scale
Sepia
Setback
Side yards
Single-line drawings
Specifications
Submittal data
Symbol
Takeoff

Ray G. Thornton

Estimator / Project Manager

Kennco Plumbing, Inc./ ABC SoCal Chapter /
Wm. S. Hart High School

Describe how you got started in the construction industry.

I got started in the business when I was in the US Air Force. I was given the choice of trades to learn and I picked plumbing because I had some prior experience in that field.

Who inspired you to enter the industry? Why?

My grandfather did. He got me my first job when he had drainage piping installed on his farm. One of the conditions was that I was to be hired as a helper by the contractor. I liked the work better than the farm work, and so when I joined the military I stayed with it.

What do you enjoy most about your job?

I enjoy everything about it. Every day is different, with far more challenges and problems to solve. Every year new products are introduced that make the job even more exciting.

Do you think training and education are important in construction? If so, why?

I totally believe in advanced education in the plumbing industry. I also believe that we are missing the problem by not reaching out to younger people to introduce them to the trade. I currently teach an introductory course in a local high school and I think this is where we need to be to get more people into the trades.

How important are NCCER credentials to your career?

They give me great credit in the trade. The different subjects that are taught and learned are a never-ending source of advancement.

How has training/construction impacted your life and your career?

Over the years, I have learned a lot of management techniques from NCCER and I have also learned even more from the students that I teach. I feel that right now I am at the top of my wage class for my age and that now is the time to return what knowledge I have learned over the years to the next generation.

I have also enjoyed the opportunities to work alongside of some of the best craftworkers in the business, both in plumbing and in other trades.

Would you suggest construction as a career to others? If so, why?

I totally believe in the plumbing construction industry and I am totally committed to the advancement of the plumbing trade. It has not only rewarded me with great opportunities, but it has also advanced my knowledge as a person, an employee, and a leader.

How do you define "craftsmanship"?

A craftsman is a person who can be handed a project and go out and complete that project; a person who can read the plans, order the material and equipment, and proceed to complete it on time and efficiently.

PLANS AND DIAGRAMS

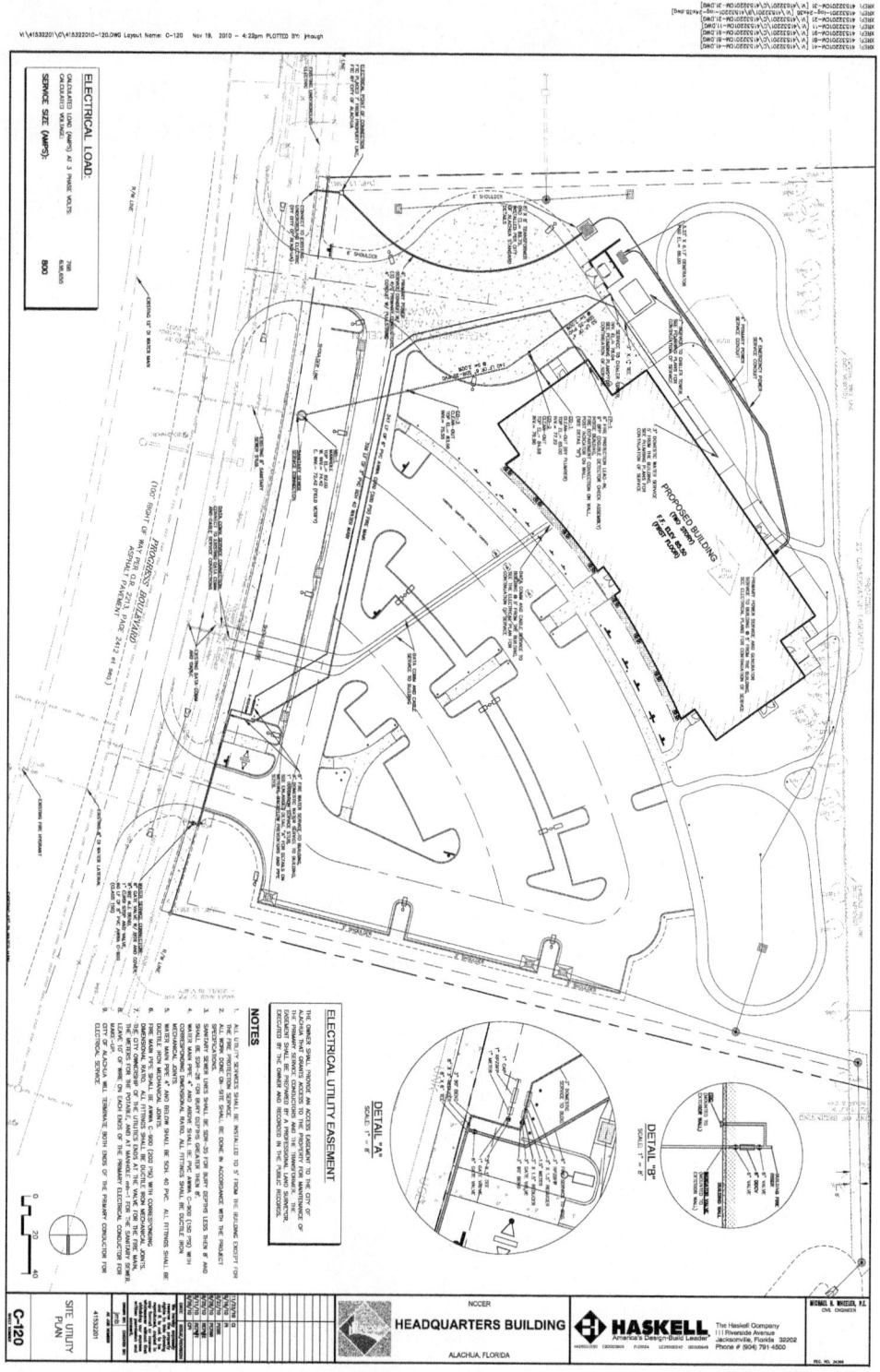

Figure A01 Site plan showing utilities.

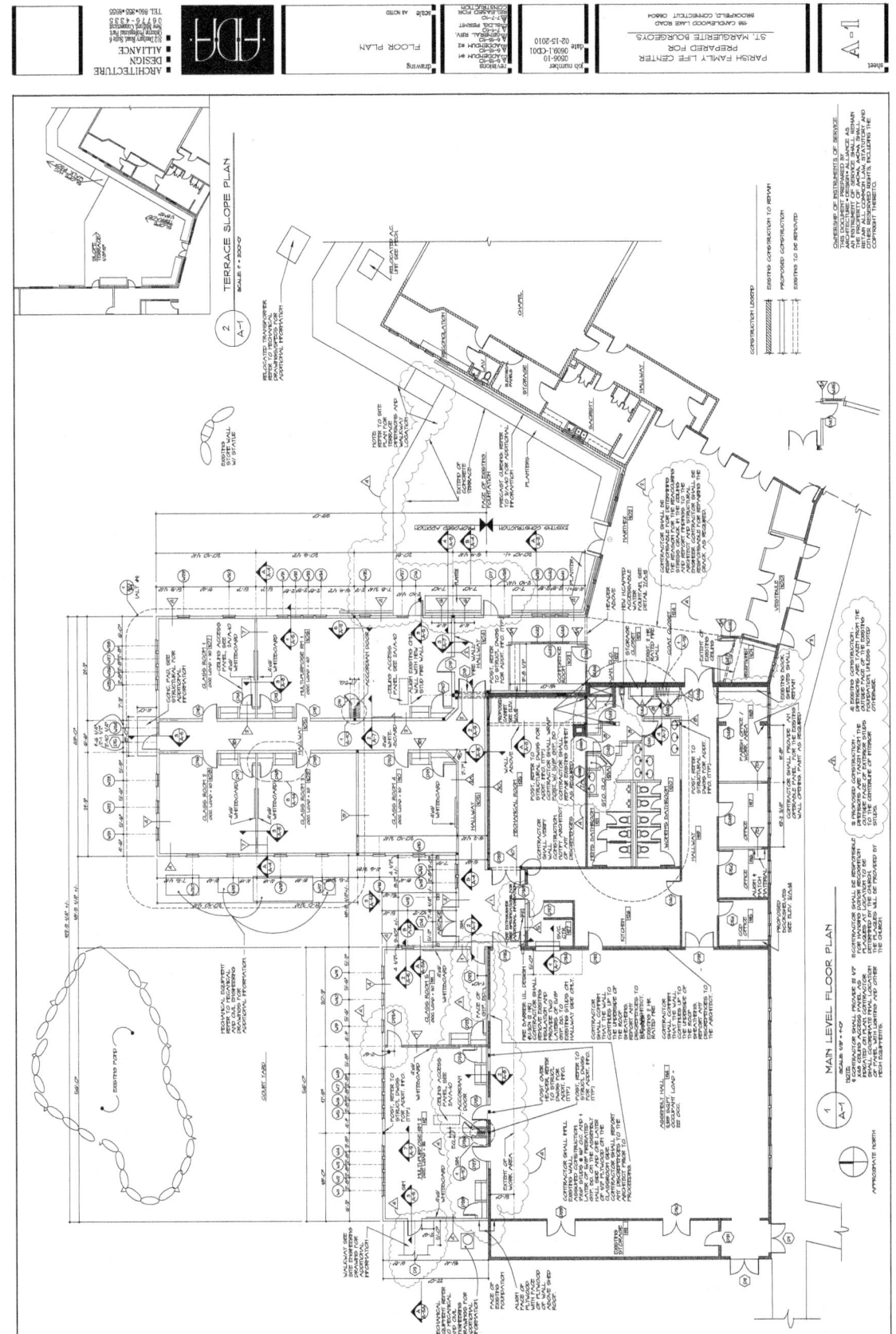

Figure A02 Floor plan.

Introduction to Plumbing Drawings

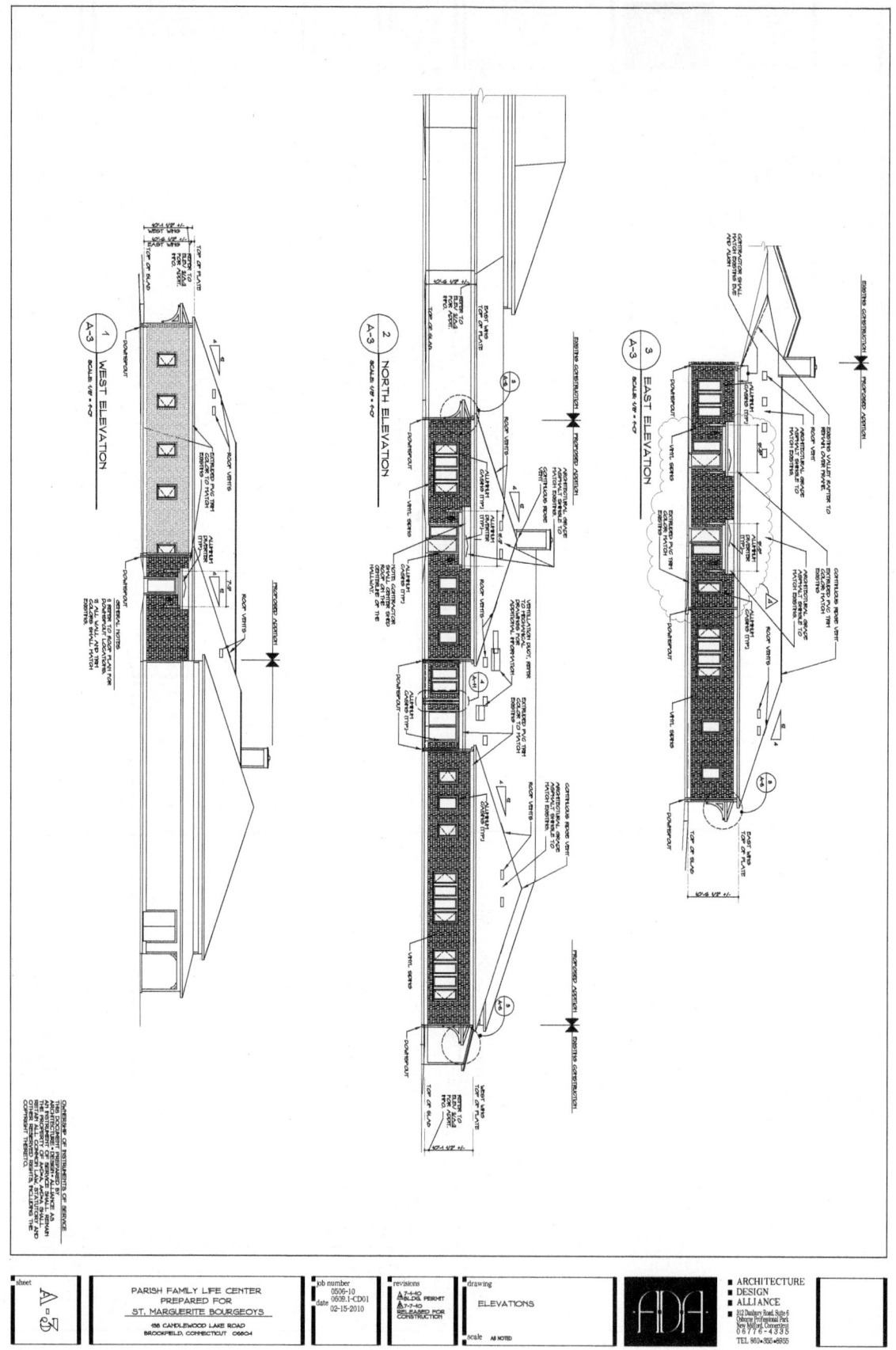

Figure A03 Elevation drawing.

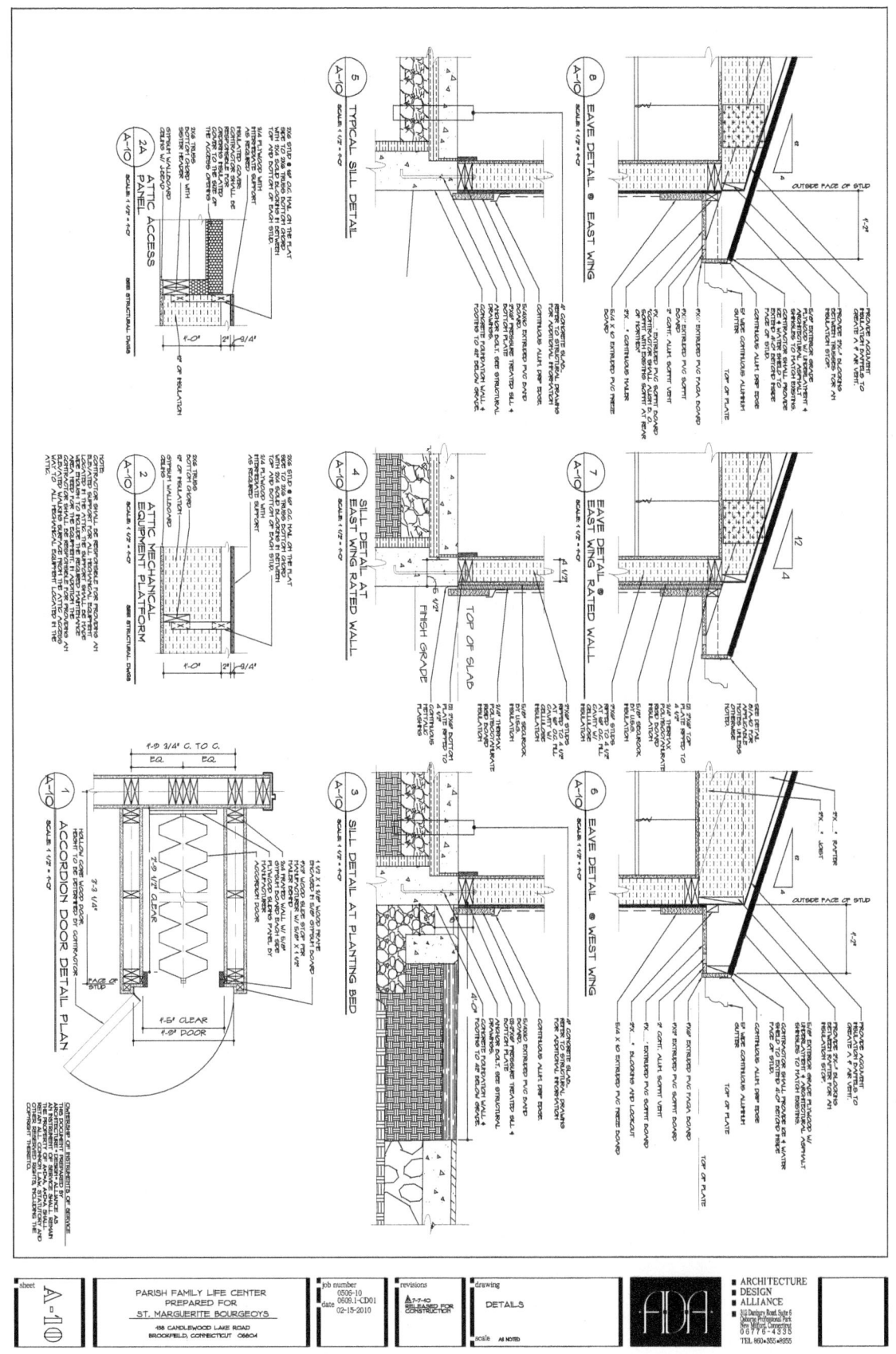

Figure A04 Detail drawing.

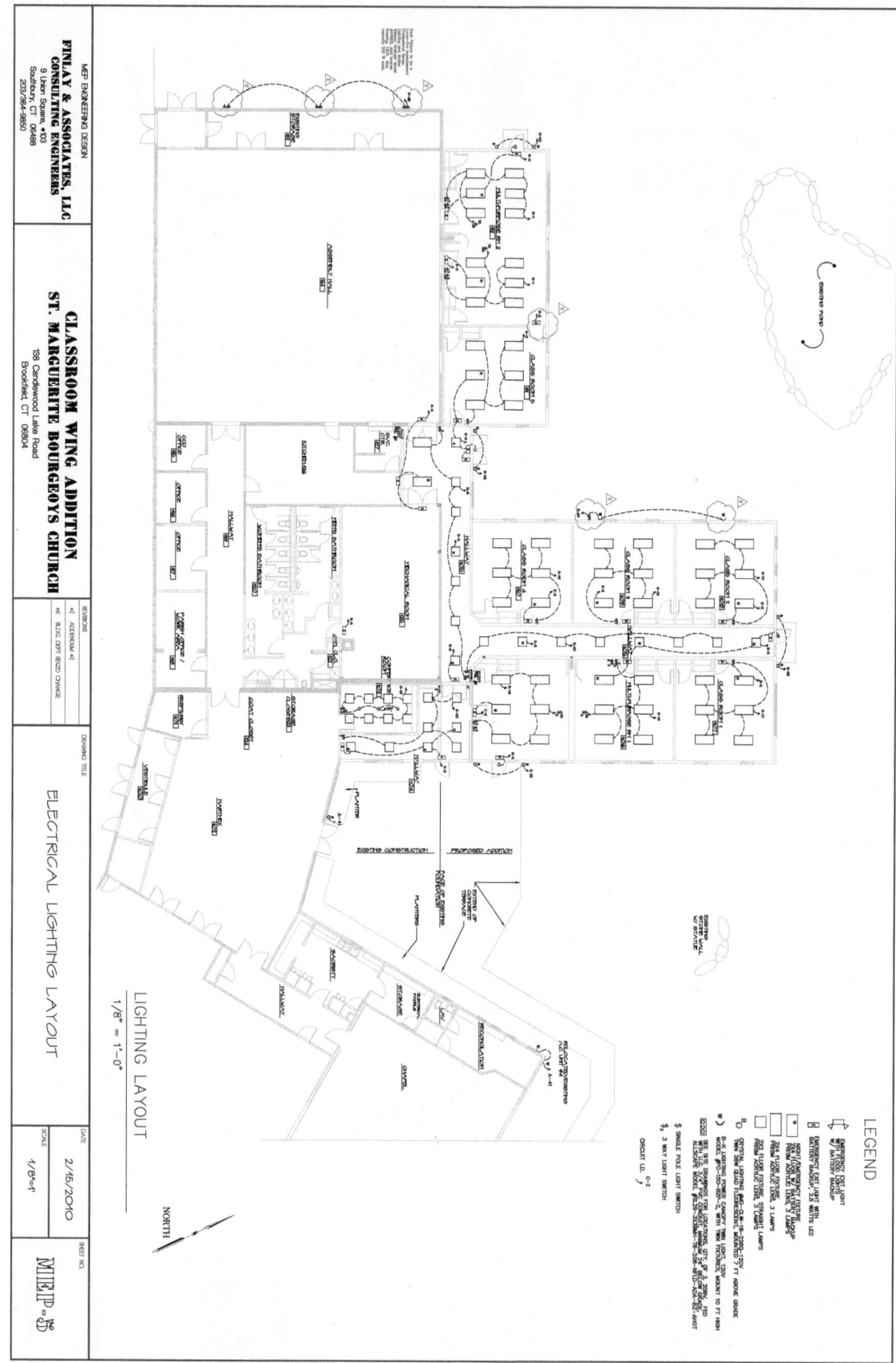

Figure A05 Electrical drawing.

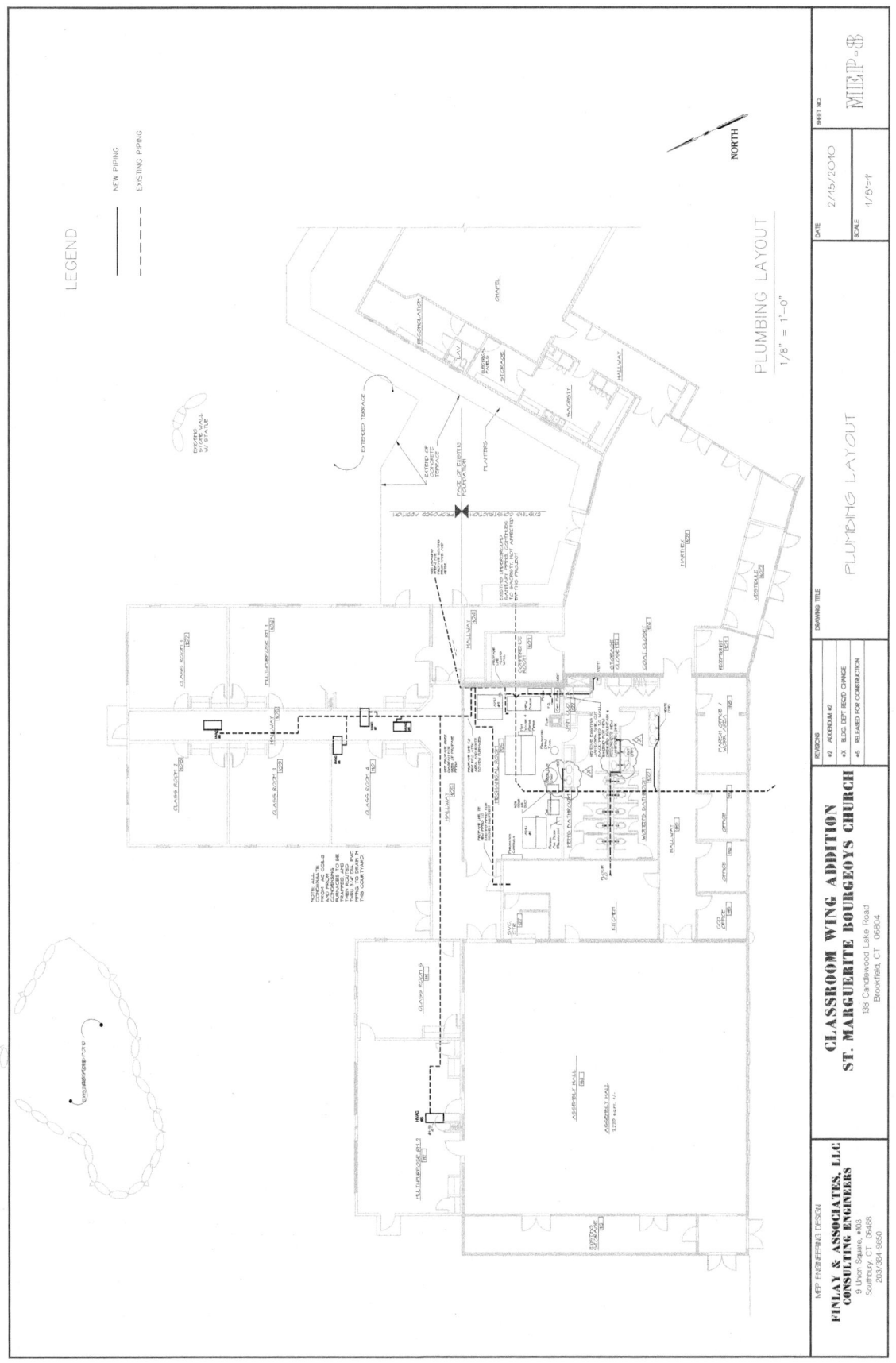

Figure A06 Plumbing drawing.

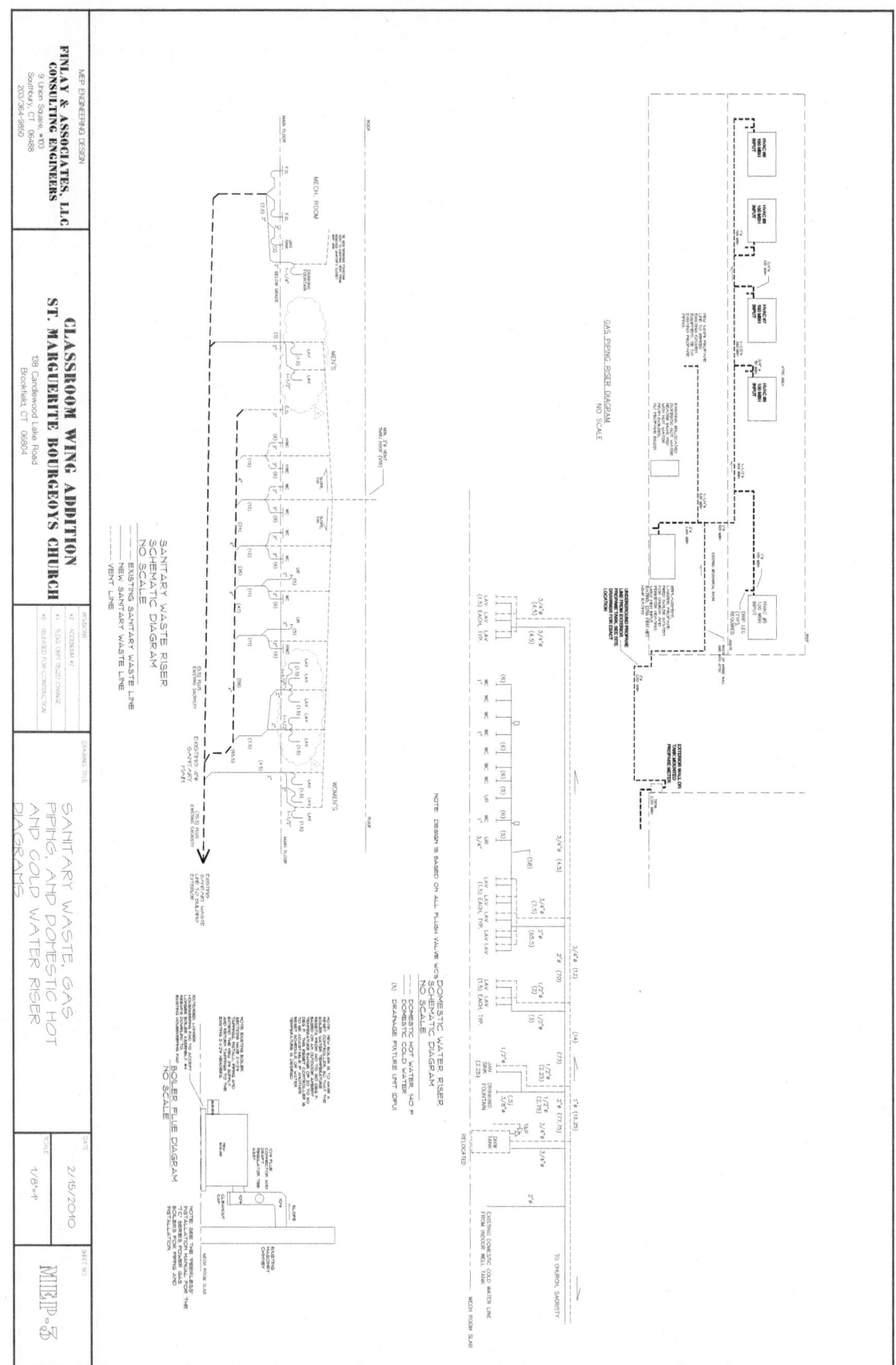

Figure A07 Riser diagram.

Approved submittal data: The fixtures and fittings in catalog drawings that have been approved by the architect or engineer.

Architect's scale: A measuring device that uses smaller units, such as $\frac{1}{2}$" or $\frac{1}{4}$", to represent 1'. Architect's scales are also issued in metric units. The units are used so that all building measurements are in proportion to their actual measurements but able to fit in the drawings.

Catalog drawing: A drawing of a plumbing fixture that is found in manufacturers' catalogs. Also referred to as submittal data.

Computer-aided drafting (CAD): A sophisticated design program used on computers. It allows designers to create drawings in two or three dimensions.

Construction drawing: A drawing that shows the design, location, and dimensions of a building and its various components.

Coordination drawing: A dimensioned drawing with elevations and sections that indicate the proposed routing of system components. The dimensioned coordination drawing includes the actual dimensioned equipment as well as the access space this equipment will need.

Cutaway drawing: A section drawing that shows the construction elements of a particular part of a building or fixture.

Details: Sections of construction drawings that are enlarged to make them clearer.

Dimension line: A line on a drawing with a measurement indicating actual length.

Easement: A designated right-of-way, such as the access guaranteed to utility companies for repair of utilities that are located on, or cross over, private land, or for vehicles to cross private property.

Electrical drawings: A drawing that shows the location of outlets, switches, and electrical fixtures. An electrical drawing may be superimposed on the floor plan.

Elevation drawings: Drawings of structures showing a side, front, or back view.

Exploded drawing: A drawing that shows how to assemble a complex product. It also shows the relationship of the individual parts to the object as a whole. Also referred to as an assembly drawing.

Extension line: A line used on a drawing to locate a dimension away from the actual points of the dimension. This method is used when a drawing would be too crowded or cluttered if the dimension were shown within the two points.

Fixture drawing: A drawing that shows the components of a fixture in detail.

Floor plan: A construction drawing of a building looking down at the floor from above (bird's-eye view). The plan shows at least the outline of the wall locations and lengths to scale. Normally, this drawing is oriented such that the top of the page represents north.

Foundation plan: A construction drawing showing the placement and dimensions of a building foundation.

HVAC (heating, ventilating, and air conditioning) drawings: Construction drawings that show the placement of the furnace and air conditioning equipment and the location of ducts and registers or pipes and radiators.

Isometric drawings: Pictorial drawings that create the illusion of a three-dimensional object. All horizontal lines are projected at a 30-degree angle.

Oblique drawing: A pictorial drawing that shows the shape of an object. It shows the front of the object with the body of the object at a slight angle.

Orthographic drawing: A construction drawing that shows straight-on views of the different sides of an object. Orthographic drawings are used for elevation drawings.

Pictorial drawings: Drawings that show a three-dimensional view of an object.

Plot plan: A drawing of a structure that includes the dimensions of the building site, location of the structure in relation to the property boundaries, elevation of key points, existing and finish contour lines, utility services, and compass directions. Also referred to as a site plan.

Plumbing drawing: A construction drawing that shows the location of fixtures and pipe runs and gives the size and type of pipe to be installed.

Riser diagram: A drawing that shows vertical and horizontal piping along with sizes and a riser number that refers back to the full set of plumbing drawings.

Scale: The relationship of the dimensions on a drawing to the actual dimensions of the structure. For example, in a ¼ scale, 0.25" represent 1'. Scale is often provided in both English and metric units.

Sepia: A print or construction drawing with dark reddish-brown lines on a light background.

Setback: The distance a code requires between a building and a property line, such as the street.

Side yards: The spaces along the sides of a structure that provide access to rear yards, reduce the possibility of fire jumping from one building to the next, and promote ventilation around the structure.

Single-line drawings: Plumbing drawings that use a single line to represent the centerline of a pipe. Single-line drawings can be used to represent pipe of any diameter. Also referred to as schematic drawing.

Specifications: Written requirements included with the drawings or blueprints of a construction project. They provide more details or descriptions of the technical standards that must be met during construction. Specifications usually override drawings but are overridden by the contract. Also referred to as specs.

Symbols: Marks or drawings used to indicate a specific object, material, class, or entity. A legend shows the symbols used on a drawing and their meanings.

Takeoffs: Detailed lists that are compiled, based on drawings and specifications, of all the material and equipment necessary to construct a project. Such a list is also called a material takeoff.

Additional Resources

This module presents thorough resources for task training. The following resource material is suggested for further study.

Core Curriculum, Latest Edition. NCCER. New York, NY. Pearson.

Hammer's Blueprint Reading Basics, 2017. Charles A. Gillis and Warren Hammer. New York: Industrial Press.

Technical Drawing, 15th Edition, 2016. Frederick E. Giesecke, et al. New York, NY. Pearson.

Section Review Answer Key

Section 1.0.0

Answer	Section Reference	Objective
1. b	1.1.1	1a
2. a	1.2.1	1b
3. b	1.3.1	1c
4. b	1.4.1	1d

Section 2.0.0

Answer	Section Reference	Objective
1. c	2.1.1	2a
2. d	2.2.0	2b
3. b	2.3.0	2c
4. a	2.4.2	2d

NCCER CURRICULA — USER UPDATE

NCCER makes every effort to keep its textbooks up-to-date and free of technical errors. We appreciate your help in this process. If you find an error, a typographical mistake, or an inaccuracy in NCCER's curricula, please fill out this form (or a photocopy), or complete the online form at **www.nccer.org/olf**. Be sure to include the exact module ID number, page number, a detailed description, and your recommended correction. Your input will be brought to the attention of the Authoring Team. Thank you for your assistance.

Instructors – If you have an idea for improving this textbook, or have found that additional materials were necessary to teach this module effectively, please let us know so that we may present your suggestions to the Authoring Team.

NCCER Product Development and Revision

13614 Progress Blvd., Alachua, FL 32615

Email: curriculum@nccer.org
Online: www.nccer.org/olf

❏ Trainee Guide ❏ Lesson Plans ❏ Exam ❏ PowerPoints Other _____

Craft / Level: _____ Copyright Date: _____

Module ID Number / Title: _____

Section Number(s): _____

Description: _____

_____

_____

_____

_____

Recommended Correction: _____

_____

_____

_____

_____

Your Name: _____

Address: _____

_____

Email: _____ Phone: _____

This page is intentionally left blank.

Plastic Pipe and Fittings

OVERVIEW

With the introduction of plastic pipe, the work of plumbers greatly changed. Plastic pipe is strong, durable, and requires little maintenance. Every conceivable plumbing fitting is available in plastic. Plumbers must be familiar with the various types of plastic pipe, each with its own properties and applications. Plumbers must also be experienced using the methods and special tools for measuring, cutting, and installing plastic pipe and fittings.

Module 02106

Trainees with successful module completions may be eligible for credentialing through the NCCER Registry. To learn more, go to *www.nccer.org* or contact us at 1.888.622.3720. Our website, *www.nccer.org*, has information on the latest product releases and training.

Your feedback is welcome. You may email your comments to *curriculum@nccer.org*, send general comments and inquiries to *info@nccer.org*, or fill in the User Update form at the back of this module.

This information is general in nature and intended for training purposes only. Actual performance of activities described in this manual requires compliance with all applicable operating, service, maintenance, and safety procedures under the direction of qualified personnel. References in this manual to patented or proprietary devices do not constitute a recommendation of their use.

02106 V4.5

Objectives

Successful completion of this module prepares trainees to:

1. Identify the types, uses, and properties of plastic pipe and fittings.
   a. Identify the types of plastic pipe and their uses.
   b. Describe the sizing and labeling of plastic pipe.
   c. Describe the different types of fittings used on plastic pipe.
   d. Identify storage and handling requirements for plastic pipe.
2. Describe the different methods for joining plastic pipe.
   a. Describe how to measure and cut plastic pipe.
   b. Describe how to join PVC and CPVC pipe.
   c. Describe the installation procedures for PVC bell-and-spigot pipe.
   d. Describe the methods for joining PEX and PE tubing.
3. Describe the methods used to support and test plastic pipe.
   a. Describe the hangers and fasteners used to support pipe.
   b. Explain methods of pressure testing plastic pipe.

Performance Tasks

Under the supervision of your instructor, you should be able to do the following:

1. Select correct types of materials for plastic piping systems.
2. Identify types of fittings and valves and their uses.
3. Select the appropriate personal protective equipment for working with plastic piping.
4. Properly measure, cut, and join plastic piping.
5. Select the correct support and spacing for the application.

Trade Terms

ABS (acrylonitrile-butadiene-styrene)
Bell-and-spigot pipe
Cellular-core wall
Chamfer
Chemically inert
Compression collars
CPVC (chlorinated polyvinyl chloride)
Elastomeric
Fusion fittings
Hydronic
Hydrostatic pressure test
Inside diameter (ID)
Interference fit
Outside diameter (OD)
PB (polybutylene)
PE (polyethylene)

PEX (cross-linked polyethylene)
Pipe riser clamps
Pounds per square inch (psi)
Pressure rating
Primer
PVC (polyvinyl chloride)
Ring-tight gasket fittings
Schedules
Size dimension ratios (SDR)
Solid wall
Solvent weld
Thermoplastic pipe
Thermoset pipe
Transition fitting
Water hammer

Industry Recognized Credentials

If you are training through an NCCER-accredited sponsor, you may be eligible for credentials from NCCER's Registry. The ID number for this module is 02106. Note that this module may have been used in other NCCER curricula and may apply to other level completions. Contact NCCER's Registry at 888.622.3720 or go to *www.nccer.org* for more information.

> **CODE NOTE**
>
> Codes vary among jurisdictions. Because of the variations in code, consult the applicable code whenever regulations are in question. Referencing an incorrect set of codes can cause as much trouble as failing to reference codes altogether. Obtain, review, and familiarize yourself with your local adopted code. Safety codes are developed by the US Occupational Safety and Health Administration (OSHA).

Contents

Figures and Tables

1.0.0 PLASTIC PIPE AND FITTINGS

Objective

Identify the types, uses, and properties of plastic pipe and fittings.

   a. Identify the types of plastic pipe and their uses.
   b. Describe the sizing and labeling of plastic pipe.
   c. Describe the different types of fittings used on plastic pipe.
   d. Identify storage and handling requirements for plastic pipe.

Performance Tasks

1. Select correct types of materials for plastic piping systems.
2. Identify types of fittings and valves and their uses.

Trade Terms

ABS (acrylonitrile-butadiene-styrene): Plastic pipe and fittings used extensively in drain, waste, and vent (DWV) systems.

Cellular-core wall: Plastic pipe wall that is low-density, lightweight plastic containing entrained (trapped) air in solid foam.

Chemically inert: Does not react with other chemicals.

CPVC (chlorinated polyvinyl chloride): Durable plastic pipe and fittings used extensively in hot and cold water distribution systems.

Elastomeric: Rubberized. Made of an elastic substance such as a polyvinyl elastomer.

Hydronic: A system that heats and cools by circulating water or steam through a closed piping system.

Inside diameter (ID): The distance between the inner walls of a pipe; approximates the nominal sizes of piping used in heating and plumbing.

Outside diameter (OD): The distance between the outer walls of a pipe.

PB (polybutylene): Plastic piping that was formerly used for plumbing pipe; it is no longer used but is still found in some residences.

PE (polyethylene): Flexible plastic pipe, tubing, and fittings, usually used for water distribution, that do not deteriorate when exposed to sunlight.

PEX (cross-linked polyethylene): Tubing and fittings made with heat and high pressure that resist high temperatures, pressure, and chemicals.

Pipe riser clamps: A type of clamp-on pipe support that is designed to support the weight of vertical pipe and tubing.

Pounds per square inch (psi): A common measure of liquid and gas pressure in the United States.

Pressure rating: The maximum pressure at which a component or system may be operated continuously.

PVC (polyvinyl chloride): Plastic pipe and fittings used for cold water distribution and for industrial water and chemicals, as well as for drain, waste, and vent (DWV) systems.

Schedules: Tables relating pipe wall thicknesses to pipe pressure ratings. Higher schedule numbers equate to thicker pipe walls for a given nominal pipe size.

Size dimension ratios (SDR): Ratios of the outer diameters of pipes to their pipe wall thicknesses.

Solid wall: Plastic pipe wall that does not contain trapped air in the form of solid foam.

Thermoplastic pipe: Pipe that can be repeatedly softened by heating and hardened by cooling. When softened, thermoplastic materials can be molded into desired shapes during manufacturing.

Thermoset pipe: Pipe that changes chemically when heated, so that once hardened by heat or chemicals, it is hardened permanently.

Water hammer: An extreme change in water pressure within a pipe that occurs when suddenly stopping a moving mass of water, as when quickly shutting a valve. It can cause a loud, banging sound as pipes flex and even damage the system.

Plastic has revolutionized plumbing, a trade that was based for hundreds of years on metal pipe and fittings. Not widely used by the plumbing industry until the 1950s, plastic is the newest of plumbing materials and has changed the way plumbers work. Plumbers use plastic today in a wide variety of ways, including DWV piping, water distribution, chemical waste systems, and fuel-gas piping.

Plastic pipe is strong, durable, and requires little maintenance. Although it is easier to use than metal pipe, it does have some disadvantages. It can be affected by temperatures, give off toxic fumes when heated or burned, and is flammable.

Plumbers must learn how to safely and properly handle plastic pipe and fittings.

Plumbers must be familiar with several types and sizes of plastic pipe, each with its own properties. Each type of plastic pipe is designed for different applications and has different installation methods and requirements. All plastic pipe must be handled and stored properly to avoid deterioration and damage. A pipe's application determines the type of fittings and joints that are compatible. Specific fittings are used for water supply and DWV systems. Plumbers use special tools to measure, cut, and join plastic pipe and fittings. While measuring and cutting techniques are standard, different types of plastic pipe have their own methods for joining.

Plastic pipe can be supported using a variety of methods. The material, size, applications, and installation affect the type of support. Hangers are used for horizontal support, while pipe riser clamps provide vertical support. Plans and specifications dictate codes that must be followed. After completing an installation, the system must be pressure-tested for leaks.

1.1.0 Plastic Pipe

Plastic pipe has both advantages and disadvantages. Plumbers must understand the different properties of plastic pipe and how the industry measures plastic piping. The following sections discuss these characteristics and properties, as well as the standard labeling practice for plastic piping and the most common plastic pipe and fitting manufacturers. The specific types of plastic pipe that you will use in plumbing applications are also introduced.

1.1.1 Uses and Applications of Plastic Pipe

Most plastic pipe and fittings are chemically inert, meaning they do not react with other substances or materials. These plastics can withstand most chemicals that are used in the home, office, or factory. Strong and durable, plastic pipe and fittings also require little maintenance because they are generally resistant to corrosion and do not pit or scale. In addition, plastics are easier to handle than metals because of their relative light weight, which can make them more cost-effective (and safer) to install.

The use of plastic pipe helps to eliminate one of the most common hazards associated with installation—fire. During the installation process, no flame is required for welding or soldering. Therefore, it is less likely that fires will occur. In the past, when plumbers joined pipe using molten lead, accidents involving spilled lead and tipped-over heating furnaces were a constant concern. Similarly, there are no soldered joints as with copper piping, so the hazard of igniting construction materials while joining pipe is avoided.

Plastic does have some drawbacks. Temperature changes cause it to expand and contract more than metal. Plastic pipe is also flammable and can give off toxic fumes when it burns. In addition, fumes from solvent chemicals used to join plastic pipes are harmful. Plumbers must be familiar with the types of safety hazards each variety of pipe and chemical solvent presents. It is your responsibility to know how to protect yourself on the job site. Always consult the manufacturer's safety data sheets (SDS) and specifications regarding the properties of the pipe you are using.

> **CAUTION**
>
> Do not use foam-core pipe in underground installations, unless explicitly allowed in your local applicable plumbing code.

Did You Know?

On November 21, 1980, an electrical fire in the MGM Grand hotel in Las Vegas caused the deaths of 85 people and injured 1,500 other people. It was the third-worst hotel fire in US history. Investigators discovered that most of the people died from inhaling toxic gases emitted by burning PVC used in wiring and piping, not from the heat or flames. The fire led to significant changes in hotel design and building codes in order to prevent such a disaster from happening again. In the 1980s, model codes addressed the use of plastic pipe in fire-resistive construction to prevent similar problems in the future.

1.1.2 Properties and Types of Plastic Pipe

Plastics contain polymers, long chains of molecules that manufacturers can mold, cast, or force through a die into desired shapes. Manufacturers can heat and shape resins in the form of pellets, powders, and solutions in various combinations to give the finished product the desired properties.

Each type of plastic pipe has its own unique properties. These properties determine where and how you can install it, how you can join it to other piping, and how you should support it. The two general categories of plastic are thermoplastic and thermoset. Thermoplastic pipe can be repeatedly softened by heating and hardened by cooling. When softened, thermoplastics can be molded into desired shapes. Thermoset pipe changes chemically when heated, so that once hardened by heat or chemicals, it is hardened permanently.

In addition, plastic pipe can be either rigid or flexible. Rigid pipe is straight and maintains its shape. Flexible pipe bends, so although it requires fewer fittings and joints, it requires more support. Flexible pipe is often referred to as tubing, since flexible pipe is practical only in small diameters. Manufacturers sell flexible pipe in coils.

The construction of plastic pipe can vary as well. Pipes can be solid wall or cellular-core wall construction. Solid wall does not contain trapped air in the form of solid foam. A pipe with cellular core construction, also called foam-core construction, has walls that contain trapped air in the form of solid foam, so it is lighter weight. Cellular-core wall piping consists of three layers of plastic: solid inner and outer layers and a foam middle layer. Because it contains of less material, it is also less expensive than solid-wall pipe. Plumbers use solid-wall and cellular-core wall pipe interchangeably for DWV applications.

There are many types of plastic from which plumbing pipe and fittings can be made. In general, only pipes and fittings of the same-type plastic can be joined using the solvent welding methods described later in this module. Some mechanical joining systems may be interchangeable among different types of plastics. Always follow the manufacturer's instructions and applicable codes when assembling and installing plastic pipe. *Table 1* summarizes the typical applications for the various types of plastic. The following sections provide descriptions of the common types of plastic used in plumbing pipe and fittings.

Table 1 Types of Plastic Pipe and Their Applications

Pipe	Applications
ABS	DWV sanitary systems, corrosive waste
PVC	DWV sanitary systems, cold water service
CPVC	Hot and cold water distribution
PE	Cold water service, corrosive waste, gas service
PEX	Hot and cold water distribution
PB	Hot and cold water distribution

1.1.3 ABS

ABS (acrylonitrile-butadiene-styrene) pipe and fittings (*Figure 1*) are made from a thermoplastic resin. ABS is the standard material for many types of DWV systems. It is available in diameters ranging from $1\frac{1}{2}$ to 6 inches. The pipe comes in both solid-wall and cellular-core wall construction, which are interchangeable in plumbing applications.

Because ABS is light, it is easy to handle and install. A 3-inch-diameter, 10-foot-long section weighs less than 10 pounds. ABS performs well at extreme temperatures because it absorbs heat slowly and also cools slowly, an important feature for a system that handles both hot and cold wastes.

ABS is highly resistant to household chemicals. In tests, it showed no effect from such common products as detergent, bleach, and household drain cleaners. Sewage treatment plants use ABS because it stands up to the highly corrosive and abrasive liquids commonly found in such systems.

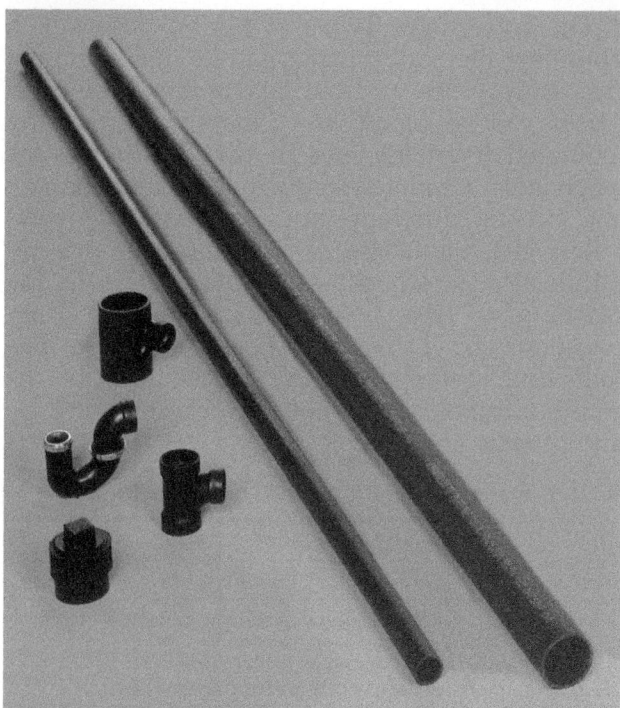
Figure 1 ABS pipe and fittings.

ABS is strong and long lasting. In 1959, ABS pipe was used in an experimental residence. Twenty-five years later, an independent research firm dug up and analyzed a section of the pipe and found no evidence of rot, rust, or corrosion. The pipe also withstands earth loads, slab foundations, and high surface loads without collapsing.

1.1.4 PVC

PVC (polyvinyl chloride) is a rigid pipe with high impact strength that is manufactured from a thermoplastic material (*Figure 2*). The material has an indefinite lifespan under most conditions. Plumbers frequently use PVC in cold water systems. Like ABS, PVC is commonly used in DWV systems. It is also used to transport many chemicals because of its chemical-resistant properties. PVC is available in solid-wall and cellular-core wall construction.

> **NOTE**
> An "indefinite lifespan" simply means that under test conditions, PVC shows no evidence of deterioration during the timespan of the testing. This prevents estimating a working lifespan of the material.

Solid-wall PVC pipe can be used in high-pressure systems, but only to carry low-temperature water. You must protect it from sunlight because ultraviolet light degrades the thermoplastic materials. It is lightweight, easy to handle and install, has joint flexibility that handles ground movement without leaking, and lasts a long time without maintenance as long as it is protected from sunlight.

1.1.5 CPVC

CPVC (chlorinated polyvinyl chloride) pipe and fittings (*Figure 3*) are made from an engineered vinyl polymer. Plumbers use CPVC in hot- and cold-water distribution systems. Improvements made to its parent polymer, polyvinyl chloride, added high-temperature performance and improved impact resistance to this material. CPVC is produced in standard copper-tube sizes (CTS) from $\frac{1}{2}$ to 2 inches, with a full line of fittings.

CPVC is acceptable under many model codes for indoor use. Its molecular structure practically eliminates condensation in the summer and heat loss in the winter, decreasing the likelihood of costly drip damage to walls or structures. CPVC pipe's smooth, friction-free interior surfaces result in lower pressure loss and higher flow rates and provide less opportunity for bacteria growth. CPVC does not break down in the presence of chemically-reactive water solutions. Like other types of plastic, it does not rust, pit, or scale.

CPVC is lightweight and easy to install. Recent improvements to CPVC have made the pipe stronger and more durable during installation. The strength of CPVC is a clear advantage for plumbers working in cold-weather states. In laboratory tests down to 20°F, CPVC pipe withstood water hammer that substantially damaged copper piping under the same conditions. Water hammer is an extreme change in water pressure within a pipe when a mass of flowing water suddenly stops. You can tell its presence from the loud banging sounds produced by deflecting pipes.

1.1.6 PE

PE (polyethylene) is a thermoset plastic. Plumbers commonly use PE as tubing because of its strength, flexibility, and chemical-resistant properties (*Figure 4*). PE is also corrosion-resistant, which makes it ideal for transporting chemical compounds. It will not deteriorate when exposed to ultraviolet light, so you can install it outdoors without a protective coating. PE is used for cold water and underground gas service lines (outside buildings).

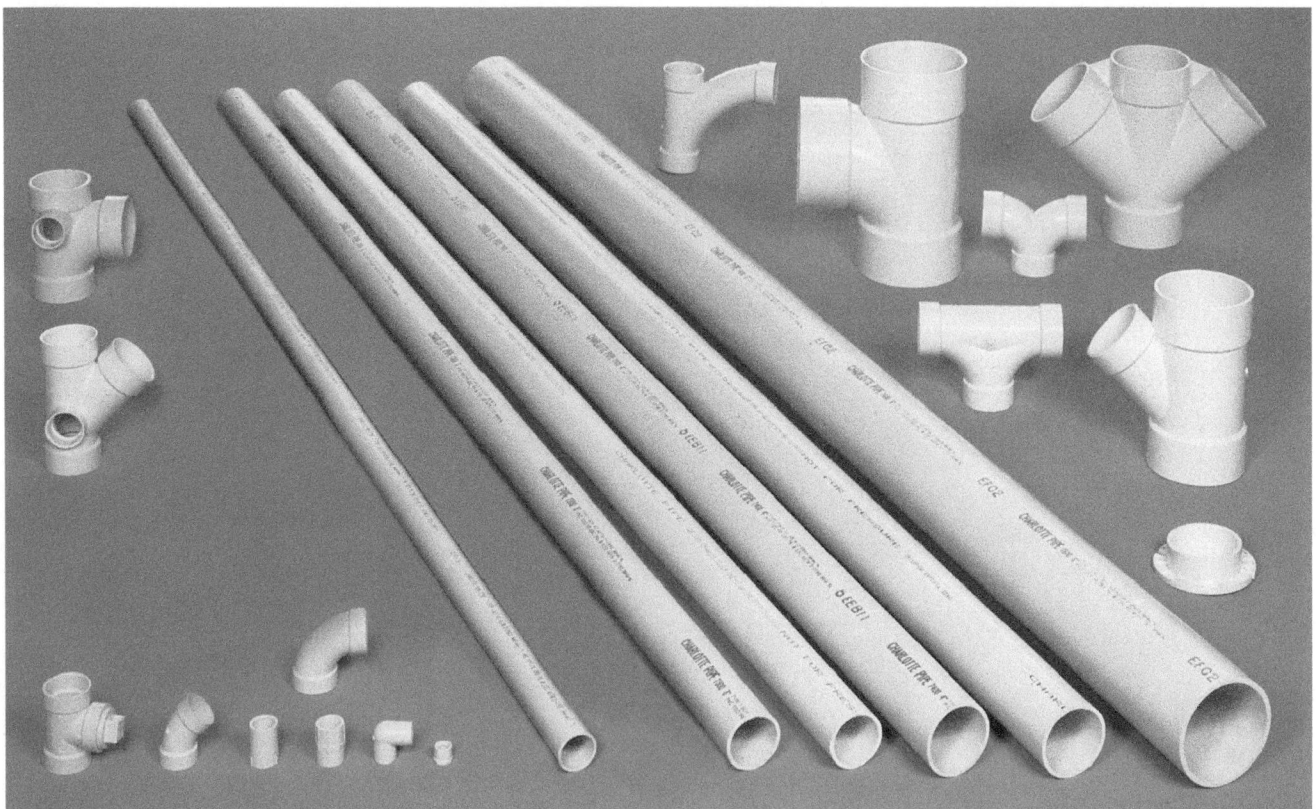

Figure 2 PVC pipe and fittings.

1.1.7 PEX

PEX (cross-linked polyethylene) tubing (*Figure* 5) is formed when high-density polyethylene is subjected to high temperature and pressure. Because it resists high temperatures, pressure, and chemicals, it is ideal for potable hot- and cold-water systems, hydronic radiant floor systems, baseboard and radiator connections, and snow-melt systems. Plumbers commonly use PEX for manifold plumbing distribution systems because of its flexibility and ease to work with.

1.1.8 PB

PB (polybutylene) is a dull gray or white thermoplastic resin pipe that plumbers used extensively for water supply piping from the late 1970s to the mid-1990s. PB is no longer available because it tended to catastrophically fail over time, and it is highly unlikely you will install PB piping. However, many buildings still contain PB piping, and repair fittings are available, so you need to be able to identify it (*Figure 6*).

1.2.0 Pipe Sizing and Labeling

Plumbers must be familiar with the sizing standards and the labeling of plastic pipe. Knowing this information allows you to understand plan specifications so that you obtain and install the correct materials. Interpreting the standard markings on pipe allows you to quickly determine whether the materials you received at the job site are what you ordered.

1.2.1 Sizing Plastic Pipe

Plastic pipe is manufactured with various wall thicknesses, based on schedules or size dimension ratios (SDR). SDR is a fixed ratio between a pipe's outside diameter (OD), or the distance between the outer walls of a pipe, and its wall thickness. An SDR number, such as SDR 35, designates this ratio. The larger the SDR number, the thinner the wall versus the outside diameter. For example, a 2-inch pipe (OD of 2.375 inches) with an SDR of 11 has a wall thickness of 0.216 inch (*Table 2*). The same pipe with an SDR of 17 has a wall thickness of 0.140 inch. Be aware that pipe can also be measured by its inside diameter (ID), the distance between its inner walls.

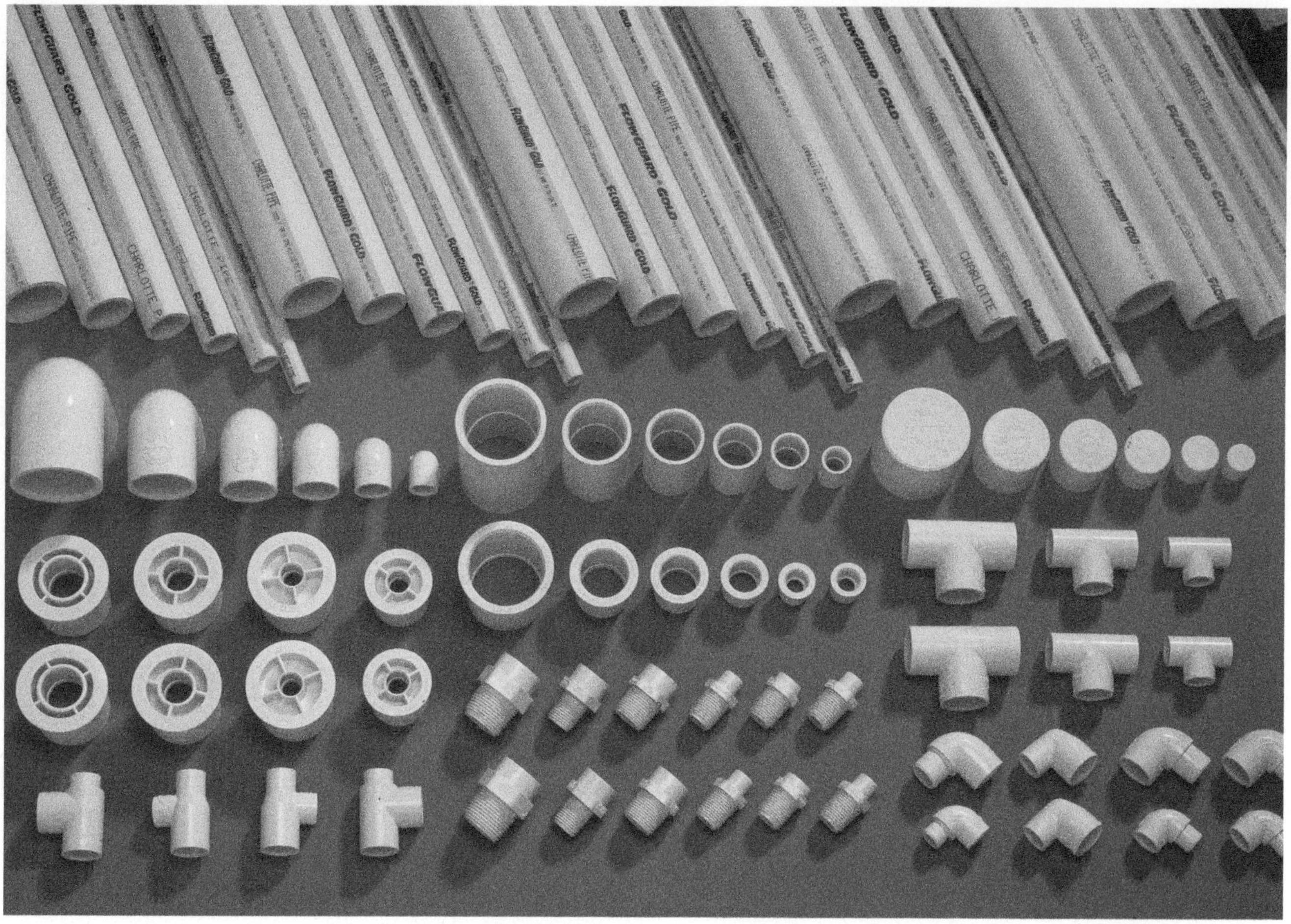

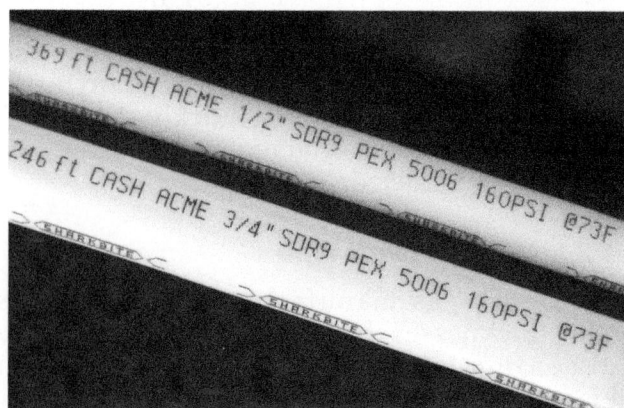

Figure 3 CPVC pipe and fittings.

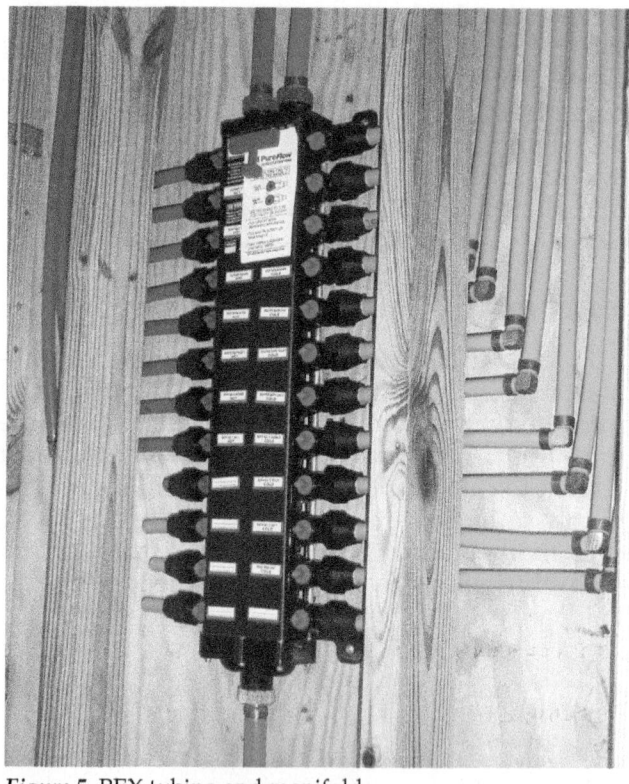

Figure 4 PE tubing.

Did You Know?

PE is the most inert of all plastic pipe materials, meaning that it is unlikely to react with other substances. That is why it is used for hospital tubing and in soda fountains.

Figure 5 PEX tubing and manifold.

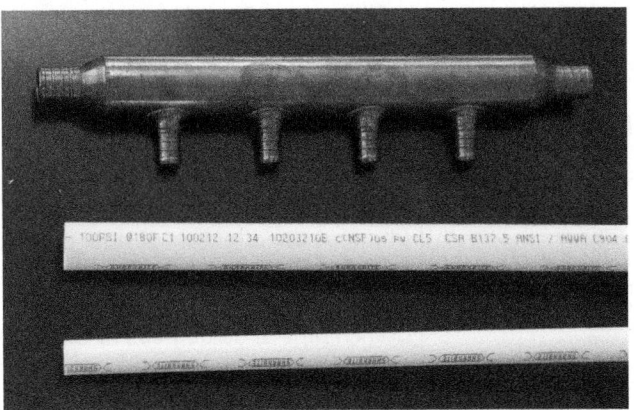
Figure 6 PB piping and manifold.

In an effort to simplify and standardize the use of plastic pipe and fittings, manufacturers fabricate pipes with SDRs based on Schedule 40 (thin wall), which is the most commonly used, and on Schedule 80 (thick wall). The schedule number is based on the design working pressure of the pipe. Most plumbing codes designate Schedule 40 as the minimum size for use in or under buildings. In some states, Schedule 80 PVC pipe can be used for gas.

Choosing the Correct Material

Be careful when choosing the material you will use for different plumbing purposes, such as PVC, which can be used for both DWV and water piping. Because PVC is white in both cases, it is possible to confuse them. However, PVC is tested and labeled differently for each use, so if you use the wrong type, as indicated by its label, your local inspecting authority could reject the installation.

1.2.2 Labeling and Markings

Plumbers must use plastic pipe and fittings that are labeled and approved for use in plumbing systems. The label on a plastic pipe includes the following:

- Nominal pipe diameter, CTS or iron-pipe size (IPS)
- Schedule or SDR
- Type of material: PVC, CPVC, and so on
- Pressure rating, usually expressed as pounds per square inch (psi) at a certain temperature
- Relevant standard for the pipe, for example, ASTM D2665
- Listing body or laboratory, such as the National Sanitation Foundation
- Name of manufacturer or brand name
- Country of origin

This label ensures that a recognized third party—the listing body or laboratory—has tested the material and approved it for the use intended (*Figure 7*). For example, the American Society for Testing Materials (ASTM) numbers on cellular core piping are different from those on Schedule 40 solid core piping. Solvent cements must also be listed and labeled for their specific use.

1.3.0 Plastic Pipe Fittings

A pipe's use determines what kinds of fittings and joints are needed. The following sections discuss the fittings used in water supply and DWV systems. While plumbers use some of the same fittings for these two applications, they also use specific fittings that differ. It is important to be familiar with the most common fittings on the market.

Table 2 Minimum Wall Thickness of 2-Inch Pipe

Plastic Pipe and Fittings Association **Minimum Wall Thicknesses of 2-Inch Pipe Based on SDR/SIDR**			
IPS-OD SDR (Outside Diameter = 2.375 inches) SDR (D 3035)	**Wall**	**IPS-ID SIDR (Inside Diameter = 2.067 inches) SIDR (D 2239)**	**Wall**
7	0.339"	5.3	0.390"
9	0.264"	7	0.295"
11	0.216"	9	0.230"
13.5	0.176"	11.5	0.180"
17	0.140"	15	0.138"
21	0.113"	19	0.109"
26	0.091"	24	0.089"
32.5	0.073"	30.5	0.071"

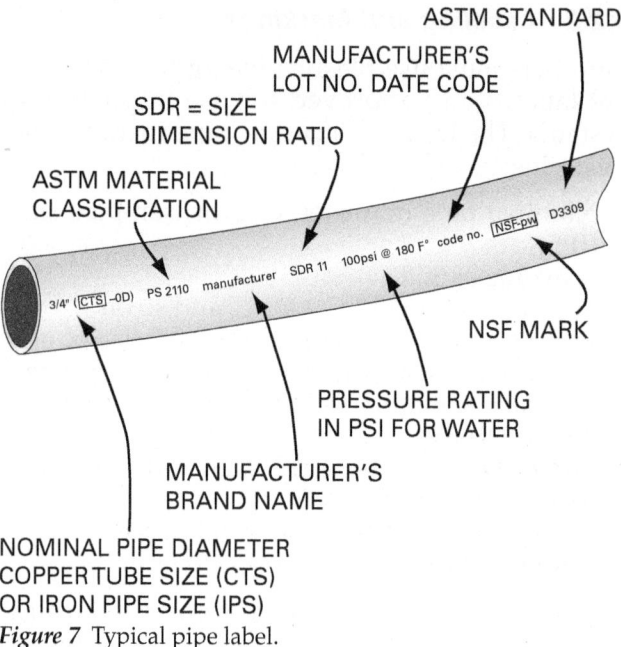

ASTM STANDARD

MANUFACTURER'S LOT NO. DATE CODE

SDR = SIZE DIMENSION RATIO

ASTM MATERIAL CLASSIFICATION

NSF MARK

PRESSURE RATING IN PSI FOR WATER

MANUFACTURER'S BRAND NAME

NOMINAL PIPE DIAMETER
COPPER TUBE SIZE (CTS)
OR IRON PIPE SIZE (IPS)

Figure 7 Typical pipe label.

1.3.1 *Water Supply Fittings*

Pressure-type fittings for use with water supply are short-turn, or radius, with ledges or shelves. The radius describes the curve or bend of a fitting that changes direction. As the radius increases, the change in flow direction becomes more gradual and less turbulent. This is particularly important for waste drain systems, where the smoother flow keeps organic waste from being deposited along the pipeline.

The most common types of water supply fittings include those listed in *Table 3* and shown in *Figure 8*.

Table 3 Water Supply Fittings and Their Uses

Fitting	Use
Union	Mechanically connects two pipes, usually at a termination downstream of a service valve
Reducer	Connects pipes of different sizes
Elbow	Changes the direction of rigid pipe by either 90 degrees or 45 degrees
Tee	Provides an opening to connect a branch pipe at 90 degrees to the main pipe run
Coupling	Joins two lengths of the same pipe size when making a straight run
Cap	Plugs water outlets when testing the system or creates an air chamber to eliminate water hammer
Plug	Closes openings in other fittings or seals the end of a pipe
Manifold	Runs several water supply lines from the main supply to different fixtures

1.3.2 *Water Supply Valves*

Valves regulate the flow of liquid. They may provide on/off service or prevent flow reversal through a line. Valves use one or more of the following methods to control flow through a tubing system:

- Move a disc or plug into or against a passageway.
- Slide a flat cylindrical or spherical surface across a passageway.
- Rotate a disc or ellipse around a shaft extending across the diameter of a tube.
- Move a flexible material into the passageway.

The Schedule System

Pipe schedules are tables of pipe sizes and wall thicknesses established by consensus standards such as those developed by the American Society of Mechanical Engineers (ASME). The schedule number is calculated from a number of factors, most importantly the safe working pressure of the pipe for a given nominal pipe size and material.

As the working pressure increases for, say, 2-inch pipe, the pipe wall thickness increases, increasing the pipe's schedule number and reducing its ID. More importantly, for a given schedule number and pipe material, as the pipe OD increases, the wall thickness increases with an increasing SDR. In other words, the pipe's OD increases faster than the wall thickness. So, while SDR is simply a ratio of pipe OD to wall thickness, the schedule number is determined by engineering calculations.

A History of Plastic Pipe

Although plastic materials became popular in the 1950s and 1960s, plastics date back more than a century to the introduction of celluloid in 1870. The 1930s and 1940s saw the introduction of Lucite™, Plexiglas™, and nylon. PVC, the first plastic used for plumbing purposes, was under development in the 1930s.

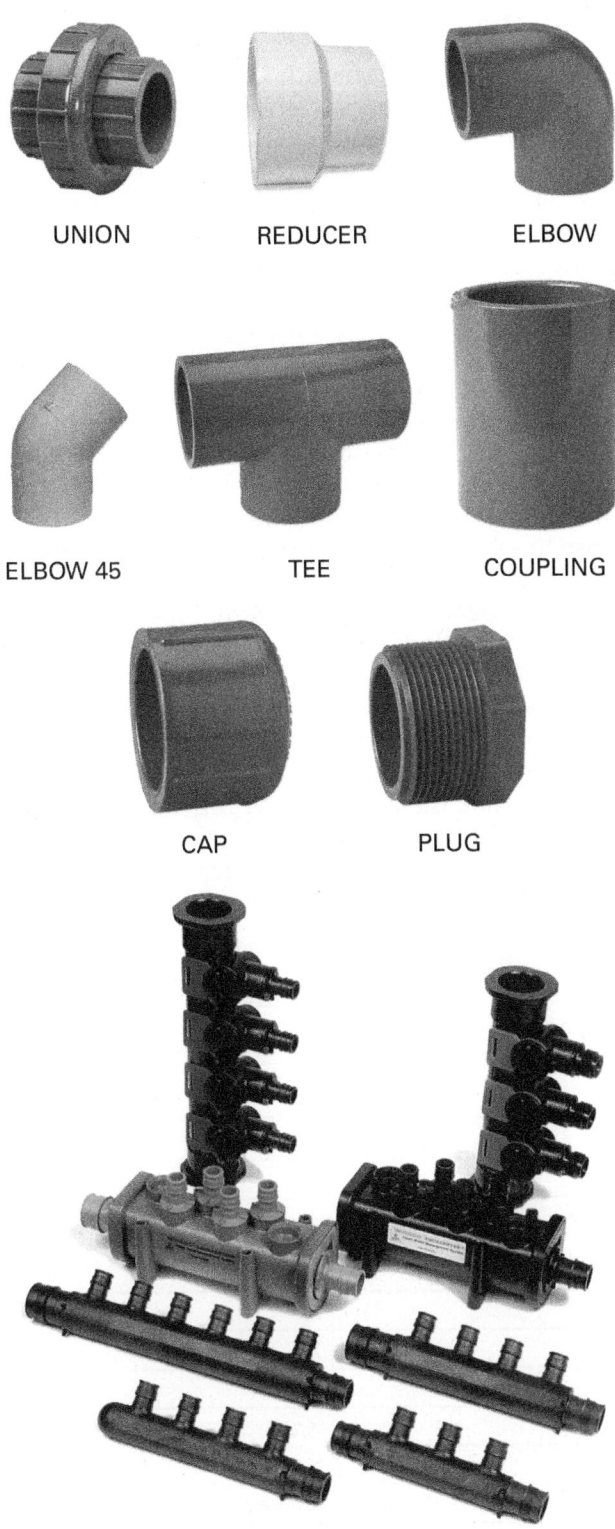

UNION   REDUCER   ELBOW

ELBOW 45   TEE   COUPLING

CAP   PLUG

MANIFOLDS

Figure 8 Water supply fittings.

Common valves for water supply systems include the following (*Figure 9*):

- *Gate*—A valve with a wedge-shaped or tapered disc that fits into a smooth-ground surface or seat with the same shape, allowing a straight-line flow with little obstruction. It is a good choice for lines that remain either completely open or completely shut most of the time.
- *Globe*—A valve that controls the flow of liquid with a conical disc, which raises or lowers to control flow through the valve seat opening. Because this valve is reliable and easy to repair, it is often used in water supply lines inside buildings.
- *Ball*—A quick-acting valve that consists of a ball with a hole bored through its diameter, mounted on a spindle. When the valve is shut, the hole is at 90 degrees to the flow passage through the valve body, so no flow can take place. When the ball is turned a quarter turn, water flows through the hole.
- *Compression, stop, and waste*—A valve that is opened or closed by raising or lowering a horizontal disc using a threaded stem. An elastomeric, or rubberized, washer on the end of the stem seals the valve seat, closing off water flow. This valve is most commonly used for draining and freeze protection above ground.
- *Check*—An automatic valve that permits the flow of liquid in one direction only. It prevents reverse flow. This valve is commonly used on domestic and irrigation wells.
- *Stop*—A valve that controls flow of liquids or gases between a building and supply source. It is also called a ground-key valve.

GATE VALVE

GLOBE VALVE

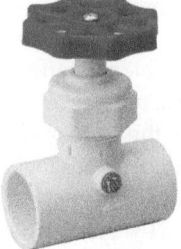
STOP & WASTE VALVE

CHECK VALVE

BALL VALVE

STOP VALVE

Figure 9 Plastic water supply valves.

1.3.3 DWV Fittings

DWV fittings have smooth interior passages with no ledges. Fittings that direct the flow of wastes also have a longer radius that makes directional changes smoother and less likely to collect solids (see *Table 4* and *Figure 10*). When selecting DWV fittings, always consult all applicable codes and manufacturers' installation recommendations.

1.4.0 Material Storage and Handling

To ensure maximum productivity on the job, use common sense when storing and handling plastic pipe. Before starting to work, always plan ahead. Organize pipe and fittings into groups by pipe size and type. In addition, store pipe and fittings close to where you are working so they are easy to access. When handling pipe and fittings, be careful not to bend or damage them. Pipe that is bent or otherwise damaged not only costs money, but also makes you less productive.

All plastic pipe is affected by ultraviolet (UV) radiation to a greater or lesser degree, unless rated for outdoors use. That means that you should store most plastic piping under opaque cover during a construction project to protect it from the sun. In general, short-term sun exposure does not harm the piping, but long-term exposure degrades the plastic material. Always consult the manufacturer's instructions regarding the proper way to select and protect the particular piping you are using.

Table 4 DWV Fittings and Their Uses

2009 International Plumbing Code®			
Table 706.3, Fittings for Change in Direction			
	Change in Direction		
Type of Fitting Pattern	**Horizontal to Vertical**	**Vertical to Horizontal**	**Horizontal to Horizontal**
Sixteenth bend	X	X	X
Eighth bend	X	X	X
Sixth bend	X	X	X
Quarter bend	X	Xa	Xa
Short sweep	X	Xa,b	Xa
Long sweep	X	X	X
Sanitary tee	Xc		
Combination wye and eighth bend	X	X	X

For SI: 1 inch = 25.4 mm

a. The fittings shall only be permitted for a 2-inch or smaller fixture drain.

b. Three inches or larger.

c. For a limitation on double sanitary tees, see Section 706.3.

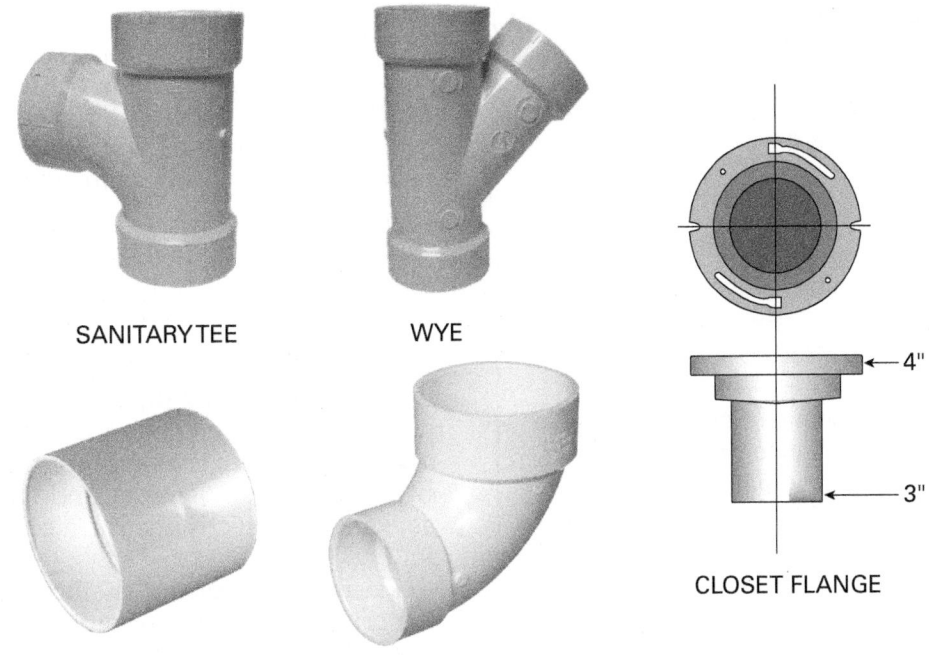

SANITARY TEE   WYE

CLOSET FLANGE

←— 4"

←— 3"

COUPLING   3×4 REDUCING CLOSET BEND

Figure 10 DWV fittings.

Additional Resources

Plumber's Handbook Revised, Howard C. Massey. 2006. Carlsbad, CA: Craftsman Book Company.

Plumbing Installation and Design, L. V. Ripka. Third Edition, 2006. Orland Park, IL: American Technical Publishers.

1.0.0 Section Review

1. The number of types of plastic pipe that the plumbing industry uses primarily is _____.

    a. 3
    b. 4
    c. 5
    d. 6

2. SDR indicates the ratio between a pipe's outside diameter and its _____.

    a. inside diameter
    b. wall thickness
    c. length
    d. flexibility

3. A check valve performs which of the following functions?

    a. Regulates water pressure downstream
    b. Quick-acting stopping and starting of flow
    c. Normally completely open or completely shut most of the time
    d. Prevents reverse flow in a pipe

4. Protect plastic pipe from long-term sun exposure because it _____.

    a. will melt
    b. will darken in color
    c. will degrade
    d. is flammable

2.0.0 WORKING WITH PLASTIC PIPE AND FITTINGS

Objective

Describe the different methods for joining plastic pipe.

  a. Describe how to measure and cut plastic pipe.
  b. Describe how to join PVC and CPVC pipe.
  c. Describe the installation procedures for PVC bell-and-spigot pipe.
  d. Describe the methods for joining PEX and PE tubing.

Performance Tasks

  3. Select the appropriate personal protective equipment for working with plastic piping.
  4. Properly measure, cut, and join plastic piping.

Trade Terms

Bell-and-spigot pipe: Pipe that has a bell, or enlargement, also called a hub, at one end of the pipe and a spigot, or smooth end, at the other end. The bell and spigot of two different pipes slide together to form a joint. Also called hub-and-spigot pipe.

Chamfer: To bevel the edge of construction material to a 45-degree angle.

Compression collars: A piece of hardware that uses compression force to connect sections of polyethylene piping.

Fusion fittings: A fitting with a butt that has the same outside diameter and inside diameter as the pipe. It is usually joined to a pipe by heat.

Interference fit: A fit that tightens as the pipe is pushed into the socket.

Primer: A liquid applied to plastic pipe prior to solvent welding in order to clean and pre-soften the pipe, and ensure a strong solvent weld.

Ring-tight gasket fittings: Fitting with a rubber O-ring or gasket in the socket.

Solvent weld: A joint created by joining two plastic pipes using solvent cement that softens the material's surface and creates a solid bond of plastic when the solvent evaporates.

Transition fitting: A special fitting used to connect plastic pipe to pipe of a dissimilar material, as specified by applicable code.

Techniques for measuring, cutting, and joining plastic pipe and fittings vary depending on the materials you are using and the function of the pipe (DWV or gas, for instance). There are many tools you need to become familiar with in order to measure, cut, and join materials properly.

2.1.0 Measuring and Cutting Plastic Pipe

Efficient pipe installation requires being able to properly measure and cut pipe so there is no wastage. Measuring pipe for cutting is more than just measuring the distance between two points. You have to take into account the dimensions of fittings, insertion lengths, and distances between sockets to correctly measure pipe before cutting it.

Plastic pipe cutting methods depend on the material and the size of pipe to be cut. You must become familiar with the tools and techniques required to properly cut and prepare pipe for installation.

2.1.1 Measuring

It is important to plan ahead when you are installing ABS and PVC pipe and fittings in DWV systems. These systems have a built-in slope, also called a pitch or fall, that is explained elsewhere in this curriculum, and you must lay them accurately, with pipe cut to exact lengths. You cannot fix mistakes later with heat or hammers. The mistake has to be cut out and corrected.

When measuring any pipe, be sure to allow for depth of joints. Take measurements to the full depth of the socket, not with pipe partly inserted into the socket. This is especially important when using solvent cement to join piping. With this method, you must dry fit the installation, then mark alignment for fittings before you make the joint.

2.1.2 Cutting

Plastic pipe requires a square cut for good joint integrity. Cut tubing as squarely as possible to create the best bonding area within a joint. If you see any indication of damage or cracking at the tubing end, cut off at least 2 inches beyond any visible crack.

> **WARNING!**
>
> Follow all manufacturer-recommended precautions when cutting or sawing pipe or when using any flame, heat, or power tools. Always wear appropriate PPE.

After cutting, you should ream the pipe. Reaming removes the fuzzy edges or burrs that result when you cut pipe. Burrs left on piping can prevent proper contact between tube and fitting during assembly. A pipe that is reamed correctly provides a smooth inner surface for better flow. You must also remove burrs on the outside of the pipe to ensure a good fit. Create a slight bevel on the end of the tube to make it easier for the tube to fit into the fitting socket and lessen the chances of pushing solvent cement to the bottom of the joint. Tools used to ream pipe ends include the reaming blade on the tubing cutter, files, pocket knives, and deburring tools (*Figure 11*).

You can cut PVC and ABS pipe with appropriate pipe cutters, a handsaw, or a power saw equipped with a carbide tip or abrasive blade. Plastic pipe cutters have one to four cutting wheels that can be rotated around a pipe to cut it. Wheels are available to fit standard cutters. You can also use ratchet shears or lightweight, quick-adjusting cutters designed exclusively for plastic piping (*Figure 12*).

To make sure you get a square cut, use a power saw with a miter gauge or fence on large jobs and a miter box on small jobs (*Figure 13*). Ensure that you use the proper plastic saw for cutting PVC pipe. If these are not available, scribe the pipe and cut to the mark. After cutting the pipe, ream it inside and chamfer the edge to remove burrs, shoulders, and ragged spots. Chamfering

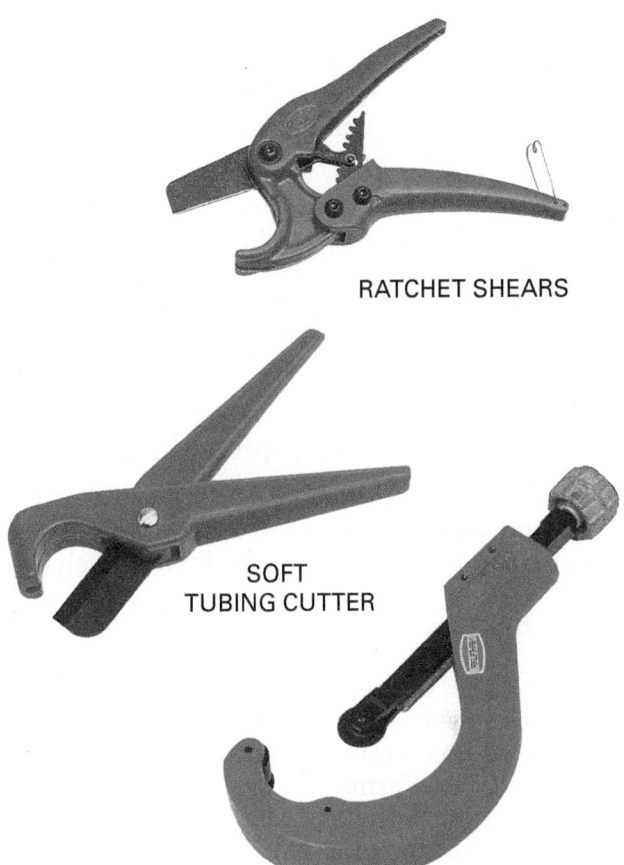

RATCHET SHEARS

SOFT TUBING CUTTER

PIPE CUTTER

Figure 12 Cutting tools used for plastic pipe.

Use Sharp Wheels

Do not use pipe cutters with dull wheels, especially wheels that have been used to cut metal, because they will exert too much pressure, causing larger shoulders and burrs that you will need to remove. Always use blade-type wheels, which actually slice through the plastic.

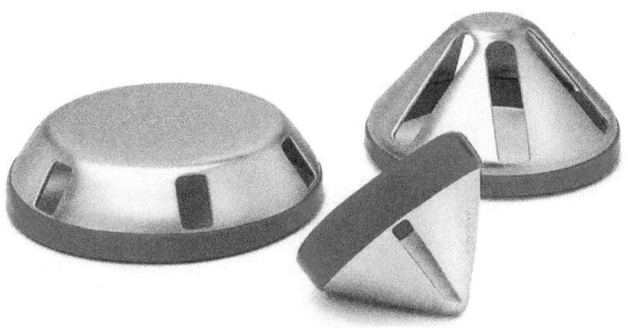

DEBURRING TOOLS

INSIDE OUTSIDE REAMER

Figure 11 Deburring tools.

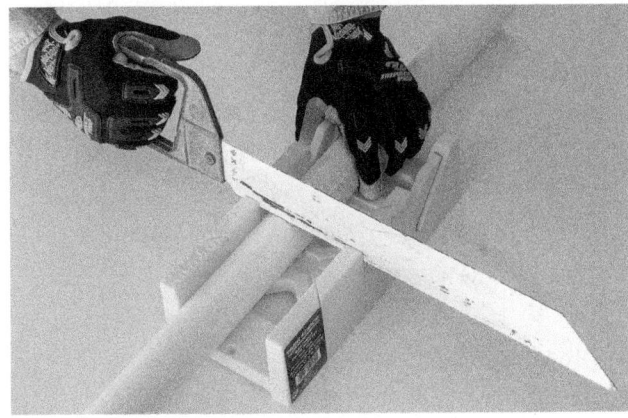

Figure 13 Using a miter box on small jobs.

involves beveling the edge of the pipe to a 45-degree angle. Chamfering is good practice because it provides for a more secure joint. When the socket and pipe fit tighter, you reduce the possibility of leakage from the pipe.

CAUTION

Mark carefully! Proper alignment in the final assembly is critical. To ensure alignment, carefully mark the positions of any fittings that will be rolled or otherwise aligned.

2.2.0 Joining Plastic Pipe and Fittings

Typically, PVC water supply and DWV fittings are joined using the same installation techniques. Among the installation methods are mechanical fittings, such as ring-tight gaskets, bell-and-spigot, threaded joints, heat fusion, and solvent weld.

Ring-tight gasket fittings have a rubber O-ring or gasket in the socket. You must bevel the pipe at the leading edge to allow it to pass the gasket that forms the seal (*Figure 14*). The bell-and-spigot pipe has a bell on one end with an internal elastomeric seal. The spigot (straight end) of the next pipe is fitted into the bell end to form a fluid-tight joint. Plumbers can also connect some plastic pipe to dissimilar pipe with a transition fitting.

There are many different types of fusion fittings, so always consult your local applicable code and the manufacturer's specifications for proper joining and installation methods (*Figure 15*). Fusion fittings and how to create a fusion weld are explained elsewhere in this curriculum.

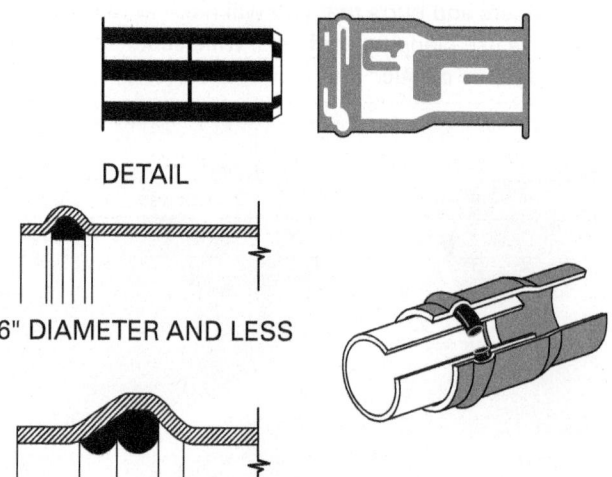

DETAIL

6" DIAMETER AND LESS

6" DIAMETER AND MORE

Figure 14 Ring-tight gasket fitting.

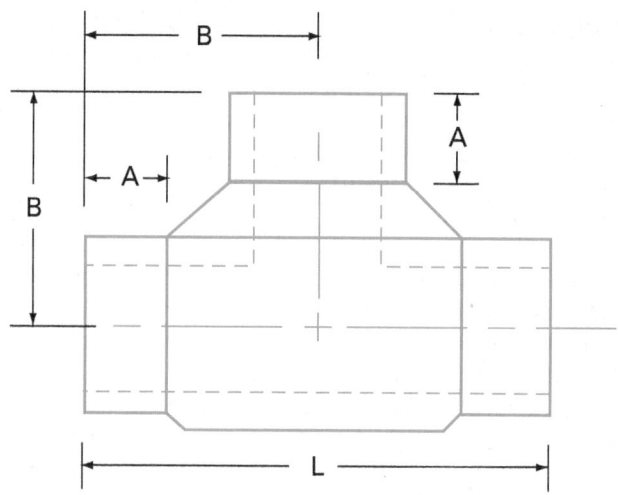

45-DEGREE ELBOW — BUTT FUSION — MOLDED

TEE — BUTT FUSION — MOLDED

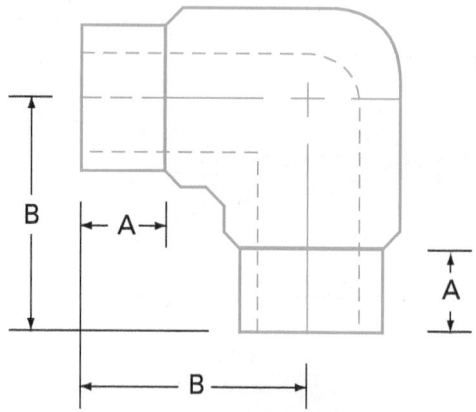

90-DEGREE ELBOW — BUTT FUSION — MOLDED
Figure 15 Fusion fittings.

2.2.1 Solvent Welding

Plumbers often join ABS, PVC, and CPVC plastic pipe and fittings with solvent cement to form a solvent weld. The process of solvent welding is also referred to as solvent cementing. Solvent-weld fittings have sockets that the pipe fits into (*Figure 16*). Plumbers apply the solvent cement to the pipe end and the inside of the fitting end, which temporarily softens the joining surfaces. This brief softening period lets you seat the pipe into the socket's interference fit—that is, the fit gets tighter as the pipe is pushed into the socket. The softened surfaces then fuse together, and joint strength develops as the solvents evaporate. The resulting joint is stronger than the pipe itself. A major advantage of the solvent-weld process is that it eliminates the need for torches or lead pots, which helps prevent the risk of fire during plumbing installations. *Table 5* shows the temperature range for a typical solvent. Always refer to the manufacturer's instructions for the appropriate temperature ranges for the type of solvent and pipe material being used.

The solvent cements you use will depend on the materials you are solvent welding. Each type of plastic has its own cement, and the cement may vary based on the size of the pipe and fittings. Be careful to choose the right cement for the job. Most codes require that you use primer, which cleans and softens the plastic, before solvent cementing pipe.

Because the cement hardens fast, you must move quickly and efficiently when joining pipe. The manufacturer's instructions will show minimum cure times for different size tubes at different temperatures before you can pressure-test the joint. Solvent cement and cure times also depend on relative humidity. Cure time is shorter for drier environments, smaller sizes, and higher temperatures.

> **WARNING!**
> Avoid eye or unnecessary skin contact with primers and cements. If contact occurs, wash immediately. Use protective eyewear and gloves during any solvent-weld procedure and consult the SDS for the cement you are using. An SDS should be available to everyone on the job site.

> **WARNING!**
> Keep primers and cements away from any sources of ignition, such as heat and open flames, and always apply these products in a well-ventilated area. Refer to the manufacturer's instructions for specific safety precautions.

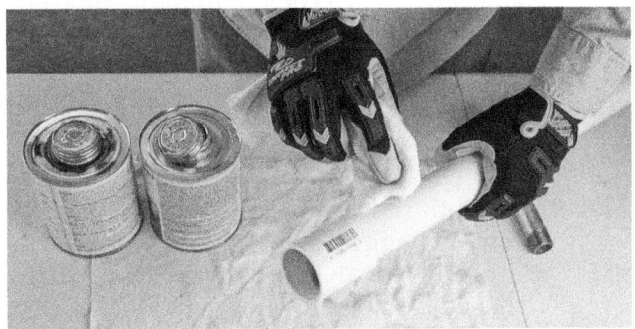

Figure 16 Solvent-welded fitting.

Table 5 Temperature Range for a Typical Solvent

Humidity: 60% or Less (180 psi)				
	Pipe Size			
Temperature Range	**Up to 1.25"**	**1.15" to 3"**	**3½" to 8"**	**10" to 18"**
60 to 100°F	1 hour	2 hours	6 hours	12 hours
40 to 60°F	2 hours	4 hours	12 hours	24 hours
20 to 40°F	6 hours	12 hours	36 hours	96 hours
Humidity: 60% or Less (350 psi)				
	Pipe Size			
Temperature Range	**Up to 1.25"**	**1.15" to 3"**	**3½" to 8"**	**10" to 18"**
60 to 100°F	6 hours	12 hours	18 hours	48 hours
40 to 60°F	12 hours	24 hours	36 hours	96 hours
20 to 40°F	36 hours	72 hours	96 hours	288 hours

Note: Figures shown should only be used as a general guide. Conditions in the field may vary, thereby affecting the times displayed.

2.2.2 Joining CPVC or PVC Pipe and Fittings

When you join CPVC or PVC pipe and fittings, use only CPVC or PVC cement or all-purpose cement conforming to *ASTM F-493* standards, or the joint may fail. Use special care when assembling CPVC and PVC systems in extremely low temperatures (below 40°F) or extremely high temperatures (above 100°F). In extremely hot environments, make sure both surfaces to be joined are still wet with cement when you put them together. If the cement has dried, the two surfaces may not adhere. Adapters are available to connect CTS CPVC pipe to PVC Schedule 40 and 80 pipe for systems requiring piping diameters larger than 2 inches.

Always store primer and cement in the shade between 40°F and 110°F and keep the containers tightly closed and covered when not in use. When applying primer and cement, remove all ignition sources and take precautions to avoid breathing the vapors.

To join CPVC or PVC pipe and fittings with solvent cement, perform the following steps (*Figure 17*). Note that these steps illustrate typical instructions, but when joining pipe, always follow the cement manufacturer's instructions.

Step 1 Cut the pipe as squarely as possible using a mechanical saw, fine-toothed saw, or wheel-type plastic pipe cutter. Use a miter box when cutting with a saw. Cut off at least 2 inches beyond any visible damage or cracks.

Step 2 To ensure proper contact between the pipe and fitting during assembly, use a chamfering tool, file, or reamer to remove burrs and filings from the inside and outside of the pipe. Bevel the end of the pipe to permit the pipe to fit smoothly into the fitting socket. This also helps reduce the chances solvent cement being scraped to the bottom of the joint during insertion.

Step 3 To ensure proper curing and adhesion, use a clean, dry rag to wipe dirt and moisture from the fitting socket and the pipe end. The pipe should make contact with the socket wall between $1/3$ and $2/3$ of the way into the fitting socket.

Step 4 Measure the depth of the fitting socket and mark this distance on the pipe. Make a second mark on the pipe 2 inches further back along the pipe to ensure that the pipe is fully inserted into the fitting. Note that the primer and solvent cement will usually erase the first mark.

Step 5 Use an applicator that is one-half the size of the pipe diameter to work the primer into the fitting socket aggressively. Re-dip the applicator in the primer as required, and ensure that the socket and applicator remain wet until the surface has been softened. Once the surfaces have been primed, use the applicator to remove any puddles of primer that have collected in the socket.

Step 6 Once the fitting socket has been correctly primed, aggressively work the primer around the end of the pipe to a depth that is approximately $1/2$ inch beyond the socket depth.

Step 7 Apply a second coat of primer to the fitting socket. You will need to begin the solvent cementing process while both surfaces are still tacky.

Step 8 While the primer is still tacky, apply the solvent cement using the same size applicator as for the primer. Aggressively work an even layer of cement onto the pipe end along a length that is equal to the depth of the fitting socket. Throughout the solvent application process, ensure that the pipe and fitting remain clean and free of dirt and moisture.

Step 9 Without re-dipping the applicator in the cement, aggressively work a medium-thick layer of cement into the fitting socket. Do not allow the cement to puddle in the fitting socket.

Step 10 Apply a second full layer of cement evenly on the pipe.

Step 11 While the cement is still wet, immediately assemble the pipe and fitting. If possible, rotate the pipe $1/4$ to $1/2$ turn until reaching the fitting stop. Using the second mark on the pipe as a reference, ensure that the pipe is fully inserted into the fitting. If applicable, rotate the fitting to the alignment mark made while dry fitting. Hold the assembly together for approximately 30 seconds to prevent the pipe from being pushed out of the fitting. You should see a continuous bead of cement around the connection of the pipe and fitting. A bead that is not continuous indicates that insufficient cement was applied, which could result in a defective joint. If this happens, discard the fitting and reassemble the joint by following Steps 1 to 11.

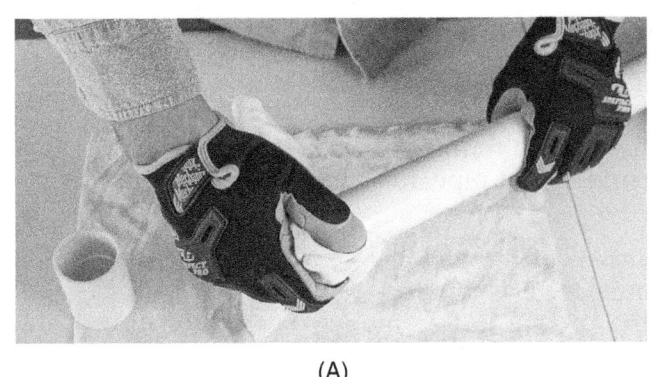

(A)

(B)

(E)

(C)

(F)

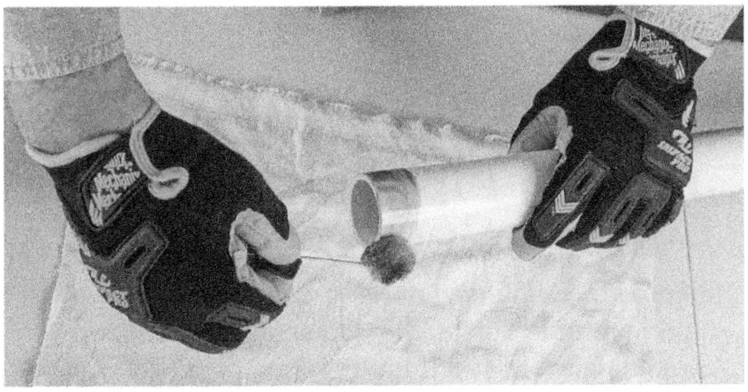

(D)

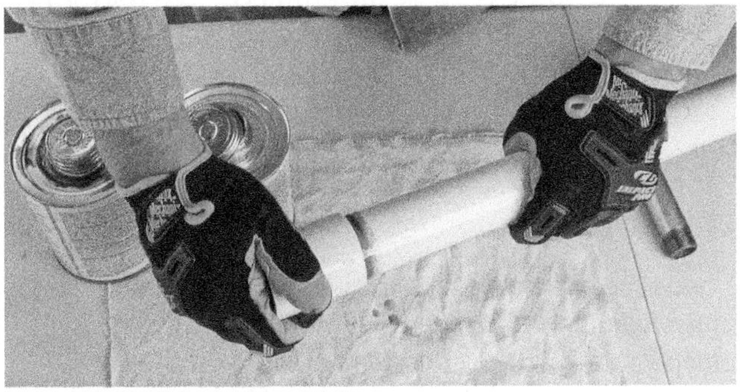

(G)

Figure 17 Joining CPVC or PVC pipe and fittings with solvent cement.

Step 12 Wipe excess cement from the pipe and fitting surfaces to ensure an attractive and professional appearance.

CAUTION

Never let solvent cement freeze. Always maintain solvent cement within the temperature range specified in the manufacturer's specifications.

2.3.0 Installing PVC Bell-and-Spigot Pipe

PVC bell-and-spigot pipe is generally used outdoors for gravity sewers. These outdoor systems are typically installed to connect with municipal utilities. To install PVC bell-and-spigot pipe, follow these steps (*Figure 18*):

Step 1 Prepare the inner surface of the bell according to the manufacturer's instructions. Ensure that the groove is free of dirt and other particles.

Step 2 Fold the gasket into a heart shape with the nose or rounded part of its cross section facing out of the mouth of the bell.

Step 3 Insert the gasket into the bell and work it into its groove until it is smooth and free from waves. You may have to snap the gasket, or wet it with clean water or a wet rag, to make sure it goes in place completely. Mark the pipe, creating a memory mark to show the proper position for alignment.

Step 4 After you place the gasket in the bell, thoroughly coat its exposed surface with lubricant. Then apply the lubricant to the entire surface of the spigot end up to the memory mark. Make especially sure that the tapered portion of the spigot is thoroughly coated. When you have finished lubricating, the pipe is ready to be joined.

Step 5 Line up the spigot with the bell and insert it straight into the bell. The spigot end of the pipe has a mark to indicate the proper depth of insertion. This mark will be about flush with the end of the bell when the joint is fully assembled. The memory mark should not be more than $3/8$ inch from the end of the bell after assembly.

CAUTION

When you are installing a ring-tight PVC gasket, you must gradually assemble the pipe either by hand or by using a bar or block. Never swing or stab the pipe to join it.

2.4.0 Joining Plastic Tubing

While solvent-welding is a common method of joining plastic pipe, modern plastic tubing, particularly PEX and PE materials, require a completely different approach. In the following sections, you will learn about the tools and methods required for joining these types of plastic tubing. The technology associated with plastic tubing is rapidly evolving, so professional plumbers need to continually familiarize themselves with the latest methods.

2.4.1 Joining PEX Tubing

Because PEX tubing resists high temperature and chemicals, you cannot join it with solvent cement or heat fusion. There are several methods for joining PEX tubing. One of the most common methods is to use an expansion and a crimp-ring system. Other tools used to join PEX tubing include the tubing cutter, hand-crimping tool, and go-no-go gauge (*Figure 19*).

NOTE

Crimping tools must be calibrated regularly. Consult and follow the manufacturer's specifications for the correct calibration intervals.

Purple versus Clear Primer

Two types of plastic pipe primer are available: purple and clear. Use purple primer; code inspectors will look specifically for the purple stain as proof that the pipes were primed before the solvent cement was applied. In some cases, with permission, you may use a clear primer on trim-out work.

Joining Pipe with Solvent Cement

Before you cement a joint, make sure the pipe and fitting are free of dust, dirt, water, and oil. Solvent cement acts fast. It is important to move quickly and efficiently when you are joining pipe using solvent welding.

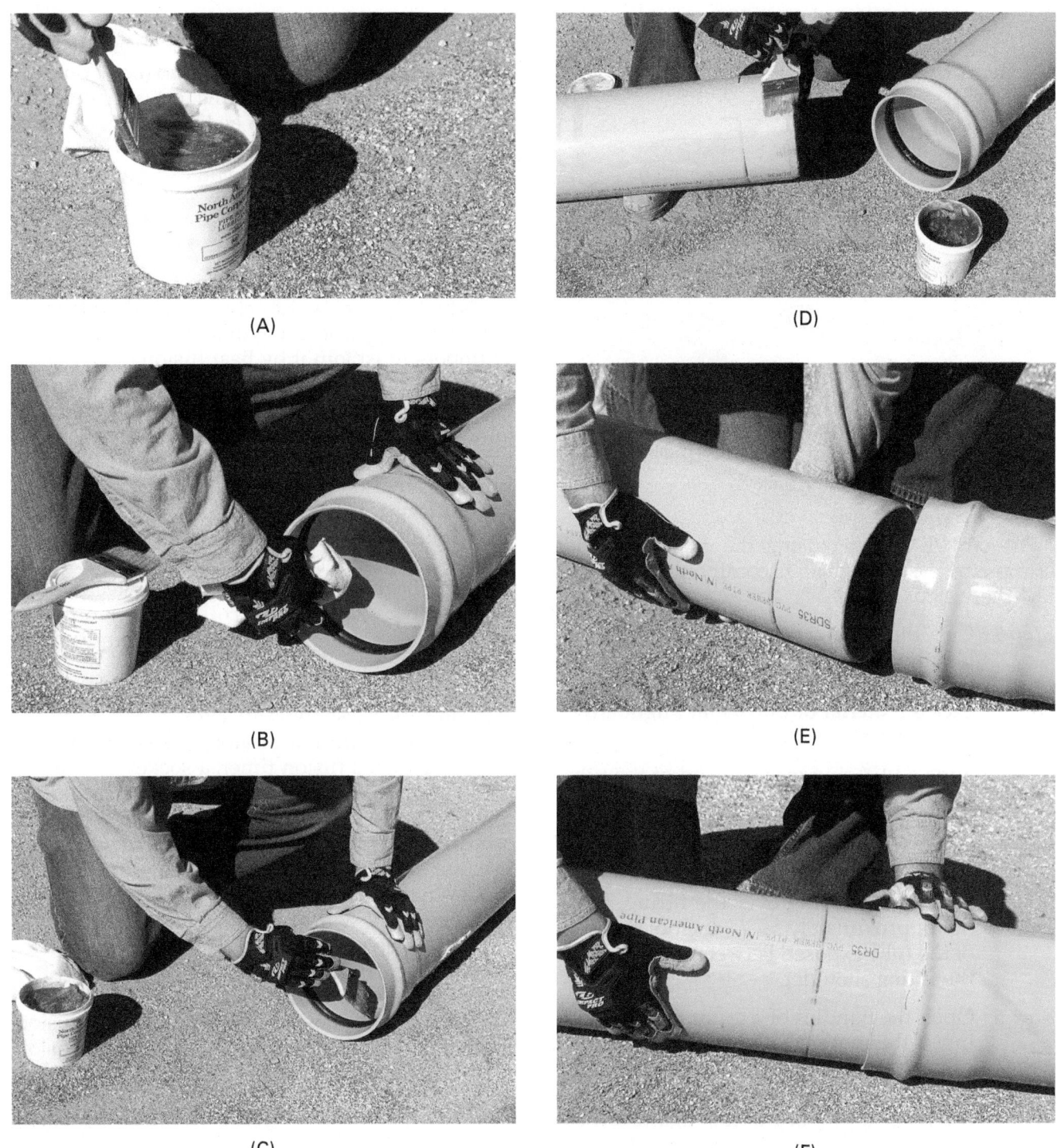

(A)

(D)

(B)

(E)

(C)

(F)

Figure 18 Installing PVC gravity sewer pipe.

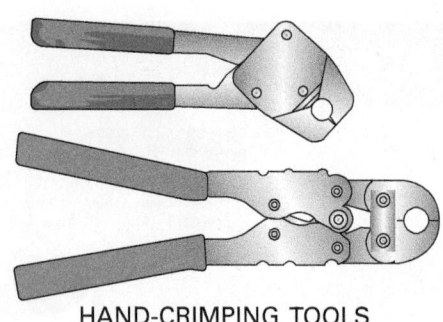

HAND-CRIMPING TOOLS

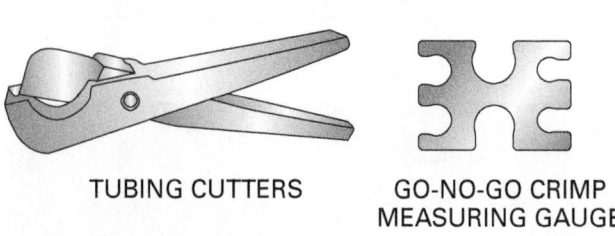

TUBING CUTTERS   GO-NO-GO CRIMP
                        MEASURING GAUGE

Figure 19 Tools for joining PEX tubing.

To join PEX tubing using the expansion and crimp-ring system, which is commonly used, follow these steps (*Figure 20*):

Step 1 Cut the tubing perpendicular to the length of the tubing, using a cutter designed for plastic tubing. Remove all excess material or burrs that might affect the fitting connection.

Step 2 Slide a PEX ring over the end of the tube about 2 inches from the end of the tube.

Step 3 Insert the fitting into the tube until its shoulder is flush with the end of the tube.

Step 4 Slide the PEX ring over the fitting, leaving approximately $\frac{1}{8}$–$\frac{1}{4}$ inch of the end of the tube exposed between the ring and the fitting shoulder.

Step 5 Open the handles of a crimping tool and insert the tool's head onto the end of the tubing until it stops. Be sure you have the correct size crimp head in the tool. Place the free handle of the tool against your hip, or place one hand on each handle when necessary.

Step 6 Make sure the crimping tool is perpendicular to the tubing. Fully separate the handles and bring them together. Repeat this process until the tubing and ring are snug against the shoulder on the crimp head. Before the final crimp, withdraw the tubing from the tool and rotate the tool one-eighth of a turn. This prevents the tool from forming ridges in the tubing.

Step 7 Check the crimp of the ring using a go-no-go tool per the manufacturer's instructions. If the crimp is incorrect, you will have to cut the fitting off and perform steps 1 through 6 again. A PEX-ring decrimping tool can be used to remove the ring and tubing so you can reuse the undamaged fitting.

2.4.2 Joining PE Tubing

PE tubing is resistant to chemicals, therefor solvent welding doesn't work for this material. Plumbers must join it by heat fusion or with mechanical joints and clamps. PE joined by fusion is similar to a weld on steel—the materials of the joined parts merge so they are indistinguishable from each other. This process gives the joint the same positive characteristics as the pipe itself.

PE fusion often requires special training and certification. Manufacturers of PE products and joining equipment often provide this training and certification free of charge. New techniques that involve compression collars for joining PE are becoming popular because they require less training.

Some of the tools that plumbers use in the fusion process are shown in *Figure 21*. These can include a temperature indicator stick or pyrometer, a heating tool, a fusion timer, a socket face, and a cold ring. Plumbers use the temperature indicator stick to make sure that piping has reached the required temperature for successful fusion. They use the stick to mark a particular area on the pipe. When the pipe reaches the desired temperature, the mark will melt. During heat fusion, the surface of the socket face comes in direct contact with the pipe or fitting.

There are several methods for joining PE tubing. One of the most common is the butt-fusion method. To join PE tubing using the butt-fusion method, follow these steps:

Step 1 Cut the ends of the tubing square with a tubing cutter.

Step 2 Mark the tubing with the proper temperature indicator stick and heat the tubing ends with a heating tool.

Step 3 When the tubing reaches the required temperature, remove the heating tool.

Step 4 Press the tubing ends together to form a tight seal at the joint.

Step 5 Allow the joint to cool before applying force.

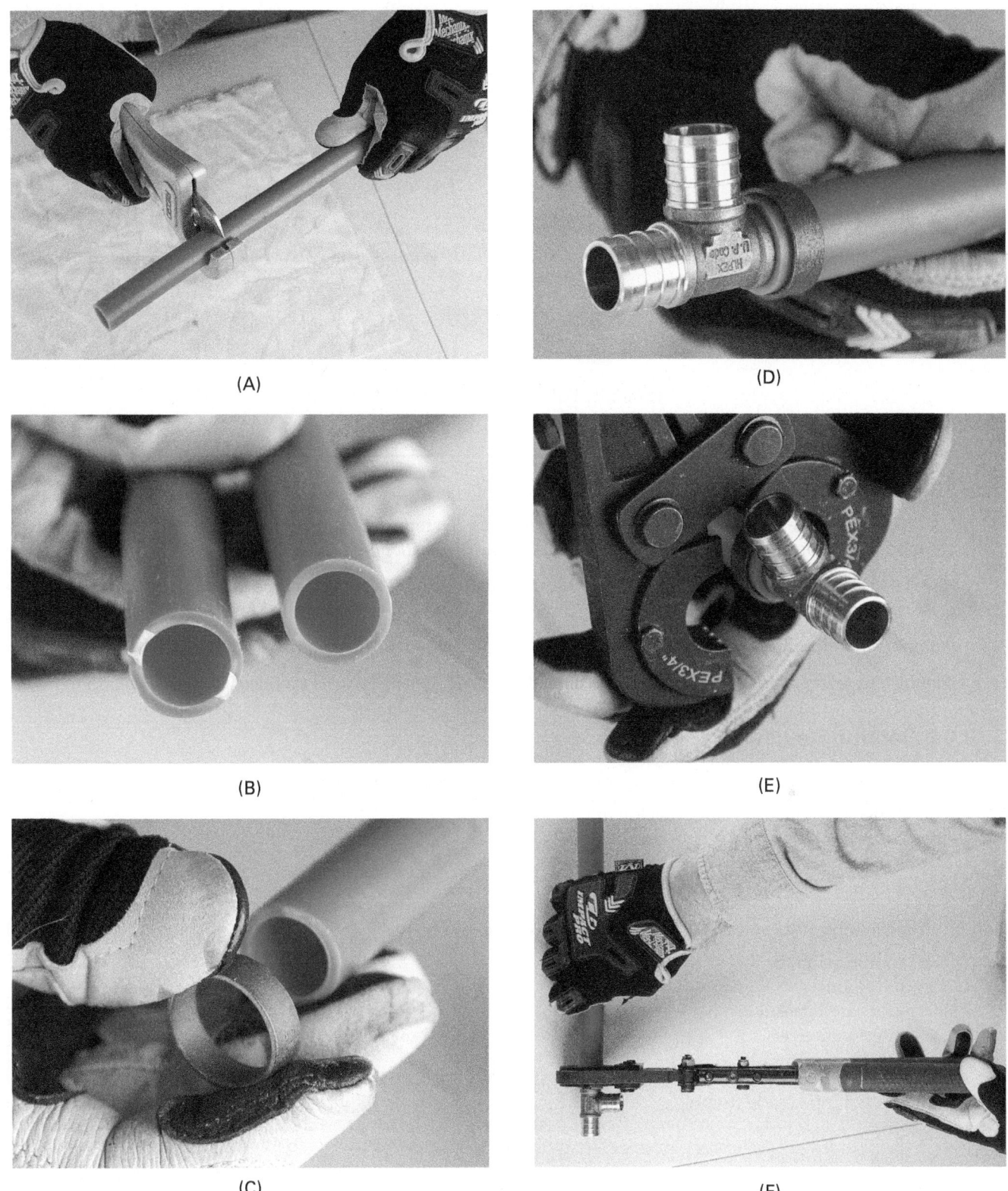

Figure 20 Joining PEX tubing.

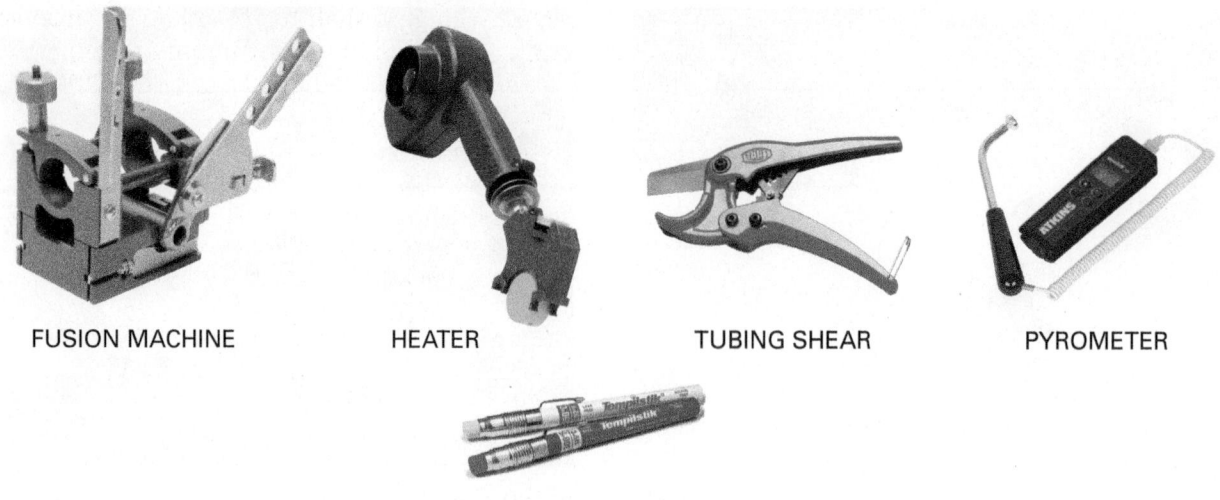

FUSION MACHINE   HEATER   TUBING SHEAR   PYROMETER

TEMPERATURE INDICATOR STICK

Figure 21 PE fusion tools.

Additional Resources

"DESIGN GUIDE, Residential PEX Water Supply Plumbing Systems," Second Edition. Revised 2013. Marlboro, MD: Home Innovation Research Labs. *http://www.plasticpipe.org*.

Handbook of Polyethylene Pipe, 2nd Edition. Revised 2009. Irving, TX: Plastics Pipe Institute. *http://www.plasticpipe.org*.

Plumber's Handbook Revised, Howard C. Massey. 2006. Carlsbad, CA: Craftsman Book Company.

Plumbing Installation and Design, L. V. Ripka. Third Edition, 2006. Orland Park, IL: American Technical Publishers.

2.0.0 Section Review

1. If you see any indication of damage or cracking at the tubing end, cut off at least _____.

    a. 1 inch beyond any visible crack
    b. 2 inches beyond any visible crack
    c. 3 inches beyond any visible crack
    d. 4 inches beyond any visible crack

2. When joining PVC and CPVC pipe, the joint bond relies on a(n) _____.

    a. compression fit
    b. butt-joint
    c. lubricated gasket
    d. interference fit

3. Outdoor gravity sewers, which are typically installed to connect with municipal utilities often use bell-and-spigot pipe made from _____.

    a. PVC
    b. PEX
    c. CPVC
    d. PE

4. Joining PEX pipe and fittings typically involves _____.

    a. solvent welding
    b. lubricated gaskets
    c. crimp rings
    d. fused joints

3.0.0 PLASTIC PIPE INSTALLATION AND TESTING

Objective

Describe the methods used to support and test plastic pipe.

  a. Describe the hangers and fasteners used to support pipe.

  b. Explain methods of pressure testing plastic pipe.

Performance Task

5. Select the correct support and spacing for the application.

Trade Term

Hydrostatic pressure test: To fill a pipe with water and bleed all air out from the highest and farthest points in the run.

Properly measuring and joining plastic pipe is only a part of the task when installing pipe systems. Piping systems must be properly supported to ensure fluids flow in the direction intended and to minimize flexing and vibration, which can cause pipe failure. This section introduces the basic types of pipe hangers and supports often used with plastic pipe.

3.1.0 Pipe Supports

Plastic pipe can be supported using several different methods. The type of support you use depends on the pipe material, its size, its use, and whether the pipe is installed in a horizontal or vertical position, as well as the system specifications and applicable plumbing codes.

When architects and engineering consultants design plumbing installations, they provide plans and specifications that completely describe the proposed system. Specifications are based on codes or ordinances and must be followed. A specification for pipe hangers, for example, may read, "All piping shall be supported with hangers spaced no more than 10 feet apart (on center). Hangers shall be the malleable iron split-ring type and shall be as manufactured by XYZ Hangers, Inc., or other approved vendor." You should always refer to applicable codes and the manufacturer's installation instructions for specific types and intervals appropriate for your particular installation. Local codes include spacing requirements for hangers for different types and sizes of plastic pipe, *Table 6* is an example of how hangers are spaced according to code.

If a plumber is installing pipe in a seismically-active area—that is, where earthquakes are a possibility—local codes require seismic restraints. The purpose of these additional requirements is to ensure that the pipe is securely fastened to the structure in the event of excessive vibration. For example, some codes require hangers and supports to be used at closer intervals than in nonseismically-active areas. In addition, they may require that a worker leave extra spacing for pipe where it meets walls and floors to allow for anticipated movement.

3.1.1 Hangers

Plumbers use hangers (*Figure 22*) for horizontal support of pipes and piping. The main purpose of hangers and brackets is to keep the piping in alignment and to prevent it from bending or distorting, but they can also prevent pipes from vibrating. Plumbers can attach horizontal hangers to wooden structures with lag screws or large nails. If vibration is a concern, follow the manufacturer's specifications to determine what type of hanger to use for your installation.

Hangers should be strong enough to support the weight of the pipe and its contents and maintain its alignment. In general, you should support horizontal PVC piping at intervals of 4 feet or less, as well as at branches and changes of direction and when using large fittings. Although supports should provide free movement, they must prevent lateral DWV runs from rising, which could create a reverse slope on branch piping and back up the system. Avoid hangers that may cut or squeeze pipe and tight clamps or straps that prevent pipe from moving or expanding. Size any holes made for pipe through framing members to allow for free movement. When working with piping in the ground, lay it on a firm bed for its entire length.

Table 6 Hanger Spacing

2003 International Plumbing Code®		
Table 308.5, Hanger Spacing		
Piping Material	**Maximum Horizontal Spacing (feet)**	**Maximum Vertical Spacing (feet)**
ABS pipe	4	10[b]
Aluminum tubing	10	15
Brass pipe	10	10
Cast-iron pipe	5[a]	15
Copper or copper-alloy pipe	12	10
Copper or copper-alloy tubing, $1\frac{1}{4}$-inch diameter and smaller	6	10
Copper or copper-alloy tubing, $1\frac{1}{2}$-inch diameter and larger	10	10
PEX pipe	2.67 (32 in)	10[b]
PEX/aluminum/PEX (PEX-AL-PEX) pipe	2.67 (32 in)	4[b]
CPVC pipe or tubing, 1 inch or smaller	3	10[b]
CPVC pipe or tubing, $1\frac{1}{4}$ inches or larger	4	10[b]
Steel pipe	12	15
Lead pipe	Continuous	4
PB pipe or tubing	2.67 (32 in)	4
PE/aluminum/PE (PE-AL-PE) pipe	2.67 (32 in)	4[b]
PVC pipe	4	10[b]
Stainless steel drainage systems	10	10[b]

For SI: 1 inch = 25.4 mm, 1 foot = 304.8 mm

a. The maximum horizontal spacing of cast-iron pipe hangers shall be increased to 10 feet where 10-foot lengths of pipe are installed.

b. Mid-story guide for sizes 2 inches and smaller.

3.1.2 Clamps

Plumbers often use beam clamps, C-clamps, and suspension clamps to fasten pipe to beams and other metal structures (*Figure 23*). Vertical hangers, also called pipe riser clamps, provide vertical support for pipe and tubing (*Figure 24*). These hangers consist of a friction clamp that you can attach to structural site components to support the vertical load of the pipe. You must use specific fasteners to attach hangers to masonry, concrete, or steel.

> **CAUTION**
> The area where the support is to be fastened should be smooth so that the item has solid footing. Uneven footing might cause the support being fastened to twist, warp, or not tighten properly.

> **CAUTION**
> Make sure the fastener is straight after working it around in the hole and installing the washer. The washer centers the fastener and holds it in place until the grout or epoxy hardens. If the grout or epoxy sets and the fastener is not straight, the fastener is unusable and has to be removed and the installation repeated.

Use supports for risers at each floor level or as required by the installation design. Mid-story guides can provide greater stability for vertical pipe that runs up through the building.

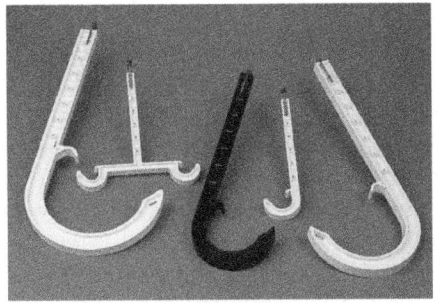
J-HOOKS

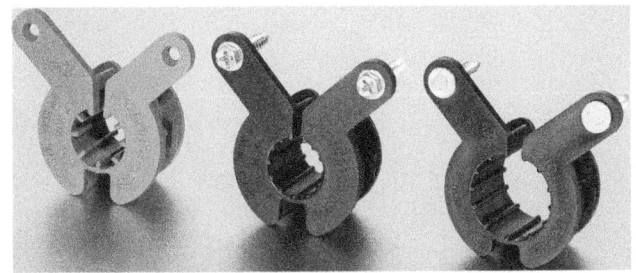
SUSPENSION CLAMPS

LOCKING TUBE STRAPS

SUPPORT BRACKET

I-BEAM CLAMP   BEAM CLAMP   C-CLAMP

Figure 23 Clamps.

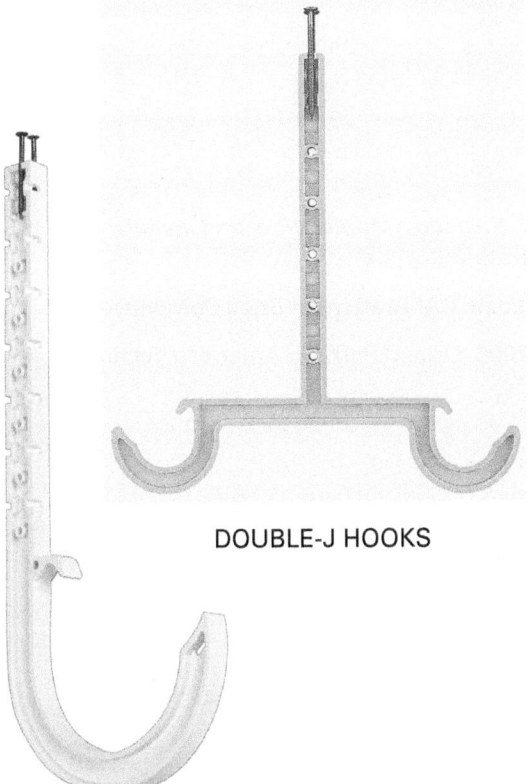
DOUBLE-J HOOKS

PIPE HOOK

Figure 22 Pipe hangers.

> **WARNING!**
> Do not use metal hangers on any plastic pipe. Always use hangers that are made of the same material as the pipe itself. For more information, refer to the MSS40 hanger standards established by the Manufacturers Standardization Society.

3.2.0 Pressure Testing

Once you have completed and cured an installation, you may be required to conduct a hydrostatic pressure test the system in accordance with applicable code requirements. This process involves filling the system with water and bleeding all air out from the highest and farthest points in the run. If you find a leak, you must remove and replace the fitting with the leaking joint. You can install a new fitting using short sections of pipe and couplings.

PIPE CLAMP

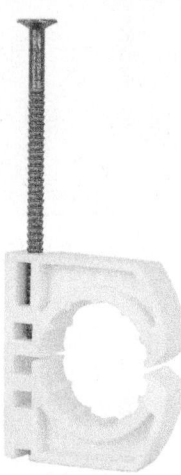

PIPE STRAP

Figure 24 Riser clamp.

During subfreezing temperatures, you should blow water out of the lines after testing to avoid possible damage to the pipes from freezing.

> **CAUTION**
>
> Never pressure-test a solvent-welded connection until the manufacturer's recommended cure times have been met. After testing a connection, thoroughly flush the system for at least 10 minutes to remove any remaining trace amounts of solvent cement.

> **WARNING!**
>
> Use air testing for plastic pipe (except PVC and CPVC) only when hydrostatic testing is not practical. If air testing must be performed, never use flammable or combustible gases. Avoid testing PVC and CPVC pipe with air. Improper air testing of these materials can cause the pipe to burst. If air testing is necessary, use a gauge that has a scale graduated in at least pounds or finer. Never test with more than 5 pounds of air pressure. When performing an air test, notify all site personnel before the test, use protective eyewear, and take precautions to protect the system from impact damage during the test. Carefully vent the system to atmospheric pressure before removing test caps and plugs.

Additional Resources

Pipe Hangers and Supports—Materials, Design, Manufacture, Selection, Application, and Installation, MSS SP-58. Current Edition. Vienna, VA: Manufacturers Standardization Society.

Plumber's Handbook Revised, Howard C. Massey. 2006. Carlsbad, CA: Craftsman Book Company.

Plumbing Installation and Design, L. V. Ripka. Third Edition, 2006. Orland Park, IL: American Technical Publishers.

3.0.0 Section Review

1. The main purpose of pipe hangers is to _____.
   a. ensure adequate pipe vibration
   b. allow pipes to flex easily
   c. maintain horizontal alignment of pipes
   d. maintain vertical alignment of pipes

2. If you find a leak when hydrostatically testing a connection, the connection must be _____.
   a. solvent welded
   b. removed and replaced
   c. epoxy patched
   d. air tested

Summary

The use of plastics in the plumbing trade has greatly increased the effectiveness and safety of indoor and outdoor plumbing systems. The introduction of different types of plastic pipe and tubing means that today's plumbers must know about new techniques, tools, and applications. Plastic may seem like the answer to all plumbing problems, but exactly which type of plastic you use makes a big difference. You must learn the special techniques for cutting and joining a variety of plastic pipe types. Recognizing the common types of materials used on a job site is a basic skill.

Developing technology affects plumbers now more than ever. Manufacturers are constantly researching and producing new products. To be competitive in the modern plumbing industry, you must keep up with innovations and improvements in installation techniques and materials.

1. ABS pipe absorbs heat _____.
   a. evenly
   b. quickly
   c. linearly
   d. slowly

2. Which of the following system conditions does CPVC pipe have more resistance compared to copper pipe?
   a. High temperatures
   b. Low temperatures
   c. Water hammer
   d. Water pressure

3. An example of a thermosetting plastic is _____.
   a. ABS
   b. PE
   c. PEX
   d. PVC

4. SDR is a fixed ratio between outside diameter and _____.
   a. wall thickness
   b. type of material
   c. fittings
   d. wyes

5. Which kind of fitting connects pipes of different sizes?
   a. Cap
   b. Reducer
   c. Elbow
   d. Manifold

6. During construction, plastic pipe should normally be stored outside under tarps in order to protect it from _____.
   a. ultraviolet light
   b. rain
   c. dirt
   d. rodents and insects

7. When measuring plastic pipe before cutting, you must take into account all of the following *except* _____.
   a. measured length between fittings
   b. insertion depth
   c. pipe diameter
   d. slope allowance

8. Before assembling plastic pipe joints involving gaskets, such as ring-tight gasket fittings, you should _____.
   a. apply primer to the joint surfaces
   b. mark the spigot end with a temperature indicator stick
   c. ream the bell end of the joint
   d. bevel the spigot end of the pipe

9. What is a major concern when joining PVC or CPVC pipe when working in hot temperatures?
   a. Pipe expansion
   b. Cement drying
   c. Pipe warping
   d. Cement running out of the joint

10. Which of the following should you avoid when assembling bell-and-spigot plastic drain pipe?
    a. Stabbing or swinging pipe sections together
    b. Lubricating the gasket
    c. Cleaning the bell-end groove
    d. Adding a memory mark to the spigot end

11. Because it resists high temperature and chemicals, you cannot use solvent cement or heat fusion when joining pipe made from _____.
    a. ABS
    b. PEX
    c. CPVC
    d. PE

12. Solvent welding doesn't work for this material, so you must use heat fusion or mechanical joints and clamps when joining _____.
    a. ABS pipe
    b. PEX tubing
    c. CPVC pipe
    d. PE tubing

13. In general, PVC piping should be supported at a maximum interval of _____.
    a. 3 feet
    b. 4 feet
    c. 5 feet
    d. 8 feet

14. The most commonly used clamps used as vertical support for pipe and tubing are _____.
    a. beam clamps
    b. C-clamps
    c. pipe riser clamps
    d. special fasteners

15. To avoid possible damage to the pipes from freezing during subfreezing temperatures, water should be blown out of pipe lines _____.
    a. before installation
    b. during installation
    c. before testing
    d. after testing

16. Plumbers commonly use PE as tubing because of its _____.
    a. strength, flexibility, and chemical-resistant properties
    b. strength, durability, and chemical resistant properties
    c. longevity, flexibility, and chemical resistant properties
    d. strength, flexibility, and ability to be used anywhere

17. Plumbers often use C-clamps to fasten pipe to _____.
    a. concrete structures
    b. metal structures
    c. thick structures
    d. thin structures

18. PEX tubing resists high temperature and chemicals, so it must be joined by _____.
    a. using solvent cement
    b. welding the parts
    c. threading and joining the pipe
    d. using a crimp-ring system

19. Reaming is a process after cutting pipe to remove _____.
    a. extra pipe
    b. burrs
    c. oxides
    d. solvents

20. All of the following are common valves used for water supply systems except for _____.
    a. gate valve
    b. ball valve
    c. butterfly valve
    d. globe valve

Trade Terms Quiz

Fill in the blank with the correct term that you learned from your study of this module.

1. A(n) _____ allows threaded plastic pipe to be connected to steel pipe.

2. _____ are tables relating pipe wall thicknesses to pipe pressure ratings.

3. Plastic is _____; it does not react with other substances or materials.

4. A(n) _____ means that the fit tightens as the pipe is pushed into the fitting socket.

5. The _____ are the ratios of pipe outer diameters to their wall thicknesses.

6. Plumbers commonly use _____ for manifold plumbing distribution systems because of its flexibility.

7. _____ changes chemically when heated, so that once hardened by heat or chemicals, it is hardened permanently.

8. After reaming cut pipe, _____ the edge to remove burrs, shoulders, and ragged spots.

9. The pressure that a pipe can withstand continuously, called the _____, is required on the pipe's label.

10. When softened, _____ can be molded into desired shapes.

11. Vertical hangers, also called _____, provide vertical support for pipes and tubing.

12. Pipe with _____ construction consists of a single layer of plastic that does not contain trapped air in the form of solid foam.

13. _____ pipe will not deteriorate when exposed to ultraviolet light, so you can install it outdoors without a protective coating.

14. Techniques for joining PE that use _____ are becoming popular because they require less training than heat fusion.

15. A(n) _____ heats and cools by circulating water or steam through a closed piping system.

16. After completing an installation, you must perform a _____ on the system by filling the system with water and bleeding all air out from the highest and farthest points in the pipe run.

17. A sudden change in water pressure due to stopping a moving mass of water is called _____.

18. Because it has been known to become weak and fail without warning in previous installations, _____ pipe is no longer available for installation today.

19. _____ has a bell on one end with an internal seal.

20. Pipe can be measured by its _____, which is the distance between its inner walls.

21. Pipe with _____ construction is more lightweight and less expensive than solid wall pipe.

22. _____ have a rubber O-ring or gasket in the socket.

23. For a given SDR, the ratio between a pipe's _____ and its wall thickness is constant for each pipe size.

24. _____ pipe performs well at extreme temperatures because it absorbs heat and cold slowly.

25. The molecular structure of _____ pipe practically eliminates condensation in the summer and heat loss in the winter, decreasing the likelihood of costly drip damage to walls or structure.

26. To _____ joints, plumbers apply solvent cement to the pipe end and the inside of the fitting end, which temporarily softens the joining surfaces.

27. A pipe's pressure rating is measured in _____.

28. To form joints using _____, you must heat the accessories to the manufacturer's specifications and press them together.

29. _____ is a rigid pipe with high-impact strength that is manufactured from a thermoplastic material and has an indefinite life span under most conditions.

30. Prior to applying solvent cement to plastic pipe, use an applicator to apply two coats of _____ to the pipe socket to ensure that it has softened thoroughly.

31. A rubberized material that is often used for gaskets is described as _____.

Trade Terms

ABS (acrylonitrile-butadiene-styrene)
Bell-and-spigot pipe
Cellular-core wall
Chamfer
Chemically inert
Compression collars
CPVC (chlorinated polyvinyl chloride)
Elastomeric
Fusion fittings
Hydronic

Hydrostatic pressure test
Inside diameter (ID)
Interference fit
Outside diameter (OD)
PB (polybutylene)
PE (polyethylene)
PEX (cross-linked polyethylene)
Pipe riser clamps
Pounds per square inch (psi)
Pressure rating
Primer
PVC (polyvinyl chloride)

Ring-tight gasket fittings
Schedules
Size dimension ratios (SDR)
Solid wall
Solvent weld
Thermoplastic pipe
Thermoset pipe
Transition fitting
Water hammer

Trade Terms Introduced in This Module

ABS (acrylonitrile-butadiene-styrene): Plastic pipe and fittings used extensively in drain, waste, and vent (DWV) systems.

Bell-and-spigot pipe: Pipe that has a bell, or enlargement, also called a hub, at one end of the pipe and a spigot, or smooth end, at the other end. The bell and spigot of two different pipes slide together to form a joint. Also called hub-and-spigot pipe.

Cellular-core wall: Plastic pipe wall that is low-density, lightweight plastic containing entrained (trapped) air in solid foam.

Chamfer: To bevel the edge of construction material to a 45-degree angle.

Chemically inert: Does not react with other chemicals.

Compression collars: A piece of hardware that uses compression force to connect sections of polyethylene piping.

CPVC (chlorinated polyvinyl chloride): Durable plastic pipe and fittings used extensively in hot and cold water distribution systems.

Elastomeric: Rubberized. Made of an elastic substance such as a polyvinyl elastomer.

Fusion fittings: A fitting with a butt that has the same outside diameter and inside diameter as the pipe. It is usually joined to a pipe by heat.

Hydronic: A system that heats and cools by circulating water or steam through a closed piping system.

Hydrostatic pressure test: To fill a pipe with water and bleed all air out from the highest and farthest points in the run.

Inside diameter (ID): The distance between the inner walls of a pipe; approximates the nominal sizes of piping used in heating and plumbing.

Interference fit: A fit that tightens as the pipe is pushed into the socket.

Outside diameter (OD): The distance between the outer walls of a pipe.

PB (polybutylene): Plastic piping that was formerly used for plumbing pipe; it is no longer used but is still found in some residences.

PE (polyethylene): Flexible plastic pipe, tubing, and fittings, usually used for water distribution, that do not deteriorate when exposed to sunlight.

PEX (cross-linked polyethylene): Tubing and fittings made with heat and high pressure that resist high temperatures, pressure, and chemicals.

Pipe riser clamps: A type of clamp-on pipe support that is designed to support the weight of vertical pipe and tubing.

Pounds per square inch (psi): A common measure of liquid and gas pressure in the United States.

Pressure rating: The maximum pressure at which a component or system may be operated continuously.

Primer: A liquid applied to plastic pipe prior to solvent welding in order to clean and pre-soften the pipe, and ensure a strong solvent weld.

PVC (polyvinyl chloride): Plastic pipe and fittings used for cold water distribution and for industrial water and chemicals, as well as for drain, waste, and vent (DWV) systems.

Ring-tight gasket fittings: Fitting with a rubber O-ring or gasket in the socket.

Schedules: Tables relating pipe wall thicknesses to pipe pressure ratings. Higher schedule numbers equate to thicker pipe walls for a given nominal pipe size.

Size dimension ratios (SDR): Ratios of the outer diameters of pipes to their pipe wall thicknesses.

Solid wall: Plastic pipe wall that does not contain trapped air in the form of solid foam.

Solvent weld: A joint created by joining two plastic pipes using solvent cement that softens the material's surface and creates a solid bond of plastic when the solvent evaporates.

Thermoplastic pipe: Pipe that can be repeatedly softened by heating and hardened by cooling. When softened, thermoplastic materials can be molded into desired shapes during manufacturing.

Thermoset pipe: Pipe that changes chemically when heated, so that once hardened by heat or chemicals, it is hardened permanently.

Transition fitting: A special fitting used to connect plastic pipe to pipe of a dissimilar material, as specified by applicable code.

Water hammer: An extreme change in water pressure within a pipe that occurs when suddenly stopping a moving mass of water, as when quickly shutting a valve. It can cause a loud, banging sound as pipes flex and even damage the system.

Additional Resources

This module is intended as a thorough resource for task training. The following reference works are suggested for further study.

"DESIGN GUIDE, Residential PEX Water Supply Plumbing Systems," Second Edition. Revised 2013. Accessed September 11, 2013. Marlboro, MD: Home Innovation Research Labs. *http://www.plasticpipe.org*.

Handbook of Polyethylene Pipe, 2nd Edition. Revised 2009. Accessed September 11, 2017. Irving, TX: Plastics Pipe Institute. *http://www.plasticpipe.org*.

Pipe Hangers and Supports—Materials, Design, Manufacture, Selection, Application, and Installation, MSS SP-58. Current Edition. Vienna, VA: Manufacturers Standardization Society.

Plumber's Handbook Revised, Howard C. Massey. 2006. Carlsbad, CA: Craftsman Book Company.

Plumbing Installation and Design, L. V. Ripka. Third Edition, 2006. Orland Park, IL: American Technical Publishers.

Figure Credits

SECTION **1.0.0**

Answer	Section Reference	Objective
1. d	1.1.2; Table 1	1a
2. b	1.2.1	1b
3. d	1.3.2	1c
4. c	1.4.0	1d

SECTION **2.0.0**

Answer	Section Reference	Objective
1. b	2.1.2	2a
2. d	2.2.1	2b
3. a	2.3.0	2c
4. c	2.4.1	2d

SECTION **3.0.0**

Answer	Section Reference	Objective
1. c	3.1.1	3a
2. b	3.2.0	3b

This page is intentionally left blank.

NCCER CURRICULA — USER UPDATE

NCCER makes every effort to keep its textbooks up-to-date and free of technical errors. We appreciate your help in this process. If you find an error, a typographical mistake, or an in-accuracy in NCCER's curricula, please fill out this form (or a photocopy), or complete the on-line form at **www.nccer.org/olf**. Be sure to include the exact module ID number, page number, a detailed description, and your recommended correction. Your input will be brought to the attention of the Authoring Team. Thank you for your assistance.

Instructors – If you have an idea for improving this textbook, or have found that additional materials were necessary to teach this module effectively, please let us know so that we may present your suggestions to the Authoring Team.

NCCER Product Development and Revision

13614 Progress Blvd., Alachua, FL 32615

Email: curriculum@nccer.org
Online: www.nccer.org/olf

❏ Trainee Guide ❏ Lesson Plans ❏ Exam ❏ PowerPoints Other _____

Craft / Level: _____ Copyright Date: _____

Module ID Number / Title: _____

Section Number(s): _____

Description: _____

_____

_____

_____

Recommended Correction: _____

_____

_____

_____

_____

Your Name: _____

Address: _____

_____

Email: _____ Phone: _____

This page is intentionally left blank.

Copper Tube and Fittings

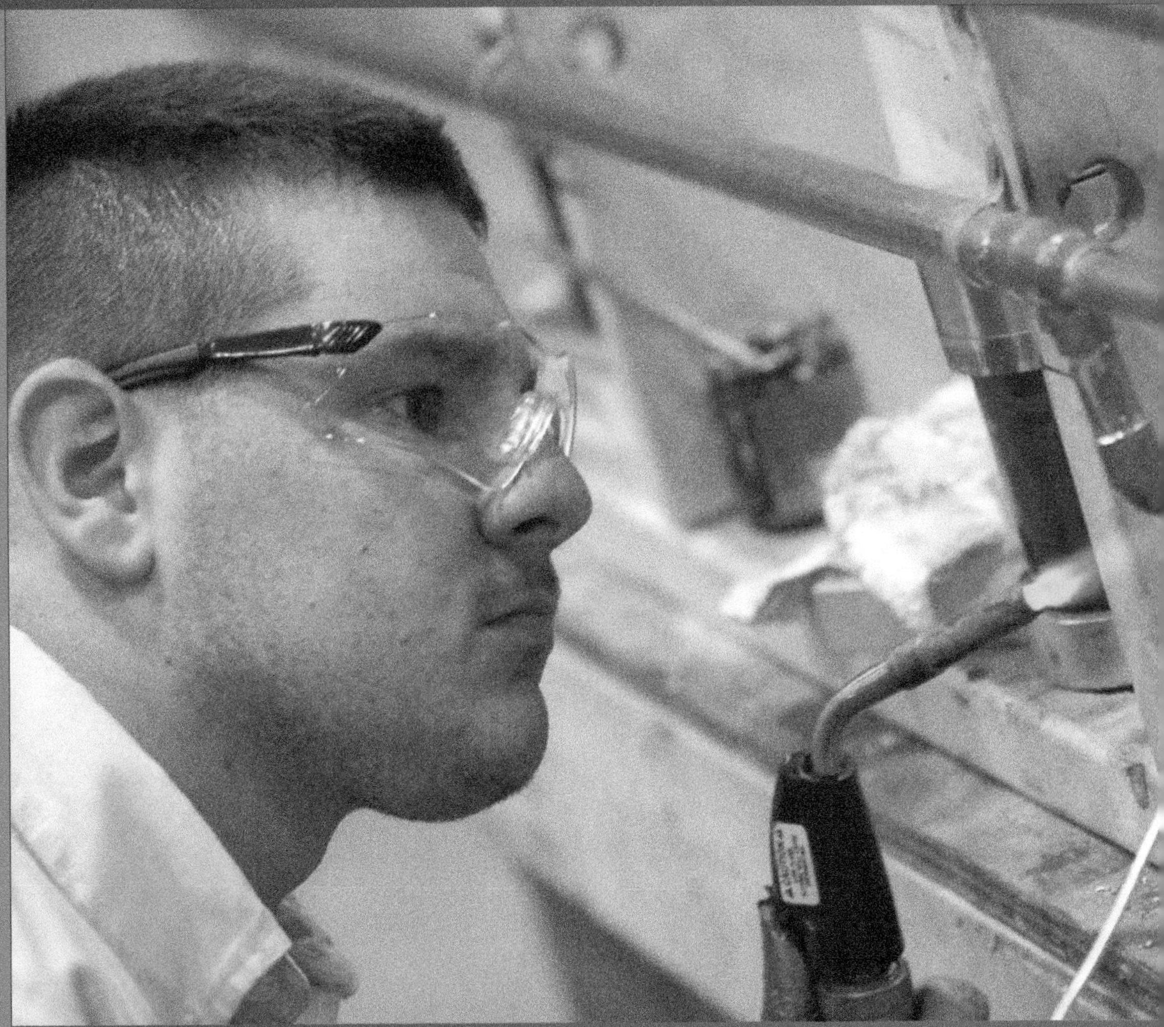

OVERVIEW

Copper tube and fittings are used in a variety of plumbing applications. Copper tube comes in a series of sizes and different wall thicknesses. Annealed copper is soft and flexible, while drawn copper is rigid and hard. Plumbers must use copper tube and fittings that have been labeled and approved by the manufacturer.

Module 02107

Trainees with successful module completions may be eligible for credentialing through NCCER's National Registry. To learn more, go to www.nccer.org or contact us at 1.888.622.3720. Our website, *www.nccer.org*, has information on the latest product releases and training.

Your feedback is welcome. You may email your comments to *curriculum@nccer.org*, send general comments and inquiries to *info@nccer.org*, or fill in the User Update form at the back of this module.

02107 V4.5

Objectives

Upon completion of this module, you will be able to do the following:

1. Identify the types, uses, and properties of copper tube and fittings.
   a. Identify the types of copper tube and their uses.
   b. Describe the sizing and labeling of copper tube.
   c. Describe the different types of fittings used on copper tube.
   d. Identify storage and handling requirements for copper tube.
2. Describe the different methods for cutting and bending copper tube.
   a. Explain the tools and methods used to measure copper tube.
   b. Explain the tools and methods used to cut copper tube.
   c. Describe the tools and methods used bend copper tube.
   d. Describe the different methods used to join copper tube.
3. Describe the methods used to install and test copper tube.
   a. Describe the hangers and fasteners used to support copper tube.
   b. Explain the insulation requirements for copper tube.
   c. Explain methods of pressure testing copper tube systems.

Performance Tasks

Under the supervision of your instructor, you should be able to do the following:

1. Select correct types of materials for copper tube systems.
2. Identify types of fittings and valves and their uses.
3. Select the appropriate personal protective equipment for working with copper tube.
4. Correctly measure, cut, and join copper tube.
5. Select the correct support and spacing for the application.

Trade Terms

ACR	Compression joint	Head
ACR tubing	Drawn copper	Insulation
Annealing	Drop-forged	Nominal size
Bullhead tee	Ferrule	Pressure drop
Capillary action	Flare joint	Sizing tool
Clevis	Formability	Sweat joint

Industry Recognized Credentials

If you're training through an NCCER-accredited sponsor you may be eligible for credentials from NCCER's Registry. The ID number for this module is 02107. Note that this module may have been used in other NCCER curricula and may apply to other level completions. Contact NCCER's Registry at 888.622.3720 or go to *www.nccer.org* for more information.

CODE NOTE

Codes vary among jurisdictions. Because of the variations in code, consult the applicable code whenever regulations are in question. Referencing an incorrect set of codes can cause as much trouble as failing to reference codes altogether. Obtain, review, and familiarize yourself with your local adopted code. Safety codes are developed by the US Occupational Safety and Health Administration (OSHA).

Contents

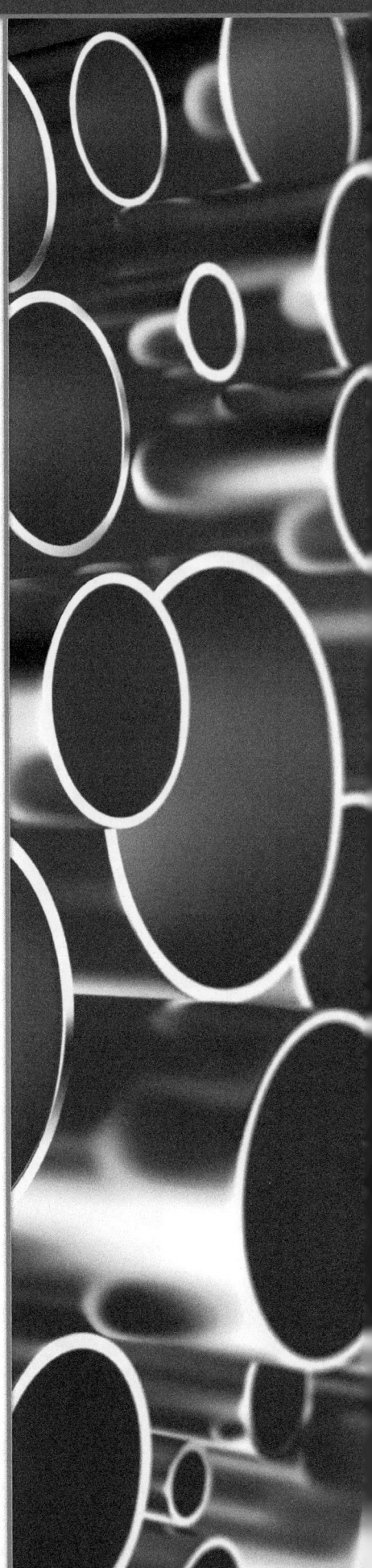

Figures and Tables

1.0.0 TYPES OF COPPER TUBE AND FITTINGS

Objective

Identify the types, uses, and properties of copper tube and fittings.

a. Identify the types of copper tube and their uses.
b. Describe the sizing and labeling of copper tube.
c. Describe the different types of fittings used on copper tube.
d. Identify storage and handling requirements for copper tube.

Performance Tasks

1. Select correct types of materials for copper tube systems.
2. Identify types of fittings and valves and their uses.

Trade Terms

ACR: Air conditioning and refrigeration system.

ACR tubing: Annealed copper tubing that is manufactured specifically for use in air conditioning and refrigeration work.

Annealing: Heating a material and slowly cooling it to relieve internal stress. This process reduces brittleness and increases toughness.

Bullhead tee: A tee fitting used on a branch that is larger than the main line, or that has one outlet larger than the run openings.

Compression joint: A method of connection in which tightening a threaded nut squeezes a compression ring to seal the joint.

Drawn copper: Tubing produced by pulling the tube through dies to reduce its diameter. The drawing process hardens the copper and makes it very rigid.

Ferrule: A brass compression ring used for joining copper tube.

Flare joint: A fitting in which one end of each tube to be joined is flared outward using a special tool. The flared tube ends mate with the threaded flare fitting and are secured to the fitting with flare nuts.

Formability: The ease with which a material can bend.

Nominal size: Approximate measurement in inches of the inside diameter (ID) of pipe for most copper tubes. However, nominal size of ACR tubing is based on the outside diameter (OD).

Pressure drop: A decrease in pressure from one point to another caused by friction losses in a fluid system, such as a water system.

Sweat joint: A pipe joint made by applying solder to the joint and heating it until it flows into the joint.

Copper is a mineral that is mined from the ground. Pure copper can be melted and molded into various sizes, lengths, and angles. Copper was first used in plumbing in the early 1800s and has been widely used in hot- and cold-water systems since the 1930s.

Copper tubes and fittings are sometimes more expensive than other types of pipe materials, but they are used in a nearly endless variety of tubing systems. They have a wide range of applications: for hot- and cold-water supply; for drain, waste, and vent (DWV) systems; for fuel gas supplies; for medical gas systems; and for transporting refrigerant in air conditioning systems.

Based on a tube's application, plumbers select appropriate fittings, which change the direction or size of a tube run, and valves, which control flow. Plumbers use specific methods to measure, cut, bend, join, and groove copper tube. These methods depend on the type of tube and its function, and the various special tools that are needed. Copper tube can be joined by soldering a sweat joint, creating a compression joint, or making a flare joint. Plumbers must learn to perform each method safely and use all tools properly.

Hangers and supports secure horizontal and vertical runs of copper tube and are used to prevent leaks and damage. Plumbers select the type of hanger depending on job specifications. Applicable codes dictate appropriate tube attachments and connectors for each installation, as well as whether the tube needs to be insulated. After completing an installation, plumbers pressure-test the system for leaks and secure connections.

1.1.0 Types of Copper Tube

Copper tubing comes in five different types: K, L, M, DWV, and ACR (air conditioning and refrigeration). Each type represents a series of sizes with different wall thicknesses. The tubing can be hard drawn copper, or soft and flexible annealed copper. Drawn copper is produced by pulling the tube through dies to reduce its diameter. Drawing hardens the copper and makes it very rigid. Annealed copper is produced by the annealing process, in which the material is heated and slowly cooled to relieve internal stress. This process reduces brittleness and increases toughness. All types of tube are available in a hard form. Hard forms come in 10' and 20' lengths and in diameters ranging from ¼" to 12". Drawn copper tubing is widely used in commercial refrigeration and air conditioning systems. Types K, L, and ACR also are available in soft coils in 40' to 100' lengths in diameters ranging from ⅛" to 2".

> **NOTE**
>
> Copper is also manufactured as pipe, which has thicker walls than copper tubing. Copper pipe size is specified using schedules, like pipe made from other materials.

Manufacturers fill lengths of drawn tubing with nitrogen and plug them at each end to maintain a clean, moisture-free internal condition. This tubing is intended for use with formed fittings to make the necessary bends or changes in direction. It is more self-supporting than annealed copper tubing, such as tubing for air conditioning and refrigeration systems (ACR tubing); therefore, it needs fewer supports.

> **WARNING!**
>
> Always check for electrical grounding when working with copper tube. Copper water supply lines are often used as the electrical grounding. Always shut off electrical power if you break the grounding. When you shut off electrical power, always use approved lockout/tagout procedures to avoid electrical shock. Check with your immediate supervisor before proceeding whenever electrical power is applied.

1.2.0 Copper Tube Sizing and Labeling

Tube sizing varies depending on the type of copper tube. Types K, L, M, and DWV use nominal, or standard, sizing. This means that the outside diameter (OD) is always ⅛" larger than the nominal size (see *Figure 1*). The nominal size is the approximate measurement in inches of the inside diameter (ID) for most copper tubes.

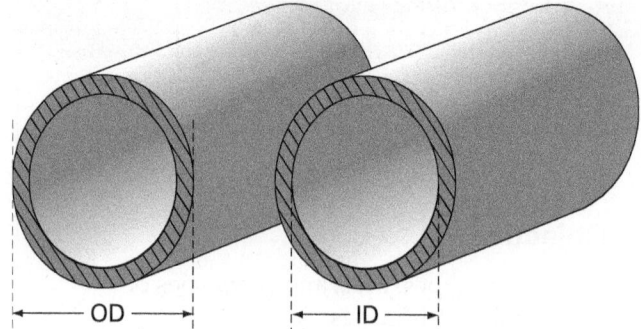

Figure 1 Copper tube sizing.

For example, the OD of a ¾" nominal type M tube measures ⅞". This allows the same fittings to be used with the different wall thicknesses and IDs of different types of copper tube. The ID is usually close to the stated size. A tube with a ½" ID has approximately ½" between the inside walls of the tube.

Type ACR tube uses actual OD sizing. A ⅞" OD ACR copper tube is actually the same OD as ¾-inch K, L, or M copper tube. This means that you can use the same fittings with these tubes.

Plumbers must use copper tube and fittings that are labeled and approved for use in plumbing systems. The labels on copper tube contain very important information. Manufacturers must permanently mark Types K, L, M, and DWV to show the tube type, the name or trademark of the manufacturer, and the country of origin. Manufacturers also print this information on the hard tubes in a color that identifies the tube type. Manufacturers label most soft copper by stamping the information into the product. *Table 1* shows five types of copper tubing and the corresponding color code for labeling.

Table 1 Copper Tubing Color Codes

Type	Color Code	Application:
K	Green	
L	Blue	
M	Red	
DWV	Yellow	
ACR	Blue	

Exercise: Using *Table 1,* fill in the blanks to indicate the applications that correspond to each type of copper tubing based on applicable codes in your area.

Consider strength and formability (ease of bending) when selecting copper tube. Consult your local plumbing and mechanical codes when selecting tube; these codes govern which types of tube you can use in particular applications.

1.3.0 Fittings

The types of fittings and valves used with copper tube depend on how the tube is used. Fittings allow you to join tubes to each other and change the direction or size of a tube run, for example. Fittings are designed so they do not block or slow the flow of materials in the tube. This section discusses water supply fittings, water supply valves, and DWV fittings, as well as relatively new alternatives.

1.3.1 *Water Supply Fittings*

Individual fitting shapes serve specific purposes, depending on whether a tube runs horizontally or vertically. Use as few fittings as possible. Fewer fittings mean fewer chances for a leak and pressure drop, which are decreases in pressure from one point to another caused by friction losses in a water system. Common copper fittings for use with water supply systems include those listed in *Table 2* and shown in *Figure 2.*

Solder fittings, also called sweat fittings, are special copper or brass fittings that are used for soldering or brazing copper tube. Solder fittings are made slightly larger than the tubes to be joined, leaving only enough room for solder to flow into the joint. Adapter fittings allow copper tubing to be joined with threaded tube on one end while the other end is soldered.

Dielectric Unions

If you attach unlike metals, such as copper and galvanized pipe, it results in a process called electrolysis. Transition fittings, such as dielectric unions, prevent electrolysis. Dielectric unions isolate the different materials by using a mixture of brass, plastic, or rubber. However, they can cause other problems. For example, calcium deposits can build up, which eventually will lead to blockages. Therefore, you should never use dielectric unions unless they are plastic lined.

Table 2 Common Copper Fittings and Their Descriptions

Fitting	Description
90-degree ell and 45-degree ell	Elbow used to change the direction of the pipeline by 90 or 45 degrees.
Drop ear ell	An elbow that allows you to attach the pipeline to the building frame; frequently used at the last joint before the pipe comes through the wall to be attached to a fixture.
Street 90	An elbow with a male end and a female end that is used to change the direction of the pipeline by 90 degrees.
Street 45	An elbow with a male end and a female end that is used to change the direction of the pipeline by 45 degrees.
Tee	A fitting with three openings used to make branches at 90-degree angles to the main pipe. Reducing tees are used to make branches at 90-degree angles from the main pipe to smaller outlet pipes.
Coupling	A fitting used to connect lengths of pipe on straight runs.
Sweat cap	A fitting used to close the end of a copper pipe.
Female and male adapters	Fittings soldered onto the end of a length of copper tubing to provide a threaded end for attaching a pipe to another threaded pipe.
Sweat-to-compression adapter	A fitting used to adapt a soldered copper tube to a compression joint by means of a ferrule, which is a brass compression ring used for joining. A compression joint consists of a threaded nut that has been squeezed over a compression ring to seal the joint.
Sweat flange	A fitting used to adapt soldered copper tube to iron pipe flange.
Reducer coupling and reducer bushing	A fitting used to connect pipe to pipe or pipe to fitting with different sizes of pipe.
Sweat union	A fitting used to connect pipe to pipe or pipe to fitting with different sizes of pipe.
Flare fittings	Fittings with a flared end that can be joined with a male cone-shaped tubing end or union. Flare fittings used in refrigeration and air conditioning include a variety of elbows, tees, and unions. They are drop-forged brass and are accurately machined to form the 45-degree flare face. The fittings used are based on the size of the tubing. Flare nuts are hexagon-shaped—they have six sides. An adjustable, or open-end, wrench is used with these fittings.
Grooved copper	A mechanical coupling material for rigidly connecting copper tubing that has been roll-grooved. Grooved copper fittings are made for connecting copper tubing in sizes from 2 to 6 inches. These fittings have grooved ends, so they can be installed using a wrench. This eliminates the need for soldering or brazing.
SharkBite® Fittings	Push-fit connections that use stainless steel teeth and a compressible O-ring seal.
ProPress® Fittings	Pressed fittings that do not require sweating.

Figure 2 Fittings for copper tubing.

1.3.2 Water Supply Valves

Valves regulate the flow of liquid. They may provide on/off service or prevent flow reversal through a line. Valves use one or more of the following methods to control flow through a tubing system:

- Move a disc or plug into or against a passageway.
- Slide a flat cylindrical or spherical surface across a passageway.
- Rotate a disc or ellipse around a shaft extending across the diameter of a tube.
- Move a flexible material into the passageway.

Common valves for water supply systems include the following (see *Figure 3*):

- *Gate*—A valve with a wedge-shaped or tapered metal disc that fits into a smooth-ground surface or seat with the same shape, allowing a straight-line flow with little obstruction. Gate valves are a good choice for lines that will remain either completely open or completely closed most of the time.
- *Globe*—A valve that controls the flow of liquid with a movable spindle, which descends to restrict flow through the valve opening. Because a globe valve is reliable and easy to repair, it is often used in water supply lines inside buildings.
- *Ball*—A valve that consists of a ball with a hole bored through its diameter, mounted on a spindle. When the ball valve is closed, the hole is at 90 degrees to the valve body so no flow can take place. When the valve is turned a quarter turn or opened completely, water flows through the hole.
- *Compression, stop, and waste*—A valve that is opened or closed by raising or lowering a horizontal disc using a threaded stem. An elastomeric, or rubberized, washer on the end of the stem seals the valve seat, closing off water flow. This valve is most commonly used for draining and freeze protection above ground.
- *Check*—An automatic valve that permits the flow of liquid in one direction only. It prevents reverse flow. Check valves are commonly used on domestic and irrigation wells.
- *Stop*—A valve that controls flow of liquids or gases between a building and supply source. A stop valve is also called a ground-key valve.

Solder Fittings

Two kinds of solder fittings are available with copper tubing. The first is a wrought copper fitting, which is made from copper tubing that is shaped into different types of fittings. Wrought copper fittings are generally lightweight, are smooth on the outside, and have thin walls. The second type is a cast solder fitting. This type of fitting is made using a mold. The first cast fittings had holes in the sockets for the solder. Today, the heated copper is poured into the mold and allowed to cool. Cast solder fittings have a rough exterior and come in a wide variety of shapes. They are heavier than wrought solder fittings.

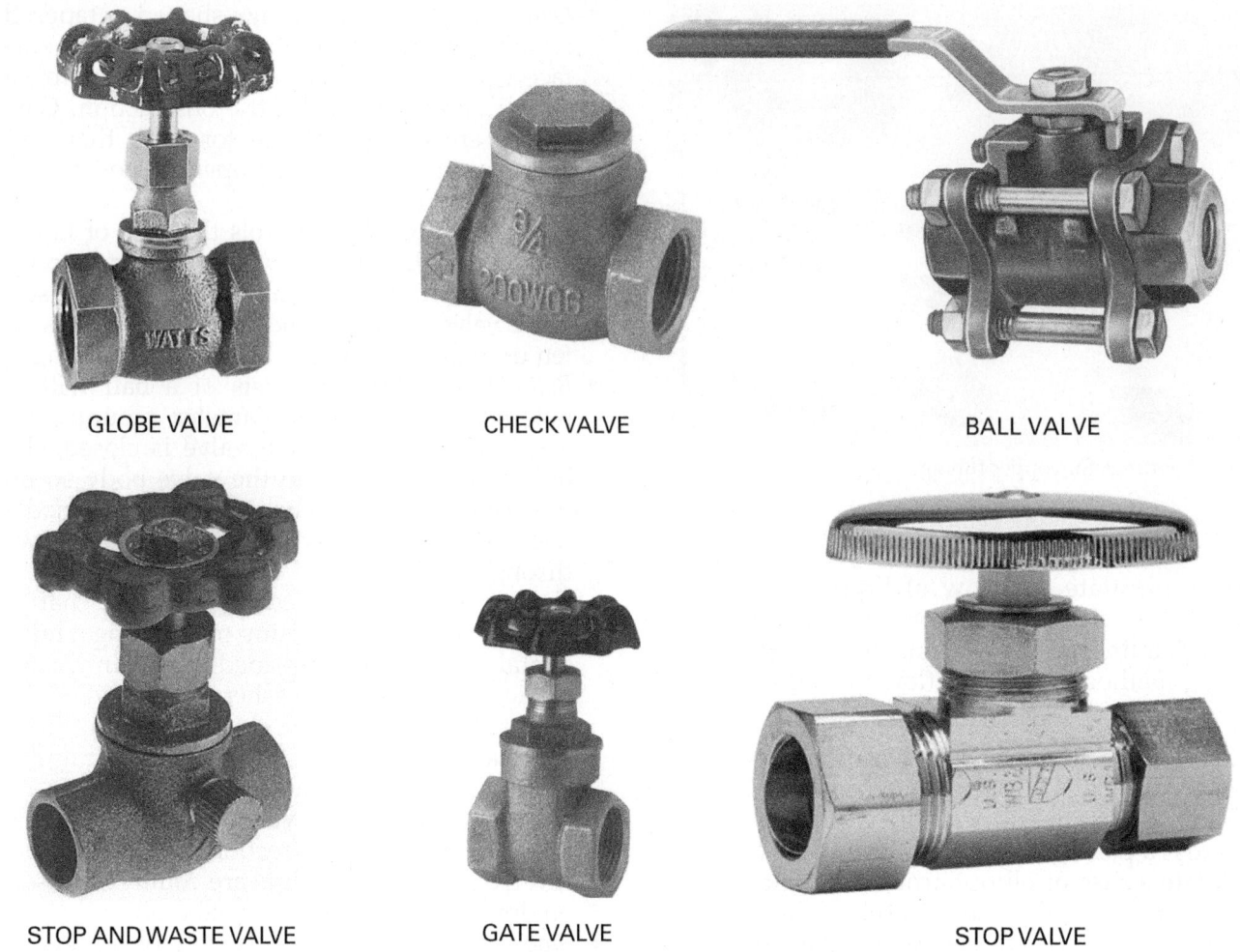

GLOBE VALVE   CHECK VALVE   BALL VALVE

STOP AND WASTE VALVE   GATE VALVE   STOP VALVE

Figure 3 Common valves for copper water supply systems.

1.3.3 DWV Fittings

DWV fittings are designed to allow liquids and other materials to flow smoothly through them. Common fittings used in DWV systems are elbows (90 degrees, 45 degrees, 22-1/2 degrees), male adapters, **ferrule** adapters, sanitary tees, cleanout tees, reducing tees, wyes, and reducers.

1.3.4 Alternative Fittings

In addition to the common fittings discussed above, you can use some alternative fittings to join copper tube. These alternatives include press fittings and mechanically formed tee connections.

Press fittings, also called press-connect fittings, are mechanical fittings that connect tube by means of a cold press fit system. First introduced in the 1970s in Germany and more recently in the United States, they are used primarily for potable water systems. Press fittings are created by an electric press tool or a hand pressing tool.

The tool mechanically presses the fitting around the tube against an O-ring inside the fitting. This ensures that the connection is strong and leakproof.

A **bullhead tee** is a tee fitting used on branches that are larger than the main line. The term is also used to describe a tee fitting in which the outlet is larger than the opening of the straight run. If tee fittings are not properly installed, they can cause a condition known as bullheading, which results from the larger outlet opening. When the flow of liquid hits the back wall of the tee, it causes turbulence. This adds to the pressure drop caused by liquid moving from a smaller tube to a larger tube. This turbulence may also cause a banging in the line. If more than one tee is installed in the line, a straight piece of tube with a length between tees of 10 times the tube's diameter is recommended to reduce turbulence. For example, a tube that is 4" in diameter should have 40" of tube between each tee (10 × 4" = 40").

Mechanically formed tee connections are joints created by a tee-pulling tool. The tool allows you to drill into a section of tube and create a tee connection (*Figure 4*). You must use brazing to join a branch line to the tube. This method increases productivity because you create only one brazed joint rather than three soldered joints to form the tee connection (see *Figure 5*). This method is commonly used to create manifolds as well as copper fire sprinkler installations. Some codes do not permit tee-pulling tools for drainage because of the possibility of leakage. Follow the manufacturer's instructions and always consult applicable codes.

SharkBite® fittings (*Figure 6*) are push-fit connections that use stainless steel teeth and a compressible O-ring seal to connect copper tube. Viega ProPress® fittings (*Figure 7*) are pressed fittings that do not require sweating in order to join copper tubing. Viega's patented Smart Connect® feature identifies unpressed connections during pressure testing so that they are not missed. They are available in ½" to 4" in copper, bronze, and stainless steel.

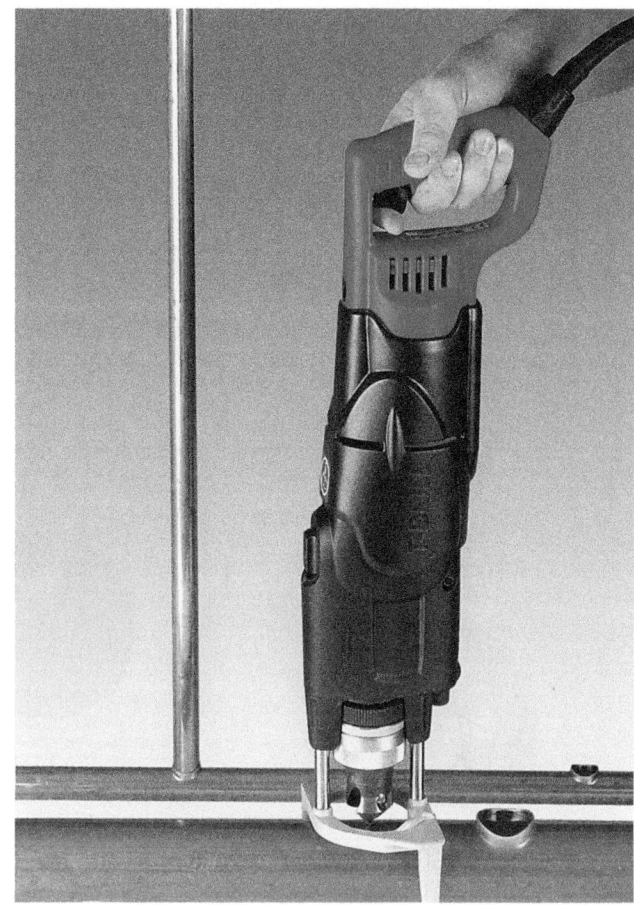

TEE PULLER

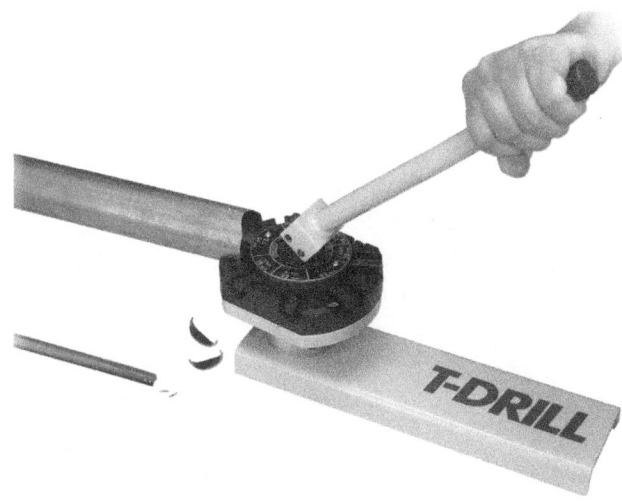

TUBE TRIMMER

Figure 4 Tee-pulling tools.

Figure 5 Mechanical tee connection.

STRAIGHT COUPLING

ELBOW

STRAIGHT CONNECTOR

TEE

MANIFOLD

FLEXIBLE WATER
HEATER CONNECTOR

Figure 6 SharkBite® fittings.

ELBOW

TEE

FEMALE ADAPTER

REDUCER

UNION

MANIFOLD

Figure 7 Viega ProPress® fittings.

1.4.0 Material Storage and Handling

To ensure maximum productivity on the job, it is important to use common sense when storing and handling copper tube. Before starting to work, always plan ahead. Always store copper tube in a secure area. Store tube and fittings near where you will be working so they are easy to access. Organize tube and fittings into groups by tube size and type. When handling tube and fittings, be careful not to bend or damage them. Tube that is bent or otherwise damaged not only costs money, but also makes you less productive. Take extra care when working in cold temperatures. Cold copper tube can stick to and injure bare hands.

> **NOTE**
>
> Because copper is so valuable, it is often stolen from job sites, even after it has been installed. Take proper steps to store copper securely and ensure the building is locked at the end of the work day.

GOING GREEN

Recycling Copper

US industries use close to 3 million tons of copper each year. About two-thirds of the copper is mined; the rest comes from recycled copper. Discarded objects, dismantled buildings, and worn-out machinery are sources of scrap copper. This scrap copper can be melted down and reused. Many recycling stations pay for scrap copper.

Additional Resources

The Copper Tube Handbook, 2016. New York: Copper Development Association (also available as a free pdf and an app for use on a smartphone or tablet at the Copper Development Association website, *www.copper.org*).

1.0.0 Section Review

1. Drawn copper is produced by a _____.

    a. heating process
    b. cooling process
    c. pulling process
    d. pressurization process

2. Copper tube used for DWV is marked with the color code _____.

    a. orange
    b. blue
    c. yellow
    d. green

3. Which of the following is true with regard to fittings?

    a. Fittings can be used to reduce pressure drops.
    b. Use as few fittings as possible.
    c. Solder fittings are slightly smaller than the tubes to be joined.
    d. Sweat fittings are used with threaded tube.

4. Take extra care when working with copper tube in cold temperatures because it can _____.

    a. break
    b. shrink
    c. injure bare hands
    d. corrode

2.0.0 MEASURING, CUTTING, AND BENDING

Objective

Describe the different methods for cutting and bending copper tube.
 a. Explain the tools and methods used to measure copper tube.
 b. Explain the tools and methods used to cut copper tube.
 c. Describe the tools and methods used bend copper tube.
 d. Describe the different methods used to join copper tube.

Performance Tasks

3. Select the appropriate personal protective equipment for working with copper tube.
4. Correctly measure, cut, and join copper tube.

Trade Terms

Capillary action: The process during soldering in which the molten solder flows into the narrow gap between the pipe and the joint, regardless of whether the solder is flowing up, down, or horizontally.

Drop-forged: A characteristic of a product made when heated metal is pounded or shaped between dies with a drop hammer or press.

Sizing tool: A tool consisting of a plug and a sizing ring that is used to reshape a deformed pipe back to roundness.

Some measuring, cutting, reaming, bending, joining, and grooving techniques are related specifically to copper. Different techniques may be used depending on the type and function of the copper tube you are using. You need to become familiar with many tools to install copper tubing properly. The following sections explain methods for measuring, cutting, bending, joining, and grooving copper tubing.

> **WARNING!**
> Before you begin a job, think through the potential hazards and wear the appropriate PPE, including, but not limited to, safety glasses, a hard hat, and gloves. In addition, beware of hair creams and shave creams, which can be flammable. Always make sure you have quick access to a fire extinguisher on the job.

2.1.0 Measuring

It is extremely important to measure tube carefully. Several methods of measuring copper tube are described in the following list:

- *End-to-end*—Measure the full length of the tube.
- *End-to-center*—Use for tube that has a fitting joined on one end only; tube length is equal to the measurement minus the end-to-center dimension of the fitting.
- *Center-to-center*—Use with a length of tube that has fittings joined on both ends; tube length is equal to the measurement minus the sum of the end-to-center dimensions of the fittings.
- *End-to-face*—Use for tube that has a fitting joined on one end only; tube length is equal to the measurement.
- *Face-to-face*—Use for same situation as center-to-center measurement; tube length is equal to the measurement.
- *Face-to-back*—Use with a length of tube that has fittings joined on both ends; tube length is equal to the measurement plus the distance from the face to the back of one sweated-on fitting.
- *Center-to-face*—Use with a length of tube that has fittings on both ends; tube length is equal to the measurement from the center of one of the fittings to the face of the opposite fitting, plus twice the insertion length.
- *End-to-back*—Use for tube that has a fitting joined on one end only; tube length is equal to the measurement plus the length of the sweated-on fitting.

2.2.0 Cutting

You can cut copper tubing with a handheld tube cutter, a hacksaw, or a midget cutter. The handheld tube cutter (*Figure 8*) is preferred because it makes a cleaner joint and leaves no metal particles. Use a tube cutter that is the right size for the copper you are cutting, and make sure that the proper cutting wheel is in place. Plumbers use internal tube cutters for trimming extended ends of installed water closet bowl and shower waste lines below the level of the flange.

After cutting, ream all cut tube ends to the full inside diameter of the tube. Reaming removes the small burr (rough inside edge) created when you cut the tube. Burrs left on tubing can cause the tube to corrode. A tube that is reamed correctly provides a smooth inner surface for better flow. You must also remove burrs on the outside of the tube to ensure a good fit. Tools used to ream tube ends include the reaming blade on the tube cutter, files (round or half-round), a pocketknife, and a deburring tool. If your tube becomes deformed, you can use a sizing tool, which consists of a plug and a sizing ring, to bring the tube back to roundness. Refer to *Figure 8* for an example of a deburring blade on a tube cutter. A variety of models are available, and the cutting sizes range from $\frac{1}{8}$" to $4\frac{1}{8}$" OD.

To use the handheld tube cutter, follow these steps (see *Figure 9*):

Step 1 Place the tube cutter on the tube at the point where you want to cut. Tighten the knob, forcing the cutting wheel against the tube.

Step 2 Make the cut by rotating the cutter around the tube under constant pressure.

Step 3 Use the built-in deburring blade to remove any burrs from inside the tube.

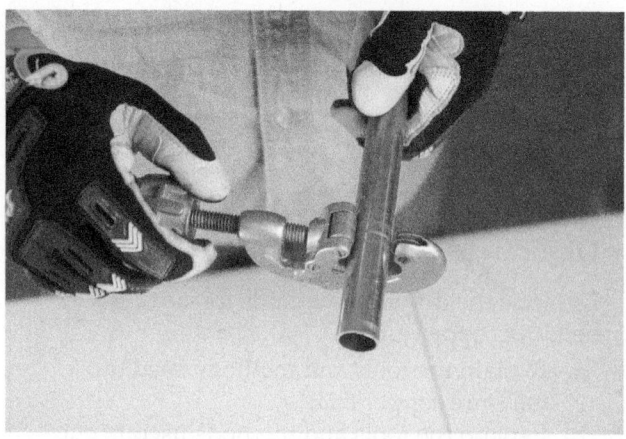

(A)

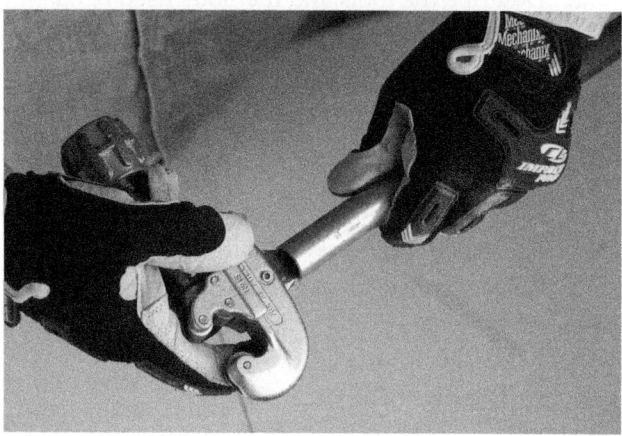

(B)

Figure 9 Using a handheld tube cutter.

To cut larger size drawn tubing, you can use a hacksaw and a vise. A vise is a gripping tool that secures an object while you work on it. Using a vise helps you square the ends and allows more accurate cuts. The blade of the hacksaw should have at least 32 teeth per inch. Avoid leaving saw cuttings inside the tubing. File the end to produce a smooth surface, and carefully clean the inside of the tubing with a cloth. To cut large quantities of tubing, use portable power tools with abrasive wheels to cut, clean, and buff the tubing.

2.3.0 Bending

Unlike other types of rigid metal tube, such as cast iron, certain kinds of soft copper tube can be bent. In fact, tests have shown that the bursting strength of bent tube is greater than that of regular tubing. Bending reduces the number of joints and fittings in a plumbing system. This can reduce leaks, and ultimately installation time and costs as well. When you are using soft tubing, it is best to bend rather than cut it. You can easily bend small-diameter tubing by hand.

Figure 8 Handheld tube cutter.

Take care not to flatten the tube with a bend that is too sharp.

You can place tube-bending springs on the outside of the tube to prevent it from collapsing while you are bending it by hand (see *Figure 10*). You can use tube-bending equipment to make accurate and reliable bends in both soft and hard tubing. This equipment has various sizes of forming attachments for use in bending tubing with diameters up to $7/8$ inch at any angle up to 180 degrees (see *Figure 11*).

When using a mechanical tube bender, the minimum bending radius depends on the tube's outside diameter as well as the type (drawn or annealed). For example, a $1/4$" annealed tube may have a minimum bend radius of $3/4$" (three times the tube's diameter), while a $1/2$" drawn tube has a minimum bend radius of $2\frac{1}{2}$" (five times the tube's diameter), and a 1" annealed tube has a minimum bend radius of 4" (4 times the diameter). Consult the manufacturer charts for the tube bender in use when selecting the appropriate bend radius.

> **NOTE**
> The bending radii for tubing bent by hand is much larger than when using a mechanical bender due to the risk of flattening.

2.4.0 Joining Copper Tube

Copper can be joined in many ways, including by a sweat joint, a compression joint, or a flare joint. The method depends on the plumbing application and environmental factors. Sweat joints, for example, require a heating process. You should not use this method when a fire hazard exists at the job site. Press-fit copper tubing can be used in fire hazard environments. As always, follow all applicable codes and manufacturer's instructions when joining copper tube. Copper tube should never be joined using a threaded joint.

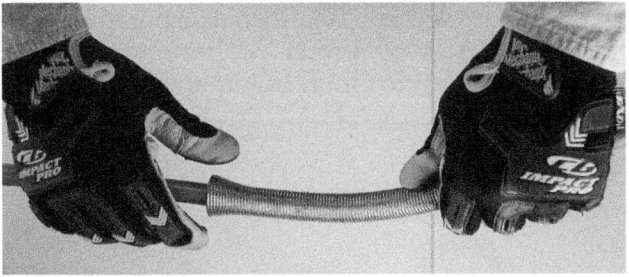

Figure 10 Tube-bending spring.

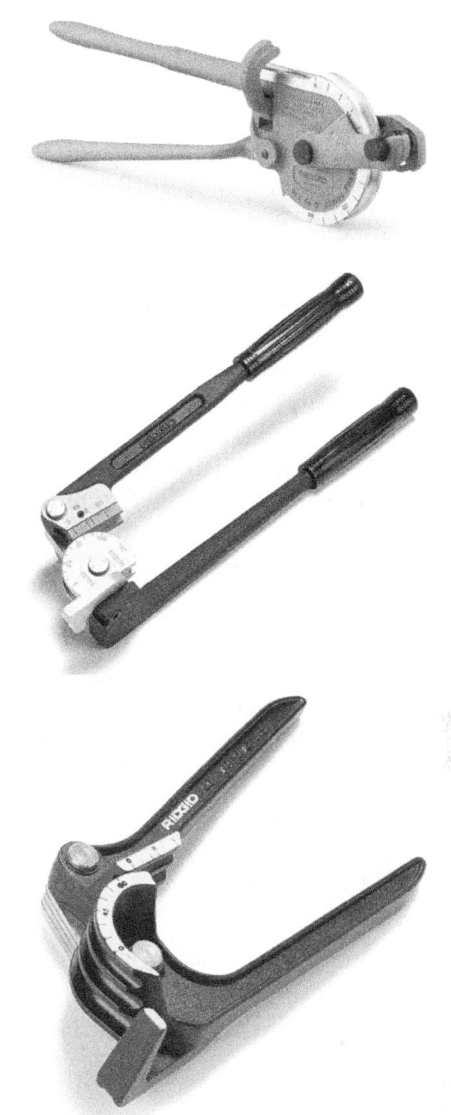

Figure 11 Tube-bending equipment.

2.4.1 Soldering

Soldering is a type of heat bonding in which copper tube is joined when a soft filler metal is melted in the joint between the two tubes. Solder fittings are made slightly larger than the tubes to be joined, leaving only enough space for solder to flow into the joint. This space is called the capillary action gap. Solder melts and flows into the gap in a process known as capillary action. Capillary action occurs regardless of whether the molten solder is flowing up, down, or horizontally. Adapter fittings allow copper tubing to be joined with threaded tube on one end while the other end is soldered.

A sweat joint is made by measuring, cutting, reaming, cleaning, and applying flux to the copper tube, then adding solder and heating to a certain temperature until the solder flows into the joint. Always use lead-free solder. Flux is a paste that acts as a wetting and cleaning agent and aids the soldering process by preventing oxidation of the joint, which would damage the copper. Flux should be lead-free and water-soluble. The heat required for soldering (350°F to 550°F) can be generated by electricity or various kinds of gases. Use sweat joints with hard or soft copper with nominal sizes from 1/8 to 4 inches.

Tools used to solder copper tubing to fittings include a tube cutter, fitting brush, solder, flux brush, and soldering torch (see *Figure 12*). In place of a fitting brush, you can use any abrasive other than steel wool. Steel wool contains oil, which contaminates the tube.

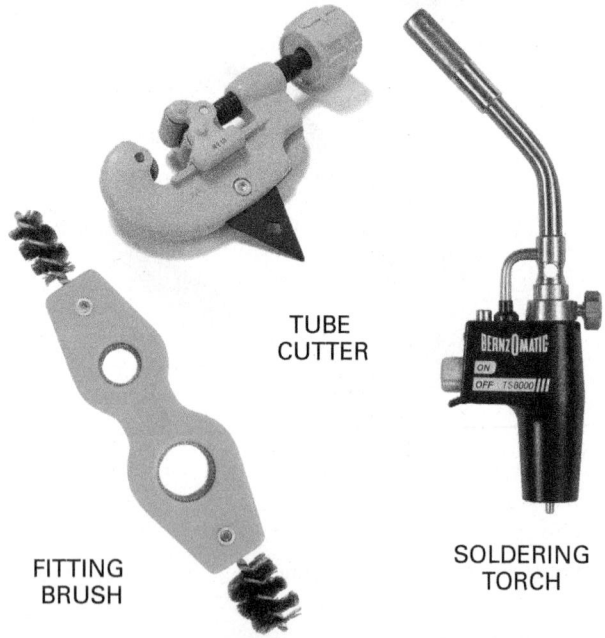

FITTING BRUSH
TUBE CUTTER
SOLDERING TORCH

WIRE SOLDER, FLUX BRUSHES AND SOLDER PASTE
Figure 12 Tools for soldering copper tube to a fitting.

To solder copper tubing to a fitting, follow these steps (see *Figure 13*):

Step 1 Measure and mark tubing to length, including the fitting allowance for the portion of the tube that extends inside fittings.

Step 2 Use a tube cutter to cut the tubing.

Step 3 Use a reaming tool on the end of the tube cutter to remove the burr from the inside of the tubing.

Step 4 Clean the inside of the tubing. Clean the inside of each fitting by scouring it with a fitting brush.

Step 5 Apply a thin layer of flux to the end of each tube using a flux brush.

Step 6 Push the tubing into the fitting and turn it a few times to spread the flux evenly.

Step 7 Hold the tip of a soldering torch flame against the fitting. When the flux starts to sizzle, move the flame around to the other side of the fitting to heat it evenly. Some plumbers use a special soldering torch fired from a 20-pound tank of gas called a B-tank.

Step 8 When the flux starts to bubble, remove the torch, and touch the wire solder to the point where the tubing enters the fitting. The solder will melt and be drawn into the joint. Once a line of solder shows completely around the joint, the connection is filled with solder.

Step 9 Allow the joint to cool until the solder solidifies.

Step 10 Once the solder has solidified, wipe it clean with a soft, wet cloth to remove any flux.

Brazing is a heating process very similar to soldering, but it requires filler metals that melt at much higher temperatures (between 1,100°F and 1,500°F) than solders. Plumbers use brazing when joints must be very strong. For example, brazing is often required to join refrigeration tube.

Did You Know?

To solder copper fittings, you must heat the fitting until the soldering paste starts to melt. This melting paste looks like beads of sweat on the pipe, which led to the term sweat fittings.

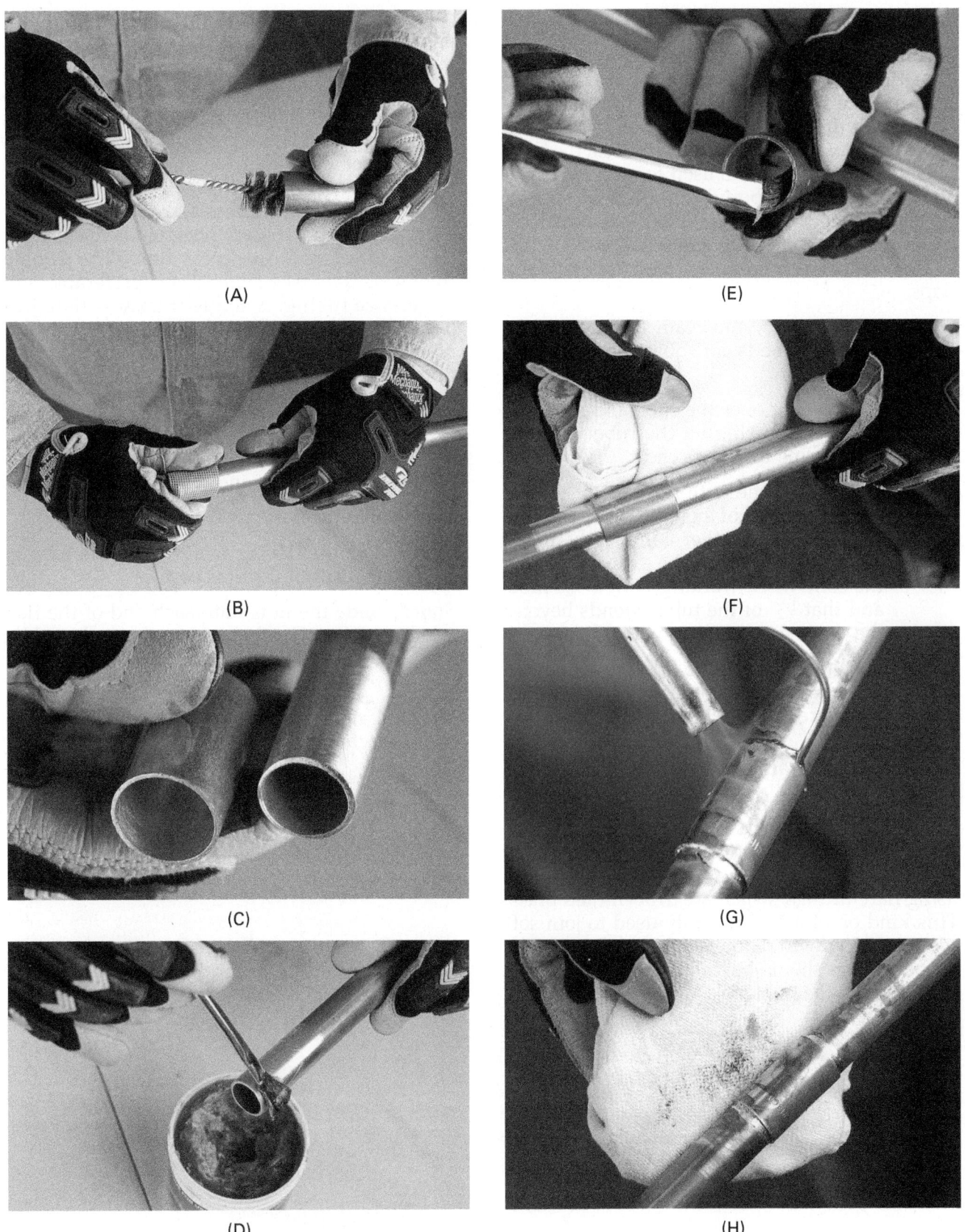

Figure 13 Soldering copper tubing to a fitting.

2.4.2 Creating Compression Joints

A compression joint (*Figure 14*) is a mechanical joint that is made by measuring, cutting, and reaming the tubes and using compression fittings. With this method, you tighten a threaded nut to squeeze a compression ring to seal the joint. This kind of joint is often used for joining refrigerant tubing. It is a popular method because it uses a threaded fitting and takes less time than making soldered or flared joints.

To create a compression joint, follow these steps:

Step 1 Measure, cut, and ream the tube. Make sure the tube is cut square and is free of burrs.

Step 2 Slip the nut over the tube and slide the compression ring on the tube with the teeth facing the tube's end.

Step 3 Install the cone with the convex surface toward the end of the tube. To be sure the fitting goes together completely, make sure that ¼" of the tube extends beyond the cone when working with a ½" tube, and that ½" of the tube extends beyond the cone when working with a ¾" tube.

Step 4 Push the nut onto the fitting and tighten. When the fitting squeaks, turn the nut one more full turn.

2.4.3 Creating Flare Joints

Flare joints may be required in an installation in which a fire hazard exists and a torch for soldering or brazing is not permitted. A flare joint is made by measuring, cutting, and reaming the tubes and using flare fittings, which are drop-forged brass. This kind of joint is commonly used to join soft copper tubing with diameters from ¼" to 2". Soft copper fittings should be leakproof and easily dismantled with the right tools.

The two kinds of flare fittings that are popular are the single-thickness flare and the double-thickness flare. For both types, you use a special flaring tool (*Figure 15*) to expand the end of the tube outward into the shape of a cone, or flare. The single-thickness flare forms a 45-degree cone that fits against the face of a flare fitting (see *Figure 16*). In a single operation, the single-thickness flare is formed, and then the lip is folded back and compressed to make a double-thickness flare. The double-thickness flare (see *Figure 17*) is preferable with larger-size tubing. A single-thickness flare may be weak when used under excessive pressure or expansion. Double-thickness flare connections are easier to dismantle and reassemble without damage than single-thickness flare connections.

To make a flare joint, follow these steps:

Step 1 Measure, cut, and ream the tubing.

Step 2 Slip the flare nut over the tubing.

Step 3 Use the flaring tool to flare the tubing's ends until the fit is perfect.

Step 4 Slide the nuts onto each end of the flare tubing. Gently bend or shape the tubing by hand over the male thread of each fitting.

Step 5 Use a smooth-jaw adjustable wrench to tighten until the fitting is snug.

Step 6 Test the joint for leaks.

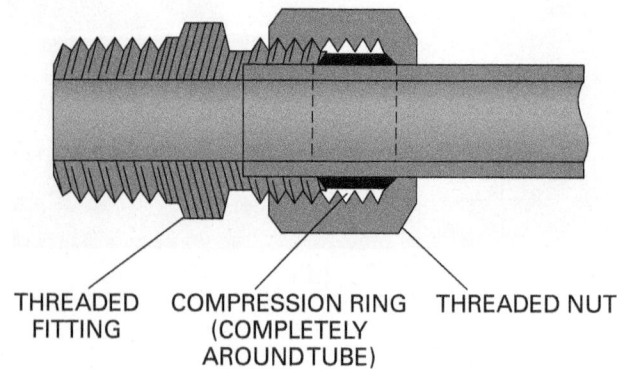

THREADED   COMPRESSION RING   THREADED NUT
FITTING   (COMPLETELY
                  AROUND TUBE)

Figure 14 Compression joint.

Figure 15 Flaring tool.

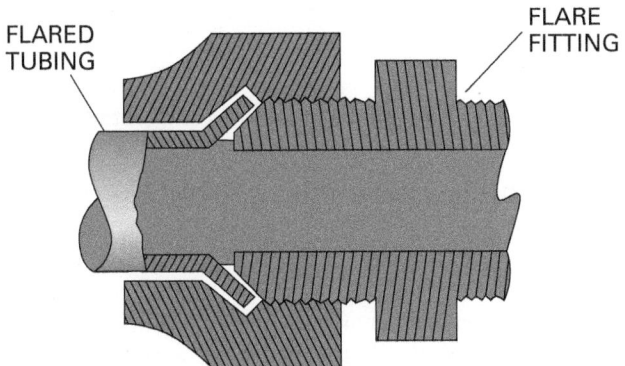

Figure 16 Single-thickness flare.

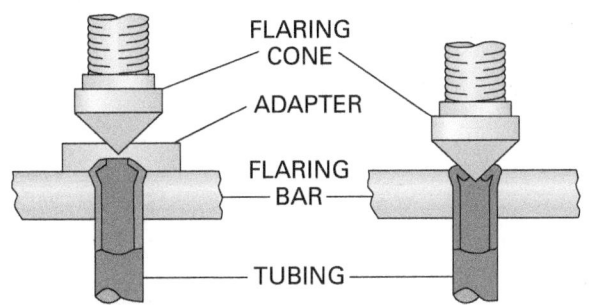

Figure 17 Double-thickness flare.

HAMMER FLARE AND HAMMER

Figure 18 Hammer flare and hammer.

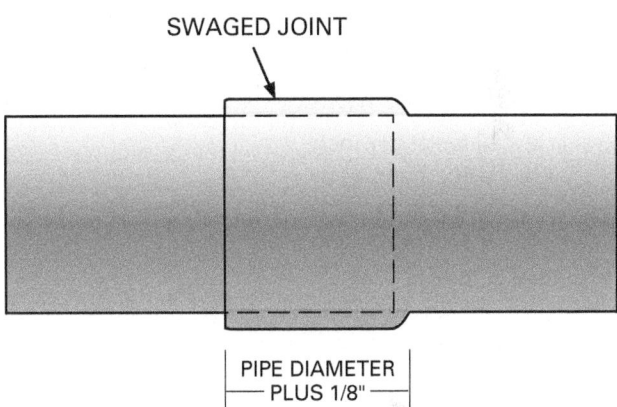

Figure 19 Swage joint.

Another commonly used type of flare is the hammer flare (*Figure 18*). This flare tool is used to make 45° flares on Type K soft copper water service tubing. Hammer flares are made from hardened steel and are driven into the tube end using a soft-face brass hammer.

2.4.4 Creating Swage Joints

Instead of using a coupling, you can also join lengths of copper tube by creating a swage joint. This is done by expanding one end of a length of copper tubing to allow another piece of tubing of the same size to fit inside it (*Figure 19*). To make a swage joint, secure the length of tube in a flaring block. The length of tube that extends above the flaring block should equal the diameter of the tube plus $\frac{1}{8}$". Use a hammer to drive the correct size swaging tool into the end of the tube to the proper shape and dimensions. This will cause the tube to expand up to the flaring block. Once the other length of tube is inserted into the swaged joint, the joint can then be soldered or brazed.

Additional Resources

This module presents thorough resources for task training. The following resource material is suggested for further study.

Plumber's Handbook Revised, Howard C. Massey. 2006. Carlsbad, CA: Craftsman Book Company.

Plumbing Installation and Design, L. V. Ripka. Third Edition, 2006. Orland Park, IL: American Technical Publishers.

The Copper Tube Handbook, 2016. New York: Copper Development Association (also available as a free pdf and an app for use on a smartphone or tablet at the Copper Development Association website, *www.copper.org*).

2.0.0 Section Review

1. To measure the full length of a line of tube, plumbers use a(n) _____.
   a. end-to-end measurement
   b. face-to-face measurement
   c. center-to-center measurement
   d. back-to-back measurement

2. The blade of a hacksaw used to cut drawn tubing should have _____.
   a. 10 teeth per inch
   b. 14 teeth per inch
   c. 24 teeth per inch
   d. 32 teeth per inch

3. When using a mechanical bender, the minimum bend radius of ½" drawn copper tubing is _____.
   a. ½"
   b. 1"
   c. 2.5"
   d. 5"

4. The type of joint that can be safely made in a fire hazard environment is a(n) _____.
   a. flare joint
   b. soldered joint
   c. brazed joint
   d. sweat fitting

3.0.0 INSTALLING AND TESTING COPPER TUBE

Objective

Describe the methods used to install and test copper tube.

a. Describe the hangers and fasteners used to support copper tube.
b. Explain the insulation requirements for copper tube.
c. Explain methods of pressure testing copper tube systems.

Performance Task

5. Select the correct support and spacing for the application.

Trade Terms

Clevis: An iron, or link in a chain, bent into the form of a horseshoe, stirrup, or letter U, with holes in the ends to receive a bolt or pin.

Insulation: A substance that retards the flow of heat.

Head: The height of a water column, measured in feet. One foot of head is equal to 0.433 pounds per square inch (psi).

All tubes must be installed and supported so that both the tube and its joints remain leakproof. Improper support can cause the tubing system to sag. This causes stress on the tube and fittings and, over time, increases the chance of breaks or leaks in the tubing system. Without proper support, the drainage tube can shift from its proper angle and form traps. These traps fill with liquid and solid wastes that block the tubeline. Copper tube is supported in a variety of ways and with various sizes of hangers and supports. Be sure to follow the manufacturer's instructions for specific installation instructions and consult applicable codes.

3.1.0 Hangers and Fasteners

Hangers and supports are designed to hold and support tube in either a horizontal or a vertical position. Which hanger to use depends on the job specifications, the documents that describe the quality of the required materials and work.

Specifications for hangers and supports are determined by the following factors:

- The combined weight of the tube fittings and valves
- The maximum weight of the contents that the tube might carry
- The material (wood, concrete, steel) the hanger will be attached to
- The distance from the anchor point to the tube
- The potential for corrosion between the hanger and the tubes, fittings, or valves
- The expansion and contraction of the tubing system
- The vibration of equipment attached to the tubing system

Different types of hangers and supports are designed to hold and support tube in either a horizontal or vertical position. They are manufactured in various materials, including carbon steel, malleable iron, cast iron, and plastics. They are available with different finishes, including copper plate, black, galvanized, chrome, and brass. To reduce corrosion, hangers and supports should be made of the same material as the tube. If another material is used, the tube must be isolated.

> **NOTE**
>
> If you are installing tube in a seismically active area—that is, where earthquakes are a possibility—local codes require seismic restraints. The purpose of these restraints is to ensure that the tube is securely fastened to the structure in the event of excessive vibration. For example, some codes require hangers and supports to be used at closer intervals than in non-seismically active areas. In addition, they may require you to leave extra spacing for tubes where they meet walls and floors to allow for anticipated movement.

The basic components of hangers and supports can be placed into the following three major categories:

- Tube attachments
- Connectors
- Structural attachments

3.1.1 Tube Attachments

A tube attachment is the part of the hanger that touches or connects directly to the tube. It may be designed for either heavy duty or light duty, for covered (insulated) tube or plain tube. Applicable fire codes may restrict the use of hangers made of material other than copper. In some cases, you can use plastic-coated hangers. For specific requirements, consult applicable codes. Examples of tube attachments are shown in *Figure 20*.

Hangers are used for horizontal or vertical support of tubes. The main purpose of hangers and brackets is to keep the tubing in alignment and prevent it from bending or distorting. Examples of hangers are shown in *Figure 21*.

Tube can be supported on wood-frame construction with several styles of tube attachments. These include tube hooks, J-hooks, tube straps, and plumber's tape (also called strap iron or band iron (see *Figure 22*). In these examples, the tube attachment, connector, and structural attachment are all one unit. Note that plumber's tape may not be allowed in your area. When choosing hangers of any type, always consult applicable codes.

Tube hangers are used mainly to support tube, but they can also be used as a support for vibration isolators (*Figure 23*). If no vibration problems are expected, use commonly accepted plumbing practices. If vibration may be a problem, the specifications should tell you what materials to use.

To fasten tubes to beams and other metal structures, one-hole clamps, steel brackets, beam clamps, or C-clamps are used. Vertical hangers, also called tube riser clamps, consist of a friction clamp that can be attached to structural site components to support the vertical load of the tube (see *Figure 24*). Special fasteners are used to attach hangers to masonry, concrete, or steel.

Other tube attachments include universal tube clamps and standard 1⅝" or 1½" channels (see *Figure 25*). The notched steel clamps are inserted by twisting them into position along the slotted side of the channel. The tubes can be aligned as close to one another as the couplings allow.

3.1.2 Connectors

The connector section of the hanger is the part that links the tube attachment to the structural attachment. Connectors can be divided into two groups: rods and bolts, and other rod attachments.

Rod attachments include eye sockets, extension pieces, rod couplings, reducing rod couplings, hanger adjusters, turnbuckles, clevises, and eye rods. A clevis is an iron bent into the form of a U, with holes in the ends to receive a bolt or pin.

To reduce the risk of corrosion, use hangers and supports that are made from the same material as the tube or pipe being hung or supported. If you must use hangers or supports made from a different material, ensure that the tube is properly isolated. If you are joining pipe and fittings of different metals (for example copper and galvanized steel), use a dielectric union to prevent galvanic corrosion.

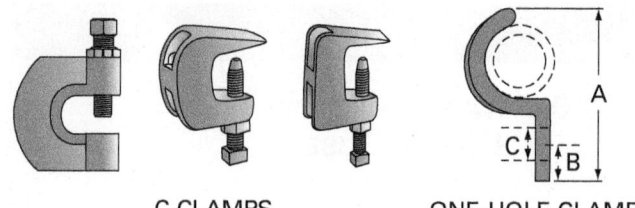

C-CLAMPS **ONE-HOLE CLAMP**

Figure 20 Tube attachments.

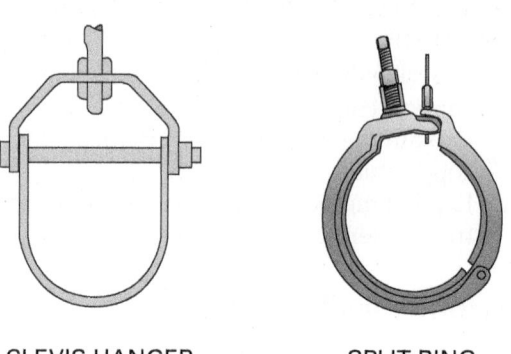

CLEVIS HANGER **SPLIT RING**

Figure 21 Tube hangers.

3.1.3 Structural Attachments

Structural attachments are used to anchor the tube hanger assembly securely to the structure. Structural attachments include threaded drop-in anchors, plastic or lead mollies, toggles, C-beam clamps, pound-in nail anchors, epoxy anchors, threaded rods, and strut channels. Water lines that are run through walls and floors should be sleeved. Refer to your local applicable code to determine whether the sleeve should be secured to the water line.

What you use to install hangers and supports depends on the item and the type of material to which you are attaching it (see *Table 3* and *Figure 26*).

3.2.0 Insulating Copper Tube

Under certain temperature and humidity conditions, condensation forms on cold water tube and may drip into equipment or occupied areas. Similarly, heat can escape from hot water tubing. To prevent these occurrences, some tube is insulated (*Figure 27*).

Insulation is a material that prevents the transfer of heat. Cork, glass fibers, mineral wool, and polyurethane foams are examples of insulating materials. Insulation should be fire-resistant, moisture-resistant, and vermin-proof.

If the insulation cannot be installed before the tubing is connected, it must be split lengthwise to fit onto the tube. Split seams and connecting seams must then be sealed. Insulation should not be stretched, because its effectiveness is then

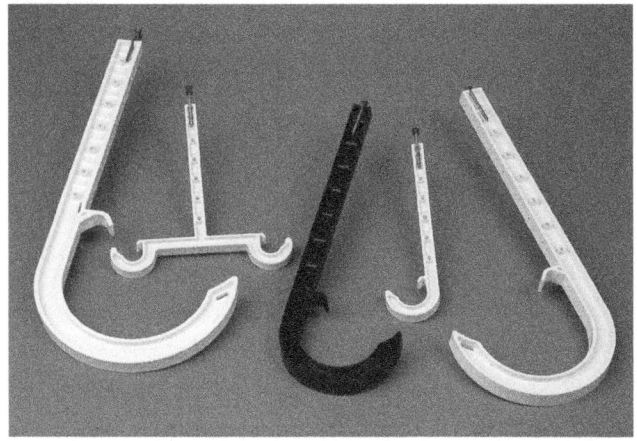

J-HOOKS

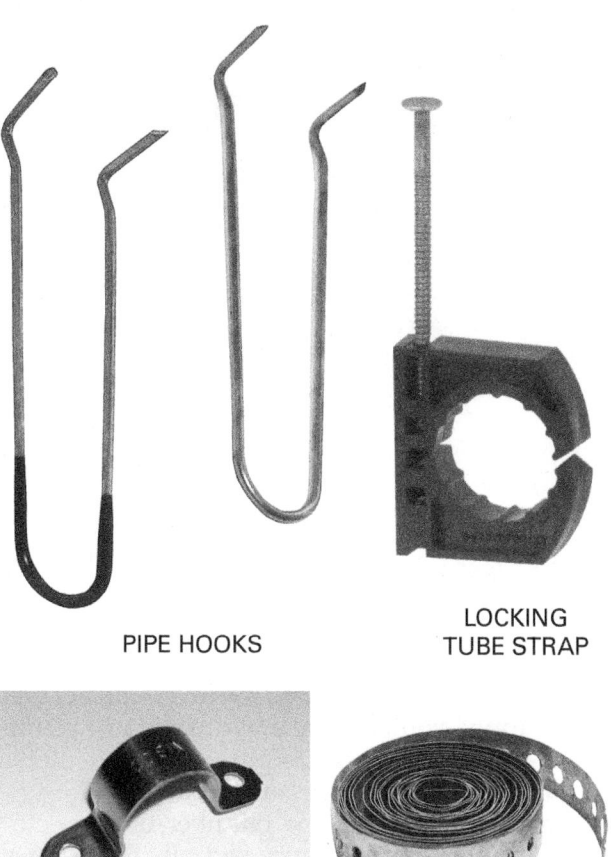

PIPE HOOKS   LOCKING
                             TUBE STRAP

PIPE STRAP   PLUMBER'S TAPE

Figure 22 Tube hangers for wood-frame construction.

reduced. You generally do not put the insulation on the tube until after pressure testing is performed. However, if you must do so before testing takes place, you must keep the joints exposed for the testers.

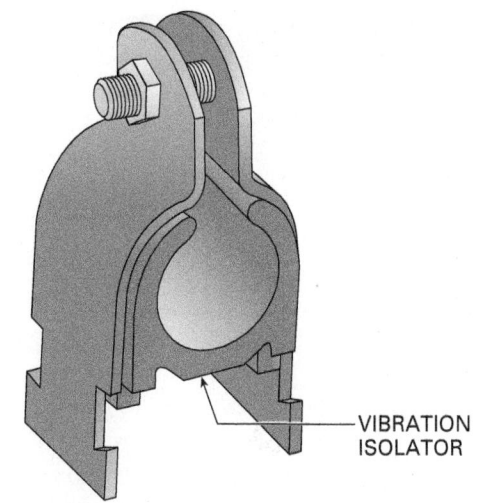

VIBRATION
ISOLATOR

Figure 23 Vibration isolator installed in hanger.

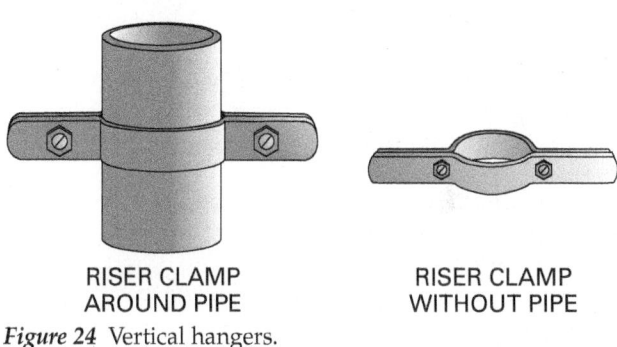

RISER CLAMP   RISER CLAMP
AROUND PIPE   WITHOUT PIPE

Figure 24 Vertical hangers.

Some tubes are always insulated, and others are insulated only under certain conditions. Local building codes and job specifications describe insulation requirements. You will learn more about which tubes to insulate and under what conditions elsewhere in this curriculum.

> **WARNING!**
> You may encounter asbestos in old buildings during renovation work. If you encounter asbestos, do not handle it or attempt to remove it. Tell your supervisor about the presence of this material immediately. Asbestos is linked to long-term illnesses. Only trained personnel who have the proper equipment for handling and disposing of asbestos can deal with such situations.

3.3.0 Pressure Testing

Once an installation is complete, you must pressure test the system for leaks to be sure that all connections are secure. The system must also be inspected to make sure that the work has been done according to the job specifications and applicable codes. Always test to the pressure that is specified in your local applicable code.

Figure 25 Universal pipe clamps and channels.

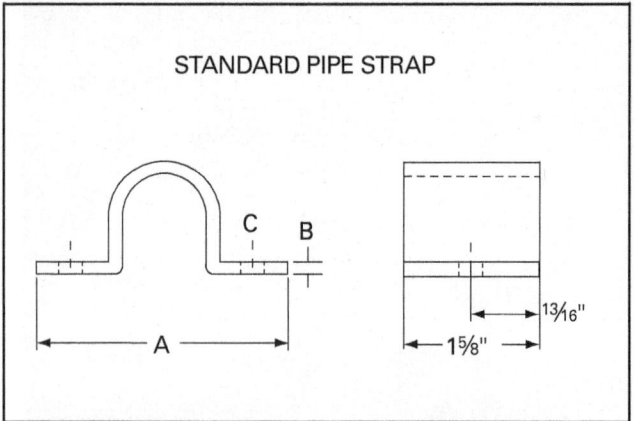

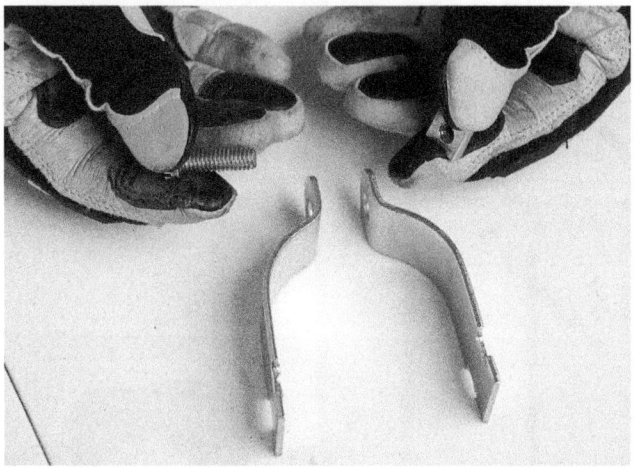

You can pressure test copper tube with water or air. You are required to supply all the equipment needed for testing, including test plugs, caps, plugs, and a source of compressed air (if air testing is required). Water testing can be done with test plugs and a hose. To test, close all outlets to the tubing system with test plugs, caps, and plugs. Fill the system with water or air under pressure and look for leaks.

3.3.1 Pressure Testing DWV Systems

To water test a DWV system, refer to your local code for the required pressure and water head. Consult all applicable codes to determine the appropriate pressure for testing. To do this, fill the vent stack completely with water. Once you have filled the DWV tubing, inspect the whole tubing system for leaks. This inspection is required before any part of the system can be covered. The plumbing inspector also checks for cross-connections, defective or inferior materials, and poor work.

3.3.2 Pressure Testing Water Supply Systems

To water test the water supply system, close all outlets and fill the system with water from the main. Inspect the system for leaks or other problems. Where codes specify water testing at pressures higher than those available from the normal water supply, you can use a hydraulic test pump (*Figure 28*) to produce the required pressure. Refer to code for the correct pressures to use.

To air test a system, the plumber or inspector fills the tubing with compressed air. By using a pressure gauge at the test plug, the plumber or inspector can determine if any pressure is lost somewhere in the tubing. Another useful method for identifying leaks is to brush soapsuds onto joints. If there are leaks in a joint, bubbles will form there.

WARNING! Never use oxygen, acetylene, or other gases to pressure test a refrigerant system. Oxygen will cause an explosion if it comes into contact with refrigerant oil or another inert gas. Acetylene is highly flammable. The only gas that should be introduced into a system is nitrogen.

Table 3 Types of Structural Attachments and Their Descriptions

Structural Attachment	Description
Threaded drop-in anchors	Most commonly used in concrete ceilings, walls, or floors. Different sizes of anchors are rated for different weights or spreads between anchors.
Toggles	Used for any hollow wall or ceiling support, including drywall and block.
Pound-in nail anchors	Used in concrete, brick, and stone.
Threaded rod hangers	Used in wood. Various lengths of all-thread rod can be cut to a specified length. Common sizes of all-thread rod are $\frac{3}{8}$ inch to $\frac{1}{2}$ inch.
C-clamps or beam clamps	Usually used to hook to iron or to a beam. These include set-bolts that must be tightened to ensure that they will not come loose.
Channel (strut)	Available in various lengths, widths, and heights and can be bolted or welded together. Various floor, wall, and overhead strut brackets are available.

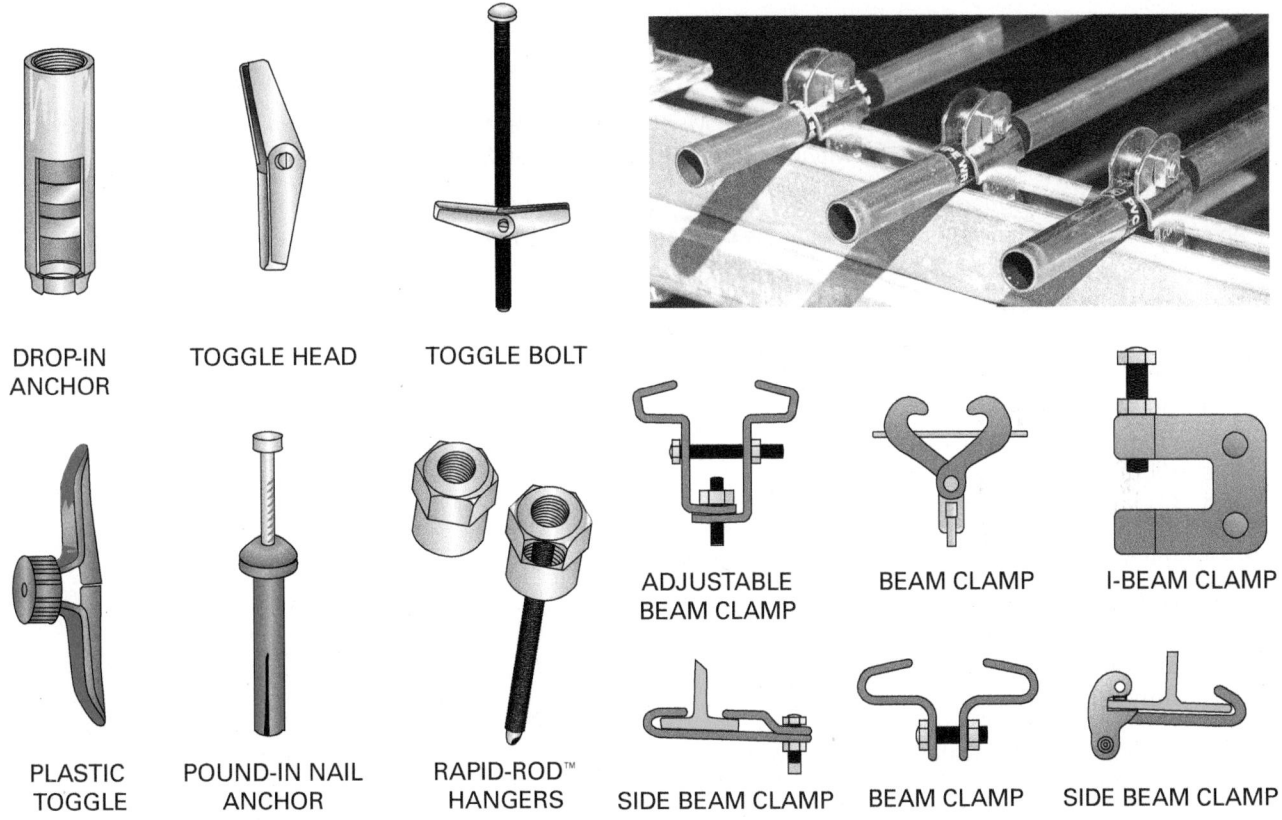

DROP-IN ANCHOR   TOGGLE HEAD   TOGGLE BOLT

PLASTIC TOGGLE   POUND-IN NAIL ANCHOR   RAPID-ROD™ HANGERS   SIDE BEAM CLAMP   BEAM CLAMP   SIDE BEAM CLAMP

ADJUSTABLE BEAM CLAMP   BEAM CLAMP   I-BEAM CLAMP

Figure 26 Structural attachments.

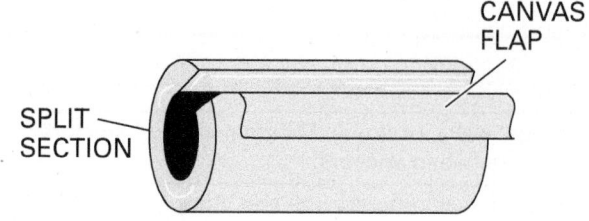

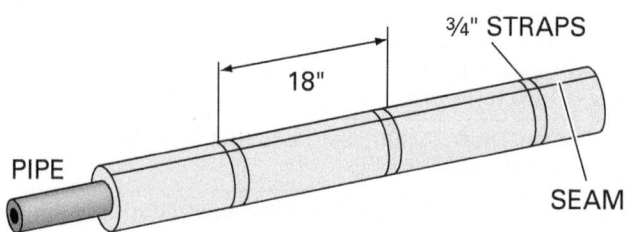

Figure 27 Insulated tube.

Figure 28 Hydraulic test pumps.

Additional Resources

Pipe Hangers and Supports—Materials, Design, Manufacture, Selection, Application, and Installation, MSS SP-58. Current Edition. Vienna, VA: Manufacturers Standardization Society.

The Copper Tube Handbook, 2016. New York: Copper Development Association (also available as a free pdf and an app for use on a smartphone or tablet at the Copper Development Association website, *www.copper.org*).

3.0.0 Section Review

1. Hangers are primarily used to support tube, but they can also be used to support _____.
   a. pressure checks
   b. channel struts
   c. vibration isolators
   d. C-clamps

2. A type of insulating material that requires special training for removal is _____.
   a. cork
   b. spray foam
   c. asbestos
   d. mineral wool

3. Where codes specify water testing at pressures higher than those available from the normal water supply, use a(n) _____.
   a. higher head
   b. hydraulic test pump
   c. hand pump
   d. pressure gauge

Summary

A plumber must be able to work with several kinds of pipe, including copper. This module introduced you to copper tube and its uses. You have also learned about the various fittings and valves that you can use to join copper tube for water distribution systems and DWV systems. You now know that tube must be properly supported and insulated, and you know the different kinds of hangers and supports that you can use. The requirements for tube hangers and insulation are usually specified in the job specifications or by local building codes.

Copper can be joined in many ways, including sweat joints, compression joints, or flare joints. This module has introduced you to all three methods of joining copper tube, and how to measure, cut, ream, and bend tube that you are going to join.

A completed piping installation must be inspected and tested. At this point, you will not be doing the testing, but you should be familiar with the procedures. Throughout this module you have read about special safety practices to prevent injury to yourself and others or damage to equipment. Follow these practices as you go into the field and work with copper tube and fittings.

1. Compared to other types of piping, copper tube and fittings are generally _____.

   a. less expensive
   b. more expensive
   c. difficult to install
   d. easy to install

2. Tubes that are measured in nominal or standard sizing have an OD that is _____.

   a. $1/8$" larger than the nominal size
   b. $1/4$" larger than the nominal size
   c. $1/2$" larger than the nominal size
   d. 1" larger than the nominal size

3. Types K, L, M, and DWV copper tubing are measured using _____.

   a. ID sizing
   b. OD sizing
   c. nominal sizing
   d. copper tube sizing

4. Type K copper tube is marked with the color code _____.

   a. orange
   b. blue
   c. yellow
   d. green

5. The valve most often used for draining and freeze protection above ground is the _____.

   a. ball valve
   b. compression, stop, and waste valve
   c. globe valve
   d. gate valve

6. The type of valve also known as a ground-key valve is the _____.

   a. stop valve
   b. globe valve
   c. check valve
   d. ball valve

7. In order to prevent turbulence when more than one tee is installed in a line, a 3" tube should have a minimum of _____.

   a. 3" of tube between tees
   b. 6" of tube between tees
   c. 18" of tube between tees
   d. 30" of tube between tees

8. If tee fittings are not properly installed, they can cause a condition known as _____.

   a. bullheading
   b. ferrulization
   c. sweating
   d. spouting

9. To measure tube that has a fitting on one end only, use a _____.

   a. center-to-face measurement
   b. face-to-back measurement
   c. end-to-face measurement
   d. center-to-center measurement

10. The preferred tool for cutting tubing is a _____.

    a. chop saw
    b. hacksaw
    c. midget cutter
    d. handheld tube cutter

11. When using a mechanical bender, the minimum bend radius of $1/4$" annealed tubing is _____.

    a. $1/4$"
    b. $1/2$"
    c. $3/4$"
    d. 1"

12. Compared to the tubes they are joining, solder fittings are _____.

    a. slightly smaller
    b. slightly larger
    c. significantly smaller
    d. significantly larger

13. A joint that is commonly used to join soft copper tubing from $1/4$" to 2" is a(n) _____.

    a. grooved joint
    b. sweat joint
    c. flare joint
    d. angle joint

14. Structural site components to support the vertical load of the tube are attached using _____.

    a. J-hooks
    b. eye sockets
    c. wedge anchors
    d. tube riser clamps

15. The part of the hanger that links the tube itself to the structural attachment is called a _____.

    a. connector
    b. clevis
    c. clamp
    d. bracket

16. The _____ of the hanger links the tube attachment to the structural attachment.

    a. c-clamps
    b. toggles
    c. rod coupling
    d. lead mollies

17. Which of the following is not a requirement of insulation for copper tubing?

    a. Moisture-resistant
    b. Fire-resistant
    c. Corrosion-resistant
    d. Vermin-proof

18. Flare joints are required in an installation where a _____ hazard exists.

    a. fire
    b. freezing
    c. corrosive
    d. weight

19. A hacksaw blade should have how many teeth per inch?

    a. 31
    b. 32
    c. 33
    d. 34

20. To pressure test a DWV system, a plumber should fill the vent stack with _____.

    a. compressed air
    b. essential oil
    c. colorful liquid
    d. water

Trade Terms Quiz

Fill in the blank with the correct term that you learned from your study of this module.

1. _____ is the approximate measurement in inches of the inside diameter of tube for most copper tubes. The exception is the measurement for ACR tubing, which is based on the outside diameter.

2. A brass compression ring used for joining is called a(n) _____.

3. A(n) _____ is commonly used to join soft copper tubing with diameters from ¼ to 2 inches.

4. Copper tubing that is described as soft and flexible is created through a process called _____.

5. Flare fittings are made of _____ brass.

6. If tube becomes deformed, use a(n) _____ to work the tube back to roundness.

7. _____, or air conditioning and refrigeration systems, is one of many plumbing applications in which copper tubing is used.

8. The use of a(n) _____, a fixture in which the main line is smaller than the branch, can prevent turbulence in the system.

9. A(n) _____ is a rod attachment made of iron that is commonly in the form of a U.

10. To prevent the transfer of heat, install _____. Materials used for this purpose include cork, glass fibers, mineral wool, and polyurethane foams.

11. Annealed copper tubing that is manufactured specifically for use in air conditioning and refrigeration work is called _____.

12. A(n) _____ consists of a threaded nut that has been squeezed over a compression ring to join tubes.

13. To water test a DWV system, you must have at least 10 feet of water _____ in the system.

14. _____ tubing is widely used in commercial refrigeration and air conditioning systems, because it is very rigid.

15. A(n) _____ is created by soldering.

16. Using fewer fittings reduces the chance for _____ or decreases in pressure from one point to another caused by friction losses in a water system.

17. The process that occurs during soldering in which solder melts and flows into the gap between the tube and the fitting is known as _____.

18. You should consider strength and _____, or ease of bending, when selecting copper tube.

Trade Terms

ACR	Clevis	Flare joint	Pressure drop
ACR tubing	Compression joint	Formability	Sizing tool
Annealing	Drawn copper	Head	Sweat joint
Bullhead tee	Drop-forged	Insulation	
Capillary action	Ferrule	Nominal size	

Terry Lunt
Instructor
ABC Southern California Chapter

Describe how you got started in the construction industry.

I started with the Santa Fe Railroad, building bridges for them for two years until a plumbing apprenticeship opened up. I started the apprentice-ship in 1979 and worked as a plumber with the railroad for eight years. Then after a big lay-off, I started working with other plumbing shops all over Southern California.

Who inspired you to enter the industry? Why?

My dad and my brother both were electricians for the railroad. My dad was an electrician for 35 years with them, and my brother worked for them for about 15 years.

What do you enjoy most about your job?

I'm at a place in my career where I can do pretty much what I want to do, so I get to do a lot of projects that other people with less experience can't do. There's a lot of variety, and I get to see a lot of Southern California too!

Do you think training and education are important in construction? If so, why?

For sure. As a plumbing instructor for over five years, I think it is crucial. One of the biggest problems I used to have was finding qualified people. It's so nice to have qualified people now. They're actually way better mechanics because they're getting hands-on experience. The combination of school and hands-on experience, you can't beat it.

For apprentices, I think there's an awakening at some point when they recognize that they can be a foreman someday. In fact, a lot of the guys I've graduated are already foremen. Even if they don't know it at first, they eventually realize that they can use this apprenticeship as a fast track. They see that they are better equipped for the job because they get a well-rounded education.

How important are NCCER credentials to your career?

For the people who have gone through the training, the credentials give them bragging rights that they have the apprenticeship behind them. For me, being a teacher is not only a boost for my company but also has made me an expert. Now I have people who I've always considered to be experts calling and asking me questions about plumbing. It's fun to be the expert now.

How has training/construction impacted your life and your career?

With a vocational teaching credential, it is the stepping stone for getting jobs. When I got a teaching credential from the State of California, I think my experience of having taken the assessment tests helped them to see that I was qualified.

Would you suggest construction as a career to others? If so, why?

For sure, I've tried to get qualified people into the program. I tell them that it serves you well throughout your life. It doesn't have to be plumbing; it can be electrical, or framing, or anything else. It can help you get employed, especially if you're not going to college.

How do you define craftsmanship?

I think craftsmanship is about doing the work and doing it right. It's a learned process. A lot of the apprentices I work with are being run hard every day, so I try to remind them to stop and think about the importance of the final product while they're working. To get them to look at the finish, how things line up, and think to themselves, "I can do that a little better next time."

Craftsmanship is also about pride. I can drive around the area and point to the buildings that I've helped build, and there's a real pride in that.

Trade Terms Introduced in This Module

ACR: Air conditioning and refrigeration system.

ACR tubing: Annealed copper tubing that is manufactured specifically for use in air conditioning and refrigeration work.

Annealing: Heating a material and slowly cooling it to relieve internal stress. This process reduces brittleness and increases toughness.

Bullhead tee: A tee fitting used on a branch that is larger than the main line, or that has one outlet larger than the run openings.

Capillary action: The process during soldering in which the molten solder flows into the narrow gap between the pipe and the joint, regardless of whether the solder is flowing up, down, or horizontally.

Clevis: An iron, or link in a chain, bent into the form of a horseshoe, stirrup, or letter U, with holes in the ends to receive a bolt or pin.

Compression joint: A method of connection in which tightening a threaded nut squeezes a compression ring to seal the joint.

Drawn copper: Tubing produced by pulling the tube through dies to reduce its diameter. The drawing process hardens the copper and makes it very rigid.

Drop-forged: A characteristic of a product made when heated metal is pounded or shaped between dies with a drop hammer or press.

Ferrule: A brass compression ring used for joining copper tube.

Flare joint: A fitting in which one end of each tube to be joined is flared outward using a special tool. The flared tube ends mate with the threaded flare fitting and are secured to the fitting with flare nuts.

Formability: The ease with which a material can bend.

Head: The height of a water column, measured in feet. One foot of head is equal to 0.433 pounds per square inch (psi).

Insulation: A substance that retards the flow of heat.

Nominal size: Approximate measurement in inches of the inside diameter (ID) of pipe for most copper tubes. However, nominal size of ACR tubing is based on the outside diameter (OD).

Pressure drop: A decrease in pressure from one point to another caused by friction losses in a fluid system, such as a water system.

Sizing tool: A tool consisting of a plug and a sizing ring that is used to reshape a deformed pipe back to roundness.

Sweat joint: A pipe joint made by applying solder to the joint and heating it until it flows into the joint.

Additional Resources

This module presents thorough resources for task training. The following resource material is suggested for further study.

Plumber's Handbook Revised, Howard C. Massey. 2006. Carlsbad, CA: Craftsman Book Company.

Pipe Hangers and Supports—Materials, Design, Manufacture, Selection, Application, and Installation, MSS SP-58. Current Edition. Vienna, VA: Manufacturers Standardization Society.

Plumbing Installation and Design, L. V. Ripka. Third Edition, 2006. Orland Park, IL: American Technical Publishers.

The Copper Tube Handbook, 2016. New York: Copper Development Association (also available as a free pdf and an app for use on a smartphone or tablet at the Copper Development Association website, *www.copper.org*).

Figure Credits

Section Review Answer Key

SECTION 1.0.0

Answer	Section Reference	Objective
1. c	1.1.0	1a
2. c	1.2.0; Table 1	1b
3. b	1.3.0	1c
4. c	1.4.0	1d

SECTION 2.0.0

Answer	Section Reference	Objective
1. a	2.1.0	2a
2. c	2.2.0	2b
3. c	2.3.0	2c
4. a	2.4.3	2d

SECTION 3.0.0

Answer	Section Reference	Objective
1. c	3.1.1	3a
2. c	3.2.0	3b
3. b	3.3.2	3c

NCCER CURRICULA — USER UPDATE

NCCER makes every effort to keep its textbooks up-to-date and free of technical errors. We appreciate your help in this process. If you find an error, a typographical mistake, or an inaccuracy in NCCER's curricula, please fill out this form (or a photocopy), or complete the online form at **www.nccer.org/olf**. Be sure to include the exact module ID number, page number, a detailed description, and your recommended correction. Your input will be brought to the attention of the Authoring Team. Thank you for your assistance.

Instructors – If you have an idea for improving this textbook, or have found that additional materials were necessary to teach this module effectively, please let us know so that we may present your suggestions to the Authoring Team.

NCCER Product Development and Revision
13614 Progress Blvd., Alachua, FL 32615

Email: curriculum@nccer.org
Online: www.nccer.org/olf

❏ Trainee Guide   ❏ Lesson Plans   ❏ Exam   ❏ PowerPoints   Other _____

Craft / Level: _____   Copyright Date: _____

Module ID Number / Title: _____

Section Number(s): _____

Description: _____

_____

_____

_____

_____

Recommended Correction: _____

_____

_____

_____

_____

Your Name: _____

Address: _____

_____

Email: _____   Phone: _____

This page is intentionally left blank.

Cast-Iron Pipe and Fittings

OVERVIEW

Cast-iron pipe is used for drain, waste, and vent (DWV) systems in residential, commercial, and industrial plumbing. This material is strong, durable, and resistant to corrosion. Manufacturers supply a full line of cast-iron pipe fittings to meet nearly every requirement in plumbing waste systems. Modern methods make working with cast-iron pipe much easier than in the past century. In this module, you will be introduced to the key materials and methods for measuring, cutting, installing, and testing cast-iron piping systems.

Module 02108

Trainees with successful module completions may be eligible for credentialing through the NCCER Registry. To learn more, go to *www.nccer.org* or contact us at 1.888.622.3720. Our website, *www.nccer.org*, has information on the latest product releases and training.

Your feedback is welcome. You may email your comments to *curriculum@nccer.org*, send general comments and inquiries to *info@nccer.org*, or fill in the User Update form at the back of this module.

This information is general in nature and intended for training purposes only. Actual performance of activities described in this manual requires compliance with all applicable operating, service, maintenance, and safety procedures under the direction of qualified personnel. References in this manual to patented or proprietary devices do not constitute a recommendation of their use.

02108 V4.5

02108
CAST-IRON PIPE AND FITTINGS

Objectives

Successful completion of this module prepares trainees to:

1. Identify the types, uses, and properties of cast-iron pipe.
   - a. Describe hub-and-spigot and no-hub pipe, and their uses.
   - b. Describe the sizing and labeling of cast-iron pipe.
   - c. Describe the different types of fittings used on cast-iron pipe.
   - d. Identify storage and handling requirements for cast-iron pipe.
2. Describe the different methods for measuring, cutting, and joining cast-iron pipe.
   - a. Explain the tools and methods used to measure and cut cast-iron pipe.
   - b. Describe how to join hub-and-spigot pipe.
   - c. Describe how to join no-hub pipe.
3. Describe the methods used to install and test cast-iron pipe.
   - a. Describe the hangers and fasteners used to support cast-iron pipe.
   - b. Explain how to support vertical and horizontal runs.
   - c. Explain installation methods used in wood, masonry, and concrete structures.
   - d. Describe the methods used in testing cast-iron pipe.

Performance Tasks

Under the supervision of your instructor, you should be able to do the following:

1. Select correct materials for cast-iron piping systems.
2. Identify types of fittings and their uses.
3. Select the appropriate personal protective equipment for cast-iron piping.
4. Correctly measure, cut, and join cast-iron pipe.
5. Select the correct support and spacing for the application.

Trade Terms

Bends	Ductile iron	Long bends	Soil pipe cutter
Closet bends	Extra-heavy	Main stack	Sweeps
Closet flange	Heel inlets	No-hub cast-iron pipe	Torque wrench
Compression joints	Hub-and-spigot cast-iron	Rotary hammer drill	Traps
Couplings	pipe	Sanitary tee	Wyes
Double sanitary tee	Increaser	Service	
Double wyes	Insertion length	Slope	

Industry Recognized Credentials

If you are training through an NCCER-accredited sponsor, you may be eligible for credentials from NCCER's Registry. The ID number for this module is 02108. Note that this module may have been used in other NCCER curricula and may apply to other level completions. Contact NCCER's Registry at 888.622.3720 or go to *www.nccer.org* for more information.

CODE NOTE

Codes vary among jurisdictions. Because of the variations in code, consult the applicable code whenever regulations are in question. Referencing an incorrect set of codes can cause as much trouble as failing to reference codes altogether. Obtain, review, and familiarize yourself with your local adopted code. Safety codes are developed by the US Occupational Safety and Health Administration (OSHA).

Contents

Figures and Tables ──────────

1.0.0 CAST-IRON PIPE

Objective

Identify the types, uses, and properties of cast-iron pipe.

   a. Describe hub-and-spigot and no-hub pipe, and their uses.
   b. Describe the sizing and labeling of cast-iron pipe.
   c. Describe the different types of fittings used on cast-iron pipe.
   d. Identify storage and handling requirements for cast-iron pipe.

Performance Tasks

   1. Select correct materials for cast-iron piping systems.
   2. Identify types of fittings and their uses.

Trade Terms

Bends: Fittings that change the direction of flow in piping.

Closet bends: Elbow-like fittings that connect water closets to the main drainage piping.

Closet flange: A fitting used to mate a water closet to a closet bend or other DWV pipe.

Double sanitary tee: A fitting with a main pipe and two branch lines that changes the direction of flow 90 degrees from the horizontal to the vertical. Also called a sanitary cross.

Double wyes: Fittings with a main pipe and two branches at 45 degrees. These connect horizontal waste pipes and branch drains to the building main.

Ductile iron: A cast-iron alloy containing magnesium to reduce brittleness.

Extra-heavy: A kind of cast-iron pipe. Extra-heavy refers to the pipe's wall thickness, which is thicker than service pipe of the same nominal pipe size.

Heel inlets: Bends that have an inlet to connect a smaller pipe to the main line or a branch line. The inlet is located at the base of the outside curve or heel of the bend.

Hub-and-spigot cast-iron pipe: Pipe that has a bell or enlargement at one end where the spigot (smooth end) of the next pipe slides in to form a joint. Also called bell-and-spigot pipe.

Increaser: A fitting used to increase the size of a straight-through line of pipe. It is often used for the vent stack before it goes through the roof to reduce the chance of frost clogging the vent opening in very cold climates.

Long bends: Bend fittings with one end (typically the spigot end) longer than the other.

Main stack: The principal DWV riser to which branches may be connected.

No-hub cast-iron pipe: A pipe with no enlargement or bell on either end.

Sanitary tee: A tee fitting for DWV systems that consists of a main pipe and a branch has a curved 90-degree side-inlet; this curve helps to channel the flow of wastewater or sewage from a branch line to the main line. This fitting is always used in the vertical position.

Service: A lightweight kind of cast-iron pipe. Service refers to the pipe's wall thickness, which is thinner for a nominal pipe size compared to extra heavy.

Sweeps: Fittings having a radius of the curve greater than those for standard 90-degree bends. Used for a smooth change of direction.

Traps: DWV fittings or devices that provide a liquid seal to prevent the emission or escape of sewer gases from the system without materially affecting the flow of sewage or wastewater through the fitting.

Wyes: Fittings consisting of a main pipe and a 45-degree branch for connecting a waste branch to a horizontal building main drain.

Cast-iron pipe, often called soil pipe, is used for drain, waste, and vent (DWV) systems in residential, commercial, and industrial plumbing. Cast iron is a strong and durable piping material that resists corrosion and abrasion, and withstand extreme temperature changes.

Cast iron refers to products made from molten iron that is formed using a mold-casting system of production. Most large water lines are made from ductile iron, which is made from an alloy of gray iron and magnesium. Ductile iron is less brittle than gray cast iron, which until the middle of the twentieth century was the most widely used form of cast-iron pipe. Because ductile iron is less brittle, it can better support its weight when installed in a building.

NOTE

Before ductile iron was introduced in 1955, gray cast iron was used for piping in water main and sanitary waste systems. Gray cast iron has a controlled carbon-content, but it is brittle. This meant that it was prone to breaking under heavy loads.

1.1.0 Hub-and-Spigot and No-Hub Pipe

The two types of cast-iron pipe are hub-and-spigot pipe and no-hub pipe. Hub-and-spigot cast-iron pipe has a bell shape or enlargement on one end called a hub, or bell. The smooth spigot end of each pipe in the run fits into the adjacent hub. Hub-and-spigot pipe is available in single hub or double hub (*Figure 1*). Double-hub pipe can be cut to create two single-hub pipes, which can be useful for a job that requires many joints.

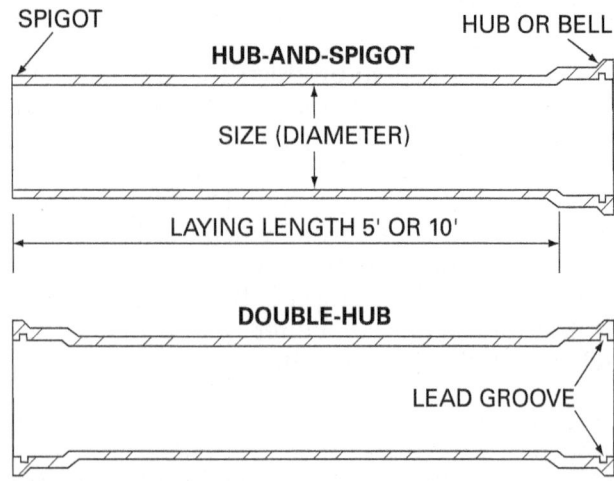

Figure 1 Hub-and-spigot and double-hub cast-iron pipe.

No-hub cast-iron pipe, as its name suggests, has no hub on either end (*Figure 2*). No-hub cast-iron pipe and fittings can be used in all areas of plumbing where cast iron is recommended. Introduced in the 1960s, no-hub pipe and fittings provide an easier method of joining cast iron in areas where space is limited. No-hub pipe has become very popular because of this feature. Both no-hub and hub-and-spigot cast-iron pipe are widely used by plumbers today.

1.2.0 Sizes and Labeling

Single hub-and-spigot cast-iron pipe is available in a variety of lengths and diameters. The smaller diameters are used in residential construction. The length of the pipe, referred to as the laying length, is the measurement from the base of the hub to the end of the spigot.

1.2.1 Sizing

Double-hub pipe is available in 30-inch, 60-inch, and 120-inch lengths. Pipe in 30-inch lengths can be cut into two short single-hub pipes. No-hub and hub-and-spigot pipe are generally made to the same thickness.

Fewer types of fittings are available for pipe sizes larger than 8 inches. Generally, when sizes above 15 inches are required under a building, the specifications will require you to use ductile iron pipe, as permitted by applicable codes.

Cast-iron pipe is classified by its wall thickness. Manufacturers make two thicknesses of pipe: service and extra-heavy. The latter is available only for hub-and-spigot pipe. Service pipe is marked SV, and the term *service* indicates the thickness of the pipe wall, which for a given

Figure 2 No-hub cast-iron pipe and fittings.

nominal pipe size is slightly thinner than an extra heavy pipe. Extra-heavy pipe is marked XH. Service is the most commonly-used hub-type pipe. However, if a structure is subject to vibration, settling, or excessive corrosion, the specifications or codes may require extra-heavy pipe and fittings. You should always use fittings that are made to work specifically with the class of pipe you are using.

1.2.2 Labeling

All cast-iron pipe and fittings with the "C$_I$" (*Figure 3*) trademark are manufactured according to the standards set by the Cast Iron Soil Pipe Institute (CISPI). The CISPI standards require that all pipe must be measured before the manufacturer can distribute it. In addition, the manufacturer must label all pipe and fittings with its name or trademark and with the CISPI trademark; fittings are frequently also labeled with the fitting's diameter (*Figure 4*). On no-hub cast-iron pipe, the label appears about $1\frac{1}{2}$ inches from the end of the pipe.

> **CAUTION**
>
> If no-hub plumbing materials are not labeled correctly, it may be an indication that they were not manufactured according to the standards of the CISPI. In most cases, plumbing codes do not permit unlabeled connections to be installed. With the growth of the global market, piping will often include the country of origin in the label.

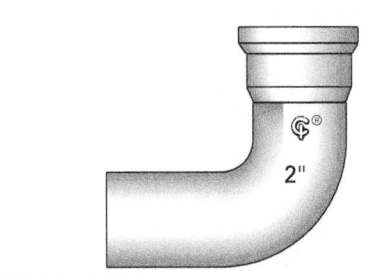

Figure 4 Fitting label.

1.3.0 Fittings

Fittings are used to join pipe sections, which often involves changing the direction of flow in a system. Almost any kind of pipe assembly is possible with the wide variety of pipe fittings made today. Plumbing codes require that appropriate types of fittings be used in different types of plumbing systems. This section identifies the common types of cast-iron fittings.

1.3.1 Bends

Bends are elbow-like fittings that change the direction of flow in a single DWV pipe. Bend type, or the amount of directional change, is expressed as a fraction of a complete circle. To determine the number of degrees a given bend turns, multiply the fraction by 360 degrees. For example, a $\frac{1}{8}$ bend turns the pipe:

$$\frac{1}{8} \times 360° = 45°$$

Bends are available in most standard pipe sizes, the smallest being $1\frac{1}{2}$ inches in diameter. Refer to *Figure 5* for common bends.

Long bends (*Figure 6*) are similar to standard bends, except that long bends consist of one leg that is longer than the other. Long bends are typically sized with a two-number system. The first number indicates the diameter of the bend; the second number indicates the length of the longer leg of the fitting. For example, a 4 × 12, $\frac{1}{4}$ bend is 4 inches in diameter, and its longer leg is 12 inches in length. Long bends are expensive and should be used only at the base of plumbing stacks.

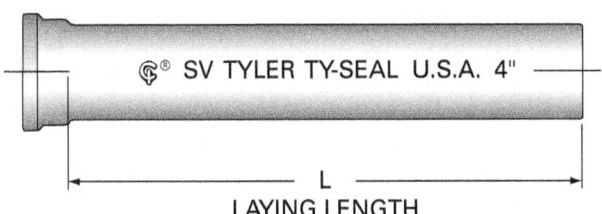

SV TYLER TY-SEAL U.S.A. 4"

L

LAYING LENGTH

Figure 3 Pipe label.

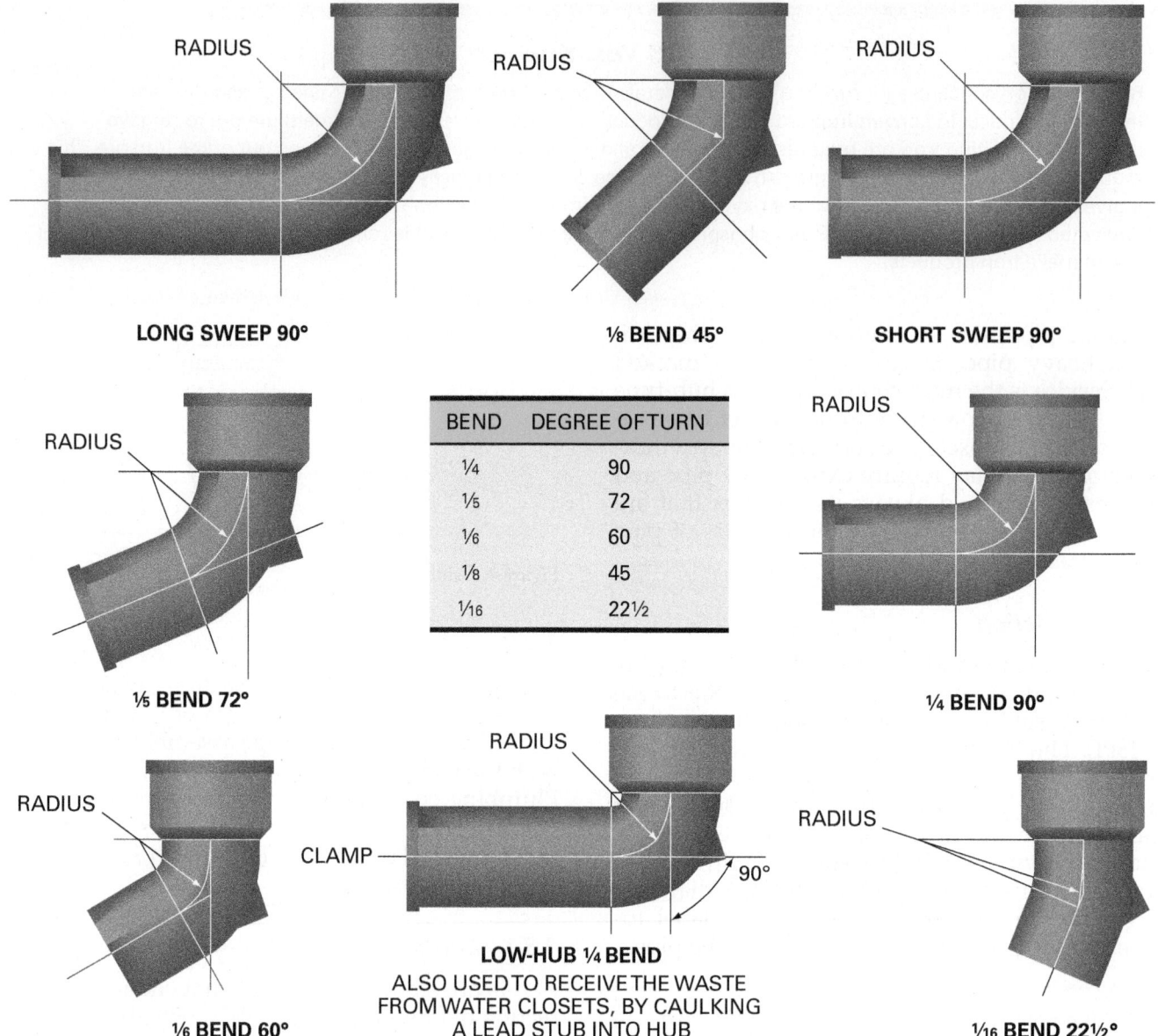

BEND	DEGREE OF TURN
¼	90
⅕	72
⅙	60
⅛	45
¹⁄₁₆	22½

LONG SWEEP 90°

⅛ BEND 45°

SHORT SWEEP 90°

⅕ BEND 72°

¼ BEND 90°

⅙ BEND 60°

LOW-HUB ¼ BEND
ALSO USED TO RECEIVE THE WASTE
FROM WATER CLOSETS, BY CAULKING
A LEAD STUB INTO HUB

¹⁄₁₆ BEND 22½°

Figure 5 Common bends.

Codes permit two methods of flow reduction: with a closet bends or with a closet flange. Always refer to the manufacturer's specifications and the plumbing plans before installing DWV system components.

Closet bends (*Figure 7*) connect the water closet to the main drainage piping. They are also available with left-hand or right-hand side inlet openings. Closet bends have a reduction of pipe size in the direction of flow. To accommodate restrictions at the water closet site, closet bends can be obtained with various configurations of the vertical leg position and diameter relative to the pipe inlet:

- Regular
- Offset
- Reducing

Both the inlet and outlet ends of the fitting may be manufactured with break-off grooves to permit easy cutting. The closet flange (*Figure 8*) is used to mate the water closet to the closet bend. These flanges contain slotted screw holes to fasten the water closet securely to the flange.

Sweeps are similar to ¼ bends, but the radius of the curve is much greater than standard bends. The greater radius allows for smoother flow of wastes through the piping system. Two types of sweeps exist: long sweeps and short sweeps. They differ in the radius of the curve made by the fittings and by their laying length. Long sweeps have a longer radius and a longer laying length; short sweeps have a shorter radius and a shorter laying length.

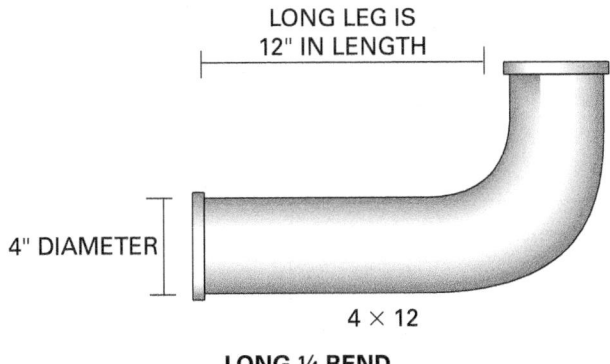

LONG LEG IS
12" IN LENGTH

4" DIAMETER

4 × 12

LONG ¼ BEND

Figure 6 Long bend.

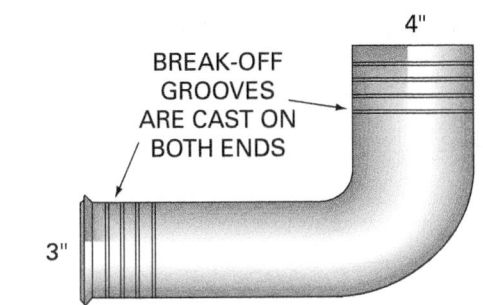

4"

BREAK-OFF
GROOVES
ARE CAST ON
BOTH ENDS

3"

Figure 7 Closet bend.

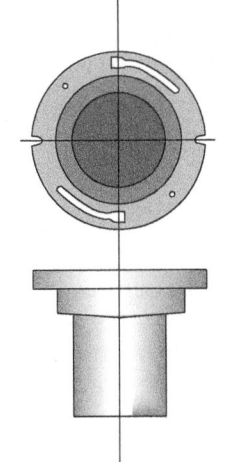

Figure 8 Closet flange.

A number of bends have side or heel inlets. Any fittings with side inlets are called either right-hand or left-hand fittings, depending on where the opening appears (*Figure 9*). To determine the handedness of the inlet, place the main bell or inlet end of the fitting toward you and any other inlet or outlet, or the spigot end, pointing down. The bend heel inlet allows small vents or drains of smaller sizes to be connected in line with the bend outlet at the heel.

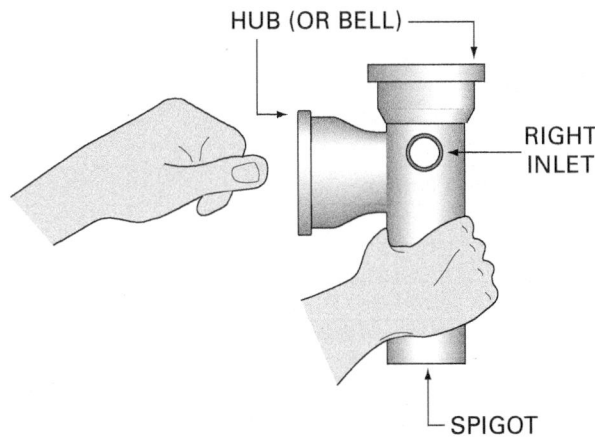

HUB (OR BELL)

RIGHT
INLET

SPIGOT

Figure 9 A fitting with a right-hand inlet.

1.3.2 Branches

A variety of fittings allow you to connect pipe to create branches in the piping system. Wyes and double wyes are used to join two or three drains into one pipe (*Figure 10*). Wyes can connect branches at 45-degree angles to the main stack, the largest riser that runs between the building drain and the highest horizontal drain in the system. Sometimes the branch fitting has a threaded or tapped opening in either the main or branch hub. You can connect a cleanout plug (used to close the fitting or pipe) or a threaded pipe from a plumbing fixture to this opening. Use an inverted wye in the venting system (*Table 1*).

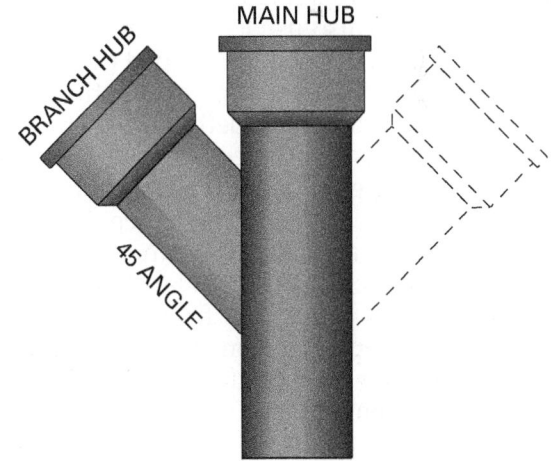

BRANCH HUB

MAIN HUB

45 ANGLE

SINGLE OR DOUBLE

Figure 10 Wye.

Table 1 DWV Fittings and Their Uses

	2009 International Plumbing Code®		
	Table 706.3, Fittings for Change in Direction		
	Change in Direction		
Type of Fitting Pattern	Horizontal to Vertical	Vertical to Horizontal	Horizontal to Horizontal
Sixteenth bend	X	X	X
Eighth bend	X	X	X
Sixth bend	X	X	X
Quarter bend	X	Xa	Xa
Short sweep	X	Xa,b	Xa
Long sweep	X	X	X
Sanitary tee	Xc		
Wye	X	X	X
Combination wye and eighth bend	X	X	X

For SI: 1 inch = 25.4 mm
a. The fittings shall only be permitted for a 2-inch or smaller fixture drain.
b. Three inches or larger.
c. For a limitation on double sanitary tees, see Section 706.3.

The sanitary tee allows you to connect pipe at right-angle intersections in branches that run from horizontal to vertical (*Figure 11*). Model codes restrict the use of sanitary tees to sanitary drainage systems where the flow of materials is from the horizontal to the vertical runs. The double sanitary tee, also termed a *sanitary cross*, allows you to connect two branch lines entering the system from opposite directions. This fitting allows you to place the horizontal runs and vertical runs centrally. You can also use tees on vent lines. Tees are available with threaded or tapped openings as well as side openings. Like bends, these fittings are designed with right-hand or left-hand openings.

The dimensions of the tee indicate the diameter of the openings. If all openings are the same diameter, only one dimension is given. When the branch openings are smaller than the straight-through openings, the fitting is labeled with two numbers, such as 4 × 3. The first number is the diameter of the openings on the main, straight-through run; the second number is the diameter of the branch openings.

1.3.3 Increasers

An increaser (*Figure 12*) is a fitting that connects two pipes of different diameters. It is used to increase the size of a straight-through line of pipe. Increasers are commonly used to connect a larger pipe to the top of a vent stack. This provides for a larger-sized pipe above the heated area of a building, which keeps frost buildup from closing the opening in the vent stack.

1.3.4 Traps

Traps are fittings designed to provide a liquid seal that prevents sewer gas from escaping into a building; they are a simple solution to the plumbing problem of separating a living space from a sewage system. The water seal in a trap prevents gas and fumes from entering the building through the fixtures. Traps also act as a barrier to vermin that may enter a building's living space through the plumbing piping.

NO-HUB SANITARY TEE

NO-HUB SANITARY
TAPPED TEE

NO-HUB SANITARY
TAPPED CROSS

SERVICE WEIGHT
STRAIGHT TEE

SERVICE WEIGHT
SANITARY TEE

SERVICE-WEIGHT
SANITARY TEE WITH LEFT-
HAND SIDE TAP

SERVICE-WEIGHT
SANITARY TEE WITH RIGHT-
HAND SIDE TAP

Figure 11 Tees.

Plumbing Codes and Cast-Iron Pipe

Plumbing codes are designed to protect the health and safety of the public. While codes vary from location to location, they are all based on sound principles of sanitation and safety.

Plumbing codes regulate every aspect of cast-iron pipe and fitting selection and installation. Take the time to become familiar with the state, city, or local plumbing codes that apply to the area where you work. Some fittings may be allowed in one area but may be strictly forbidden in another area.

**SERVICE WEIGHT NO-HUB
INCREASER (REDUCER)**

SERVICE WEIGHT INCREASER

Figure 12 Increasers.

Traps are installed between the drainage line and fixtures or drains. Many traps are designed with cleanout plugs that you can remove to clean out the trap. Various P-traps (*Figure 13*) are available to suit different applications. P-traps are commonly used at sinks and lavatories.

Note that S-traps are no longer installed but are still found in older buildings. Running traps may be illegal in one area, but not in another.

1.4.0 Material Storage and Handling

To ensure maximum productivity on the job, use common sense when storing and handling cast-iron pipe. Before starting to work, always plan ahead. Organize pipe and fittings into groups by pipe size and type. Store pipe and fittings near where you are working so they are easy to access. When handling pipe and fittings, be careful not to break, chip, or otherwise damage them. Pipe that is damaged not only wastes money, but also makes you less productive. No special protection from weather is needed for cast-iron pipe. However, when storing and handling cast-iron pipe, ensure that dirt, insects, rodents, and animals are prevented from entering the pipe.

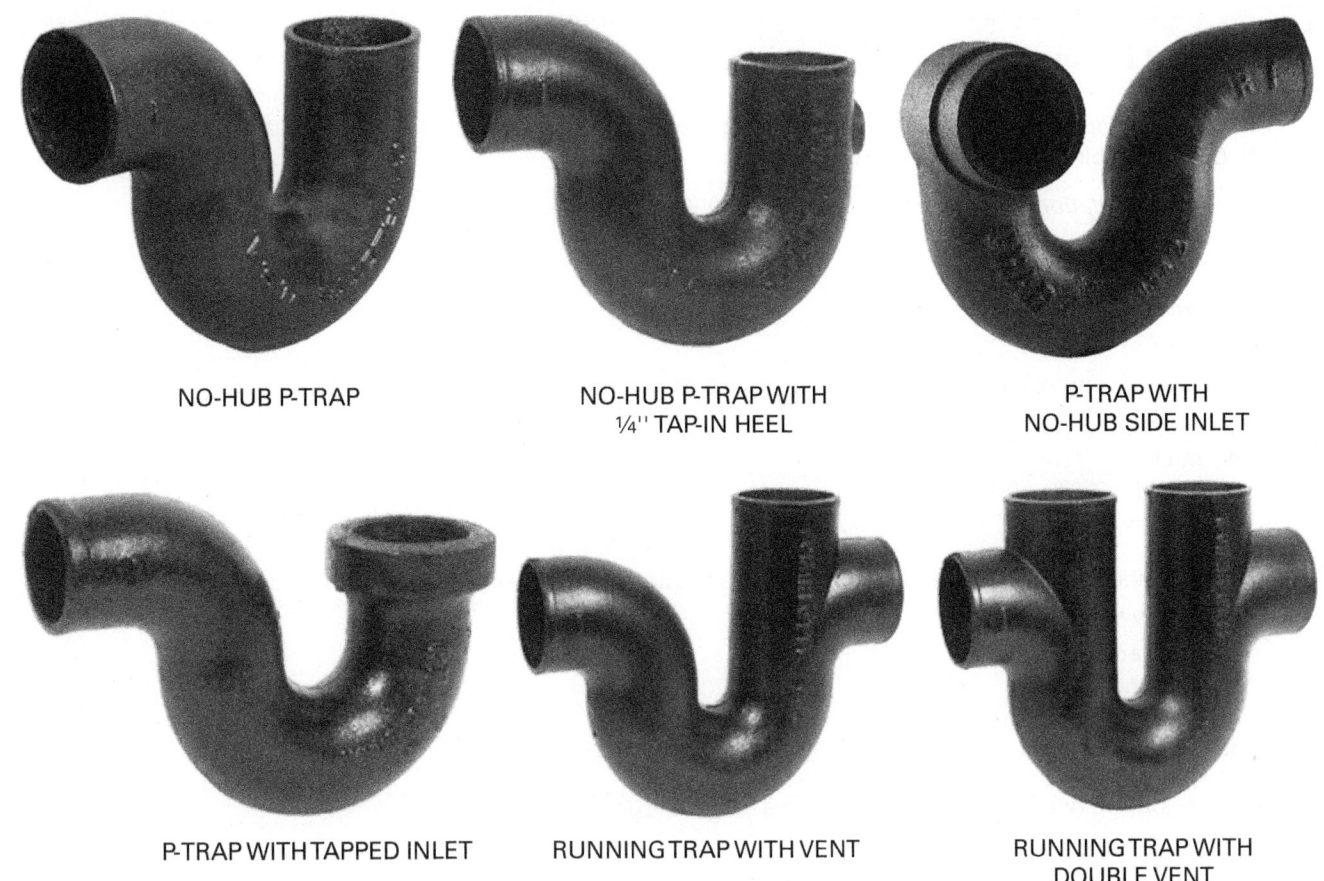

NO-HUB P-TRAP

NO-HUB P-TRAP WITH
¼'' TAP-IN HEEL

P-TRAP WITH
NO-HUB SIDE INLET

P-TRAP WITH TAPPED INLET

RUNNING TRAP WITH VENT

RUNNING TRAP WITH
DOUBLE VENT

Figure 13 Traps.

1.0.0 Section Review

1. The smooth end of cast-iron pipe manufactured with a bell or enlarged end is called the _____.

   a. boss
   b. inlet
   c. spigot
   d. trap

2. When installing cast-iron pipe and fittings in structures that are subject to vibration, settling, or excessive corrosion, use _____.

   a. standard pipe
   b. extra-heavy pipe
   c. service-weight pipe
   d. vibration-dampening pipe

3. When joining two horizontal branches to a single vertical pipe, use a _____.

   a. closet bend
   b. wye
   c. sweep
   d. sanitary cross

4. Which of the following statements is true about storage and handling of iron pipe?

   a. Contractors should order pipe and fittings as they are needed to save time and money.
   b. Carefully stage pipe and fittings into groups to avoid damage and the entry of pests.
   c. Carefully cover exposed pipe and fittings with loose-fitting polyethylene tarps to protect them from the weather.
   d. Cast-iron pipe and fittings are strong and tough, so no special care is needed when handling them.

2.0.0 MEASURING, CUTTING, AND JOINING

Objective

Describe the different methods for measuring, cutting, and joining cast-iron pipe.

   a. Explain the tools and methods used to measure and cut cast-iron pipe.
   b. Describe how to join hub-and-spigot pipe.
   c. Describe how to join no-hub pipe.

Performance Tasks

   3. Select the appropriate personal protective equipment for cast-iron piping.
   4. Correctly measure, cut, and join cast-iron pipe.

Trade Terms

Compression joints: Joints formed using a neoprene gasket inserted into the bell end of a hub-and-spigot pipe. The pressure applied between the two joined pipes forces the gasket to compress and fill in any air gaps in the joint.

Couplings: Fittings used to connect two lengths of no-hub cast-iron pipe.

Insertion length: The length of a pipe that fits into the fitting or other joint when assembling a pipe run. The measurement must be figured into the calculation before cutting a piece of pipe for joining.

Soil pipe cutter: A heavy-duty tool used for cutting cast-iron pipe. Soil pipe cutters can be of the snap or ratchet type.

Torque wrench: A wrench with a gauge or other means to indicate the amount of rotating force applied to a fastener, such as a nut, as it is turned.

This section describes procedures for measuring, cutting, joining, and assembling cast-iron pipe. Be sure to measure and cut pipe carefully. Inaccurate measuring and improper or sloppy cutting cost a business money. If a pipe is cut too short, it is useless. If a pipe is cut improperly, it cannot be used to make a secure joint. Measuring and cutting hub-and-spigot pipe is more involved than working with no-hub pipe.

2.1.0 Measuring and Cutting Cast-Iron Pipe

Hub-and-spigot cast-iron pipe has a standard laying lengths of 5 feet or 10 feet, measured from the base of the hub to the end of the spigot. Some manufacturers also offer $2\frac{1}{2}$ and $3\frac{1}{2}$-foot laying lengths in single-hub pipe. When connecting long runs containing many sections of pipe, you must add the insertion length to the measurement between the spigot at one end of the run and the hub at the other end to get the correct measurement of pipe needed. The insertion length is the length of the spigot end that is inserted into the hub (*Figure 14*). Insertion lengths for the most commonly used pipe sizes are $2\frac{1}{2}$ inches for 2-inch pipe; $2\frac{3}{4}$ inches for 3-inch pipe; and 3 inches for 4-, 5-, or 6-inch pipe.

2.1.1 Measuring Hub-and-Spigot Pipe

To measure hub-and-spigot pipe correctly, you must also know how to locate the center of a fitting. The center of a fitting is the point where the center lines of the inlet(s) and outlet of the fitting intersect (*Figure 15*). For example, in a drainage system, if you are connecting a $\frac{1}{4}$-bend to a $\frac{1}{8}$-bend fitting with a length of soil pipe, and if the center-to-center measurement must be 36 inches, the length of pipe is actually less than 36 inches, because the fittings take up part of the overall measurement (*Figure 16*).

To make a proper measurement:

1. Place the fittings together (*Figure 17*)

2. Measure the distance between the center marks on the fittings and subtract that number from the overall measurement

3. Use a folding rule to measure pipe length

4. Begin by inserting the fittings

5. Establish the centers of the fittings and mark these points on their surfaces

6. Place the 36-inch mark of the rule on the center mark of the right fitting (*Figure 18*)

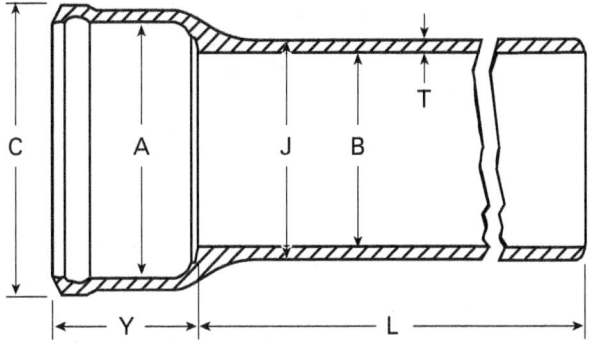

INSERTION LENGTH   LAYING LENGTH

Figure 14 Measuring hub and spigot pipe.

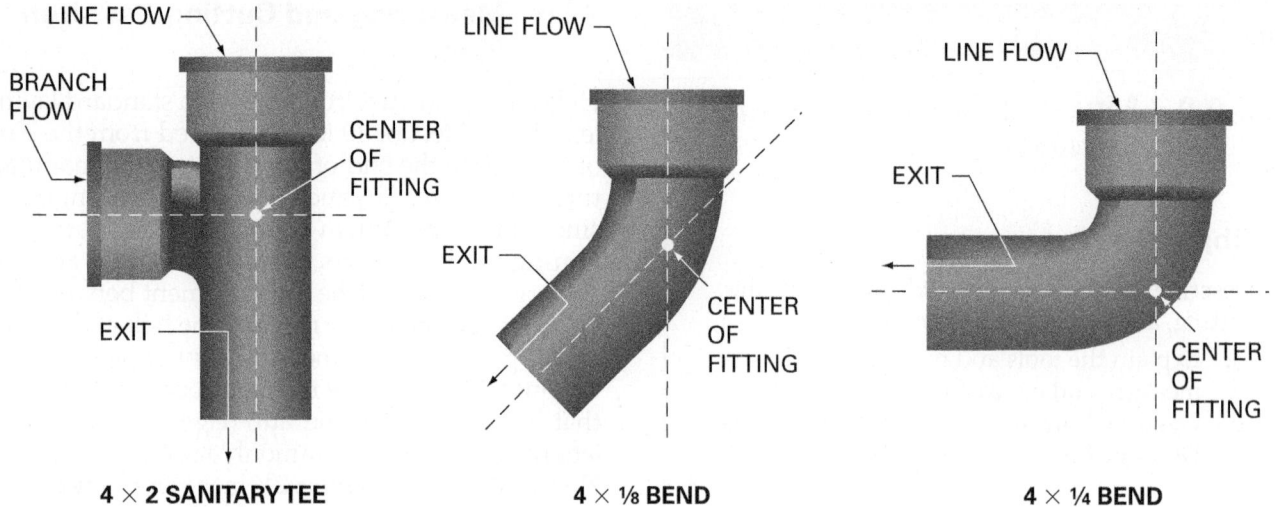

4 × 2 SANITARY TEE

4 × ⅛ BEND

4 × ¼ BEND

Figure 15 Finding the center of a fitting.

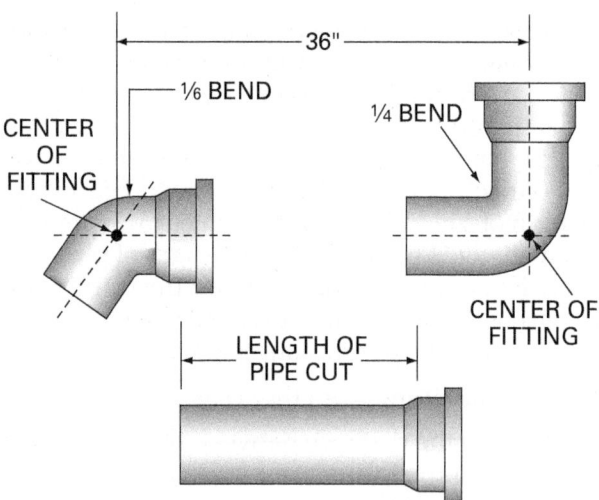

Figure 16 Determining the cutting length.

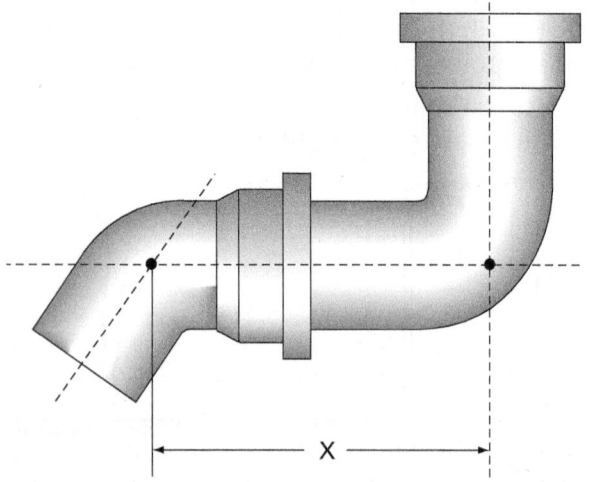

NOTE: SUBTRACT MEASUREMENT X FROM CENTER-TO-CENTER MEASUREMENT TO FIND LENGTH OF THE CUT.

Figure 17 Placing fittings together.

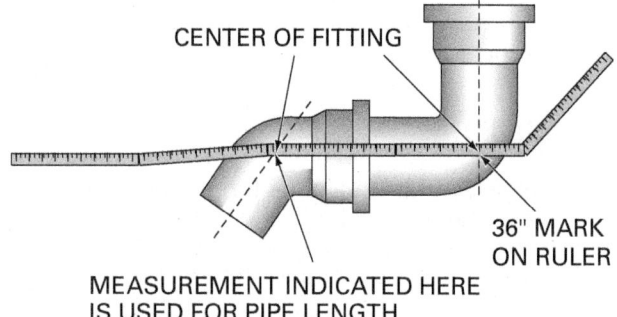

MEASUREMENT INDICATED HERE IS USED FOR PIPE LENGTH

Figure 18 Using a folding rule.

The mark on the rule that falls over the center mark of the fitting on the left is the length to subtract from the actual distance between fittings when measuring the pipe to cut.

2.1.2 *Measuring No-Hub Pipe*

Measuring no-hub cast-iron pipe is rather simple. Using a folding rule or tape measure, measure the exact distance between the fitting connection and the pipe, or from pipe to pipe, and subtract the allowance for the ridge on the inside of the neoprene gasket.

2.1.3 *Cutting Cast-Iron Pipe*

After you measure the pipe, cut it to the desired length with a soil pipe cutter. Soil pipe cutters are available in a variety of models (*Figure 19*). Each cutter can be used to cut a variety of pipe sizes. The soil pipe cutter uses a length of chain, each link of which contains a cutting wheel. When the chain is tightened around the pipe, the cutters press into the metal. Most manufacturers recommend scoring the pipe before tensioning the cutter to break it. A straight score leads to a clean break.

SOIL PIPE RATCHET CUTTER

MINI SOIL PIPE CUTTER

SOIL PIPE SNAP CUTTER

ELECTRICALLY OPERATED
SOIL PIPE CUTTER

ELECTRICALLY OPERATED SOIL PIPE CUTTER
ATTACHED TO POWER TOOL

Figure 19 Soil pipe cutters.

2.2.0 Joining Hub-and-Spigot Pipe

Hub-and-spigot pipe can be joined by compression joints (*Figure 20*). Compression joints are used more often because they are less labor-intensive and less expensive than lead and oakum. When joining pipe, you must ensure proper alignment of the piping to ensure watertight joints. Proper alignment means that the center line in every straight run of pipe is perfectly straight and that pipe and fittings meet at the correct angle.

Compression joints are a fast method of joining cast-iron pipe and fittings. They can absorb vibrations and be deflected or bent up to 5 degrees. They require the use of pipe and fittings that do not have a bead on the spigot end. Always follow the manufacturer's specifications when assembling and tightening hub-and-spigot pipe.

To create a compression joint, insert a neoprene gasket into the hub end of the pipe. Use a chain puller, lead hammer, or pushing bar to push or draw the spigot into the hub. This process causes the neoprene gasket to compress and fill in any air space between the two sections of pipe.

To make a compression joint, follow these steps:

Step 1 Clean the neoprene gasket and pipe ends.

Step 2 Insert the gasket in the hub.

Step 3 Apply a rubber lubricant to the inside of the gasket and the spigot end of the pipe.

Step 4 Force the spigot end of the pipe into the gasket using a puller tool (*Figure 21*).

Figure 21 Puller tool.

2.3.0 Joining No-Hub Pipe

To join no-hub pipe, use couplings (*Figure 22*). Couplings have three components: a gasket, a stainless-steel shield, and clamps. These parts are manufactured to conform to the standards of the CISPI. The gaskets are made of an elastomeric material, usually neoprene. They must be labeled with the gasket material, and must have CISPI's trademark and patent number.

The gasket is flexible, with a ridge on the inside diameter to control the distance the gasket can be slipped onto the end of no-hub pipe. It seals the joint between no-hub components. The shield and clamps are made of stainless steel. They must also be labeled with the CISPI's trademark and patent number as well as with the marking

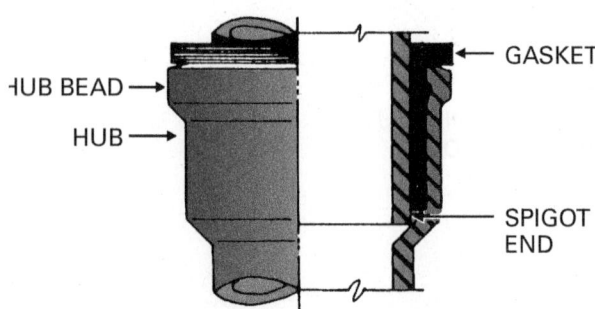

Figure 20 Compression joint.

GASKET

HUB BEAD

HUB

SPIGOT END

Figure 22 No-hub couplings.

"all stainless" or other recognized abbreviations. The corrugated stainless-steel shield surrounds the neoprene gasket, with one or more pairs of clamps around the gasket and shield. Transition couplings are available to allow you to join cast-iron pipe to other types of pipe, such as plastic. Heavy-duty couplings have two or three pairs of clamps and are available with either a $5/16$-inch or $3/8$-inch nut.

Before you join the pipe with the coupling, always clean the area of the pipe and fitting that the gasket is to cover. Dirt, mud, sand, or any other material between the gasket and the pipe can cause a leaky connection.

Joining no-hub pipe requires a torque wrench (*Figure 23*). The torque wrench, which you use to tighten the coupling clamps, has either a $5/16$-inch or a $3/8$-inch nut driver. You must tighten each clamp to the correct torque. For drivers with a $5/16$-inch driver, the correct torque is approximately 60 inch-pounds of force. For drivers with a $3/8$-inch driver, the correct torque is approximately 80 inch-pounds of force or more, depending on the brand of coupling and its size. Use a torque wrench that is designed to disengage at the correct torque; you hear a clicking sound when the wrench disengages.

Proper pipe alignment is necessary to guarantee tight joints. Before tightening the clamps, ensure that you have proper alignment and that the coupling is in place. When you are tightening the coupling, be sure that you apply the torque evenly. For couplings with more than two clamps, manufacturers specify a staggered torque sequence to prevent puckering of the coupling. Switch the wrench from one clamp to the other per the manufacturer's torque sequence. Also tighten all the clamps in steps; do not fully tighten one clamp before going to the next. Tightening one clamp before you begin to tighten the other clamp could cause the gasket to become misaligned, and the coupling might leak. Always follow the manufacturer's instructions regarding the appropriate torque for the torque wrench you are using.

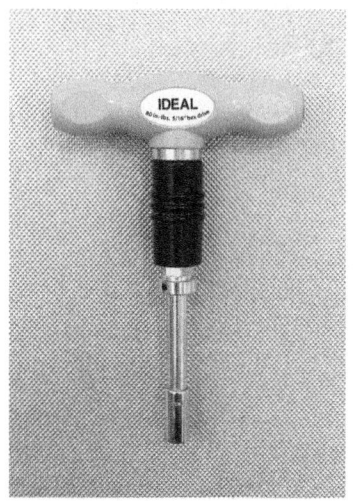

Figure 23 Torque wrench.

Figure 24 Impact driver.

Many plumbers use an impact driver (*Figure 24*) to quickly drive fasteners into hard surfaces. This tool combines bit rotation and blows per minute—a combination of drilling and hammering—to drive the screw in more rapidly than conventional drills. This drill increases productivity with less fatigue for the operator because it delivers torque directly to the screw and not the operator's wrist.

Additional Resources

Cast Iron Soil Pipe and Fittings Handbook. Last revised 2006. Accessed September 7, 2017. Chattanooga, TN: Cast Iron Soil Pipe Institute. *www.cispi.org*.

Plumbing Installation and Design, Third Edition, 2006. L. V. Ripka. Orland Park, IL: American Technical Publishers.

2.0.0 Section Review

1. The point at which the center lines of the inlet and outlet of a fitting intersect is called the _____.

   a. bend
   b. sweep
   c. center
   d. joint

2. Plumbers join hub-and-spigot pipe using _____.

   a. flanges
   b. couplings
   c. expansion joints
   d. compression joints

3. The component of a no-hub coupling that seals the joint between pipes and fittings is _____.

   a. the lead sheet gasket
   b. the neoprene gasket
   c. the corrugated stainless steel shield
   d. one or more pairs of clamps

3.0.0 HANGERS, SUPPORTS, AND TESTING

Objective

Describe the methods used to install and test cast-iron pipe.

a. Describe the hangers and fasteners used to support cast-iron pipe.
b. Explain how to support vertical and horizontal runs.
c. Explain installation methods used in wood, masonry, and concrete structures.
d. Describe the methods used in testing cast-iron pipe.

Performance Task

5. Select the correct support and spacing for the application.

Trade Terms

Rotary hammer drill: A drill with a pounding action that lets you drill into concrete, brick, or block. It rotates and hammers at the same time and drills much faster than regular drills.

Slope: Measurement of the fall of a length of pipe from level. The change in level is expressed in degrees of an angle or distance of fall per foot of run. Also referred to as percent of grade.

Hangers are used for horizontal or vertical support of pipe and pipe fittings. The primary purpose of hangers is to keep the piping in alignment and prevent it from bending, swaying, or distorting. Without proper alignment, waste may not flow through the system properly, and leaks may develop. Plumbing codes require pipe to be securely fastened horizontally and vertically. Never hang one pipe from another or allow one pipe to support another.

> **NOTE**
> The construction drawings and specifications for your job may have unusual pipe-support requirements. Always check the construction drawings and specifications before starting your installation.

If you are installing pipe in a seismically active area—that is, where earthquakes are likely—local codes require seismic restraints. The purpose of these requirements is to ensure that the pipe is securely fastened to the structure in the event of excessive vibration. For example, some codes require hangers and supports to be used at closer intervals than in non-seismically active areas. They may require seismic dampers to absorb vibrations. In addition, codes may require extra spacing for pipe where it penetrates or passes near walls, floors, and ceilings to allow for anticipated movement.

3.1.0 Types of Hangers and Supports

The hangers and supports you use depend on the size of the pipe and the type of material to which the hanger or support is attached. When selecting the types of hangers or supports, you must consider whether the pipe runs horizontally or vertically. All hangers and supports are designed to keep pipe in alignment, but you must also ensure that horizontal pipes do not move back and forth. When selecting supports for vertical pipe runs, you must also take into account the vertical load, or total weight, of the pipe. A variety of hangers are available for horizontal and vertical runs. Proper spacing of hangers and supports is very important.

> **NOTE**
> When installing soil pipe above food-handling areas, you use special installation techniques.

3.2.0 Supporting Horizontal and Vertical Pipe Runs

Plumbing runs in two general orientations—horizontal and vertical. Only in special circumstances may you find pipe installed at some large angle to the horizontal. The design engineers will provide special instructions for these situations. Horizontal DWV pipe isn't truly level, but has a slight incline as discussed below. In this section you will learn about the methods used for supporting horizontal and vertical pipe.

3.2.1 Horizontal Runs

For horizontal pipe runs, you can use many different types of horizontal support hangers and brackets (*Figure 25*). Your choice depends on the conditions and the specifications set forth by the contractor, architect, or engineer. Most authorities and plumbing codes specify that pipe must have hangers to support it at each joint. For maximum support, hangers must be no more than 18 inches from each joint (*Figure 26*). No matter what kind of support you use on a horizontal run of pipe, be sure to maintain a proper slope.

Horizontal runs of pipe must be installed to provide a slope, which allows the waste to drain away. Slope, also called fall, is the downward slant of the pipe. DWV piping systems depend on gravity to move waste out of the system. Without the correct slope, gravity is not effective. Branch pipe generally requires a slope of $\frac{1}{8}$ inch to $\frac{1}{4}$ inch per foot on all horizontal runs. At this slope, the waste in the branch moves fast enough to clean the pipe.

If the slope is more than $\frac{1}{4}$ inch per foot, liquid and solid wastes may separate. When separation occurs, the heavier solid waste is deposited in the bottom of the pipe instead of flowing away with the liquid. If the slope is less than $\frac{1}{8}$ inch per foot, the waste flow slows, and the branch tends to become blocked. When the pipe is perfectly aligned, it must be secured to the building structure to ensure that it remains aligned. You will learn more about determining the slope or grade required for DWV pipes elsewhere in this curriculum.

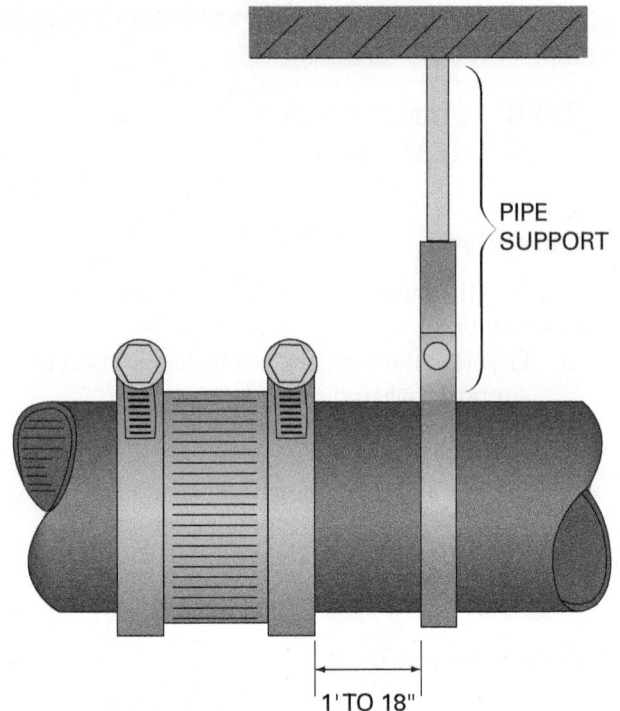

Figure 26 Hanger distance.

CAUTION	If you are installing pipe in a seismically active area—that is, where earthquakes are a possibility—local codes require seismic restraints.

To attach horizontal hangers to wooden structures, use screws, lag screws, or large nails. To fasten pipe to I-beams, bar joists, and other metal structures, use beam clamps or C-clamps (*Figure 27*).

3.2.2 Vertical Runs

Vertical hangers are used to support vertical runs of pipe and to maintain pipe alignment. Proper pipe alignment is critical to ensure a leak-proof system. To support the vertical load (or total weight) of the pipe, use a riser clamp (*Figure 28*). Support vertical pipe at each floor level. Mount the riser clamp so the bracket adequately supports the pipe weight.

3.3.0 Installation

You can attach pipe hangers or supports to wood, masonry, concrete, and steel. Because of the heavy weight of cast-iron pipe, it is extremely important for you to choose a fastener with the appropriate holding power for your installation.

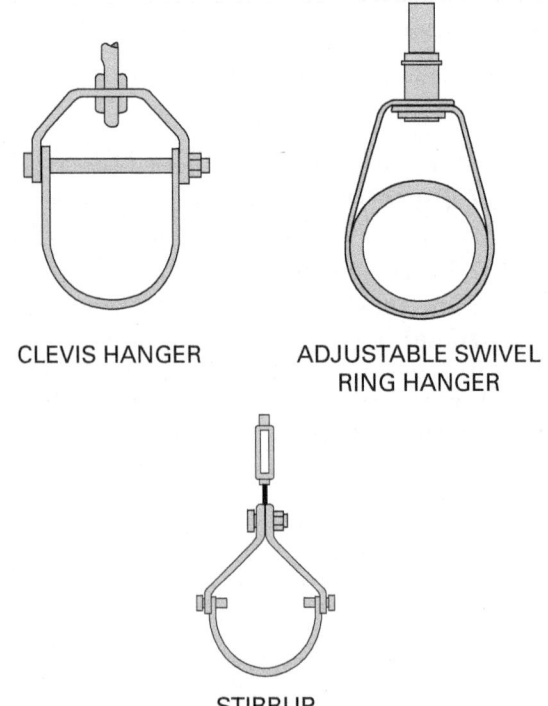

CLEVIS HANGER   ADJUSTABLE SWIVEL RING HANGER

STIRRUP

Figure 25 Horizontal pipe hangers.

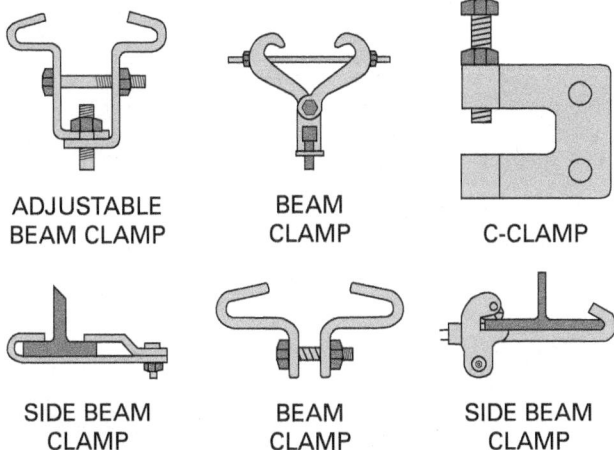

ADJUSTABLE
BEAM CLAMP

BEAM
CLAMP

C-CLAMP

SIDE BEAM
CLAMP

BEAM
CLAMP

SIDE BEAM
CLAMP

Figure 27 Beam clamps.

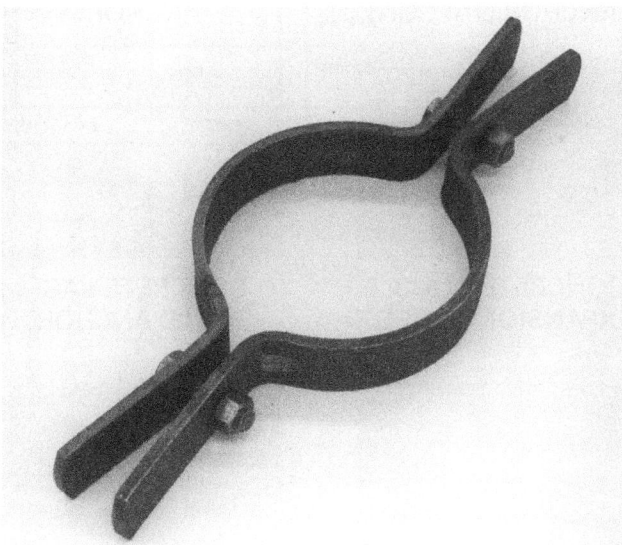

Figure 28 Riser clamp.

3.3.1 Wooden Structures

Use nails, screws, and bolts to fasten pipe to wooden structures. Screws and bolts have more holding power than nails. For both horizontal and vertical pipe runs, use nails only if vibration and strength are not factors in your installation.

Seismic Codes

The International Building Code (IBC) is considered the leading source for information on seismic code. The IBC was a combined effort of the Building Officials and Code Administrators International (BOCA), the Southern Building Code Congress International (SBCCI), and the Uniform Building Code (UBC). The IBC includes seismic hazard maps, seismic design loads, and seismic code criteria.

Plumbing Codes for Supports and Fasteners

As the plumber, you must select and use the supports and fasteners best suited for the job. You are also responsible for installing hangers or fasteners at the correct intervals along the length of the pipe run. If unusual circumstances exist, the project specifications may direct you to make changes from generally accepted plumbing practices. Never work without knowing the applicable codes and detailed project specifications.

3.3.2 Masonry and Concrete Structures

You can use various expansion anchors or threaded masonry fasteners to fasten support hangers to masonry and concrete. Use a rotary hammer drill (*Figure 29*) with a special masonry bit to drill into masonry. After drilling the hole, insert a screw anchor, a plastic expansion anchor, or a threaded masonry anchor in the hole (*Figure 30*). Turn the screw into the insert to expand the insert, allowing it to hold the screw tightly in the masonry. Use toggle bolts to attach pipe supports to hollow masonry units, such as concrete block.

> **CAUTION**
>
> Manufacturers usually provide two types of load ratings for masonry fasteners—axial (lengthwise) and shear (sideways). Examine the way the pipe load will be applied to the pipe hangers' fasteners to determine what load rating will be needed for the fasteners. Shear load rating for a given type of fastener in masonry is almost always higher than axial ratings. Project specifications usually have detailed instructions for selecting and installing pipe hanger fasteners.

> **CAUTION**
>
> Powder-actuated fastening tool can be used to anchor pipe supports to concrete, masonry, and steel. However, you must be certified by the manufacturer to use a powder-actuated fastening tool. Manufacturers' representatives visit shops and job sites to conduct training on their equipment and to certify people in its use.

Figure 29 Rotary hammer drill.

3.4.0 Testing Installations

Cast-iron piping installations of venting and drainage systems can be tested by a hydrostatic test (water test), air pressure test, smoke test, and peppermint test. Air and water testing can be performed at the rough-in stage. Smoke and peppermint testing must be performed after final installation. If possible, test only one floor at a time.

Codes have different requirements for the water test. Some require a 10-foot head of water; others a 5-foot head of water. First, plug all openings except for one vertical pipe. Extend the vertical pipe at least 5 feet higher than the horizontal drain piping. If the code requires a 10-foot head of water, extend the pipe at least 10 feet. Fill this pipe with water. This allows you to check for any leaking joints in the system. An inspector may require you to remove cleanout plugs or caps at different points along the system piping to make sure that the water has reached everywhere in the drainage system.

Air testing is similar to water testing except that the pipes are filled with compressed air. The system should generally be pressurized to a maximum of 5 pounds per square inch (psi). Use a pressure gauge at the test plug to determine whether any pressure has been lost. While cast-iron pipe and fittings joined with compression joints or couplings should have a reduction in air pressure during a 15-minute test, a reduction of more than 1 psi indicates failure of the system.

A smoke test involves filling the traps with water and adding a thick smoke into the DWV system. The stack should be closed when the smoke appears at the roof opening. A pressure of 1 inch water column is needed throughout the smoke test. The system passes the test if no smoke is visible at any connection in the system.

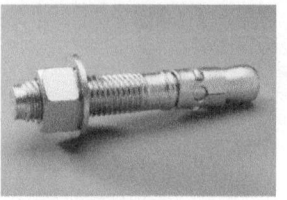

CONCRETE WEDGE ANCHOR

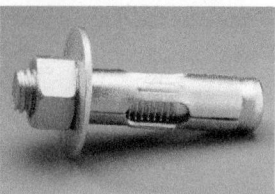

CONCRETE SLEEVE ANCHOR

CONCRETE SCREW ANCHOR (TAPCON)

CONCRETE DROP-IN ANCHOR

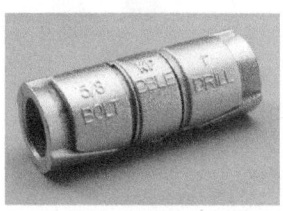

CONCRETE DOUBLE EXPANSION ANCHOR

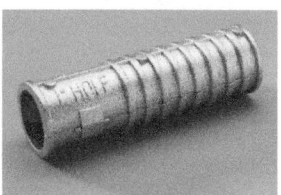

CONCRETE LAG SHIELD ANCHOR

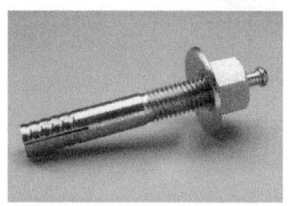

CONCRETE STRIKE ANCHOR

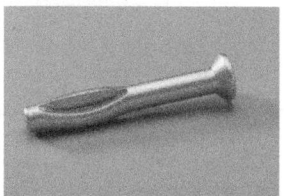

CONCRETE SPLIT-DRIVE ANCHOR

Figure 30 Anchors.

If plumbing inspectors decide that a smoke test cannot be used, use a peppermint test. To conduct a peppermint test, add 2 ounces of peppermint oil into each roof opening of the system. After adding the peppermint oil, add 10 quarts of hot water (160°F). Seal each terminal. If you can smell peppermint at any point in the system, you have identified a leak.

Adhering to all codes is extremely important to the success of a plumbing installation. An inspector will test your installation to ensure that you have complied with all codes. Inspectors are especially careful to check for cross-connections, defective or inferior materials, or poor craftsmanship. Because it is important to inspect the joints in the piping system visually, the inspection must be done before you cover the pipes.

Additional Resources

Cast Iron Soil Pipe and Fittings Handbook. Last revised 2006. Accessed September 7, 2017. Chattanooga, TN: Cast Iron Soil Pipe Institute. *www.cispi.org*.

Plumbing Installation and Design, Third Edition, 2006. L. V. Ripka. Orland Park, IL: American Technical Publishers.

3.0.0 Section Review

1. In seismically active areas, the following pipe support characteristics are dictated by seismic codes except _____.

   a. diameter of the clevis pipe hangers
   b. intervals of pipe hangers
   c. the use of seismic dampers
   d. clearances around pipes at penetrations

2. To provide maximum support for horizontal pipe runs, the maximum distance that a hanger should be installed from a joint is _____.

   a. 12 inches
   b. 16 inches
   c. 18 inches
   d. 24 inches

3. Nails can be used to secure pipe hangers in wooden structures if they will not be subject to _____.

   a. moisture
   b. heat
   c. vibration
   d. cold

4. When conducting an air test of a cast-iron piping installation, a failure in the system is indicated if test pressure drops more than _____.

   a. 0.5 psi
   b. 1 psi
   c. 1.5 psi
   d. 2 psi

SUMMARY

Cast iron refers to iron products that are formed using a mold. Cast-iron pipe and fittings are available in two types of cast iron: gray iron and ductile iron. Either type can be used to form hub-and-spigot and no-hub cast-iron pipes in standard lengths of 5 or 10 feet. The pipes are made in two wall thicknesses: service and extra-heavy.

Hub-and-spigot pipes are joined using a compression joint to produce a watertight seal. No-hub cast-iron pipes are joined using a coupling consisting of a neoprene gasket, a stainless-steel shield, and clamps.

Planning ahead saves time and money when you are installing cast-iron pipe. Accurate measurements reduce the chances that you will waste pipe by making the wrong cut. Several techniques are used to measure pipe correctly before the pipe is cut and joined. Always measure and mark before you cut.

You must support cast-iron pipe to prevent it from bending, sagging, or otherwise losing its alignment. Hangers or supports such as brackets, clamps, and pipe trapezes secure the pipe in either a horizontal or vertical position. Supports can be fastened to wooden, concrete, and masonry frames using anchors. The type of anchor you use depends on the building frame material.

As with all aspects of plumbing, the selection and installation of cast-iron pipe is strictly regulated by building and plumbing codes.

1. When creating cast-iron pipe and fittings, manufacturers use a process involving _____.
   a. 3D printing
   b. molten pours in molds
   c. hand tooling
   d. additive manufacturing

2. Which of the following statements about cast-iron pipe is true?
   a. For a given nominal pipe size, SV and XH pipes have the same wall thickness.
   b. For a given nominal pipe size, SV pipes have the thicker walls than XH pipes.
   c. For a given nominal pipe size, SV pipes have the thinner walls than XH pipes.
   d. Both SV and XH pipes are available in hub-and-spigot as well as no-hub products.

3. How many degrees does a 1/5 bend change the direction of flow?
   a. 90
   b. 72
   c. 60
   d. 45

4. Tees are typically dimensioned with two numbers, such as $4 \times 2\frac{1}{2}$. The first number indicates the _____.
   a. length of the straight-through opening
   b. length of the branch
   c. diameter of the branch opening
   d. diameter of the straight-through opening

5. Which of the listed fittings is permitted only in horizontal-flow to vertical-flow applications?
   a. Eighth bend
   b. Wye
   c. Sanitary tee
   d. Long sweep

6. To connect a length of pipe to another of a different size, use a(n) _____.
   a. closet flange
   b. wye
   c. increaser
   d. sanitary tee

7. When preparing for a cast-iron plumbing job, you should _____.
   a. organize pipe and fittings into groups by pipe size and type
   b. leave all the fittings and pipes banded on the pallets where they were delivered until ready to use
   c. precut all the pipes to length before starting installation
   d. lay out the pipes and fittings in sequence throughout the building before starting the work

8. When measuring the length of a long, straight run of cast-iron hub-and-spigot pipe, the total length of pipe required is equal to the run _____.
   a. measurement
   b. measurement plus the insertion length
   c. measurement minus the insertion length
   d. measurement plus the insertion length times the number of laying lengths required

9. When inserting the spigot end of cast-iron pipe into a hub, you should _____.
   a. twist the pipe both ways by hand as you insert it into the gasket
   b. position the pipe in the hub then tap the gasket into place
   c. pull the pipe into the hub using a puller tool
   d. position the pipe in the hub, then pour quick-setting epoxy sealant into the gap to finish the joint

10. When using a torque wrench, do not tighten the clamps before ensuring the correct placement of the _____.
    a. coupling
    b. bushing
    c. anchor
    d. gasket

11. When using a 5/16-inch torque wrench, the final torque on each coupling clamp is typically approximately _____.
    a. 30 inch-pounds
    b. 60 inch-pounds
    c. 80 inch-pounds
    d. 120 inch-pounds

12. All hangers and supports are designed to _____.
    a. bear the same amount of weight
    b. fasten to any kind of building structure
    c. work in any seismic condition
    d. keep pipe in alignment

13. Vertical runs of pipe must be supported by _____.
    a. beam clamps
    b. stirrups
    c. riser clamps
    d. clevis hangers

14. Which of those listed is the best tool or method for making anchor holes for pipe hangers in masonry and concrete?
    a. Rotary hammer drill
    b. Standard drill with a masonry bit
    c. Hammering powder-actuated anchors into masonry
    d. Hammering nails into mortar joints

15. The amount of water-column pressure needed for a smoke test is _____.
    a. 1 inch
    b. 2 inches
    c. 3 inches
    d. 4 inches

16. A plumber must use a _____ to drill into masonry.
    a. rotary hammer drill
    b. torque wrench
    c. impact driver
    d. hammer

17. To maximize the support hangers offer for horizontal runs of pipe, they should be no further than _____ apart.
    a. 17 inches
    b. 18 inches
    c. 19 inches
    d. 20 inches

18. Plumbers should use _____ to join no-hub pipe.
    a. compression joints
    b. couplings
    c. oakum joints
    d. traps

19. A plumber will need _____ to cut cast-iron pipe.
    a. torque wrench
    b. hammer drill
    c. hacksaw
    d. soil pipe cutter

20. _____ are seals designed to prevent sewer gas from escaping into a building.
    a. Increasers
    b. Reducers
    c. Traps
    d. Branches

Trade Terms Quiz

Fill in the blank with the correct term that you learned from your study of this module.

1. _____ are small openings in-line with the curve of bends that permit connecting smaller pipes to the main lines at the bends.

2. _____ are fittings designed to hold a water seal, which prevents sewer gas from escaping into the building.

3. The _____ used to join no-hub pipes have three components: a gasket, a stainless steel shield, and clamps.

4. The fitting that mates the water closet to the closet bend is a(n) _____.

5. Cast-iron pipe that has no bell on either end is called _____.

6. When fastening support hangers to masonry or concrete with anchors, use a(n) _____ to drill into the masonry.

7. _____ connect water closets to the main drainage piping.

8. Manufacturers of cast-iron pipe make two thicknesses of pipe: _____ and _____.

9. Horizontal runs of pipe must be installed with the proper _____ to allow waste to drain away.

10. _____ are used to join branches at a 45-degree angle to the main pipe.

11. A(n) _____ is a fitting that changes the direction of flow from horizontal to vertical and has openings for two branch lines.

12. Cast-iron pipe that has a bell at one end of the pipe where the smooth end of the next pipe slides in to form a joint is called _____.

13. In connecting long runs containing many sections of hub-and-spigot pipe, you must add the _____ to the measured length of the run to determine the amount of pipe to use.

14. The _____ has a length of chain, each link of which contains a cutting wheel.

15. Using fittings called _____ allows for smoother flow of wastes through turns in the piping system.

16. The _____ is the largest DWV riser that runs between the building drain and the highest horizontal drain in the system.

17. _____ are elbow-like fittings that change the direction of flow within piping.

18. _____ provide a fast method of joining hub-and-spigot cast-iron pipe and fittings.

19. To connect two DWV pipes at a right-angle intersection, use a(n) _____.

20. A(n) _____ is often used above the heated area of a building to prevent frost buildup from closing the opening in the vent stack.

21. _____ permit joining two branch drains into to a main pipe at 45-degree angles.

22. _____ are fittings that change the direction of flow in piping and have one leg longer than the other.

23. A(n) _____ is used to tighten the coupling clamps when you are joining no-hub cast-iron pipe.

24. _____, which is an alloy of gray iron and magnesium, is less brittle than pure gray iron.

Trade Terms

Bends	Ductile iron	Long bends	Soil pipe cutter
Closet bends	Extra-heavy	Main stack	Sweeps
Closet flange	Heel inlets	No-hub cast-iron pipe	Torque wrench
Compression joints	Hub-and-spigot cast-iron	Rotary hammer drill	Traps
Couplings	pipe	Sanitary tee	Wyes
Double sanitary tee	Increaser	Service	
Double wyes	Insertion length	Slope	

Bends: Fittings that change the direction of flow in piping.

Closet bends: Elbow-like fittings that connect water closets to the main drainage piping.

Closet flange: A fitting used to mate a water closet to a closet bend or other DWV pipe.

Compression joints: Joints formed using a neoprene gasket inserted into the bell end of a hub-and-spigot pipe. The pressure applied between the two joined pipes forces the gasket to compress and fill in any air gaps in the joint.

Couplings: Fittings used to connect two lengths of no-hub cast-iron pipe.

Double sanitary tee: A fitting with a main pipe and two branch lines that changes the direction of flow 90 degrees from the horizontal to the vertical. Also called a sanitary cross.

Double wyes: Fittings with a main pipe and two branches at 45 degrees. These connect horizontal waste pipes and branch drains to the building main.

Ductile iron: A cast-iron alloy containing magnesium to reduce brittleness.

Extra-heavy: A kind of cast-iron pipe. Extra-heavy refers to the pipe's wall thickness, which is thicker than service pipe of the same nominal pipe size.

Heel inlets: Bends that have an inlet to connect a smaller pipe to the main line or a branch line. The inlet is located at the base of the outside curve or heel of the bend.

Hub-and-spigot cast-iron pipe: Pipe that has a bell or enlargement at one end where the spigot (smooth end) of the next pipe slides in to form a joint. Also called bell-and-spigot pipe.

Increaser: A fitting used to increase the size of a straight-through line of pipe. It is often used for the vent stack before it goes through the roof to reduce the chance of frost clogging the vent opening in very cold climates.

Insertion length: The length of a pipe that fits into the fitting or other joint when assembling a pipe run. The measurement must be figured into the calculation before cutting a piece of pipe for joining.

Long bends: Bend fittings with one end (typically the spigot end) longer than the other.

Main stack: The principal DWV riser to which branches may be connected.

No-hub cast-iron pipe: A pipe with no enlargement or bell on either end.

Rotary hammer drill: A drill with a pounding action that lets you drill into concrete, brick, or block. It rotates and hammers at the same time and drills much faster than regular drills.

Sanitary tee: A tee fitting for DWV systems that consists of a main pipe and a branch has a curved 90-degree side-inlet; this curve helps to channel the flow of wastewater or sewage from a branch line to the main line. This fitting is always used in the vertical position.

Service: A lightweight kind of cast-iron pipe. Service refers to the pipe's wall thickness, which is thinner for a nominal pipe size compared to extra heavy.

Slope: Measurement of the fall of a length of pipe from level. The change in level is expressed in degrees of an angle or distance of fall per foot of run. Also referred to as percent of grade.

Soil pipe cutter: A heavy-duty tool used for cutting cast-iron pipe. Soil pipe cutters can be of the snap or ratchet type.

Sweeps: Fittings having a radius of the curve greater than those for standard 90-degree bends. Used for a smooth change of direction.

Torque wrench: A wrench with a gauge or other means to indicate the amount of rotating force applied to a fastener, such as a nut, as it is turned.

Traps: DWV fittings or devices that provide a liquid seal to prevent the emission or escape of sewer gases from the system without materially affecting the flow of sewage or wastewater through the fitting.

Wyes: Fittings consisting of a main pipe and a 45-degree branch for connecting a waste branch to a horizontal building main drain.

Additional Resources

This module is intended as a thorough resource for task training. The following reference works are suggested for further study.

Cast Iron Soil Pipe and Fittings Handbook. Last revised 2006. Accessed September 7, 2017. Chattanooga, TN: Cast Iron Soil Pipe Institute. *www.cispi.org.*

Plumbing Installation and Design, Third Edition, 2006. L. V. Ripka. Orland Park, IL: American Technical Publishers.

Figure Credits

SECTION **1.0.0**

Answer	Section Reference	Objective
1. c	1.1.0	1a
2. b	1.2.1	1b
3. d	1.3.2	1c
4. b	1.4.0	1d

SECTION **2.0.0**

Answer	Section Reference	Objective
1. c	2.1.1	2a
2. d	2.2.0	2b
3. b	2.3.0	2c

SECTION **3.0.0**

Answer	Section Reference	Objective
1. a	3.0.0	3a
2. c	3.2.1	3b
3. c	3.3.1	3c
4. b	3.4.0	3d

This page is intentionally left blank.

NCCER CURRICULA — USER UPDATE

NCCER makes every effort to keep its textbooks up-to-date and free of technical errors. We appreciate your help in this process. If you find an error, a typographical mistake, or an inaccuracy in NCCER's curricula, please fill out this form (or a photocopy), or complete the online form at **www.nccer.org/olf**. Be sure to include the exact module ID number, page number, a detailed description, and your recommended correction. Your input will be brought to the attention of the Authoring Team. Thank you for your assistance.

Instructors – If you have an idea for improving this textbook, or have found that additional materials were necessary to teach this module effectively, please let us know so that we may present your suggestions to the Authoring Team.

NCCER Product Development and Revision
13614 Progress Blvd., Alachua, FL 32615

Email: curriculum@nccer.org
Online: www.nccer.org/olf

❏ Trainee Guide   ❏ Lesson Plans   ❏ Exam   ❏ PowerPoints   Other _____

Craft / Level: _____ Copyright Date: _____

Module ID Number / Title: _____

Section Number(s): _____

Description: _____

_____

_____

_____

_____

Recommended Correction: _____

_____

_____

_____

_____

Your Name: _____

Address: _____

_____

Email: _____ Phone: _____

This page is intentionally left blank.

Steel Pipe and Fittings

OVERVIEW

Steel pipe is used for hot- and cold-water distribution; steam and water heating systems; gas and air piping systems; and drain, waste, and vent (DWV) systems. Plumbers commonly use black-steel pipe for gas or air pressure pipe, and galvanized pipe for venting. Steel pipe is manufactured with threads on both ends, according to standards established by the American Standard Pipe and Pipe Thread Dimensions, and is categorized by size and strength.

Module 02109

Trainees with successful module completions may be eligible for credentialing through the NCCER Registry. To learn more, go to *www.nccer.org* or contact us at 1.888.622.3720. Our website, *www.nccer.org*, has information on the latest product releases and training.

Your feedback is welcome. You may email your comments to *curriculum@nccer.org*, send general comments and inquiries to *info@nccer.org*, or fill in the User Update form at the back of this module.

This information is general in nature and intended for training purposes only. Actual performance of activities described in this manual requires compliance with all applicable operating, service, maintenance, and safety procedures under the direction of qualified personnel. References in this manual to patented or proprietary devices do not constitute a recommendation of their use.

Objectives

Successful completion of this module prepares trainees to:

1. Identify the types, uses, and properties of steel pipe.
   a. Describe the sources, applicable material standards, and storage and handling of steel pipe.
   b. Explain methods used to identify pipe threads.
   c. Describe the sizing and labeling of steel pipe.
   d. Describe the different types of fittings and valves used on steel pipe.
2. Describe the methods for measuring, cutting, and joining steel pipe.
   a. Explain the tools and methods used to measure and cut steel pipe.
   b. Describe the tools and methods used to thread steel pipe.
   c. Describe how to join threaded pipe.
   d. Describe how to join grooved pipe.
3. Describe the methods used to install and test steel pipe.
   a. Describe the hangers and fasteners used to support steel pipe.
   b. Explain how to support vertical and horizontal piping runs.
4. Explain the properties and uses of corrugated stainless-steel tubing (CSST).
   a. Describe the uses of CSST.
   b. Describe the sizing and labeling of CSST.
   c. Explain the use of CSST regulators and valves.
   d. Describe CSST installation.

Performance Tasks

Under the supervision of your instructor, you should be able to do the following:

1. Identify the common types of materials, schedules, sizes, and labels used for steel piping.
2. Identify the common fittings and valves used with steel piping.
3. Properly measure, cut, and join steel piping.
4. Identify the hazards and safety precautions when working with steel piping.
5. Identify the various techniques used in hanging and supporting steel piping.

Trade Terms

Ball valve
Black-steel pipe
Chain vise
Chain wrenches
Channels
Corrugated stainless steel tubing (CSST)
Crosses
Dies
Double-extra strong

Drainage fittings
Elevated pressure systems
Extra strong
Flange union
Galvanized pipe
Gas cocks
Gate valve
Globe valve
Ground-joint union
Hangers

Manifolds
Nipples
Pipe tongs
Pipe-joint compound
Pounds per square inch gauge (psig)
Power-threading machine
Pressure fittings
Regulators
Sealant tape

Standard weight
Stock
Stop-and-waste valve
Strap wrench
Striker plates
Thread engagement
Water column (w.c.)
Yoke vise

Industry Recognized Credentials

If you are training through an NCCER-accredited sponsor, you may be eligible for credentials from NCCER's Registry. The ID number for this module is 02109. Note that this module may have been used in other NCCER curricula and may apply to other level completions. Contact NCCER's Registry at 888.622.3720 or go to *www.nccer.org* for more information.

> **CODE NOTE**
>
> Codes vary among jurisdictions. Because of the variations in code, consult the applicable code whenever regulations are in question. Referencing an incorrect set of codes can cause as much trouble as failing to reference codes altogether. Obtain, review, and familiarize yourself with your local adopted code. Safety codes are developed by the US Occupational Safety and Health Administration (OSHA).

Contents

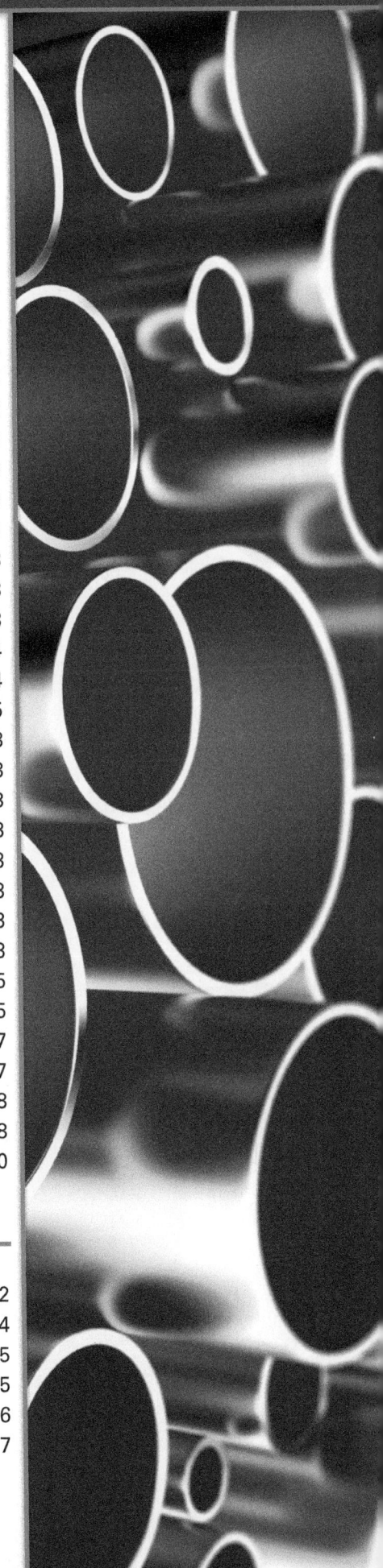

Figures and Tables

Figures and Tables (continued)

1.0.0 STEEL PIPE

Objective

Identify the types, uses, and properties of steel pipe.

a. Describe the sources, applicable material standards, and storage and handling of steel pipe.
b. Explain methods used to identify pipe threads.
c. Describe the sizing and labeling of steel pipe.
d. Describe the different types of fittings and valves used on steel pipe.

Performance Tasks

1. Identify the common types of materials, schedules, sizes, and labels used for steel piping.
2. Identify the common fittings and valves used with steel piping.

Trade Terms

Ball valve: A valve used to control the flow of gases and liquids. It is installed in piping systems where quick shutoffs for operation or in-line maintenance may be necessary.

Black-steel pipe: Steel pipe with a black color used for gas and air-pressure pipe. It should not be used where corrosion affects its uncoated surface.

Crosses: Fittings that connect four lengths of pipe in a cross-like shape in a distribution system.

Dies: Devices used to cut the threads on steel pipe.

Double-extra strong: An industry standard measurement of the weight or strength of steel pipe. Also referred to as Schedule 120 and double extra-heavy.

Drainage fittings: Fittings designed with smooth inside surfaces and shaped for easy flow to form an unbroken, internal contour for use only on drainage systems. Also called recessed fittings.

Extra strong: An industry standard measurement of the weight or strength of steel pipe. Also referred to as Schedule 80 and extra heavy

Flange union: A drilled fitting that connects two pipes. A flange is screwed onto the end of each pipe that needs to be joined and then bolted together with nuts and bolts. A gasket in between flanges creates a leakproof seal.

Galvanized pipe: Pipe electroplated with a zinc alloy to provide a protective coating that resists corrosion or oxidation (rust)

Gate valve: A valve in which the flow is controlled by moving a gate or disc that slides in machined grooves at right angles to the flow. This type of valve is designed to be fully open or fully shut.

Globe valve: A valve in which the flow is controlled by moving a beveled circular disc or seat washer against a circular or conical metal seat that surrounds the flow opening.

Ground-joint union: A fitting that joins two pieces of pipe by screwing the thread and shoulder pieces of the fitting onto the pipe. The collar piece is then tightened to join the sections of pipe into a watertight joint. The fitting permits later disassembly when required.

Hangers: For plumbing installations, the pipe attachments, connectors, and structural attachments that support and secure pipes.

Nipples: Short lengths of pipe with male threads at both ends, used to make extensions from a fitting or to join two fittings.

Pressure fittings: Any type of fitting used for steam, gas, or water supply piping under pressure. Not designed for use in drainage piping, which is typically at atmospheric or a low water-column pressure.

Standard weight: An industry standard measurement of the weight or strength of steel pipe. Also referred to as Schedule 40.

Stop-and-waste valve: Valve similar to a globe valve that is opened or closed by raising or lowering a disc oriented perpendicular to the flow through the valve seat using a threaded stem. An elastomeric, or rubberized, washer attached to the valve disc seals the valve seat, closing off water flow. This valve is most commonly used for water faucets.

Thread engagement: The length of pipe that gets threaded into the fitting when joining threaded pipe.

Plumbers need to know about common fittings and valves; measuring, cutting, and joining steel pipe, and become familiar with the hangers and supports used with steel pipe.

During smelting and fabrication, manufacturers add different chemical elements to steel alloy to create three basic kinds of steel: carbon steel, low-alloy steel, and high-alloy (stainless) steel. Carbon steel is created by combining iron ore with carbon. Carbon steel may also have small percentages of manganese, silicon, aluminum, copper, and other elements. Manganese helps prevent brittleness and improves strength, hardness, and flexibility. Low-alloy steel is harder, which allows it to resist corrosion better, but it can be difficult to weld depending on the amount of carbon and other metals used. Stainless steel is similar to carbon steel but includes chromium, which protects the metal against rust.

The two common types of carbon-steel pipe are black-steel pipe and galvanized pipe. Black-steel pipe, which gets its color from the carbon in the steel, is most often used as gas or air-pressure pipe. Because it is not highly resistant to corrosion, you should use black-steel pipe only in applications where corrosion does not affect its uncoated surfaces. Galvanized pipe is steel pipe that has been electroplated with molten zinc. This protects the pipe surfaces from abrasive, corrosive materials and gives the pipe a dull gray color. Although more resistant to corrosion than black-steel pipe, galvanized pipe is still less resistant to corrosion than other piping materials. Galvanized pipe is commonly used for venting systems.

1.1.0 Sources, Storage, and Handling of Steel Pipe

Steel is manufactured both in the United States and other countries, and plumbers use both. The US steel industry must comply with trade regulations that control the amount of imported steel that companies can purchase and distribute domestically.

1.1.1 Steel Pipe Sources and Standards

The safety and quality of the pipe you use for plumbing systems are critical. US pipe manufacturers must adhere to US manufacturing codes and standards that may not apply elsewhere. For example, the US government requires that pipe adhere to the standards of ASTM International (formerly known as the American Society for Testing and Materials) and piping is marked with their initials.

1.1.2 Storage and Handling of Steel Pipe

To ensure maximum productivity on the job, it is important to use common sense when storing and handling steel pipe. Before starting work, always plan ahead. Organize pipe and fittings into groups by pipe size and type. Store pipe and fittings near where you are working so that they are easy to access. When handling pipe and fittings, be careful not to bend or damage them. Pipe that is bent or otherwise damaged not only wastes money, but also makes you less productive.

1.2.0 Threads

Steel pipe for plumbing is typically manufactured with threads on both ends and connected with threaded fittings. The length of pipe that screws into the fitting is called the thread engagement. In 1886, the decision to standardize the thread size on all manufactured pipe products was adopted by manufacturers. The standard thread used on pipe is V-shaped, with an angle of 60 degrees (*Figure 1*). The taper is $\frac{1}{32}$ inch per inch of threaded length. When threaded together, tapered pipe threads produce a pressure-tight connection. In addition, after assembly, it produces a mechanically rigid piping system.

Note in *Figure 1* that the first seven threads from the end are perfect threads. They are sharp at both tops and bottoms. The next two threads are imperfect at the top and perfect at the bottom, and the remaining threads are imperfect at the top and bottom. Imperfect threads result when threads are not cut completely or when perfect threads are marred and broken. These threads can't form a tight seal. If perfect threads are marred or broken, they lose their sealing ability as well.

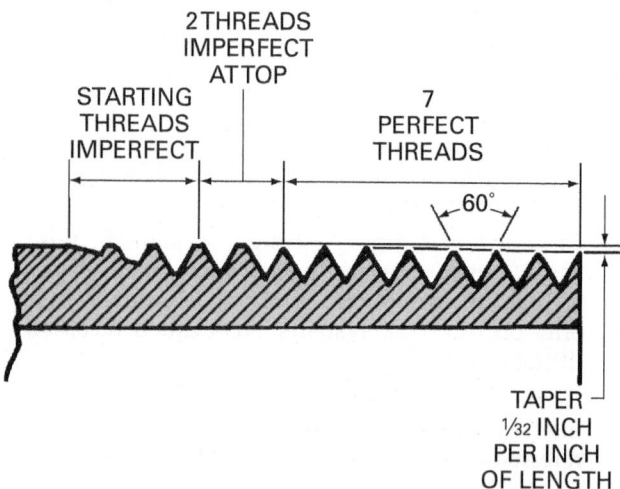

Figure 1 American standard thread.

Producing Steel

Steel is produced by one of four processes: Bessemer, open-hearth, electric furnace, or basic oxygen. The Bessemer process, developed in the 1850s, produced steel by blowing air through molten iron at high pressure. The process used oxygen to draw off impurities, but it was not efficient enough to remove all the unwanted elements. The Bessemer process is no longer used in the United States.

The open-hearth process was the main method of producing steel for many years. An open-hearth furnace is a huge, shallow tank lined with firebrick. A fuel, such as gas, oil, or powdered coal, is mixed with hot air in burners and blasted across the tank. Molten iron, recycled steel, and limestone are melted together in the tank. After the removal of impurities collected by the limestone, the steel is ready.

The electric-furnace process uses tremendous charges of electricity to produce steel. The furnace is loaded with scrap iron. An electrical arc is produced using electrodes. The resulting temperatures (6,000°F or 3,300°C) melt the scrap iron and produce the steel alloy.

The basic-oxygen process is the method most often used today. Molten iron from a blast furnace (a giant oven lined with firebrick) is poured into the basic-oxygen furnace. Pure oxygen is blown onto the metal through a tube called an oxygen lance. The oxygen combines with the impurities and is then removed from the furnace, leaving the steel.

After the liquid steel has been created in the furnaces, it must be further processed and cast into solid form before manufacturers can use it.

Table 1 lists the number of threads per inch required for each pipe size. Note that the total number of threads required is different from the number of usable threads required.

1.3.0 Steel Pipe Sizing and Labeling

Steel pipe is sized in one of two ways depending on the pipe diameter. It is labeled with sizing and other data.

Table 1 American Standard Pipe and Pipe Thread Dimensions

Nominal Pipe Size (Inches)	Threads per Inch	Number of Usable Threads*	Hand-Tight Engagement** (Inches)	Thread Makeup Wrench and Hand Engagement** (Inches)	Total Length of External Threads (Inches)
$\frac{1}{8}$	27	7	$\frac{3}{16}$	$\frac{1}{4}$	$\frac{3}{8}$
$\frac{1}{4}$	18	7	$\frac{1}{4}$	$\frac{3}{8}$	$\frac{9}{16}$
$\frac{3}{8}$	18	7	$\frac{1}{4}$	$\frac{3}{8}$	$\frac{5}{8}$
$\frac{1}{2}$	14	7	$\frac{5}{16}$	$\frac{1}{2}$	$\frac{3}{4}$
$\frac{3}{4}$	14	8	$\frac{5}{16}$	$\frac{9}{16}$	$\frac{13}{16}$
1	$11\frac{1}{2}$	8	$\frac{3}{8}$	$\frac{11}{16}$	1
$1\frac{1}{4}$	$11\frac{1}{2}$	8	$\frac{7}{16}$	$\frac{11}{16}$	1
$1\frac{1}{2}$	$11\frac{1}{2}$	8	$\frac{7}{16}$	$\frac{3}{4}$	1
2	$11\frac{1}{2}$	9	$\frac{7}{16}$	$1\frac{3}{4}$	$1\frac{1}{16}$
$2\frac{1}{2}$	8	9	$\frac{11}{16}$	$1\frac{1}{8}$	$1\frac{9}{16}$
3	8	10	$\frac{3}{4}$	$1\frac{3}{16}$	$1\frac{5}{8}$
$3\frac{1}{2}$	8	10	$\frac{13}{16}$	$1\frac{1}{4}$	$1\frac{11}{16}$
4	8	10	$\frac{13}{16}$	$1\frac{5}{16}$	$1\frac{3}{4}$
5	8	11	$\frac{15}{16}$	$1\frac{3}{8}$	$1\frac{13}{16}$
6	8	12	$\frac{15}{16}$	$1\frac{1}{2}$	$1\frac{15}{16}$
8	8	14	$1\frac{1}{16}$	$1\frac{11}{16}$	$2\frac{1}{8}$
10	8	15	$1\frac{3}{16}$	$1\frac{15}{16}$	$2\frac{3}{8}$
12	8	17	$1\frac{3}{8}$	$2\frac{1}{8}$	$2\frac{9}{16}$

* Rounded to the nearest thread.
** Rounded to the nearest $\frac{1}{16}$ inch.

1.3.1 Pipe Sizing

Steel pipe is measured by the nominal size. For pipe up to 12 inches, this size is based on the approximate inside diameter (ID) of the pipe. The nominal size is not necessarily equal to the exact ID, pipe larger than 12 inches is measured by its outside diameter (OD). Steel pipe is manufactured in a variety of sizes, measured by the pipe's ID and its length.

Steel pipe can be manufactured in lengths of 21 feet, threaded or unthreaded. A 21-foot length of steel pipe is called random length. A 42-foot length of pipe is called double random. Pipe with a nominal size of 6 inches or larger is available in random lengths between 18 and 22 feet.

In addition to size and length, you need to understand the different weights of steel pipe. Steel pipe for threading is manufactured in three weights or strengths: standard weight, extra strong, and double-extra strong. Some manufacturers refer to extra strong as extra heavy and double extra strong as double extra heavy. Because all weights for a particular size of pipe have the same OD, as shown in *Figure 2*, the same threading dies (devices used to cut the threads) fit all three weights of each size of pipe.

The wall thickness of the pipe and the ID differ with each weight. The thicker the wall, the smaller the ID, and the more pressure the pipe can withstand. Standard weight is adequate in most plumbing situations. However, the two stronger weights must be used when pressures require it. Always check applicable codes for weight requirements. See the dimensions of Schedule 40 (standard weight) and Schedule 80 (extra strong) steel pipe in *Table 2*.

1.3.2 Labeling

For ease of handling, manufacturers ship steel pipe in bundles. The number of lengths per bundle is determined by the size of the pipe. Each bundle has a tag that identifies the number of feet in the bundle. While there is no specific standard for labeling, steel pipe must be labeled with the name or brand of the manufacturer, its length, and *ASTM A53/A53M*, which indicates that the pipe meets ASTM International standards.

> **NOTE**
>
> The full title for *ASTM A53/A53M* is *Standard Specification for Pipe, Steel, Black and Hot-Dipped, Zinc-Coated, Welded and Seamless.* This standard supersedes *ASTM A120.*

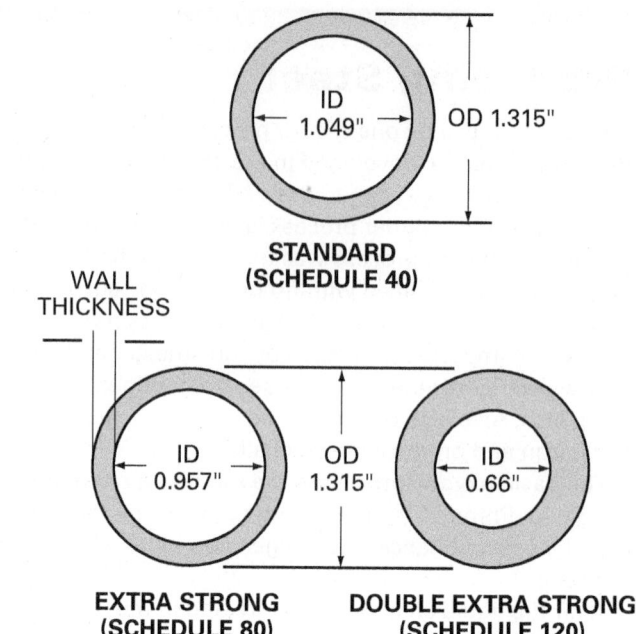

Figure 2 ID and OD of 1-inch steel pipe of different weights.

1.4.0 Fittings and Valves

The purpose of fittings is to connect lengths of straight pipe together in order to change the direction of flow in water supply and distribution systems, and in drain, waste, and vent (DWV) systems. Valves regulate the flow of water in a system. They are often attached to the ends of pipe to provide access to fluid flowing through the system. Fittings are designed so they do not block or significantly slow the flow of fluids in the pipe unless intended to do so.

Fittings for steel pipe are generally made of malleable iron or cast iron. Annealing ordinary cast iron produces malleable iron. The annealing process, which consists of heating and slowly cooling the iron, makes the iron stronger so that it can be bent or pounded to a reasonable extent without breaking. This is unlike copper, which becomes flexible when heated.

Steel pipe fittings are manufactured in two types: pressure fittings and drainage fittings, also called recessed drainage fittings (*Figure 3*). Drainage fittings have a drainage pattern design and recessed threads. They should be used only in soil and waste piping systems.

Pressure pipe fittings cannot be used in drainage systems because the inside surfaces are not smooth enough. Drainage fittings differ from pressure fittings in several ways. The insides of drainage fittings are smooth and are shaped for easy flow. Their shoulders are recessed so that when pipe is screwed in, the pipe fills the recessed area and forms an unbroken, internal contour. Ideally, a pipe screwed into a drainage

Table 2 Dimensions of Schedule 40 and Schedule 80 Steel Pipe

Nominal Diameter (Inches)	Dimensions of Schedule 40 (Standard Weight) Steel and Wrought Iron Pipe			Dimensions of Schedule 80 (Extra Strong) Steel and Wrought Iron Pipe		
	Actual Inside Diameter (Inches)	Actual Outside Diameter (Inches)	Weight per Foot (Pounds)	Actual Inside Diameter (Inches)	Actual Outside Diameter (Inches)	Weight per Foot (Pounds)
1/8	0.269	0.405	0.25	0.215	0.405	0.31
3/4	0.364	0.540	0.43	0.302	0.540	0.54
3/8	0.493	0.675	0.57	0.423	0.675	0.74
1/2	0.622	0.840	0.86	0.546	0.840	1.09
3/4	0.824	1.050	1.14	0.742	1.050	1.47
1	1.049	1.315	1.68	0.957	1.315	2.17
1 1/4	1.380	1.660	2.28	1.278	1.660	3.00
1 1/2	1.610	1.900	2.72	1.500	1.900	3.63
2	2.067	2.375	3.66	1.939	2.375	5.02
2 1/2	2.469	2.875	5.80	2.323	2.875	7.66
3	3.068	3.500	7.58	2.900	3.500	10.25
3 1/2	3.548	4.000	9.11	3.364	4.000	12.50
4	4.026	4.500	10.80	3.820	4.500	14.98
5	5.047	5.583	14.70	4.810	5.563	20.78
6	6.065	6.625	19.00	6.760	6.625	28.57

fitting completely fills the recessed cavity, but this rarely occurs. Drainage fittings are also designed so that horizontal lines entering them will have a slope of 1/4 inch per foot. Refer to *Figure 3* for illustrations of the normal and ideal conditions of most drainage joints.

The following sections discuss the different fittings and valves you can use with threaded and grooved pipe. The fittings for threaded and grooved pipe are not interchangeable. The common pressure fittings used for hot- and cold-water distribution systems, steam and hot water heating systems, and gas and air piping systems are covered. Slight obstructions on the inside surface of the pipe and fitting joint do not affect the flow of materials in any of these systems.

1.4.1 Tees

Tees are available in a number of sizes and patterns, and they are used to connect branches to a pipe at a 90-degree angle. If all three outlets are the same size, the fitting is called a regular tee (*Figure 4*). If outlet sizes differ, the fitting is called a reducing tee (*Figure 5*). Tees are identified first by the dimensions of the straight-through run, then by the side-opening dimensions. For example, a tee with a run inlet of 2 inches, a run outlet of 1 inch, and a branch outlet of 3/4 inch is a 2 × 1 × 3/4 tee. Always state the large run size first, the small run size next, and the branch size last. Tees are also available with male threads (on the outside of a pipe) on a run or branch outlet.

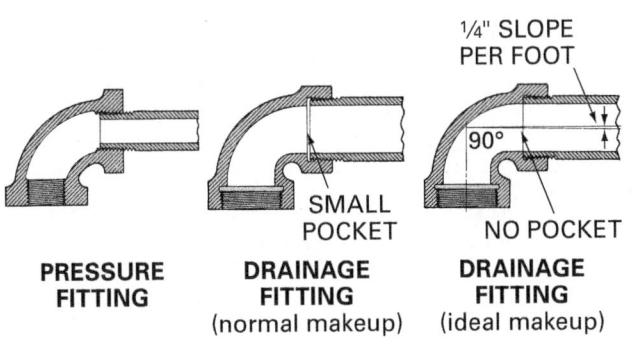

Figure 3 Pressure and drainage fittings.

Figure 4 Regular tee.

Figure 5 Reducing tee.

1.4.2 Elbows

Elbows, or ells, are installed to change the direction of the pipe. The most common elbows are the 90-degree ell; the 45-degree ell; the street ell, which has a male thread on one end; and the reducing ell, which has outlets of different sizes (*Figure 6*). Ells are also available to make $11^1/_4$-degree, $22^1/_2$-degree, and 60-degree bends.

1.4.3 Unions

Unions make it possible to disassemble a threaded piping system. After disconnecting the union, you can unscrew the length of pipe on either end of the union. The two most common steel pipe unions are the ground joint and the flange.

The ground-joint union connects two pipes by screwing the thread and shoulder pieces of the union onto the pipes. The collar then draws the shoulder and thread parts together. This union creates a gastight or watertight joint (*Figure 7*).

The flange union allows you to screw the flanges to the two pipes you are joining. You then pull the flanges together using nuts and bolts. A gasket between the flanges also makes this connection gastight or watertight (*Figure 8*).

1.4.4 Couplings and Other Fittings

Couplings are used to connect two lengths of pipe in making straight runs. Couplings can't be used in place of unions because they can't be disassembled (*Figure 9*) after installation is complete.

> **CAUTION**
>
> Cylindrical thread protectors are not tapered and may not be used as couplings. Use caution when working with thread protectors because they can easily be mistaken for couplings.

Other fittings for steel pipe include nipples, crosses, plugs, caps, and bushings. Nipples are short pieces of pipe, 12 inches long or less, that are threaded on both ends and are used to make extensions from a fitting or to join two fittings (*Figure 10*). Nipples are manufactured in many sizes, beginning with the close or all-thread nipple. Unlike most fittings, which are typically female, nipples are male pieces that connect female components.

Did You Know?

Durham fittings are a type of pressure drainage fitting that are designed so that horizontal lines connected to them have a slope of $^1/_4$ inch per foot. Durham fittings are not commonly used anymore, but you may encounter them when servicing drainage systems in older buildings.

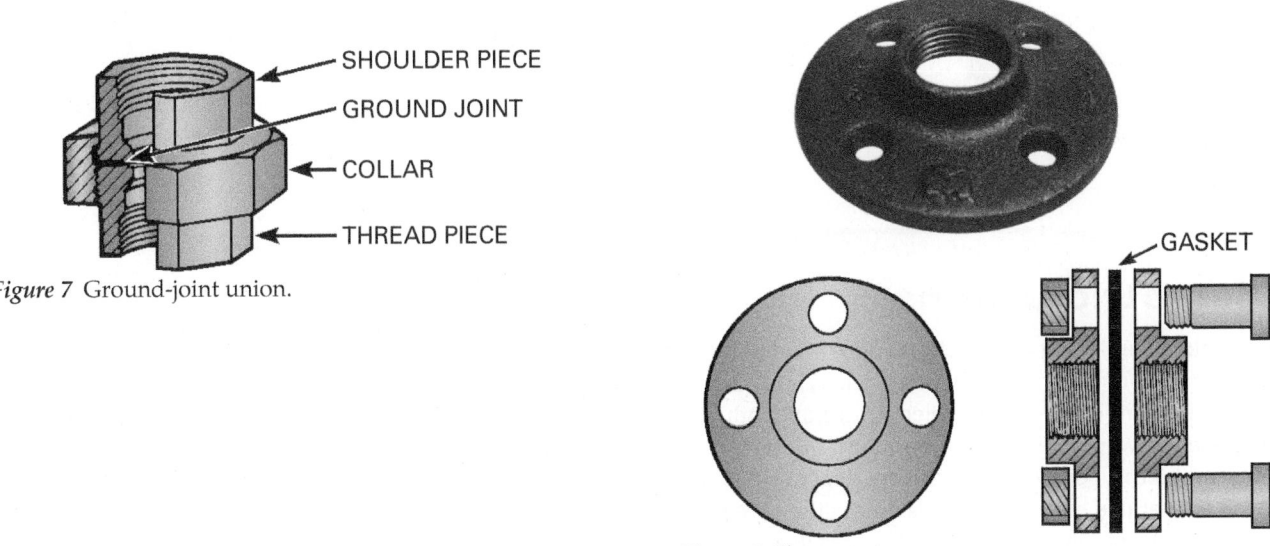

REDUCING ELL

90-DEGREE ELL

45-DEGREE ELL

STREET ELL

Figure 6 Elbows.

SHOULDER PIECE

GROUND JOINT

COLLAR

THREAD PIECE

Figure 7 Ground-joint union.

GASKET

Figure 8 Flange union.

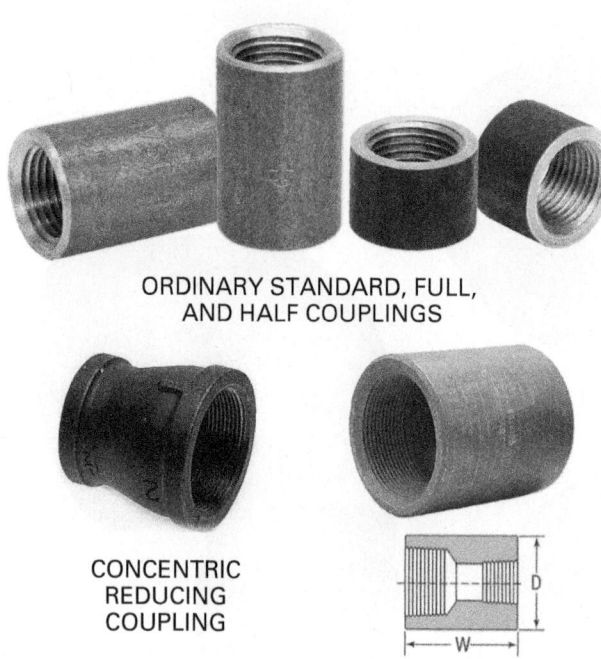

ORDINARY STANDARD, FULL, AND HALF COUPLINGS

CONCENTRIC REDUCING COUPLING

REDUCING COUPLING

Figure 9 Couplings.

pipe in a cross-like shape in a distribution system. This fitting allows you to join horizontal runs and vertical runs at a single point. Plugs close openings in pipe fittings and are available with a variety of heads, such as square, slotted, and hexagon (*Figure 11*). Caps are used to close open-ended pipes (*Figure 12*). Bushings are usually used to connect the male end of a pipe to a larger fitting. The ordinary bushing has a hexagon nut at the female end (*Figure 13*). Some couplings have extension pieces, which are fittings that have female openings on one end and male connections on the other. Flange isolation kits use a gasket to protect flanges from coming into contact with each other when joining, while also ensuring a tight seal on the joint. Insulating unions are used to protect exterior pipes against galvanic corrosion.

> **CAUTION**
>
> Only use reducers in gas pipe installations, and never use bushings as they can damage the pipe.

Crosses are fittings that connect four lengths of

BLACK AND GALVANIZED WELDED PIPE NIPPLES

SHORT OR SHOULDER NIPPLE

LONG NIPPLE

CLOSE NIPPLE

Figure 10 Nipples.

Dielectric Unions

If you attach unlike metals, such as copper and galvanized pipe, it results in a process called electrolysis (chemical breakdown caused by the flow of electricity) that produces galvanic corrosion. Transition fittings, such as dielectric unions, prevent electrolysis. Dielectric or nonconducting unions isolate the dissimilar materials by placing plastic or rubber between the two types of metals to break the electrical path. Dielectric unions, however, can cause other problems. For example, calcium deposits can build up, which eventually leads to blockages. Therefore, you should never use dielectric unions unless they are plastic lined. When installing a dielectric union, always check the temperature rating of the union before installing to ensure that it is appropriate for the installation. When installing a dielectric union between brass and steel pipe, install an isolation kit according to the manufacturer's instructions.

SQUARE AND HEXAGONAL SOCKET
COUNTERSUNK PIPE PLUGS

SQUARE HEAD PIPE PLUG

Figure 11 Plugs.

Figure 12 Cap.

Figure 13 Bushing.

1.4.5 *Valves for Threaded Pipe*

All common valves are manufactured for threaded steel pipe. However, the valves used with steel pipe are not necessarily made of steel. The most commonly used valves for threaded pipe are the globe valve, gate valve, ball valve, and stop-and-waste valve. The first three are shown in *Figure 14*.

Globe valves are recommended for general service on steam, water, gas, and oil, where frequent operation and precise control of flow are required. To close the valve, turn the handle clockwise until the valve stem firmly seats the washer or disc between the valve stem and the valve seat. This stops the flow of gas or liquid.

Gate valves are designed to be fully open or fully closed and are best suited as stop valves. They can be used in lines for steam, water, gas, oil, or air. If they are installed as service valves (that is, they are not often used), then they need to be operated periodically to prevent them from freezing shut because of corrosion, mineral, or scale buildups.

Ball valves are used as stop valves to control the flow of gases and liquids. Ball valves are best suited for terminations on main supply lines and pump lines. They are also used where quick cut-offs for line maintenance are required. The ball part of the valve with a drilled passage rotates into the open or shut position.

Stop-and-waste valves are opened or shut by raising or lowering a horizontal disc (as viewed with the stem vertical) using a threaded stem. An elastomeric, or rubberized, washer on the end of the stem seals the valve seat, closing off water flow. This valve is most commonly used as faucet fixtures, and for draining and freeze protection above ground.

> **WARNING!**
> Stop-and-waste valves are prohibited on water supply lines. Do not use stop-and-waste valves underground. There is a risk of contamination from the possibility of a cross-connection.

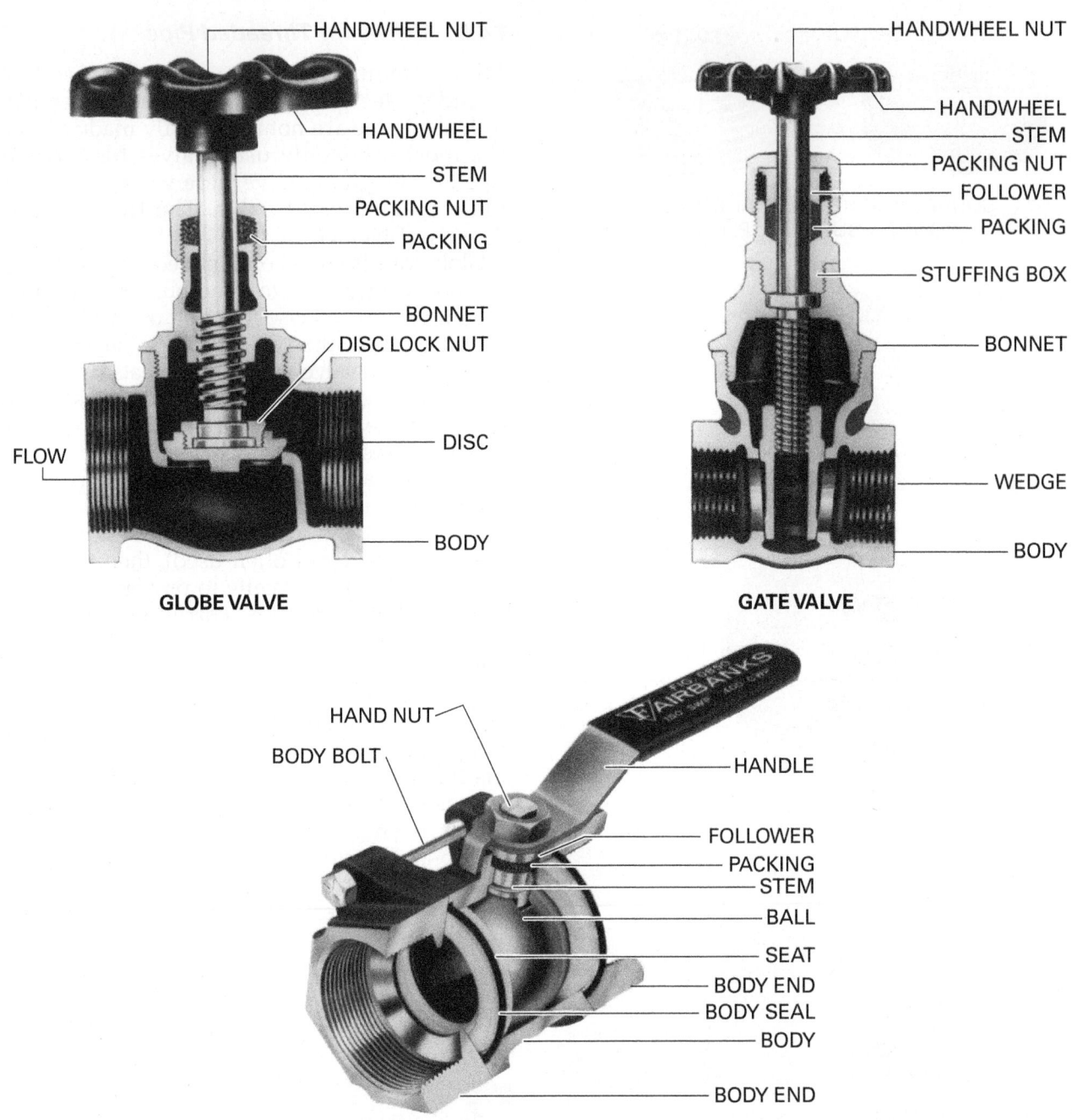

Figure 14 Valves for threaded pipe.

1.4.6 Fittings and Valves for Grooved Pipe

A wide variety of fittings and valves are available with grooved ends (*Figure 15*). Special gaskets are also available for various service conditions. It is important to consult the manufacturer's design specifications before using a grooved pipe system.

A grooved pipe joint consists of the following components:

- Specially grooved pipe ends rolled by for-the-purpose machines to exact manufacturer's standards
- A synthetic rubber gasket
- Lubricant
- A coupling consisting of two housings that mate with the grooves

Figure 15 Grooved pipe.

Did You Know?

The concept of grooved piping dates back to World War I, when armies in the field needed to join pipe quickly and efficiently. Since then, grooved piping has undergone continuous development. Grooved pipe is now a primary means of joining pipe. Steel pipe, copper pipe, cast-iron pipe, and ductile-iron pipe may have grooved joints.

ProPress® for Steel

Viega ProPress® is a system of grooved-pipe coupling that has been used for stainless piping systems in North America since 1999. ProPress® was the first press-pipe system to be introduced in this country. Pipe, valves, and fittings are available in 304 and 316 grade stainless steel.

Additional Resources

Plumbing Installation and Design, Third Edition, 2006. L. V. Ripka. Orland Park, IL: American Technical Publishers.

1.0.0 Section Review

1. The US government requires that steel pipe meets the _____.

   a. American Thread standard
   b. ANSI standard
   c. ASTM standard
   d. ADA standard

2. The taper of the threaded portion of a pipe is _____.

   a. 60 degrees
   b. 7 threads per inch of pipe length
   c. perfect
   d. $1/32$ inch per inch of threaded length

3. For a particular nominal size pipe, schedule 40, 80, and 120 have the same _____.

   a. outside diameter
   b. inside diameter
   c. thread diameter
   d. weight per foot

4. The valves that are best suited for terminals on main supply lines and pump lines are _____.

   a. gate
   b. globe
   c. ball
   d. stop-and-waste

2.0.0 MEASURING, CUTTING, AND JOINING STEEL PIPE

Objective

Describe the different methods for measuring, cutting, and joining steel pipe.

a. Explain the tools and methods used to measure and cut steel pipe.
b. Describe the tools and methods used to thread steel pipe.
c. Describe how to join threaded pipe.
d. Describe how to join grooved pipe.

Performance Tasks

3. Properly measure, cut, and join steel piping.
4. Identify the hazards and safety precautions when working with steel piping.

Trade Terms

Chain vise: A tool that aids in cutting large pipe by supporting the pipe and clamping it in an attached vise with V-jaws and secured by a chain attached to a screw tensioner.

Chain wrenches: Similar in construction and function to pipe tongs, but smaller in size for use on smaller-diameter pipes.

Pipe tongs: Wrench-like tools that use a length of chain wrapped around a pipe and linked into a cam tensioner to secure the tool to the pipe; used for threading and tightening large pipes.

Pipe-joint compound: A soft, semi-solid sealing material used when joining lengths of pipe. Also called pipe dope.

Power-threading machine: An electrically-powered machine useful for cutting threads in large quantities of pipe. The machine rotates, threads, cuts, and reams pipe.

Sealant tape: Tape used to wrap the threads of a pipe before joining to lubricate and ensure a watertight, secure fit. Also called PTFE or Teflon® tape.

Stock: The component of a hand or power threader that receives and securely holds the pipe-thread die cutter.

Strap wrench: A wrench consisting of a heavy strap wrapped around a pipe and run through a buckle at the top end of the wrench handle, which creates a cam-lock tensioner. Used to thread and tighten chrome-plated or other types of finished pipe so that there are no jaw marks or scratches left on the pipe.

Yoke vise: A pipe vise with a lower toothed V-jaw topped by a hinged attachment consisting of a jaw positioned by a threaded compression screw. When secured, the assembly creates a yoke that firmly holds a piece of pipe and prevents it from turning.

2.1.0 Working with Threaded Pipe

You will need to become familiar with many tools and processes to measure, cut, thread, and join materials. Working with threaded steel pipe is very different from working with grooved steel pipe. In addition to requiring different types of fittings, each type of pipe is joined using different methods. You do, however, cut threaded and grooved pipe using the same process. The following sections explain the methods for working with threaded and grooved steel pipe.

When you are working with threaded pipe, you must know how to measure a length of pipe for proper installation. You must take the thread engagement into account by referring to thread makeup tables such as *Table 3* or your pipe will be cut too short. You must also calculate the makeup of the fitting.

2.1.1 Measuring

It is extremely important to measure pipe carefully. If you cut a pipe improperly, it cannot make a secure joint. You can measure threaded steel pipe using a variety of methods:

- *End-to-end*—Measure the full length of the pipe, including both threaded ends.
- *End-to-center*—Use for pipe that has a fitting screwed on one end only. Pipe length is equal to the measurement minus the end-to-center dimension of the fitting, plus the length of thread engagement.
- *Center-to-center*—Use with a length of pipe that has fittings screwed onto both ends. Pipe length is equal to the measurement minus the sum of the end-to-center dimensions of the fittings, plus twice the length of the thread engagement.

Table 3 Thread Makeup

Size of Pipe (Inches)	Outside Diameter (Inches)	Number of Threads per Inch	Total Length of Threads (Inches)	Effective Length (Inches) (Approx.)	Thread Engagement (Inches) (Approx.)
$1/4$	0.540	18	$5/8$	$3/8$	$3/8$
$3/8$	0.675	18	$5/8$	$7/16$	$3/8$
$1/2$	0.840	14	$13/16$	$9/16$	$1/2$
$3/4$	1.050	14	$13/16$	$9/16$	$1/2$
1	1.315	$11\,1/2$	1	$11/16$	$9/16$
$1\,1/4$	1.660	$11\,1/2$	1	$11/16$	$5/8$
$1\,1/2$	1.900	$11\,1/2$	1	$3/4$	$5/8$
2	2.375	$11\,1/2$	$1\,1/16$	$3/4$	$11/16$
$2\,1/2$	2.875	8	$1\,9/16$	$1\,1/8$	$15/16$
3	3.500	8	$1\,5/8$	$1\,3/16$	1
4	4.500	8	$1\,3/4$	$1\,5/16$	$1\,1/16$
5	5.560	8	$1\,13/16$	$1\,3/8$	$1\,3/16$
6	6.625	8	$1\,15/16$	$1\,1/2$	$1\,1/4$

- *End-to-face*—Use for pipe that has a fitting screwed on one end only. Pipe length is equal to the measurement plus the length of the thread engagement.
- *Face-to-face*—Use for the same situation as center-to-center measurement. Pipe length is equal to the measurement plus twice the length of the thread engagement.
- *Face-to-back*—Use with pipe that has fittings screwed onto both ends. Pipe length is equal to the measurement plus the distance from the face to the back of one screwed-on fitting, plus twice the length of the thread engagement.
- *Center-to-face*—Use with pipe that has fittings screwed onto both ends. Pipe length is equal to the measurement from the center of one of the screwed-on fittings to the face of the opposite fitting, plus twice the length of the thread engagement.
- *End-to-back*—Use for pipe that has a fitting screwed on one end only. Pipe length is equal to the measurement plus the length of the screwed-on fitting, plus the length of the thread engagement.

2.1.2 Cutting

You can cut steel pipe with pipe cutters, which may have one to four cutting wheels (*Figure 16*). A single cutting wheel requires enough room for the cutter to be rotated all the way around the pipe. Use a cutter with more wheels when space is limited. The more wheels a cutter has, the fewer rotations it requires to cut the pipe. In addition to the cutting wheels, pipe cutters have an adjusting screw and at least two guiding wheels.

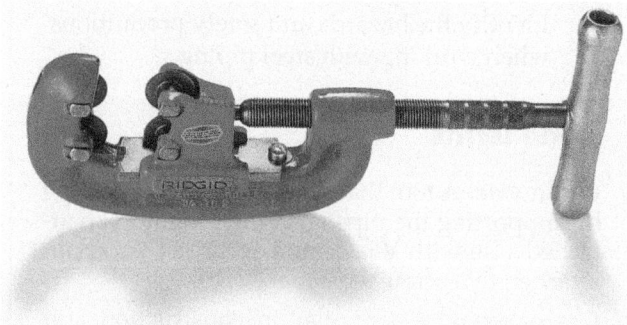

Figure 16 Pipe cutter.

To operate the pipe cutter, revolve it around the pipe and tighten the cutting wheel 1 revolution with each turn. Avoid overtightening the cutting wheel; this could damage or break the wheel.

Always use sharp cutting wheels. If the tube or pipe looks mashed after it has been cut, replace the cutting wheel(s). Apply lubricating oil periodically to all movable parts on the pipe cutter so it operates smoothly.

2.1.3 Reaming

After you cut a piece of pipe, use a reamer to remove the burr that remains on the inside of the pipe. If the burr is not reamed off, it will collect deposits and slow the flow of liquid through the pipe. Because reamers are tapered, one reamer can deburr many sizes of pipe. Reamers may be straight or spiral (*Figure 17*). Reaming is always required for pipe installations.

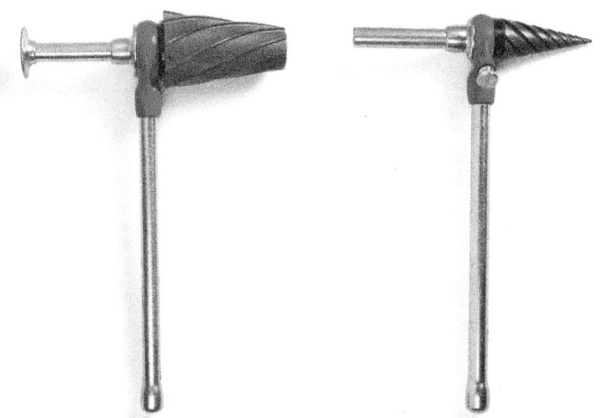

Figure 17 Reamers.

Figure 19 Power-drive threader.

2.2.0 Threading

There are two types of pipe threaders: hand and power. Hand threaders (*Figure 18*) consist of the pipe die and the stock. The stock holds the die while the die cuts the threads. Apply cutting oil to protect the die segments. Although you usually operate a hand threader manually, you can use it with a power drive, which threads automatically (*Figure 19*).

To thread large quantities of pipe, use a power-threading machine (*Figure 20*). This machine rotates, threads, cuts, and reams pipe. To use this machine, mount the pipe through the machine chuck (the piece that centers and grips the pipe), and then tighten the chuck. The oil-pumping trigger allows you to control the application of oil from the cutting oil pan. The oil lubricates the die. Direct the threading process by using the foot switch. Service the foot switch often to ensure that it always works as effectively as possible. Ensure that you use the proper oil for the foot switch. Refer to the manufacturer's specifications for the proper types of oil.

Figure 18 Hand threader.

Figure 20 Power-threading machine.

You can use a variety of vises and wrenches to hold a pipe securely during threading operations. Vises have jaws that hold the pipe firmly and prevent it from turning when threading a fitting onto it. The teeth on vises may leave marks on the pipe, so you should use a vise only on pipe that will not be visible after installation. Use the standard yoke vise to hold pipe that is small in diameter. Use the chain vise (*Figure 21*) to hold much larger pieces of pipe. Keep the chain on this vise oiled to prevent it from becoming stiff. Always make sure that you use the correct oil. A stiff chain makes the chain vise operate poorly.

Pipe wrenches have teeth that are set at an angle and should be used to grip and turn around stock. This angle allows the teeth to grip when turning the wrench in only one direction. Common pipe wrenches include the straight pipe wrench (*Figure 22*), the offset pipe wrench (*Figure 23*), and the compound-leverage pipe wrench.

Figure 21 Chain vise.

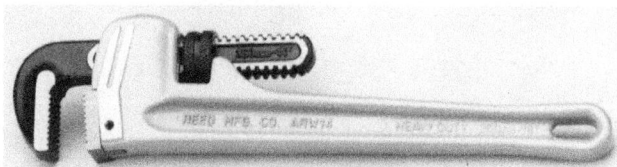

Figure 22 Straight pipe wrench.

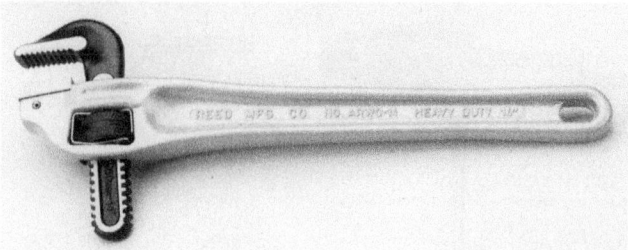

Figure 23 Offset pipe wrench.

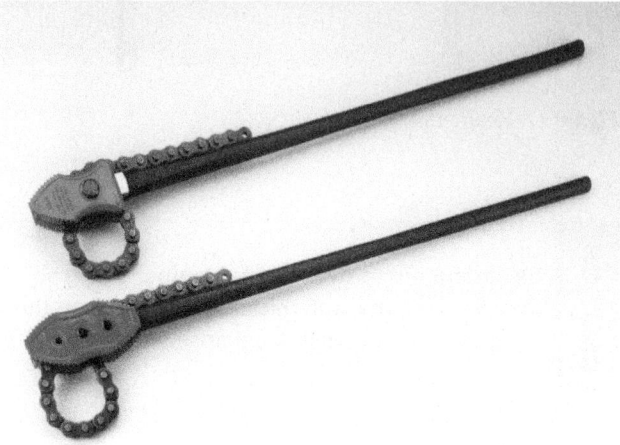

Figure 24 Pipe tongs.

Figure 25 Chain wrench.

Use an offset pipe wrench in confined spaces where a straight pipe wrench does not fit. Use a compound-leverage pipe wrench when you need extra leverage to turn pipe assemblies. As the hook jaw turns the pipe one way, an offset-chain-trunnion (pivot) assembly applies pressure in the opposite direction.

To hold large pipes, use pipe tongs, also called chain wrenches (*Figure 24* and *Figure 25*). Pipe tongs are wrench-like tools with a length of chain that holds the pipe. You must oil the chain in pipe tongs and chain wrenches often to prevent the chain from becoming stiff or rusty.

To hold chrome-plated or other types of finished pipe, use a strap wrench. The strap allows the wrench to grip the pipe without leaving jaw marks or scratches on the pipe. You can apply resin to the strap to increase the wrench's holding power and reduce slippage.

To cut threads using a hand die and stock, follow these steps (*Figure 26*):

Step 1 Select the correct size of die for the pipe you are threading.

Step 2 Inspect the die to make sure the cutters are free of nicks and wear.

Step 3 Lock the pipe securely in a vise.

Step 4 Slide the die over the end of the pipe, guide-end first.

Step 5 Push the die against the pipe with the heel of one hand. Take three or four short, slow, clockwise turns. Be careful to keep the die pressed firmly against the pipe.

Step 6 When you have cut enough thread to keep the die firmly against the pipe, apply some thread-cutting oil. This oil prevents the pipe from overheating from

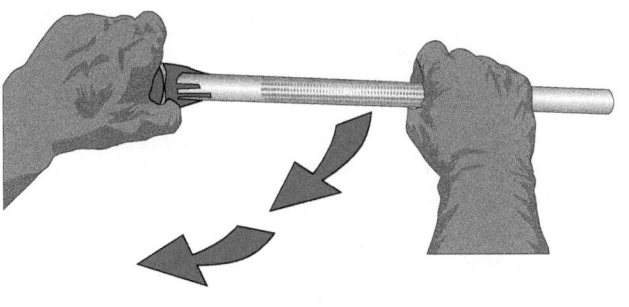

OVER-THE-SHOULDER VIEW

SIDE VIEW

Figure 26 Cutting threads using a hand die and stock.

friction, and it lubricates the die. Oil the threading dies every two or three downward strokes.

Step 7 Back off one quarter-turn after each full turn forward to clear out the metal chips. Continue until the pipe projects one or two threads from the die end of the stock. Having too few threads is as bad as having too many threads.

Step 8 Remove the die by rotating it counterclockwise.

Step 9 Wipe off excess oil and any chips.

To cut threads using a power-threading machine, follow these steps, referring to *Figure 27*:

Step 1 Make sure the machine and the foot pedal are in proper working order. Check for frayed wires and cords.

Step 2 Select and install the correct size of die, and inspect it for nicks.

Step 3 Mount the pipe into the machine chuck. Make sure to provide additional support for long pipe.

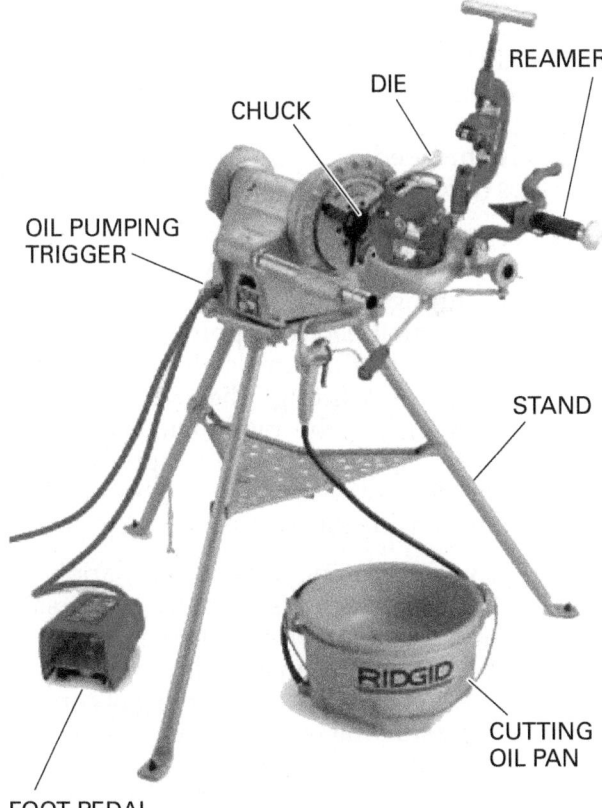

Figure 27 Cutting threads using a power-threading machine.

Step 4 Check the pipe and die alignment.

Step 5 Cut threads until two threads appear at the other end of the die. Stop the threading action. Apply the correct cutting oil to the threads and pipe during the threading operation to reduce heat and friction.

Step 6 Back off the die until it is clear of the pipe.

Step 7 Remove the pipe from the machine chuck. Be careful not to mar the threads.

Step 8 Wipe the pipe clean of oil and metal chips.

> **WARNING!**
>
> Before using any power machine, become familiar with the machine's maintenance and safety instructions. Foot pedals for safety cutoffs are now standard. A poorly maintained machine is a safety hazard. Dirt and metal chips can get stuck in the cutting dies and produce inadequate threads. Clogged components may also send debris flying—these flying objects can blind you or a co-worker.

2.3.0 Joining Threaded Pipe

To join threaded pipe, use pipe-joint compound, often called pipe dope, or sealant tape to provide lubrication for assembly.

To join threaded steel pipe, perform the following steps (refer to *Figure 28*):

Step 1 Apply pipe-joint compound or sealant tape only to the male threads of the pipe or fittings before you assemble a pipe connection. Do not apply the compound or tape to the female threads of either a pipe or a fitting. The twisting motion used to join the pipes will ball up the pipe dope or tape if it is applied to the female threads, and the debris will get stuck inside the pipe system.

Step 2 If you are using sealant tape, apply the tape in a clockwise direction, the same direction as the fitting turns. Before using sealant tape in a gas line assembly, be sure to check all applicable gas codes to make sure the use of tape is permitted.

Step 3 Start the fitting into the threaded pipe by hand, making sure to leave the starter threads free of tape and compound. Turn the fitting clockwise. Finish tightening the fitting with a pipe wrench.

Pipe-Joint Compound

Pipe-joint compound is available in a wide variety of materials. The most common pipe-joint compound is used for domestic water lines. Joint compound is designed to resist interaction with substances moving through the system. Depending on the type of gas or liquid moving through the pipe system, manufacturers recommend different kinds of joint compound. Always read manufacturer's recommendations before using the product. PTFE (polytetrafluoroethylene) plumber's tape can also be used instead of pipe-joint compound.

2.4.0 Grooving and Joining Grooved Pipe

As noted earlier in this module, grooved pipe connections have become quite common in a variety of metal pipe materials. In this section you will learn the methods of grooving and joining grooved-pipe sections. Working with grooved pipe is a specialized skill and requires training for specific grooved-pipe systems.

2.4.1 Grooving Pipe Ends

While grooved pipe is usually ordered prefabricated for large jobs, you can groove pipe in the field by rolling or cutting using a special powered machine similar to that shown in *Figure 29*. Roll-grooving involves cold-forming pipe, pressing a circumferential trough around the pipe. It does not remove any metal from the pipe. Cut-grooving removes metal from the outside diameter of the pipe, cutting a circumferential depression in the pipe wall. You will learn more about grooving pipe elsewhere in this curriculum.

2.4.2 Joining Grooved Pipe

Joining grooved steel pipe is very different from joining threaded steel pipe. To join grooved pipe, follow these steps as illustrated in *Figure 30A* and *Figure 30B*:

Step 1 Check the pipe ends. To make a leakproof seal, the ends must be free from indentations, projections, or roll marks.

Step 2 Check to make sure that the gasket is suitable for the intended use. Some manufacturers color-code their gaskets. Apply a thin coat of lubricant to the lips and the outside of the gasket.

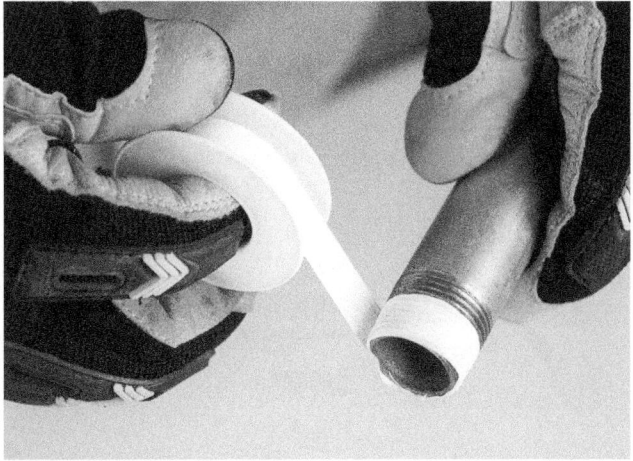

(A)

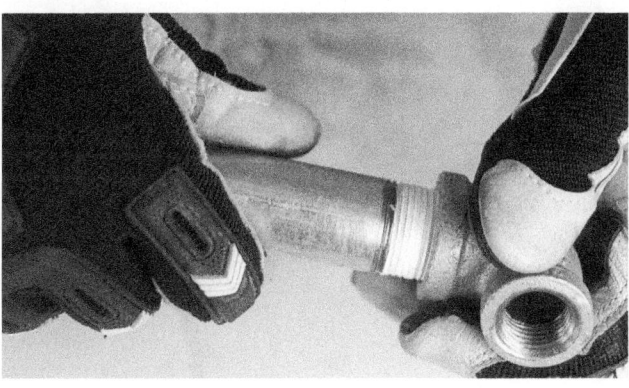

(B)

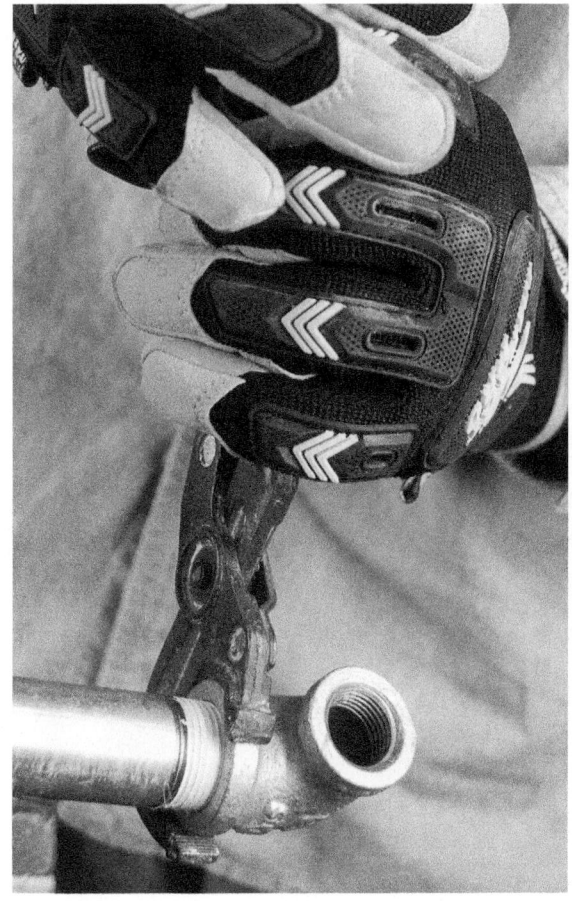

(C)

Figure 28 Joining threaded steel and pipe.

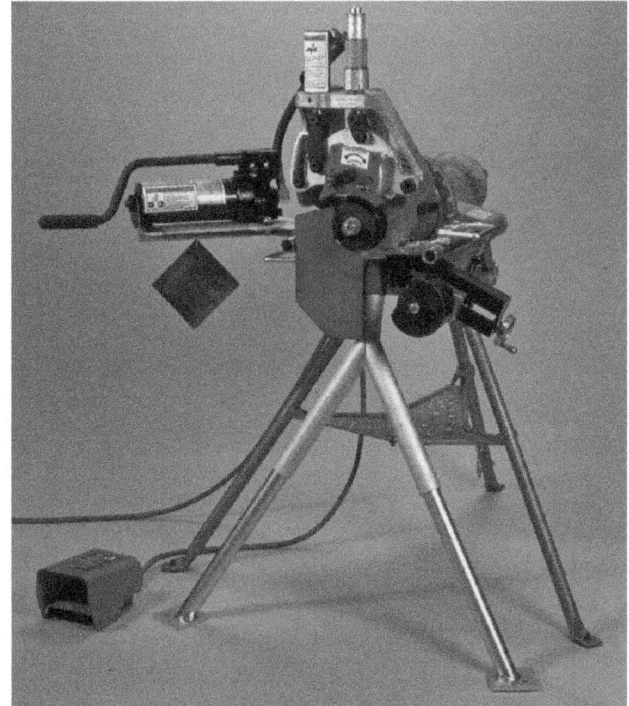

Figure 29 Roll-grooving tool.

Step 3 Install the gasket over the pipe end. Be sure the gasket lip does not hang over the pipe end.

Step 4 Align and bring the two pipe ends together. Slide the gasket into position and center it between the grooves on each pipe. Be sure that no part of the gasket extends into the groove on either pipe.

Step 5 Assemble the housing segments loosely, leaving one nut and bolt off to allow the housing to swing over the joint.

Step 6 Install the housing, swinging it over the gasket and into position in the grooves on both pipes.

Step 7 Insert the remaining bolt and nut. Be sure that the bolt track head engages into the recess in the housing (see *Figure 31* and refer to *Figure 30A* and *Figure 30B*).

Step 8 Tighten the nuts alternately and equally to maintain metal-to-metal contact at the angle bolt pads.

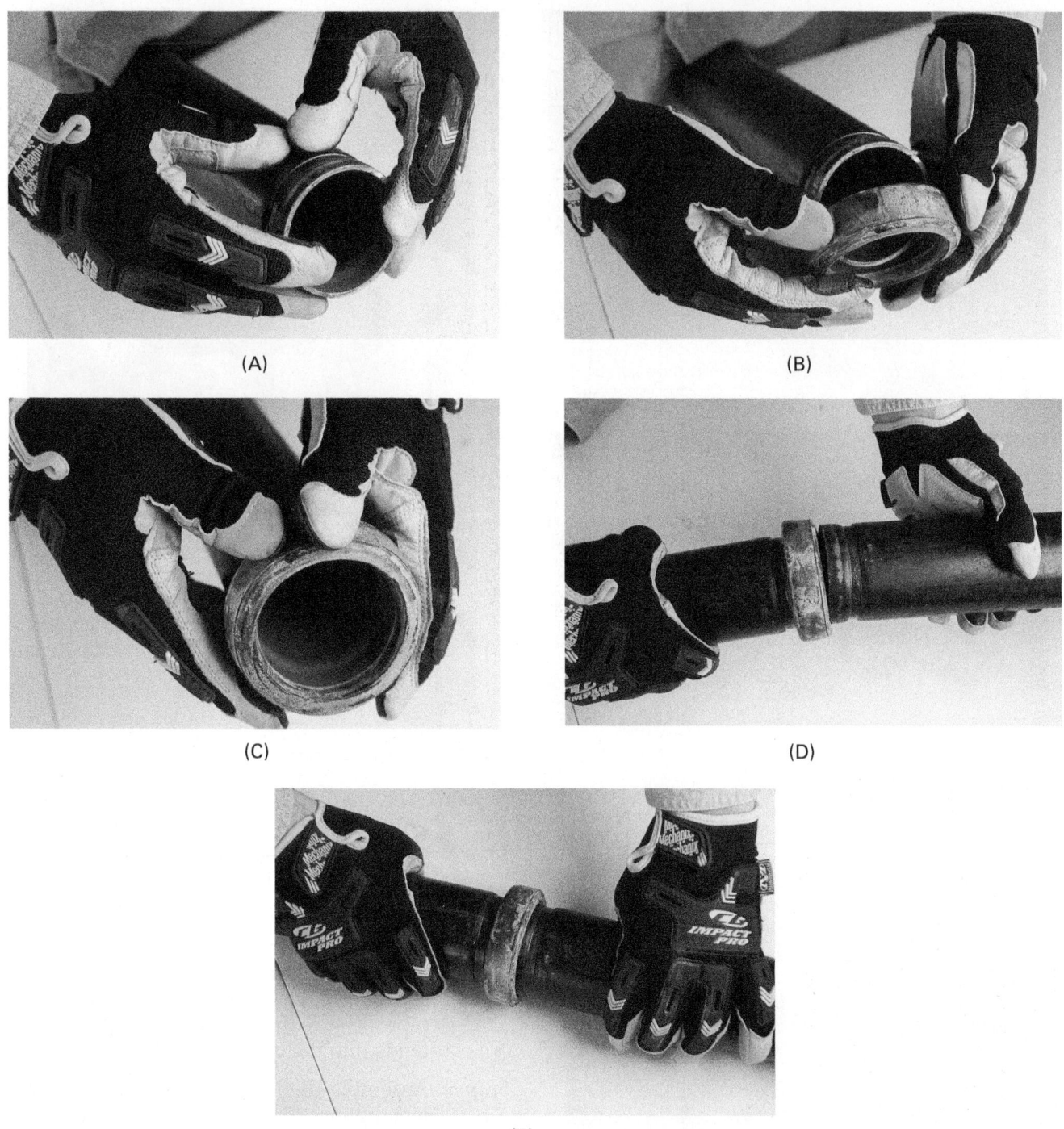

(A)

(B)

(C)

(D)

(E)

Figure 30A Joining grooved pipe.

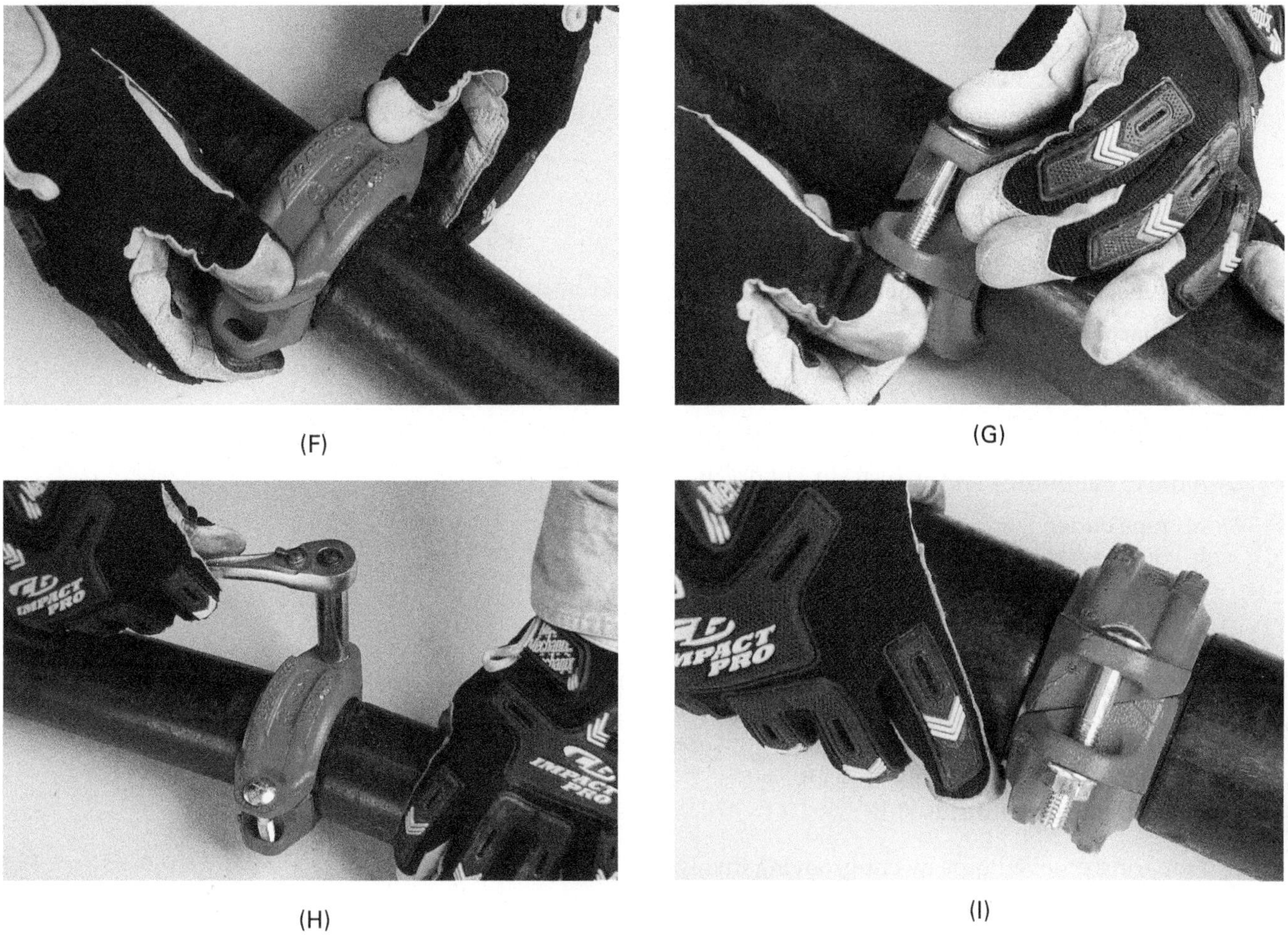

(F)

(G)

(H)

(I)

Figure 30B Joining grooved pipe.

Figure 31 Coupling installed.

Additional Resources

Plumbing Installation and Design, Third Edition, 2006. L. V. Ripka. Orland Park, IL: American Technical Publishers.

2.0.0 Section Review

1. If a tube or pipe looks mashed after it has been cut, the cutting wheel or wheels should be _____.

   a. sharpened
   b. replaced
   c. polished
   d. adjusted

2. A die to cut threads and a stock to hold the die are the two parts of a(n) _____.

   a. pipe cutter
   b. offset wrench
   c. hand threader
   d. chain vise

3. When using sealant tape to join threaded pipe, apply the tape _____.

   a. in a clockwise direction
   b. in a counterclockwise direction
   c. on both male and female threads
   d. on the female threads only

4. Preparing grooved pipe by cut-grooving involves _____.

   a. pressing a circumferential groove into the pipe
   b. removing metal in a circumferential depression around the pipe
   c. using a die to cut spiral grooves called threads around the pipe
   d. cutting longitudinal grooves in the end of the pipe

3.0.0 PIPE INSTALLATION AND SUPPORT

Objective

Describe the methods used to install and test steel pipe.

a. Describe the hangers and fasteners used to support steel pipe.
b. Explain how to support vertical and horizontal piping runs.

Performance Task

5. Identify the various techniques used in hanging and supporting steel piping.

Trade Term

Channels: Sections of C-, U-, or box-shaped extrusions of steel or aluminum used for supporting pipes and pipe hangers.

Installing plumbing systems involves connecting various appliances and fixtures with pipe. This provides the functionality of the piping system. But if the pipes don't stay where they are supposed to be, the system will soon fail. This section introduces you to some of the methods for securely supporting pipes so that they will be able to reliably fulfill their functions.

3.1.0 Supporting Pipes and Attachments

The industry has developed many ways of attaching pipes to the supporting structures around them. There are many types of components that go into pipe hangers and supports.

3.1.1 Hangers and Supports

A plumber must install and support all pipe so the pipe and joints remain leakproof. Improper support can cause joints to sag, and the added stress increases the chances that the pipe will leak, break, or crack. Plumbing codes require pipe to be securely fastened, both horizontally and vertically, using a system of supports.

If a plumber is installing pipe in a seismically active area (where earthquakes are a possibility) local codes require seismic restraints. The purpose of these restraints is to ensure that the pipe is securely fastened to the structure in the event of excessive vibration. For example, some codes require hangers and supports to be used at closer intervals than in nonseismically-active areas. In addition, they may require that the worker leave extra spacing for pipe where it meets walls and floors to allow for anticipated movement.

Plumbers need to know how to attach pipe to wood, masonry (including tile and concrete), and steel surfaces and should use one or more hangers to support steel pipe. The basic components of hangers are pipe attachments, connectors, and structural attachments. Always use hangers and supports that are made from materials that are compatible with the piping materials.

Consult local applicable codes and manufacturer's recommendations for spacing intervals and equipment needed when installing pipe hangers and supports. Spacing requirements are often found in the specifications section of the building plans. Failing to support pipe adequately will get a rejection from the plumbing inspector. Poor workmanship or misunderstandings will earn a rejection as well.

> **NOTE**
>
> If you do not agree with manufacturer's recommendations or codes, check with a supervisor about what needs to be done and what material can be used to accomplish the job.

3.1.2 Pipe Attachments

Pipe attachments touch or connect directly to the pipe. Rings, clamps, and clevises are among the attachments that can support piping horizontally from the ceiling (*Figure 32*). To support piping on wood-frame construction, a worker can use tube straps, tin straps, plumber's tape (also called strap iron or band iron), half clamps, suspension clamps, hold-down clips, and pipe clamps (*Figure 33*). Vertical pipe should be supported at each floor level using riser clamps (*Figure 34*).

Other pipe attachments include universal pipe clamps attached to pipe channels (*Figure 35*) by various means. The clamps are available in standard finishes of mild steel and electro-galvanized steel.

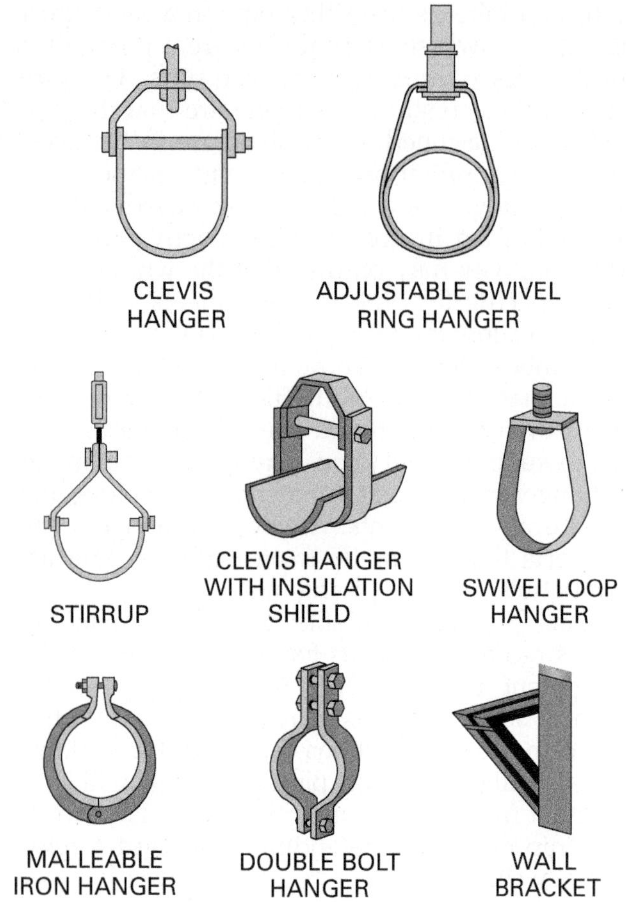

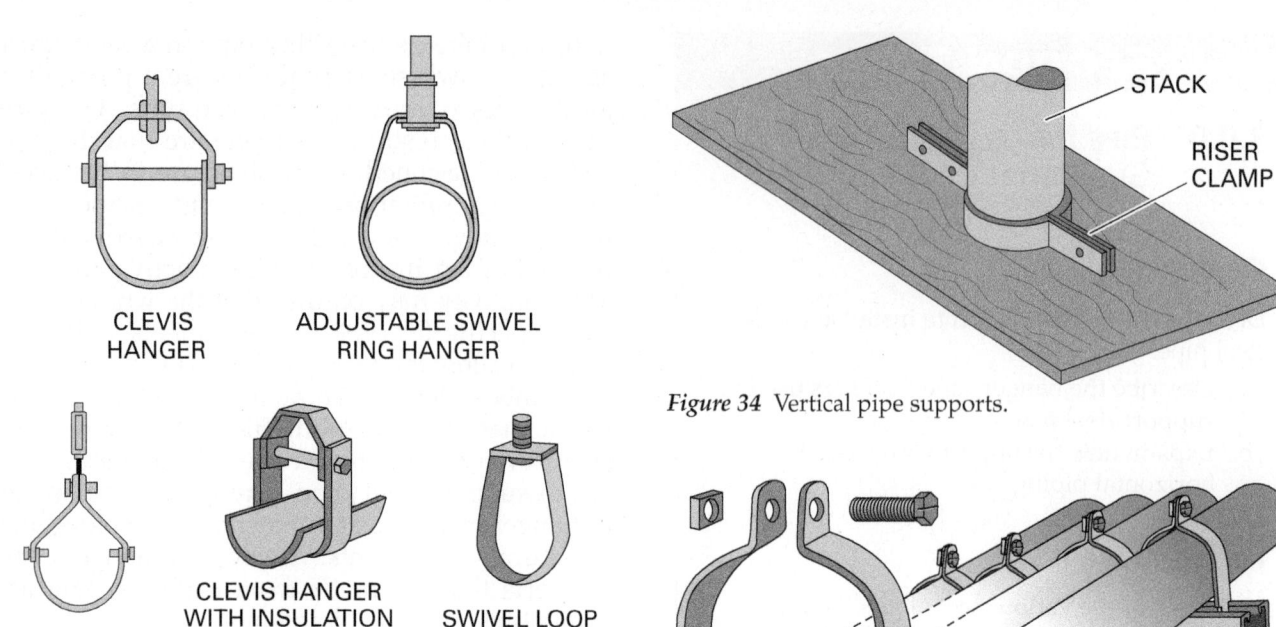

CLEVIS
HANGER

ADJUSTABLE SWIVEL
RING HANGER

STIRRUP

CLEVIS HANGER
WITH INSULATION
SHIELD

SWIVEL LOOP
HANGER

MALLEABLE
IRON HANGER

DOUBLE BOLT
HANGER

WALL
BRACKET

Figure 32 Hangers used for horizontal support.

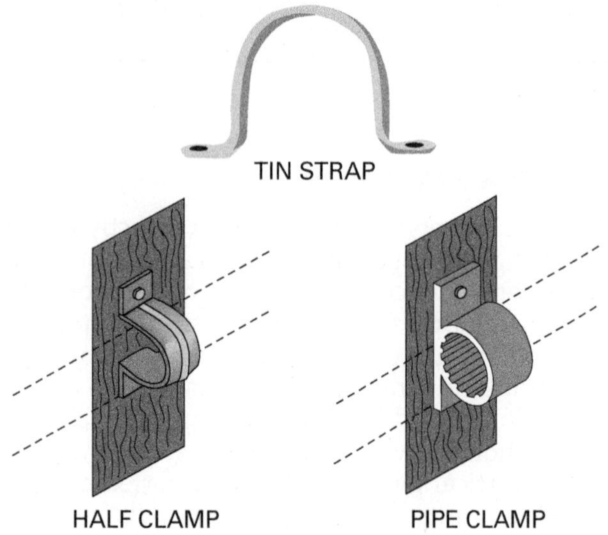

TIN STRAP

HALF CLAMP

PIPE CLAMP

Figure 33 Pipe supports for wooden frames.

STACK

RISER
CLAMP

Figure 34 Vertical pipe supports.

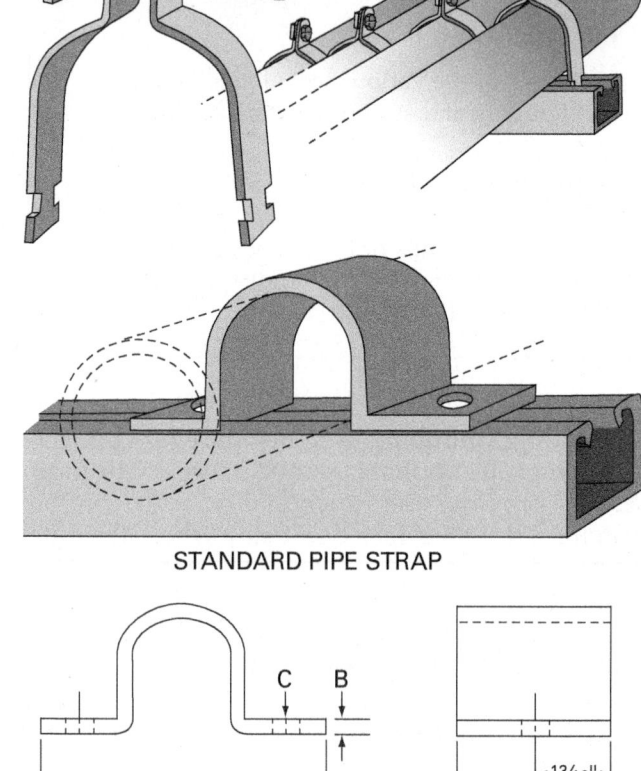

STANDARD PIPE STRAP

A

C

B

13⁄16"

15⁄8"

Figure 35 Universal pipe clamps and pipe channels.

3.1.3 Connectors

Connectors link the pipe attachment to the structural attachment. The two main groups of connectors are (1) rods and bolts and (2) other rod attachments.

Rod attachments, which are available in several sizes, include eye sockets, extension pieces, rod couplings, hanger adjusters, turnbuckles, clevises, and eye rods. Eye sockets are used with hanger rods. Extension pieces are used to attach hanger rods to beam clamps and other types of structural attachments. Rod couplings and reducing rod couplings support pipe runs where it is possible to connect to an existing stud.

3.2.0 Structural Attachments

Structural attachments are the anchors and anchoring devices that hold the pipe hanger assembly securely to the building structure. They include powder-actuated anchors, concrete inserts, beam clamps, C-clamps, beam attachments, various brackets, ceiling flanges and plates, plate washers, and lug plates.

You can use wedge anchors, sleeve anchors, drop anchors, and stud anchors to fasten the piping system to concrete or brick (*Figure 36*). You can also use non-drilling anchors and self-drilling anchors in concrete.

CAUTION	Check with your supervisor to determine the proper curing time for fresh concrete, plumbers do not make this judgment and working with the concrete before it cures can cause serious problems.

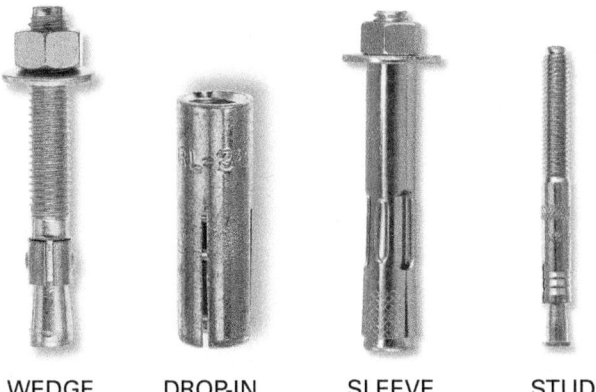

WEDGE   DROP-IN   SLEEVE   STUD

Figure 36 Anchors for concrete or brick.

In steel-framed buildings, you should use clamp-like devices to attach hangers for smaller pipe. These devices include beam clamps, I-beam clamps, and steel C-clamps (*Figure 37*).

In wood-framed buildings, use nails and screws to attach pipe hangers and supports to the structure. For vertical pipe runs, place riser clamps over the pipe, and secure the pipe to the structural frame (see *Figure 38*). If using plumber's tape, nail it directly onto the floor joists.

Vertical piping must be supported at sufficient intervals to keep the pipe in alignment. Supporting vertical piping depends on the size of pipe and whether it needs to be supported at or between the floor lines. Support vertical pipe at each floor line with riser clamps (refer to *Figure 37*). As always, follow applicable code requirements.

Figure 37 Clamp.

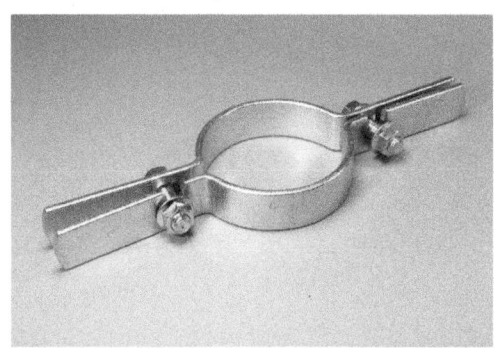

Figure 38 Riser clamp.

Additional Resources

Plumbing Installation and Design, Third Edition, 2006. L. V. Ripka. Orland Park, IL: American Technical Publishers.

3.0.0 Section Review

1. Preventing pipe from sagging, bending, or moving out of alignment requires the use of a system of _____.

   a. anchors
   b. attachments
   c. supports
   d. slopes

2. To support vertical pipes passing between floors, the attachments best suited for this purpose are _____.

   a. C-clamps
   b. riser clamps
   c. concrete inserts
   d. hanger rods

SECTION FOUR

4.0.0 CORRUGATED STAINLESS-STEEL TUBING

Objective

Explain the properties and uses of corrugated stainless-steel (CSST) tubing.
 a. Describe the uses of CSST.
 b. Describe the sizing and labeling of CSST.
 c. Explain the use of CSST regulators and valves.
 d. Describe CSST installation.

Trade Terms

Corrugated stainless steel tubing (CSST): A gas piping material made from flexible steel tubing, usually but not always supplied with a polyethylene jacket.

Elevated pressure systems: CSST systems that operate above the normal gas mains pressure of 7 inches w.c.

Gas cocks: Valves designed mainly for gas service that provide positive, quick flow control on gas piping systems.

Manifolds: Devices that allow the supplies to multiple gas devices to branch off of one CSST gas pipe.

Pounds per square inch gauge (psig): A unit of pressure that measures system pressure above atmospheric (i.e., 0 psig is atmospheric pressure). Used to measure the operating pressures for gas and other plumbing systems.

Regulators: Devices used in elevated-pressure systems to reduce and maintain a lower gas pressure at the pipe outlet.

Striker plates: A heavy manufactured metal plate of a required thickness that is installed behind a wall to protect CSST from penetration by nails, screws, and drill bits.

Water column (w.c.): A unit of pressure based on the pressure exerted by the height of water in a tubular column or manometer (e.g., 12 inches w.c. = 0.43 psig). Used for measuring low pressures, such as those in a standard gas supply.

Fuel gases such as natural gas and liquid petroleum (LP) gas are commonly used for cooking, heating, and power generation throughout the United States. Plumbers are responsible for installing safe and efficient fuel-gas systems. One of the most popular methods among plumbers for delivering fuel gas is to use corrugated stainless steel tubing (CSST).

CSST is a flexible stainless-steel tube, typically encased in a jacket made of polyethylene (PE) plastic. As an alternative to rigid gas pipe, CSST has been steadily gaining in popularity over the past 20 years. As of 2003, all national model codes recognize CSST as an approved gas piping material. In 2002, manufacturers sold almost 50 million feet of CSST, which represents nearly 40 percent of the current gas piping market. In this section, you will be introduced to CSST pipe and fittings and learn the basics of cutting, connecting, and hanging this type of pipe.

4.1.0 Introduction to CSST

CSST has several advantages over other types of gas piping, such as black steel, PE, or copper. It is delivered in precut lengths coiled on drums, which allows the pipe to be loaded and carried more easily (*Figure 39*). It is lighter than rigid pipe and does not require special tools for cutting or installing. Because it can be bent, CSST does not require angle fittings, such as elbows, to go around corners. These factors all reduce the labor time and costs required to install CSST to as much as half the costs of other types of pipe. CSST costs more per foot than other pipe, but labor and time cost savings offset at least some of this expense.

Unlike other types of gas pipe, CSST is sold as a complete system, including pipe, fittings, and connections. The components of CSST systems, however, are not interchangeable among manufacturers. *Table 4* lists the major manufacturers of CSST systems and the trade names of their pipe.

Figure 39 CSST is delivered in coils.

Table 4 CSST Names and Manufacturers

CSST System Trade Name	Manufacturer
Gastite®	Titeflex Corporation
TracPipe®	OMEGAFLEX®, Inc.
Parker Gas Piping (PGP™)	Parker Hannifin Corporation
Pro-Flex™	Tru-Flex™ Metal Hose, LLC

4.2.0 Sizes and Labeling

CSST is manufactured in sizes ranging from $^3/_8$ inch to 2 inches in diameter. The designer determines the proper pipe size for each installation. Regardless of the size, each length of pipe is labeled with the manufacturer's name and part number, as well as the label of the testing organization that certified its use in gas piping installations. The maximum operating pressure and the words "Fuel Gas" are marked at regular intervals. Some manufacturers also provide measurement markings at 1-foot intervals (*Figure 40*). Refer to your local applicable code for sizing specifications.

Qualified installers are responsible for sizing and installing CSST piping systems. Either the CSST manufacturer or one of the manufacturer-certified representatives can train installers. Installers must follow the manufacturer's instructions for proper installation of the pipe and fittings. Qualified installers ensure that the pipe runs are placed safely, supported properly, and connected securely.

Figure 40 Typical markings on a length of CSST.

When bending CSST, allow a generous bending radius. A tight bending radius can stress the pipe and fittings, causing leaks. Never kink, twist, or stretch CSST. Refer to the manufacturer's instructions for the allowable bending radius of the pipe you are using.

> **WARNING!**
>
> Never connect CSST directly to a meter. Use a length of steel pipe with a minimum of a 1-inch diameter between the meter and the CSST. Attach the steel pipe securely to the exterior wall of the building to avoid accidents and injury. Regulations are based on gas company and code requirements. Consult your applicable local code and the manufacturer's specification for sizing information.

4.3.0 Regulators and Valves

Most residential, commercial, and industrial gas systems operate at a relatively low pressure of 7 inches of water column (w.c.) or 0.25 pounds per square inch gauge (psig) as measured at the system's meter. Some installations require higher operating pressures. These are called elevated pressure systems. These systems operate within a range of $^1/_2$ to 2 psig. The maximum allowable operating pressure for gas systems, according to standards developed by ANSI, is 5 psig. Elevated pressure systems require regulators where the piping connects to an appliance or other device (*Figure 41*). Regulators decrease the pressure and maintain a uniform supply of gas to the device using the gas.

How CSST Came to the United States

CSST was introduced to the United States in the late 1980s through the efforts of the Gas Research Institute (GRI) and Foster-Miller, Inc., an engineering consulting company. GRI was seeking alternatives to black-steel pipe for gas utilities. CSST systems had been introduced in Japan and Europe only a few years earlier, and the two organizations examined the technology to see whether it could be used in the United States.

Tests showed that the use of CSST could reduce the labor required for installation by anywhere from one-quarter to two-thirds. Furthermore, the material proved to be as safe and reliable as other widely used types of gas piping. The 1989 National Fuel Gas Code was the first edition of the code to include CSST as an approved gas piping material, and the use of CSST in the United States has continued ever since.

Figure 41 Regulators are used to lower gas pressure in elevated pressure systems.

Ball valves and gas cocks are installed in CSST systems to shut off gas flow quickly (*Figure 42*). A handle on the outside of the valve body rotates the valve's ball into the opened or closed position. Ball valves provide positive, quick-flow control on gas piping systems. They are widely used in liquid plumbing systems as well. Install the valve upstream from appliances or connectors. Some codes require the valve to be located within 6 feet of the appliance. Other codes require a distance of only 3 feet and require the valve to be in the same room as the appliance.

Figure 42 Manufacturer-specific example of CSST valves.

Fuel Gas Codes and Standards

The National Fuel Gas Code is approved by the American National Standards Institute (ANSI) and the National Fire Protection Association (NFPA). The code is also referred to as *ANSI Z223.1* and *NFPA 54*, the *National Fuel Gas Code*. It applies to pipe that is installed between the meter outlet and all appliance inlets in a building.

ANSI and the American Gas Association (AGA) publish another important standard related to CSST. *ANSI/AGA LC 1, Interior Fuel Gas Piping Systems Using Corrugated Stainless Steel Tubing*, sets the construction, installation, and performance requirements for CSST systems.

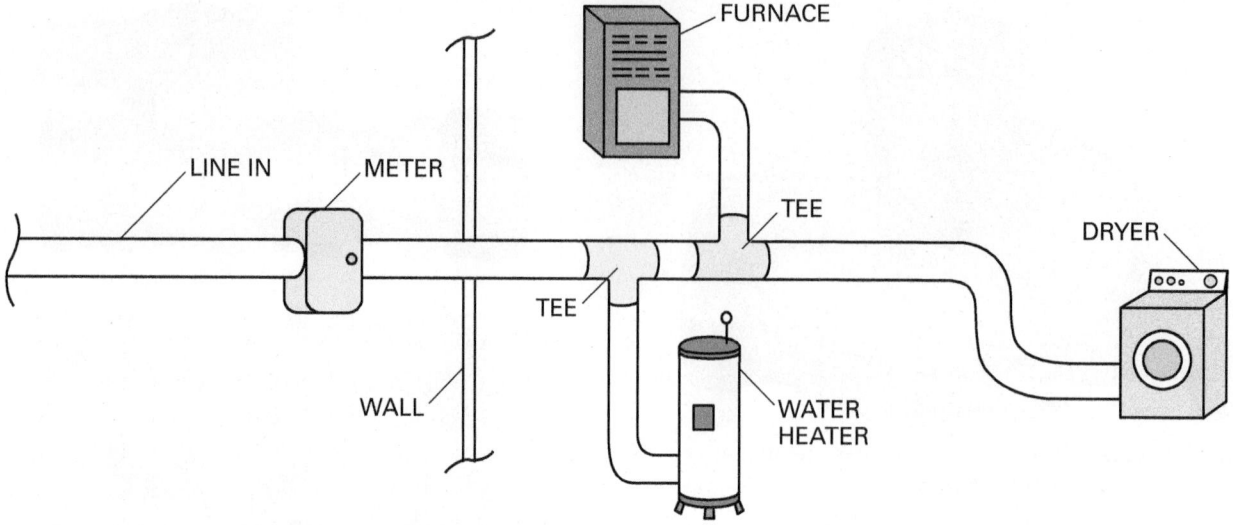

Figure 43 Simple CSST installation.

4.4.0 Types of CSST Installations

CSST can be run directly from the building meter to each appliance by using connectors. Use tees to connect branches to the main gas line (*Figure 43*). Multiple appliances, such as unit heaters, can be connected this way (*Figure 44*). Ensure all connectors and lines are the same size; otherwise the difference increases or decreases the gas pressure.

CSST can also be connected using manifolds (*Figure 45*). Manifolds connect a single gas line to multiple appliances without repeated cuts along the line for each branch. Install the manifold in a central location in the building, and then connect the CSST lines from the appliances to the manifold using approved connectors. Manifolds allow the piping for all appliances to be joined at a single, easily-accessible location (*Figure 46*). Striker plates (*Figure 47*), also called nail guards, are used to protect runs of CSST from accidental penetration by nails, screws, and other objects that could puncture or damage the pipe.

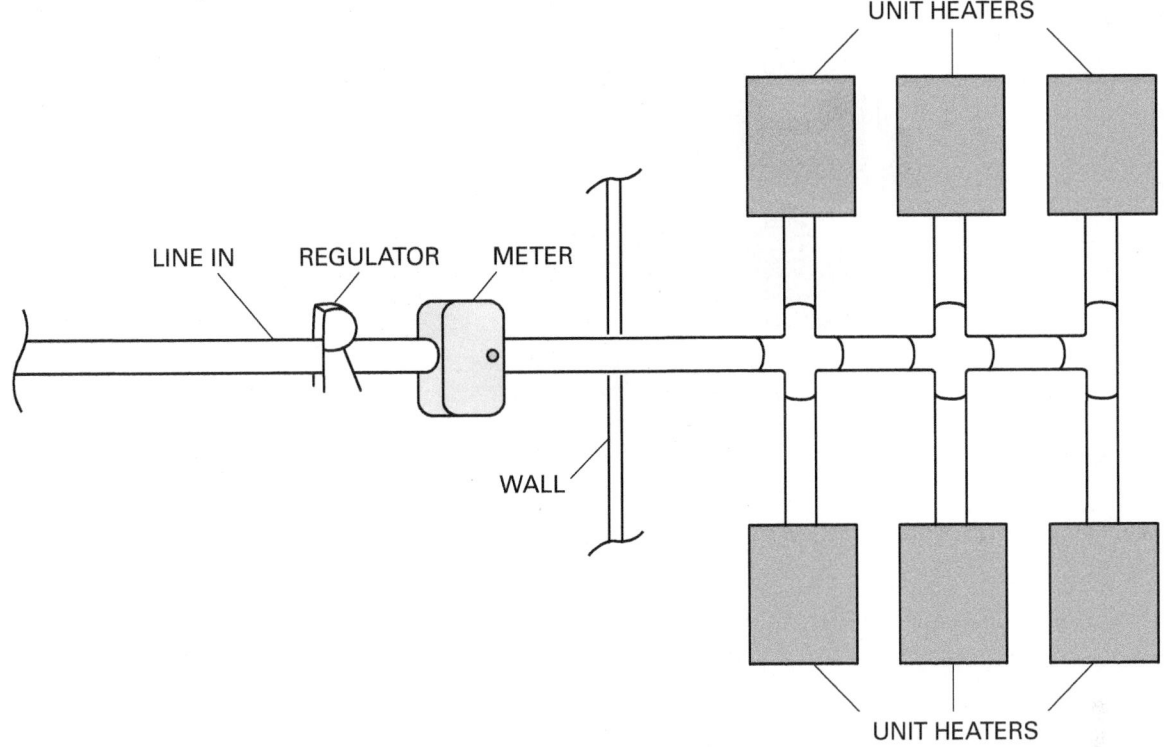

LOW-PRESSURE SYSTEM

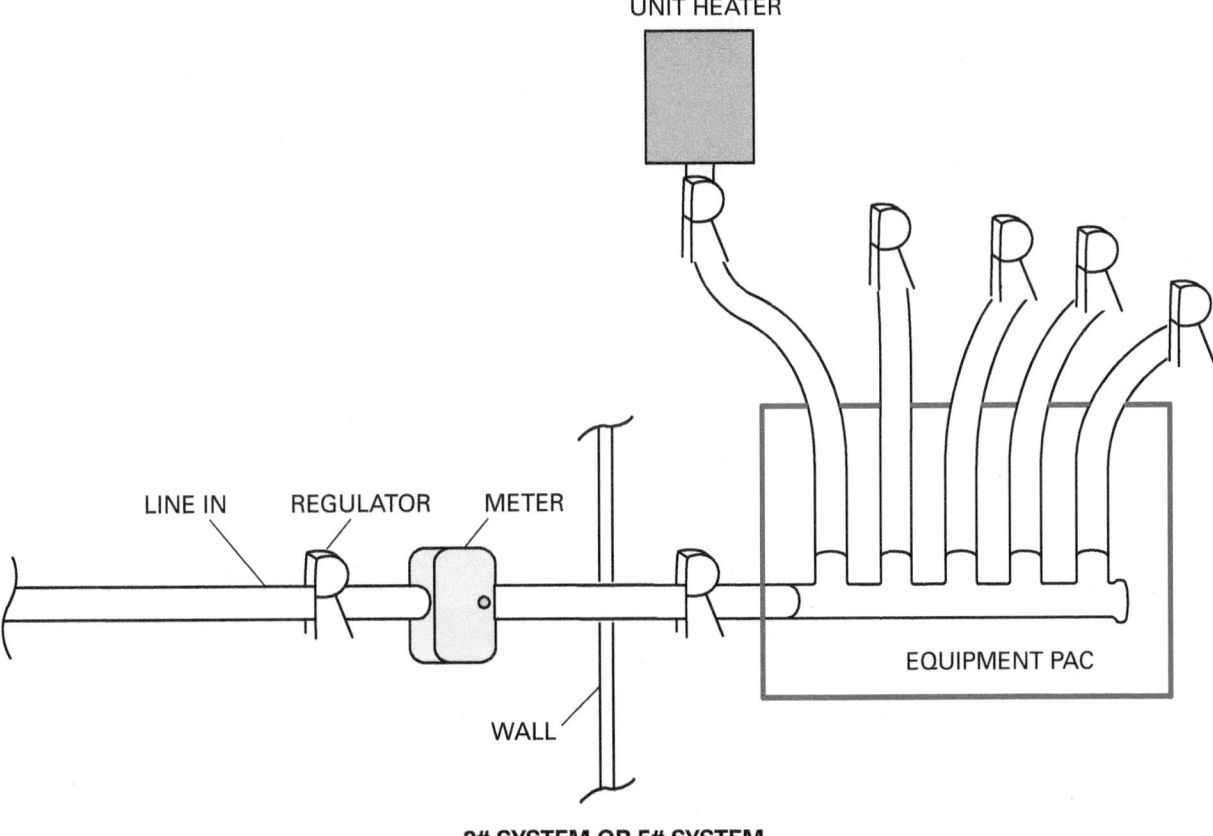

2# SYSTEM OR 5# SYSTEM

Figure 44 Typical parallel installation for multiple unit heaters.

Figure 45 Manifolds connect a single gas line to multiple appliances.

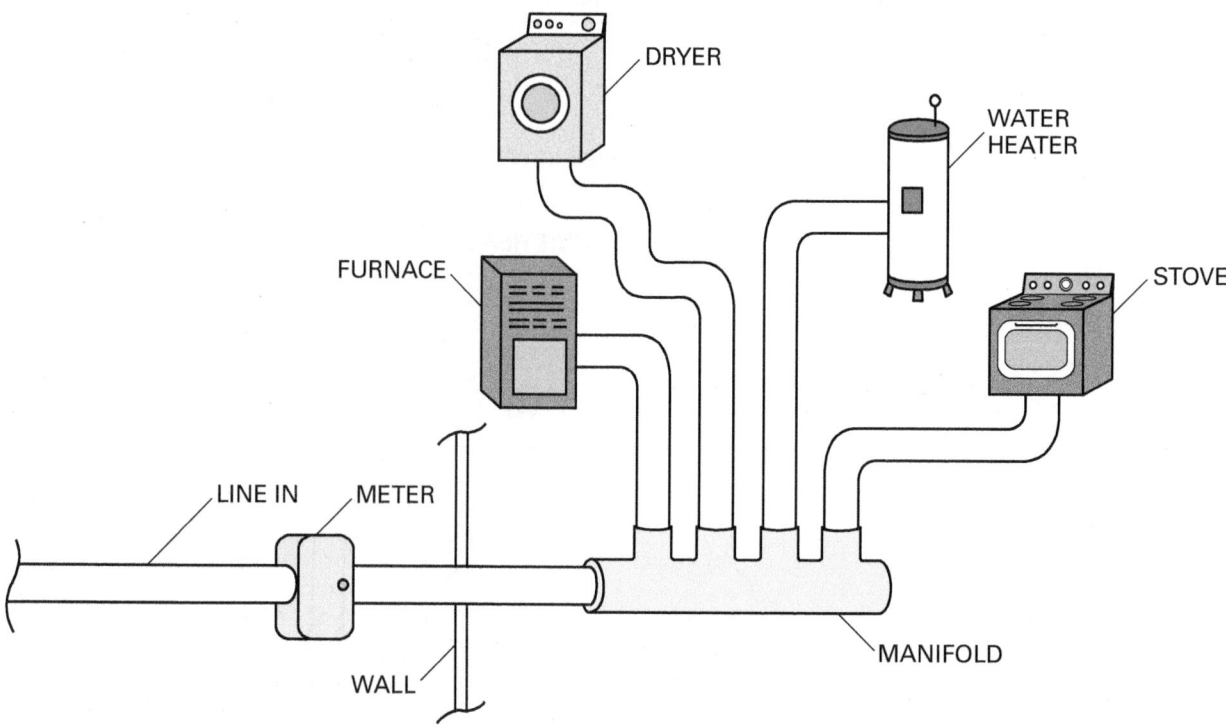

Figure 46 Typical installation with a manifold.

Figure 47 Striker plate.

Additional Resources

ANSI/AGA LC 1, Interior Fuel Gas Piping Systems Using Corrugated Stainless Steel Tubing (CSST). Current Edition. Washington DC: International Code Council (ICC).

NFPA 54, ANSI Z223.1 National Fuel Gas Code. Current Edition. Washington DC: National Fire Protection Association (NFPA).

Plumbing Installation and Design, Third Edition, 2006. L. V. Ripka. Orland Park, IL: American Technical Publishers.

4.0.0 Section Review

1. A special type of piping that is delivered in precut lengths coiled on drums is called _____.

   a. black steel
   b. PE
   c. copper
   d. CSST

2. At the time of publication, the largest pipe diameter available in CSST is _____.

   a. 3 inches
   b. 2 inches
   c. 1 inch
   d. $\frac{3}{8}$ inch

3. Which of the following is the main reason for using the water column unit when measuring gas pressures?

   a. It is useful for measuring low pressures such as those found in fuel-gas systems.
   b. Water does not present a combustion hazard when working with fuel gas.
   c. A water column provides a wider range of pressures compared to other methods.
   d. It is simply a tradition in the fuel-gas industry.

4. A convenient method of supplying multiple fuel-gas fixtures or appliances from a single CSST pipe involves using a _____.

   a. regulator
   b. manifold
   c. striker plate
   d. gate valve

SUMMARY

Plumbers work with threaded and grooved steel pipe. Pipe fittings for steel pipe are made of malleable or cast iron. Common valves for threaded pipe include globe, gate, ball, and stop and waste valves. Fittings and valves are also available for pipe with grooved ends. There are different methods for joining both types of steel pipe. To thread steel pipe, plumbers use hand and power threaders; to join threaded pipe, plumbers use pipe joint compound. Steel pipe is grooved by rolling or by cutting.

Plumbers install and support steel pipe so that the pipe and joints are leakproof and properly aligned. Pipe attachments, connectors, and structural attachments must be installed at sufficient intervals and to the appropriate surfaces. Because the installation of steel pipe is strictly regulated, plumbers always must follow applicable code requirements.

1. Which of the following statements is true about the metals used for plumbing pipes?
   a. Galvanized steel is more corrosion resistant than stainless steel.
   b. Galvanized steel is less corrosion-resistant that black steel.
   c. Manganese is often added to a carbon steel during smelting to improve strength.
   d. Stainless steel is similar to carbon steel but does not include chromium.

2. Proper storage and organization of plumbing materials at the work site significantly increases _____.
   a. the time before you can actually get to work
   b. your productivity
   c. the overall costs of the job
   d. the time needed to sort through all the parts

3. Thread engagement refers to _____.
   a. how much each thread overlaps the adjacent thread in a fitting
   b. how wide the fitting thread is compared to the pipe thread
   c. the length of pipe that screws into a fitting
   d. the ratio of imperfect threads to perfect threads on a threaded pipe

4. The inside diameter (ID) of 2-inch, nominal, Schedule 80 pipe compared to Schedule 40 of the same pipe size is _____.
   a. larger
   b. smaller
   c. the same
   d. varies by manufacturer

5. Bundles of steel pipe are labeled with tags that identify _____.
   a. the name or brand of the manufacturer
   b. the number of feet in the bundle
   c. storage and handling instructions
   d. the ASTM A53 standard

6. Fittings with recessed threads that are used only on soil and waste piping systems are called _____.
   a. drainage fittings
   b. plugs
   c. caps
   d. pressure fittings

7. Which of the listed fittings are used to close open-ended pipes?
   a. nipples
   b. plugs
   c. caps
   d. bushings

8. A globe valve is most effective for which of the following purposes in a plumbing system?
   a. Rapidly shutting off or starting flow
   b. Mainly acting as stop valves
   c. Providing precise control of flow
   d. Preventing back flow in a pipe

9. When measuring pipe, your pipe will be cut too short unless you take into account its _____.
   a. diameter
   b. weight of pipe
   c. type of pipe
   d. thread engagement

10. When measuring pipe before cutting and threading it, which method would be best when the pipe section has fittings threaded on both ends?
    a. End-to-end
    b. Center-to-center
    c. End-to-back
    d. End-to-center

11. Which of the following tools listed would be the best for turning a small-diameter (e.g., 1 inch), black-iron pipe while threading it into a fitting?
    a. Chain wrench
    b. Chain vise
    c. Pipe tong
    d. Yoke vise

12. Which of the following statements about cutting threads on a pipe is *not* true?
    a. Initially, take three or four short clockwise turns.
    b. Frequently wipe off chips with your bare hands to check for smoothness of the cut.
    c. Apply cutting oil after the die has firmly engaged the pipe to reduce friction.
    d. Finish cutting when one or two threads projects from the die.

13. The first step in joining a threaded carbon steel water pipe to a fitting is _____.
    a. grooving the end of the pipe
    b. reaming the fitting until it is smooth
    c. applying pipe-joint compound or tape
    d. hand-threading the pipe into the fitting before tightening

14. When joining grooved pipe, the plumber must _____.
    a. make sure that no part of the gasket extends into the grooves
    b. rough up the pipe ends to provide a good grip for the gasket
    c. tighten one of the bolts in housing segment to ensure a good alignment before inserting and tightening the other bolt
    d. make sure that both sides of the gasket cover the two groove equally

15. A hanger that touches or connects directly to the pipe itself is called a(n) _____.
    a. attachment
    b. support
    c. clamp
    d. channel

16. Securing a pipe directly to a concrete surface could involve _____.
    a. stirrup hangers
    b. universal pipe clamps attached to channels
    c. pipe clamps and powder-actuated anchors
    d. C-clamps

17. Among different manufacturers, the components of CSST systems are _____.
    a. interchangeable
    b. identical
    c. not interchangeable
    d. consistent

18. Which of the following statements about installing CSST pipe for fuel-gas service is true?
    a. Journey plumbers after completing first-year training are qualified to install CSST pipe and fittings.
    b. CSST piping can only be installed by gas utility workers.
    c. Any licensed plumber can install CSST piping.
    d. Plumbers certified by CSST manufactures or their certified trainers can install CSST pipe and fittings.

19. A regulator performs which of the following functions in a fuel-gas system?
    a. Raises supply pressure to a gas appliance
    b. Prevents backflow in the system
    c. Lowers supply pressure and maintains a uniform pressure
    d. Meters the flow of gas for billing purposes

20. What can protect CSST piping from penetration when it runs within hollow walls?
    a. striker plates
    b. metal conduits
    c. fiberglass sheathing
    d. manifolds

21. Which of the following factors would inform a decision to use an extra strong or double-extra strong pipe?
    a. The pipe will have to be longer than 21 feet.
    b. The pipe will have to be protected from corrosive materials.
    c. The pipe will have to withstand strong pressures
    d. The pipe only has to be threaded on one end.

22. Drainage fittings are designed so that horizontal lines connected to them will have _____.
    a. extra threads
    b. a slope
    c. flange unions
    d. thread protectors

23. When cutting a pipe, how much should the cutting wheel be turned with each turn of the pipe?

    a. 1 revolution per turn
    b. 2 revolutions per turn
    c. As slightly as it can possibly be turned.
    d. As much as it can possibly be turned.

24. Which of the following is a source for information about proper hanger spacing?

    a. The plumbing drawings
    b. The seismic mitigation plan
    c. The temperature and pressure chart
    d. The job specifications

25. Which of the following is used in CSST systems to shut off the flow of gas quickly?

    a. Regulators
    b. Striker plates
    c. Gas cocks
    d. Manifolds

Fill in the blank with the correct term that you learned from your study of this module.

1. When joining threaded pipe, use _____ or _____ to provide lubrication for assembly.

2. A _____ is best suited as a stop valve.

3. Where frequent operation and close control of flow are required, use a(n) _____.

4. The purpose of the _____ is to allow you to join two lengths of pipe using threaded flanges at their ends.

5. _____ and _____ are wrench-like tools with a length of chain that holds the pipe.

6. A(n) _____ connects two pipes by screwing the thread and shoulder pieces of the union onto the pipes.

7. You can use the same _____, or tools used to cut the threads, with all three weights of each size of pipe.

8. A _____ is most commonly used in faucets and for draining and freeze protection aboveground.

9. Hand threaders consist of two parts: the pipe die and the _____, the device that holds the die.

10. Pieces of pipe that are used to make extensions from a fitting or to join two fittings are called _____.

11. _____ allow you to connect the horizontal runs and vertical runs of pipe at a central point.

12. When threading a small-diameter pipe, use a(n) _____ to hold the pipe in place.

13. The length of pipe that gets threaded into the fitting is called the _____.

14. To avoid leaving jaw marks or scratches on plated or stainless pipe, use a(n) _____ to grip the pipe.

15. _____ is most often used as gas or air-pressure pipe.

16. Although more resistant to corrosion than black-steel pipe, _____ is still less resistant to corrosion than other piping materials.

17. To thread large quantities of pipe, use a(n) _____, which rotates, threads, cuts, and reams pipe.

18. _____ cannot be used in drainage systems because their inside surfaces are not smooth enough.

19. When threading pipe that is larger in diameter, use a(n) _____ to hold the pipe in place.

20. _____ have a drainage pattern design and recessed threads that should be used only on soil and waste piping systems.

21. Where quick shutoffs for line maintenance are required, use a(n) _____.

22. Steel pipe for threading is manufactured in three weights or strengths: _____, _____, and _____.

23. _____ and _____ are installed in CSST systems to quickly shut off gas flow.

24. Use _____ to split single CSST gas lines into multiple lines.

25. _____ protect hidden CSST from puncture damage.

26. The maximum pressure for gas systems is 5 _____.

27. Box- or C-shaped extrusions of steel used for attaching pipe hangers when securing pipes are _____.

28. _____ is widely used as gas piping because of its ease of installation.

29. _____ operate at pressures greater than standard pressures.

30. The standard pressure for gas systems is 7 inches of Water column (w.c.).

31. _____ are used to control gas pressure in fuel-gas piping systems.

32. Pipe _____ are pipe attachments, connectors, and structural attachments used for securing and supporting pipes.

Trade Terms

Ball valve
Black-steel pipe
Chain vise
Chain wrenches
Channels
Corrugated stainless steel tubing (CSST)
Crosses
Dies
Double-extra strong

Drainage fittings
Elevated pressure systems
Extra strong
Flange union
Galvanized pipe
Gas cocks
Gate valve
Globe valve
Ground-joint union
Hangers

Manifolds
Nipples
Pipe tongs
Pipe-joint compound
Pounds per square inch gauge (psig)
Power-threading machine
Pressure fittings
Regulators
Sealant tape

Standard weight
Stock
Stop-and-waste valve
Strap wrench
Striker plates
Thread engagement
Water column (w.c.)
Yoke vise

Ball valve: A valve used to control the flow of gases and liquids. It is installed in piping systems where quick shutoffs for operation or in-line maintenance may be necessary.

Black-steel pipe: Steel pipe with a black color used for gas and air-pressure pipe. It should not be used where corrosion affects its uncoated surface.

Chain vise: A tool that aids in cutting large pipe by supporting the pipe and clamping it in an attached vise with V-jaws and secured by a chain attached to a screw tensioner.

Chain wrenches: Similar in construction and function to pipe tongs, but smaller in size for use on smaller-diameter pipes.

Channels: Sections of C-, U-, or box-shaped extrusions of steel or aluminum used for supporting pipes and pipe hangers.

Corrugated stainless steel tubing (CSST): A gas piping material made from flexible steel tubing, usually but not always supplied with a polyethylene jacket.

Crosses: Fittings that connect four lengths of pipe in a cross-like shape in a distribution system.

Dies: Devices used to cut the threads on steel pipe.

Double-extra strong: An industry standard measurement of the weight or strength of steel pipe. Also referred to as Schedule 120 and double extra-heavy.

Drainage fittings: Fittings designed with smooth inside surfaces and shaped for easy flow to form an unbroken, internal contour for use only on drainage systems. Also called recessed fittings.

Elevated pressure systems: CSST systems that operate above the normal gas mains pressure of 7 inches w.c.

Extra strong: An industry standard measurement of the weight or strength of steel pipe. Also referred to as Schedule 80 and extra heavy.

Flange union: A drilled fitting that connects two pipes. A flange is screwed onto the end of each pipe that needs to be joined and then bolted together with nuts and bolts. A gasket in between flanges creates a leakproof seal.

Galvanized pipe: Pipe electroplated with a zinc alloy to provide a protective coating that resists corrosion or oxidation (rust)

Gas cocks: Valves designed mainly for gas service that provide positive, quick flow control on gas piping systems.

Gate valve: A valve in which the flow is controlled by moving a gate or disc that slides in machined grooves at right angles to the flow. This type of valve is designed to be fully open or fully shut.

Globe valve: A valve in which the flow is controlled by moving a beveled circular disc or seat washer against a circular or conical metal seat that surrounds the flow opening.

Ground-joint union: A fitting that joins two pieces of pipe by screwing the thread and shoulder pieces of the fitting onto the pipe. The collar piece is then tightened to join the sections of pipe into a watertight joint. The fitting permits later disassembly when required.

Hangers: For plumbing installations, the pipe attachments, connectors, and structural attachments that support and secure pipes.

Manifolds: Devices that allow the supplies to multiple gas devices to branch off of one CSST gas pipe.

Nipples: Short lengths of pipe with male threads at both ends, used to make extensions from a fitting or to join two fittings.

Pipe tongs: Wrench-like tools that use a length of chain wrapped around a pipe and linked into a cam tensioner to secure the tool to the pipe; used for threading and tightening large pipes.

Pipe-joint compound: A soft, semi-solid sealing material used when joining lengths of pipe. Also called pipe dope.

Pounds per square inch gauge (psig): A unit of pressure that measures system pressure above atmospheric (i.e., 0 psig is atmospheric pressure). Used to measure the operating pressures for gas and other plumbing systems.

Power-threading machine: An electrically-powered machine useful for cutting threads in large quantities of pipe. The machine rotates, threads, cuts, and reams pipe.

Pressure fittings: Any type of fitting used for steam, gas, or water supply piping under pressure. Not designed for use in drainage piping, which is typically at atmospheric or a low water-column pressure.

Regulators: Devices used in elevated-pressure systems to reduce and maintain a lower gas pressure at the pipe outlet.

Sealant tape: Tape used to wrap the threads of a pipe before joining to lubricate and ensure a watertight, secure fit. Also called PTFE or Teflon® tape.

Standard weight: An industry standard measurement of the weight or strength of steel pipe. Also referred to as Schedule 40.

Stock: The component of a hand or power threader that receives and securely holds the pipe-thread die cutter.

Stop-and-waste valve: Valve similar to a globe valve that is opened or closed by raising or lowering a disc oriented perpendicular to the flow through the valve seat using a threaded stem. An elastomeric, or rubberized, washer attached to the valve disc seals the valve seat, closing off water flow. This valve is most commonly used for water faucets.

Strap wrench: A wrench consisting of a heavy strap wrapped around a pipe and run through a buckle at the top end of the wrench handle, which creates a cam-lock tensioner. Used to thread and tighten chrome-plated or other types of finished pipe so that there are no jaw marks or scratches left on the pipe.

Striker plates: A heavy manufactured metal plate of a required thickness that is installed behind a wall to protect CSST from penetration by nails, screws, and drill bits.

Thread engagement: The length of pipe that gets threaded into the fitting when joining threaded pipe.

Water column (w.c.): A unit of pressure based on the pressure exerted by the height of water in a tubular column or manometer (e.g., 12 inches w.c. = 0.43 psig). Used for measuring low pressures, such as those in a standard gas supply.

Yoke vise: A pipe vise with a lower toothed V-jaw topped by a hinged attachment consisting of a jaw positioned by a threaded compression screw. When secured, the assembly creates a yoke that firmly holds a piece of pipe and prevents it from turning.

Additional Resources

This module is intended as a thorough resource for task training. The following reference works are suggested for further study.

ANSI/AGA LC 1, Interior Fuel Gas Piping Systems Using Corrugated Stainless Steel Tubing (CSST). Current Edition. Washington DC: International Code Council (ICC).

NFPA 54, ANSI Z223.1 National Fuel Gas Code. Current Edition. Washington DC: National Fire Protection Association (NFPA).

Plumbing Installation and Design, Third Edition, 2006. L. V. Ripka. Orland Park, IL: American Technical Publishers.

Section Review Answer Key

SECTION 1.0.0

Answer	Section Reference	Objective
1. c	1.1.1	1a
2. d	1.2.0	1b
3. a	1.3.1	1c
4. c	1.4.5	1d

SECTION 2.0.0

Answer	Section Reference	Objective
1. b	2.1.2	2a
2. c	2.2.0	2b
3. a	2.3.0	2c
4. a	2.4.1	2d

SECTION 3.0.0

Answer	Section Reference	Objective
1. c	3.1.1	3a
2. b	3.2.0	3b

SECTION 4.0.0

Answer	Section Reference	Objective
1. d	4.1.0	4a
2. b	4.2.0	4b
3. a	4.3.0	4c
4. b	4.4.0	4d

NCCER CURRICULA — USER UPDATE

NCCER makes every effort to keep its textbooks up-to-date and free of technical errors. We appreciate your help in this process. If you find an error, a typographical mistake, or an inaccuracy in NCCER's curricula, please fill out this form (or a photocopy), or complete the online form at **www.nccer.org/olf**. Be sure to include the exact module ID number, page number, a detailed description, and your recommended correction. Your input will be brought to the attention of the Authoring Team. Thank you for your assistance.

Instructors – If you have an idea for improving this textbook, or have found that additional materials were necessary to teach this module effectively, please let us know so that we may present your suggestions to the Authoring Team.

NCCER Product Development and Revision

13614 Progress Blvd., Alachua, FL 32615

Email: curriculum@nccer.org
Online: www.nccer.org/olf

❏ Trainee Guide   ❏ Lesson Plans   ❏ Exam   ❏ PowerPoints   Other _____

Craft / Level: _____   Copyright Date: _____

Module ID Number / Title: _____

Section Number(s): _____

Description: _____

_____

_____

_____

_____

Recommended Correction: _____

_____

_____

_____

_____

Your Name: _____

Address: _____

_____

Email: _____   Phone: _____

This page is intentionally left blank.

Introduction to Plumbing Fixtures

OVERVIEW

Plumbers use a variety of fixtures and faucets for plumbing installations. Understanding how fixtures and faucets operate enables plumbers to select the appropriate item for each installation. Plumbing codes determine the types of fixtures and faucets that are allowed in an area. All codes are based on the same principles of access, safety, and sanitation.

Module 02110

Trainees with successful module completions may be eligible for credentialing through the NCCER Registry. To learn more, go to *www.nccer.org* or contact us at 1.888.622.3720. Our website, *www.nccer.org*, has information on the latest product releases and training..

Your feedback is welcome. You may email your comments to *curriculum@nccer.org*, send general comments and inquiries to *info@nccer.org*, or fill in the User Update form at the back of this module.

This information is general in nature and intended for training purposes only. Actual performance of activities described in this manual requires compliance with all applicable operating, service, maintenance, and safety procedures under the direction of qualified personnel. References in this manual to patented or proprietary devices do not constitute a recommendation of their use.

02110 V4.5

Objectives

Successful completion of this module prepares trainees to:

1. Identify and describe the various plumbing fixtures.
   a. Identify and describe the various materials used in making plumbing fixtures.
   b. Identify and describe common bathroom fixtures.
   c. Explain the operating principles of water closets.
   d. Identify and describe common kitchen fixtures.Identify and describe other common plumbing fixtures.
2. Describe the different types of faucets used in plumbing systems.
   a. Describe compression and non-compression faucets.
   b. Describe kitchen and bathroom fixture faucets.
   c. Describe utility faucets.

Performance Task

Under the supervision of your instructor, you should be able to do the following:

1. Identify the most commonly installed fixtures and appliances.

Trade Terms

ADA compliant
American National Standards Institute (ANSI)
Anti-scald valve
Bathtubs
Bibb
Diverter
Faucets
Fiberglass

Flange
Flood-level rim
Flush valves
Flushometer valve
Laundry trays
Lavatory
Plumbing fixtures
Porcelain enamel
Seat washer

Sinks
Terrazzo
Valve disc
Valve seat
Valve stem
Vent pipe
Vitreous china
Waste pipe

Industry Recognized Credentials

If you are training through an NCCER-accredited sponsor, you may be eligible for credentials from NCCER's Registry. The ID number for this module is 02110. Note that this module may have been used in other NCCER curricula and may apply to other level completions. Contact NCCER's Registry at 888.622.3720 or go to *www.nccer.org* for more information.

CODE NOTE

Codes vary among jurisdictions. Because of the variations in code, consult the applicable code whenever regulations are in question. Referencing an incorrect set of codes can cause as much trouble as failing to reference codes altogether. Obtain, review, and familiarize yourself with your local adopted code. Safety codes are developed by the US Occupational Safety and Health Administration (OSHA).

Contents

Figures

Figures (continued)

1.0.0 PLUMBING FIXTURES

Objective

Identify and describe the various plumbing fixtures.

   a. Identify and describe the various materials used in making plumbing fixtures.

   b. Identify and describe common bathroom fixtures.

   c. Explain the operating principles of water closets.

   d. Identify and describe common kitchen fixtures.

   e. Identify and describe other common plumbing fixtures.

Performance Task

   1. Identify the most commonly installed fixtures and appliances.

Trade Terms

ADA compliant: Follows the accessibility guidelines of the Americans With Disabilities Act of 1990, as described in the Code of Federal Regulations (28 CFR, Part 36, Appendix A—ADA Accessibility Guidelines for Buildings and Facilities)

American National Standards Institute (ANSI): A nonprofit organization that is responsible for creating and implementing standards for many industries, including the construction trades.

Bathtubs: Low-profile plumbing fixtures designed to hold semi-reclined bathers. Bathtubs are identified by the location of the drain hole as one faces the fixture. If the drain is on the right end, it is a right-hand bathtub; if on the left end, it is a left-hand bathtub.

Faucets: Valve-like fixtures that are used to control the flow of water from pipes into sinks, bathtubs, and similar plumbing fixtures.

Fiberglass: Spun filaments of glass woven into yarn, roving (twisted) strands, and textile materials, such as cloth or mats. Fiberglass cloth saturated with a plastic (vinyl resin or epoxy), called glass-reinforced plastic or GRP, can be pressed into molds to make products such as bathtubs and shower enclosures.

Flange: A rim on one end of a length of pipe that provides a connection point to another length of pipe or to a plumbing fixture, such as a water closet or valve. Bolts or studs with nuts are used to hold two flanges together, with some type of gasket between them.

Flood-level rim: The edge of the basin, bowl, or fixture over which water flows when it is too full.

Flush valves: Devices located at the bottom of tanks for flushing water closets and similar fixtures.

Flushometer valve: A device that discharges a predetermined quantity of water to a fixture for flushing purposes and is automatically closed by direct water pressure or other mechanical means. Also referred to as simply a flushometer.

Laundry trays: Fixed tubs on legs or wall-mounted, usually installed in the laundry room or utility area of homes. They are used for washing clothes and for receiving wastewater from automatic clothes washers.

Lavatory: A basin designed for installation in bathrooms and other locations, primarily for washing the hands and face. Compare with sinks.

Plumbing fixtures: Receptacles or devices which are either permanently or temporarily connected to the water plumbing system of the premises. Also referred to as simply fixtures. They can connect to a supply of water therefrom or discharge used water, liquid-borne waste materials, or sewage either directly or indirectly to the drainage system of the premises. Fixtures may also require both a water supply connection and a discharge to the drainage system.

Porcelain enamel: A coating of vitrified porcelain (or china) that provides an attractive and protective finish to fixtures made of cast iron or steel.

Sinks: Shallow, flat-bottomed basins mainly used for the preparation of food, dishwashing, and for utility purposes. Compare with lavatory.

Terrazzo: A type of floor surface (commonly used for showers and other plumbing fixtures) made by pouring in place or precasting a mixture of small pieces of marble, other hard stone, or glass and a mortar base. When hardened, the surface can be ground and polished smooth, although it may be left slightly rough to create a nonslip surface.

Vent pipe: A pipe installed to provide a flow of air to or from a drainage system, or to provide a circulation of air within such a system to protect trap seals from siphonage and back pressure. The combination of all vent pipes in a plumbing system is referred to as the vent system.

Vitreous china: A durable ceramic material usually consisting of a fine clay mineral (kaolinite) and is often a mixture with quartz, feldspar, silica or other materials. The mixture is heated to temperatures between 2,200°F and 2,600°F (1,200°C and 1,400°C) to create a nonporous, glass-like finish. Also referred to as china, porcelain, and vitrified porcelain.

Waste pipe: A pipe in a plumbing system that carries only waste water and sewage

As a plumber, you need to be familiar with the different types of plumbing fixtures and faucets that are available for use in plumbing installations. A plumbing fixture is a receptacle or device that receives water from a water supply line and/or discharges water, waste, or sewage into a connected drainage system. Plumbing fixtures are also referred to simply as fixtures. A faucet is a fixture that is used to control the flow of water from a pipe.

Plumbing fixtures include sinks, bathtubs, toilets, domestic dishwashers, and drinking fountains, to name a few. Faucets include utility faucets, bathroom faucets, and kitchen faucets. This module discusses the materials commonly used to make fixtures, the most common types of fixtures, and the types of faucets available. In addition, this module explains how each type of fixture and faucet operates. Knowing this information will help you to choose the proper fixtures and faucets for each installation you perform.

If a plumber is installing fixtures and faucets in a seismically-active area (where earthquakes may occur) local codes specify seismic requirements. The purpose of these requirements is to ensure pipe is securely fastened to the structure in the event of excessive vibration. For example, codes may require that extra space is left around pipes where they meet walls and floors to allow for anticipated movement.

In addition to the basic tools that you normally carry, the following tools are often used when installing and servicing fixtures and faucets:

- Basin wrench
- Spud wrench
- Seat wrench
- Crescent wrench
- Basket strainer wrench
- Shower valve socket wrench
- Smooth-jawed crescent wrench

Plumbing codes determine the types of fixtures and faucets that are allowed in an area. Always follow all applicable codes when installing fixtures and faucets. While the details of plumbing construction may vary among different plumbing codes, all codes are based on the same basic principles of sanitation and safety. The principles of basic sanitation require the isolation of waste to prevent it from causing contamination and spreading disease or causing illness.

The following is a list of basic principles as stated in the National Standard Plumbing Code, and it is incorporated into most other codes:

- *Principle 1: All occupied premises shall have potable water*—All premises intended for human habitation, occupancy, or use shall be provided with a supply of potable water. Such a water supply shall not be connected with unsafe water sources, nor shall it be subject to the hazards of backflow.
- *Principle 2: Adequate water required*—Plumbing fixtures, devices, and appurtenances shall be supplied with water in sufficient volume and at pressures adequate to enable them to function properly and without undue noise under normal conditions of use.
- *Principle 3: Hot water required*—Hot water shall be supplied to all plumbing fixtures that normally need or require hot water for their proper use and function.
- *Principle 4: Water conservation*—Plumbing shall be designed and adjusted to use the minimum quantity of water consistent with proper performance and cleaning.
- *Principle 5: Safety devices*—Devices for heating and storing water shall be so designed and installed as to guard against dangers from explosion or overheating.
- *Principle 6: Use public sewer where available*—Every building with installed plumbing fixtures and intended for human habitation, occupancy, or use, and located on premises where a public sewer is on or passes said premises within a reasonable distance, shall be connected to the public sewer.
- *Principle 7: Required plumbing fixtures*—Each family dwelling unit shall have at least one water closet, one lavatory, one kitchen-type sink, and one bathtub or shower to meet the basic requirements of sanitation and personal hygiene. All other structures for human habitation shall be equipped with sufficient sanitary facilities. Plumbing fixtures shall be made of durable, smooth, non-absorbent and corrosion-resistant material and shall be free from concealed fouling surfaces.
- *Principle 8: Drainage system*—The drainage system shall be designed, constructed, and maintained to guard against fouling, deposit of

solids, and clogging, and with adequate cleanouts arranged so that the pipes may be readily cleaned.

- *Principle 9: Durable materials and good workmanship*—The piping of the plumbing system shall be of durable material, free from defective workmanship, and so designed and constructed as to give satisfactory service for its reasonable expected life.
- *Principle 10: Fixture traps*—Each fixture directly connected to the drainage system shall be equipped with a liquid-seal trap.
- *Principle 11: Trap seals shall be protected*—The drainage system shall be designed to provide an adequate circulation of air in all pipes with no danger of siphonage, aspiration, or forcing of trap seals under condition of ordinary use.
- *Principle 12: Exhaust foul air to outside*—Each vent terminal shall extend to the outer air and be so installed as to minimize the possibilities of clogging and the return of foul air to the building.
- *Principle 13: Test the plumbing system*—The plumbing system shall be subjected to such tests as will effectively disclose all leaks and defects in the work or the material.
- *Principle 14: Exclude certain substances from the plumbing system*—No substance that will clog or accentuate clogging of pipes, produce explosive mixtures, destroy the pipes or their joints, or interfere unduly with the sewage-disposal process shall be allowed to enter the building drainage system.
- *Principle 15: Prevent contamination*—Proper protection shall be provided to prevent contamination of food, water, sterile goods, and similar materials by backflow of sewage. When necessary, the fixture, device, or appliance shall be connected indirectly with the building drainage system.
- *Principle 16: Light and ventilation*—No water closet, urinal, or bidet shall be located in a room or compartment that is not properly lighted and ventilated.
- *Principle 17: Individual sewage disposal systems*—If water closets or other plumbing fixtures are installed in buildings where there is no public sewer within a reasonable distance, suitable provision shall be made for disposing of the sewage by some accepted method of sewage treatment and disposal.
- *Principle 18: Prevent sewer flooding*—Where a plumbing drainage system is subject to backflow of sewage from the public sewer or private disposal system, suitable provision shall be made to prevent its overflow in the building.

- *Principle 19: Proper maintenance*—Plumbing systems shall be maintained in a safe and serviceable condition from the standpoint of both mechanics and health.
- *Principle 20: Fixtures shall be accessible*—All plumbing fixtures shall be so installed with regard to spacing as to be accessible for their intended use and for cleaning.
- *Principle 21: Structural safety*—Plumbing shall be installed with due regard to preservation of the strength of structural members and prevention of damage to walls and other surfaces through fixture usage.
- *Principle 22: Protect ground and surface water*—Sewage and other waste shall not be discharged into surface or sub-surface water unless it has first been subjected to some acceptable form of treatment.

1.1.0 Materials to Make Fixtures

Modern plumbing fixtures are made from a variety of durable, corrosion-resistant, nonabsorbent materials that have smooth, easy-to-clean surfaces. To make plumbing fixtures, manufacturers use different materials including china (also called *vitrified porcelain*), steel coated with porcelain enamel, cast iron coated with porcelain enamel, stainless steel, acrylic plastic, and fiberglass reinforced with plastic (also called glass-reinforced plastic or GRP). Manufacturers may produce fixtures from a single material, such as kitchen sinks made of stainless steel, or a combination of materials, such as shower stalls made of fiberglass-reinforced plastic.

1.1.1 Vitreous China

A common material used in the production of toilets, urinals, bidets, and often for lavatories is vitrified clay which has been cast in molds and then fired with a glazed finish. These fixtures are often referred to by a number of common names, such as china and porcelain. In this text, the material is referred to as vitreous china. Vitreous china is made from a mixture of fine clay (called kaolin) and can include quartz, feldspar, silica, and/or other materials. The materials are fused into a glass-like or vitrified state in a kiln. The complete process of fusion, or vitrification, requires a temperature between 2,200°F and 2,600°F (1,200°C and 1,400°C). The process produces a sanitary surface that is strong, nonporous, and easy to clean.

1.1.2 Porcelain Enamel

Porcelain is also used as an enamel coating for easily corrodible materials, such as steel or cast iron. Manufacturers use a high-temperature bonding process that fuses the two materials together. This process creates a more durable and functional material for making fixtures. The resulting coating is called porcelain enamel, also referred to as *glass lining* or *vitreous enamel*.

1.1.3 Cast Iron

Some plumbing fixtures, such as bathtubs and slop sinks, are made from a type of cast iron called gray iron, a metal that is ideally suited for plumbing fixtures because of its ability to be molded into a variety of shapes. Cast iron is an iron alloy made from molten iron that is formed using a mold. Cast iron can be coated with an enamel powder and then heated at extremely high temperatures. The enamel powder contains pigments that allow manufacturers to produce fixtures of different colors.

1.1.4 Sheet Steel

Manufacturers also produce fixtures from sheet steel. Using this low-cost process, manufacturers form the fixtures by stamping the sheet steel into the desired shapes. A porcelain enamel finish is fused to the surfaces of these plumbing fixtures, which is glossy, decorative, protective, and sanitary. While these fixtures are usually less expensive than those made of cast iron, they are also less durable, because they dent and chip easily.

1.1.5 Stainless Steel

Stainless steel is another metal commonly used to make plumbing fixtures, particularly kitchen sinks. As a corrosion-resistant steel alloy, stainless steel is an excellent material for plumbing fixtures. Although it is difficult to form, stainless steel has several advantages over enameled surfaces. It does not require an additional coating to produce a sanitary surface and it has a durable finish that does not chip.

1.1.6 Plastics

Plastic is often used for plumbing fixtures because of its versatility, light weight, and relatively low cost, although the finish is not as durable as enameled cast iron. The plastics most commonly used today are acrylic plastic and fiberglass-reinforced plastic (GRP). GRP is used to make one-piece showers and bathtubs, which include both the fixture and the adjoining walls. The strength of the product comes from the glass fibers. Fiberglass tubs and shower stalls are lightweight and easy to install. A plastic gel coating applied during manufacturing gives these fixtures a shiny, enamel-like surface. Manufacturers can now use marbled, colored acrylic plastic sheets to make decorative fixtures that appear to be made of marble. The fixtures can be just as attractive as marble, but they do not absorb water as marble does.

1.2.0 Basic Types of Bathroom Fixtures

Bathroom fixtures are available in a variety of styles. Fixtures made for commercial installations are designed to withstand heavier use than fixtures for residential installations. For example, a sink in a restaurant needs to be more durable than a sink in a home. Some commercial fixtures are even designed to be vandal-proof.

> **NOTE**
>
> Always consult submittal data to obtain the correct fixture. Shipping containers for fixtures are usually not labeled with the name of the item (for example, "corner sink"). More commonly, they only feature the product's catalog number assigned by the manufacturer.

For a public building to meet the standards of the Americans with Disabilities Act (ADA) of 1990, it must be ADA compliant. This means that it must adhere to ADA requirements regarding the location and spacing of plumbing fixtures. The ADA established explicit requirements to improve the accessibility of facilities to people with disabilities. The ADA requirements are based on space allowances and reach ranges for people with various disabilities. The technical requirements in the *ADA Standards for Accessible Design* apply to all public accommodations, such

> **Did You Know?**
>
> Many of the most popular bathtubs, shower stalls, and hot tubs that manufacturers produce today appear to be made of natural materials but are actually made from acrylic, a type of plastic. One example of such a material is DuPont Corian®, an acrylic-based, premium, solid surface material that resists damage, is easy to maintain. They can be shaped into countertops, work surfaces, sinks and bowls, shower and tub surrounds, and wainscoting and windowsills.

as restaurants, movie theaters, stores, and medical facilities. The requirements affect everything from sinks and toilets to water coolers and grab bars. For example, ADA-compliant sinks must be mounted with the counter or rim no higher than 34″ above the finish floor. As a plumber, you must have the appropriate information in order to rough-in the dimensions to meet the ADA requirements.

The American National Standards Institute (ANSI) also sets forth similar guidelines for accessible design. The ANSI guidelines are outlined in ANSI document *A117.1 (Current Edition) Accessible and Usable Buildings and Facilities.*

1.2.1 Sinks and Lavatories

A sink is a shallow, flat-bottomed basin used for food preparation and dishwashing. Kitchen sinks come in single, double, and triple compartments and are often installed in a countertop (*Figure 1*). They are manufactured from a variety of materials and are available in different shapes, sizes, colors, and styles.

People often confuse the terms sink and lavatory. A lavatory is a basin designed for the primary purpose of washing hands and face. Lavatories are available in a variety of colors, sizes, and shapes. Installing lavatories requires a small-diameter vent pipe and a waste pipe. The vent pipe allows a flow of air to escape from the drain, waste, and vent (DWV) system and provides a circulation of air within a DWV system.

Figure 1 Double-compartment sink.

Vent pipes protect trap seals from siphonage and back pressure. The waste pipe, which connects the trap to the vent pipe, conveys only wastewater to the building drain.

Sinks and lavatories are classified by their method of installation. The five major categories of sinks and lavatories are wall hung, self-rimming, built-in rim, undercounter, and pedestal (*Figure 2*). Additionally, *Figure 3* shows two examples of the roughing-in measurements that you typically find in fixture catalogs. Plumbers need roughing-in measurements to ensure that you install the correct fixture at the correct location and height.

Moisture Control Systems

Manufacturers are introducing systems that control water use in landscaping and irrigation systems more efficiently. For example, some systems monitor weather conditions to adjust the watering schedule or change the amount of water needed to offset evaporation and transpiration (water loss in plants). Other systems monitor the dryness of the surrounding soil. Many of these systems can be fitted to existing systems through the addition of a controller, which functions similarly to the way a thermostat adjusts heat and cooling in a house. Larger, more sophisticated systems use networked computers to monitor conditions in many different locations.

WALL HUNG

PHOTOGRAPH COURTESY OF KOHLER

SELF RIMMING

BUILT-IN RIM

PHOTOGRAPH COURTESY OF KOHLER

UNDERCOUNTER

PEDESTAL

Figure 2 Five major categories of lavatories and sinks.

A wall-hung lavatory is fastened to the wall by a support bracket, also called a lavatory carrier (*Figure 4*), or a concealed-arm carrier. The manufacturer supplies the bracket, which is usually made from stamped steel or cast iron. You typically install this type of lavatory with the **flood-level rim** 31" above the floor, but no more than 34" according to the ADA guidelines. The flood-level rim is the top edge of the basin, bowl, or fixture. If water overflows the rim, it spills over onto the countertop and floor. In addition, wall-hung lavatories require a knee clearance of at least 27" from the floor to the underside of the lavatory and extending at least 8" from the wall to the front edge of the lavatory (*Figure 5*).

Several different styles of wall-hung lavatories are available. They include corner, raised back, ledge or shelf back, and wheelchair accessible (*Figure 6*).

To comply with the ADA, wheelchair-accessible lavatories must be installed with the flood-level rim or counter surface no higher than 34" above the floor. This makes the lavatory accessible to people in wheelchairs (refer to *Figure 5*) and permits the wheelchair to slide easily under the lavatory. The drain fittings should be offset or moved back toward the wall rather than dropping directly down from the fixture. This placement of the fittings allows room for the user's legs to fit under the lavatory.

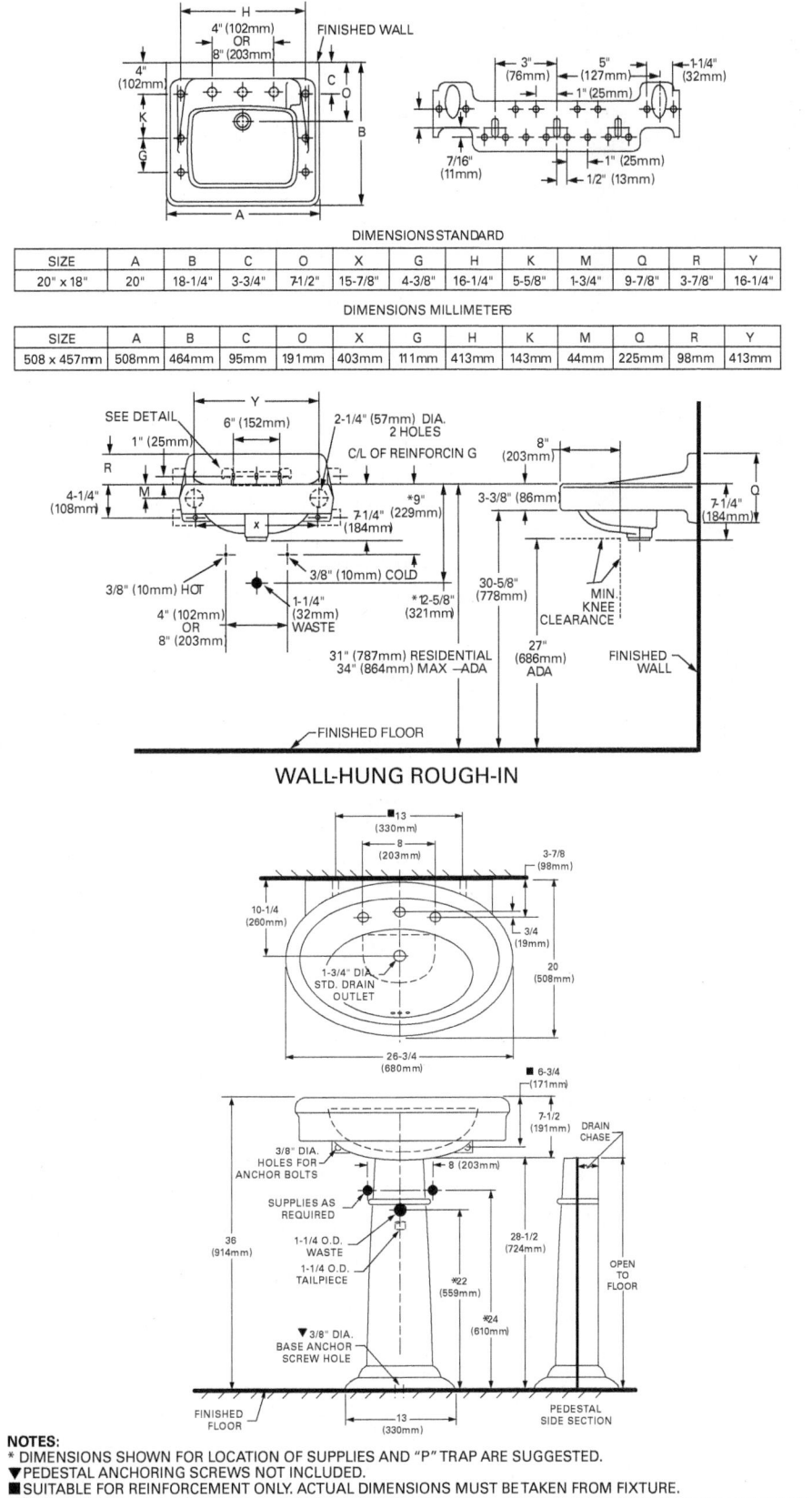

DIMENSIONS STANDARD

SIZE	A	B	C	O	X	G	H	K	M	Q	R	Y
20" x 18"	20"	18-1/4"	3-3/4"	7-1/2"	15-7/8"	4-3/8"	16-1/4"	5-5/8"	1-3/4"	9-7/8"	3-7/8"	16-1/4"

DIMENSIONS MILLIMETERS

SIZE	A	B	C	O	X	G	H	K	M	Q	R	Y
508 x 457mm	508mm	464mm	95mm	191mm	403mm	111mm	413mm	143mm	44mm	225mm	98mm	413mm

WALL-HUNG ROUGH-IN

PEDESTAL LAVATORY ROUGH-IN

NOTES:
* DIMENSIONS SHOWN FOR LOCATION OF SUPPLIES AND "P" TRAP ARE SUGGESTED.
▼ PEDESTAL ANCHORING SCREWS NOT INCLUDED.
■ SUITABLE FOR REINFORCEMENT ONLY. ACTUAL DIMENSIONS MUST BE TAKEN FROM FIXTURE.
Fittings not included and must be ordered separately. Provide suitable reinforcement for all wall supports. Installation instructions supplied with lavatory.

IMPORTANT:
Dimensions of fixtures are nominal and may vary within the range of tolerances established by ANSI Standard A112.19.9M. These measurements are subject to change or cancellation. No responsibility is assumed for use of superseded or voided pages.

Figure 3 Two methods of sink rough-in.

Figure 4 Concealed-arm carrier for a wall-hung lavatory.

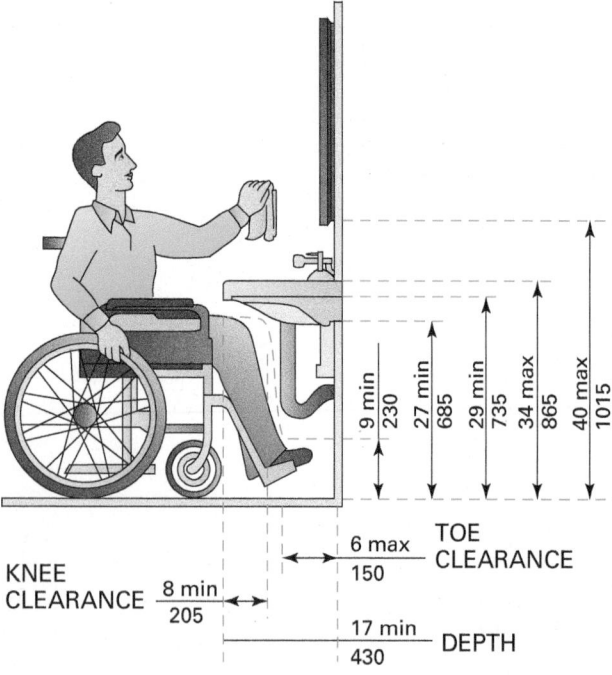

KNEE CLEARANCE

8 min 205

6 max 150

TOE CLEARANCE

17 min 430 — DEPTH

9 min 230

27 min 685

29 min 735

34 max 865

40 max 1015

Figure 5 ADA compliant wall-mounted lavatory height.

Wheelchair-accessible lavatories are outfitted with special faucets that have long blade handles that extend toward the user and do not require grasping or twisting to turn on the water (*Figure 7*). This feature is particularly necessary for users who have restricted mobility or who suffer from arthritis. As well as space for a wheelchair turnaround, ANSI also has a minimum ambulatory space requirement of 36" deep and 60" wide (*Figure 8*).

The self-rimming lavatory or sink is placed directly on the countertop. This fixture is manufactured with the mounting rim as part of the overall structure. It requires no external mounting frame to secure it to the countertop. Kitchen and bar sinks are often self-rimming designs. The built-in rim lavatory has a metal frame that is set in a stainless-steel rim. This rim supports the bowl in the countertop.

The undercounter lavatory is attached to the underside of a countertop. It has no rim. It requires brackets placed underneath the counter to support it. The vanity-top lavatory is an undercounter lavatory that has a one-piece washbasin and countertop set on top of a vanity cabinet (*Figure 9*). Vanity tops may be manufactured from marbleized acrylic plastic or artificial stone materials and are available in a variety of colors and sizes.

Pedestal lavatories are decorative lavatories that sit on pedestals rather than on countertops or vanities (*Figure 10*). They may be mounted to the floor or to the wall. They must be firmly mounted according to the manufacturer's specifications.

1.2.2 Bathtubs

Bathtubs are plumbing fixtures that are shaped to fit a semi-reclining human body and are used for bathing. A bathtub is described as either right-hand bathtub or left-hand bathtub, depending on where the drain is. A right-hand tub has the drain on the right end of the tub as you face the length of its finished side (*Figure 11*). A left-hand tub has the drain on the left end of the tub.

CORNER

RAISED BACK

WHEELCHAIR ACCESSIBLE

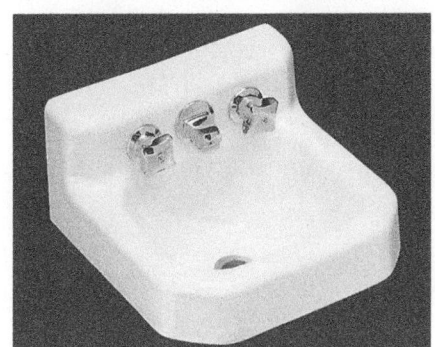

LEDGE OR SHELF BACK

Figure 6 Wall-hung lavatories.

PHOTOGRAPH COURTESY OF KOHLER

Figure 7 Faucet for wheel-chair accessible lavatory.

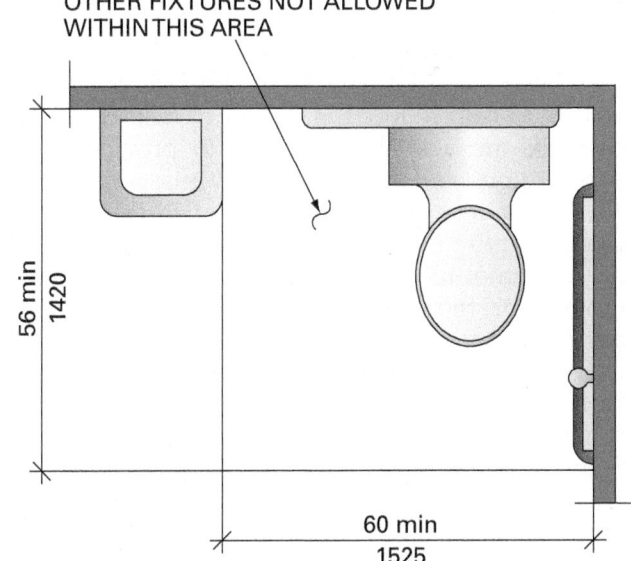

OTHER FIXTURES NOT ALLOWED WITHIN THIS AREA

56 min
1420

60 min
1525

Figure 8 Floor plan of an ADA compliant bathroom.

Figure 9 Vanity-top lavatory.

Figure 10 Pedestal lavatory.

Figure 11 Right-hand bathtub.

To provide proper sanitation and to reduce maintenance, the exterior surface of the tub must be durable enough to withstand frequent cleanings. Bathtubs are available in several sizes and shapes, but the standard bathtub is usually 5″ long. Some tubs have a special nonslip or slip-resistant safety feature to help reduce the danger of falls. Specialty bathtubs are available with jets to provide high-pressure outlets for water therapy massage (*Figure 12*). These tubs are commonly called whirlpools. Codes require the installation of removable panels on whirlpool tubs to provide access to the piping, motors for the high-pressure jets, and other required components illustrated in *Figure 13*.

Bathtubs are available in freestanding, built-in, or bath-shower module styles. The freestanding bathtub (*Figure 14A*) rests on feet that raise it off the floor. Some freestanding tubs sit directly on the floor but are not attached to adjacent walls or enclosures. The built-in bathtub (*Figure 14B*) is built permanently into the walls and floor of the bathroom. Some built-ins include an apron on one or more sides. Others are drop-in units as shown in. Bath-shower modules (*Figure 14C*) are usually manufactured as one piece, although some models are available with the walls and tub as separate units. Because of their large size, bath-shower modules are generally installed only in new, larger homes.

Figure 12 Whirlpool bathtub.

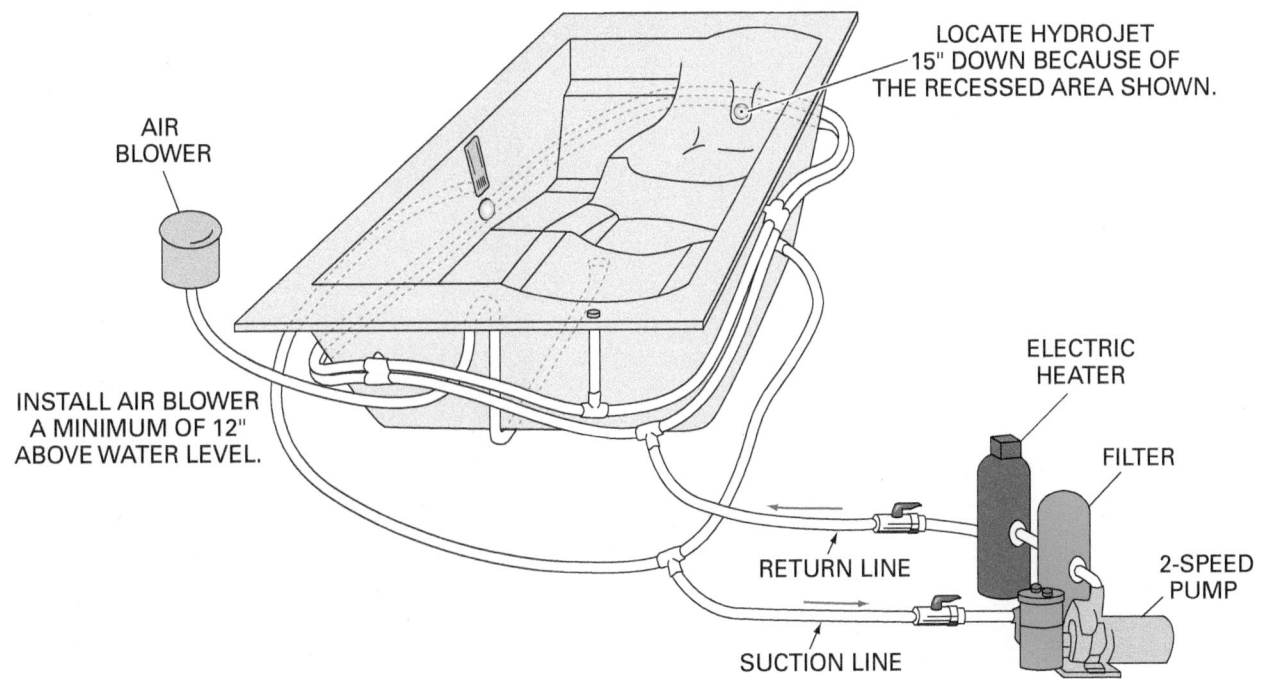

LOCATE HYDROJET 15" DOWN BECAUSE OF THE RECESSED AREA SHOWN.

AIR BLOWER

INSTALL AIR BLOWER A MINIMUM OF 12" ABOVE WATER LEVEL.

ELECTRIC HEATER

FILTER

RETURN LINE

2-SPEED PUMP

SUCTION LINE

Figure 13 Whirlpool bathtub components.

FREESTANDING

Figure 14A Freestanding bathtub.

BUILT-IN

Figure 14B Built-in bathtub.

BATH-SHOWER MODULE

Figure 14C Bath-shower module bathtub.

Figure 15 Premade built-in shower.

Bath-shower modules are manufactured from GRP. The walls are usually high enough to eliminate the need for conventional wall material such as ceramic tile. Because these modules have no joints, cleaning the tub after installation is easy.

1.2.3 Shower Stalls

Shower stalls, also called shower baths, are stalls or baths with faucets that spray water from above onto the user's body. They are available in many sizes, shapes, and colors. Residential shower baths usually consist of one showerhead installed in a small, enclosed space. Shower stalls with seats are also available. Showers in industrial buildings, schools, gymnasiums, and similar facilities usually have a series of showerheads installed in a large shower room.

The walls of shower stalls consist of waterproof materials such as gel-coated GRP, enameled steel, or glazed tile. They can be either freestanding or built-in stalls. Built-in showers can consist of a premade shower enclosure or have walls constructed of glazed ceramic tile (*Figure 15*). In either case, plumbing codes require a safety pan under the installation. You should always insulate the shower stall. You can use nonexpanding foam when installing a fiberglass shower. Simply inject the foam between the wall and the floor and the outside of the shower enclosure. Check applicable plumbing code requirements regarding size, installation of shower stalls and floors, and materials allowed in your area.

Shower enclosures are available as one-piece modules or with the walls and base as separate units. You must assemble shower enclosures that consist of separate units on the job site. The dimensions of the shower base are usually 36" by 36" or 36" by 48", with a slope of $\frac{1}{4}$" per foot toward a center drain. Bases for shower stalls are usually made of fiberglass, cast stone, enameled steel, or terrazzo.

1.3.0 Water Closets and Urinals

Water closets, commonly referred to as toilets or WCs, are water-flushed plumbing fixtures designed to carry away solid human waste. They are usually manufactured from vitreous china and are available in both round-front and elongated shapes. Elongated water closets are used in public facilities. Both round-front and elongated shapes are used in residences.

The ADA sets forth specific requirements for accessible water closets, including the height of the fixture and the size and location of grab bars, flush controls, and toilet paper dispensers. If you are installing a water closet for individuals with disabilities, you must adhere to these requirements. An accessible water closet must be between 17″ and 19″ high, measured to the top of the toilet seat. The grab bar behind the water closet must be a minimum of 36″ long. Controls for flush valves and flushometers must be mounted on the wide side of toilet areas, no more than 44″ above the floor. Flush valves are used to flush water closets and similar fixtures.

Water closets are available in two styles: floor mounted and wall hung (*Figure 16*). A floor-mounted water closet is secured directly to the floor. Its outlet is mated to the drainage system pipe provided with a closet flange. Wall-hung water closets are suspended from the wall by supporting chair carriers (*Figure 17*).

New toilets must comply with the *Energy Policy Act of 1992*, as amended. Compliant toilet models must use less than 1.6 gallons of water per flush. High-efficiency toilets use 1.28 gallons or less per flush.

Figure 17 Carrier for wall-hung water closet.

1.3.1 Basic Operating Principles

The water closet is designed to carry away solid and liquid human waste. When you flush a toilet, the water moves through passages in the bowl under pressure or gravity. The flow of the water cleans the bowl. The water then moves the solid waste through the trap and into the drain. A siphoning action removes the solid waste from the bowl. This siphoning action occurs when the water moves through the downward leg of the trap.

The flowing mass of water, along with the head of water from the bowl, sweeps out the wastes.

The siphoning action continues to draw water out of the bowl even after it slows to a trickle. As the water flow through the trap lessens, air is allowed to rush into the trap. This causes a break in the siphon water column and equalizes the pressure on both sides of the trap. The flush valve delivers enough additional water to reseal the trap.

FLOOR MOUNTED

WHEELCHAIR ACCESSIBLE

PHOTOGRAPH COURTESY OF KOHLER

WALL HUNG

Figure 16 Water closets.

The flush valve controls the flow of water from the water closet flush tank into the bowl (*Figure 18*). When you depress the tank lever, the valve lifts above the tank outlet and floats there. This allows the water in the tank to flow rapidly into the bowl. When the water level in the tank drops to the point where the flush valve no longer floats, the flush valve reseats on the tank drain and the tank refills.

1.3.2 Older-Style Water Closets

Generally, you are not permitted to sell or install water closets that do not conform to the *Energy Policy Act of 1992*. However, you may still find older styles of water closets in commercial and residential buildings, so you need to be familiar with them. You may even come across such old styles as the wall-mounted chain-pull tank and the corner tank. Older-style toilets that you are most likely to encounter are washdown water closets and siphon water closets.

The washdown water closet is inexpensive and simple in construction (*Figure 19*). It is the least efficient and the noisiest of all water closets. The waste passageway is located at the front of the bowl, which causes the front portion of the closet to protrude over the passageway. It is more susceptible to staining and contamination because the flat surface at the front of the bowl area is above the water level.

Siphon water closets use a jet of water to increase the speed of the siphon action and have a downstream leg that is longer than the upstream leg. This design enhances the siphon effect, which helps pull the waste from the bowl.

1.3.3 Newer-Design Water Closets

Although the reverse-trap water closet looks and flushes like the siphon-action water closet, its water surface area, passageway, and water seal are smaller than those of the other siphon water closet models (*Figure 20*). The waste passageway is 2" in diameter, the water surface area is 10¼" × 10", and the water seal is 2½" deep. These features allow the reverse-trap water closet to operate on less water than the other models.

The siphon-jet water closet has a larger passageway and greater water surface area that reduce the tendency for waste articles to become stuck and clog the passageway (*Figure 21*). It has a waste passageway that measures either 2¼" or 2⅝" in diameter depending on design specifications, a water area of 12" × 10½", and a 3" deep water seal. This type of water closet does not require a head of water in the bowl to flush out solids. Instead, the stream of water that the closet delivers spins to the outlet of the trap.

Figure 18 Flush valve.

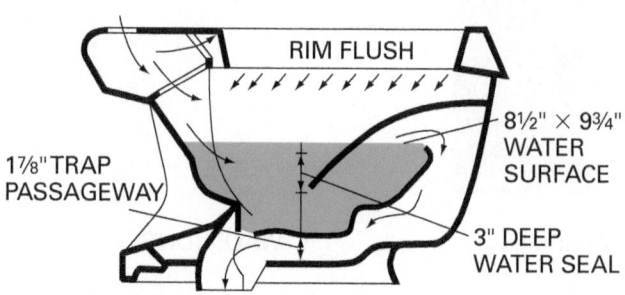

Figure 19 Washdown water closet.

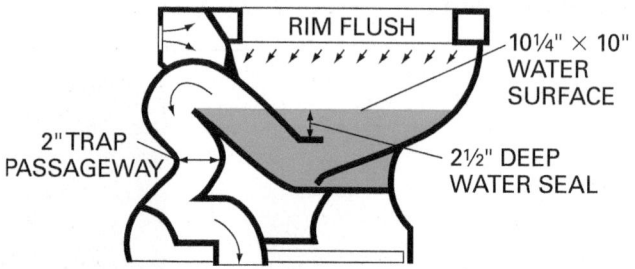

Figure 20 Reverse-trap water closet.

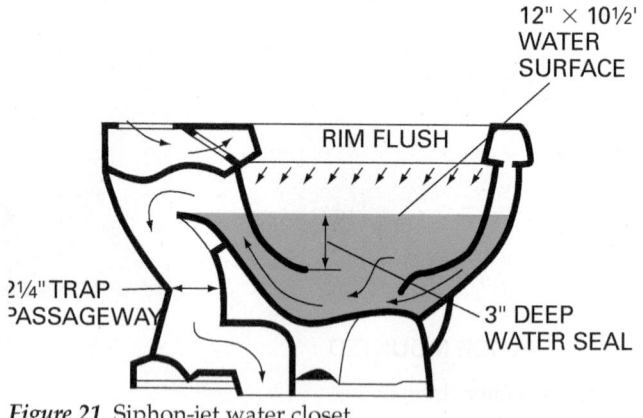

Figure 21 Siphon-jet water closet.

The siphon-action water closet is usually manufactured as a one-piece closet that combines the closet and flush tank in one unit, which makes it more compact than other types (*Figure 22*). It has a water surface area that is 12" × 11½", and it flushes quietly. It has a trap passageway diameter of 2" and a 3" deep water seal.

The blowout water closet, which is usually a wall-hung bowl, is found only in commercial plumbing installations (*Figure 23*). It is flushed by the action of a large jet of water directed to the inlet of the trap passageway. The force of the jet of water draws the contents of the bowl into the inlet trap passageway. The jet then forces or blows the waste into the outlet passageway to flush the bowl. The blowout water closet's main advantage is its ability to flush large amounts of toilet paper without becoming clogged. The blowout water closet has a waste passageway that is 2¾" in diameter, a water surface area of 11¼" × 14⅛", and a 3¼" deep water seal.

1.3.4 Washout Urinals

Unlike water closets, urinals are water-flushed plumbing fixtures designed to receive only liquid human waste directly. They are manufactured from vitreous china or stainless steel. Urinals are commonly installed in public restrooms for men. To comply with the ADA, urinals must have an elongated rim at a maximum of 17" above the floor. The two main types are washout and siphon-jet urinals.

The washout urinal washes waste out of the trap rather than flushing it out. The washing water enters the top of the urinal, flows out through openings in the top rim of the urinal into the fixture, spreads across the back of the urinal, and flows by gravity out through the urinal trap at the base and into the sanitary drainage system.

Washout urinals are characterized by the restricted opening over the trap inlet at the base of the urinal (*Figure 24*). The purpose of these small openings over the trap is to prevent the trap from becoming plugged by debris in the urinal. These restricted openings may be of two designs: either a beehive or small openings cast in the china.

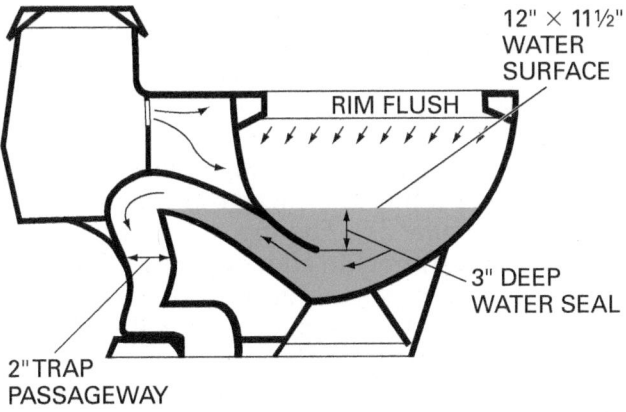

Figure 22 Siphon-action water closet.

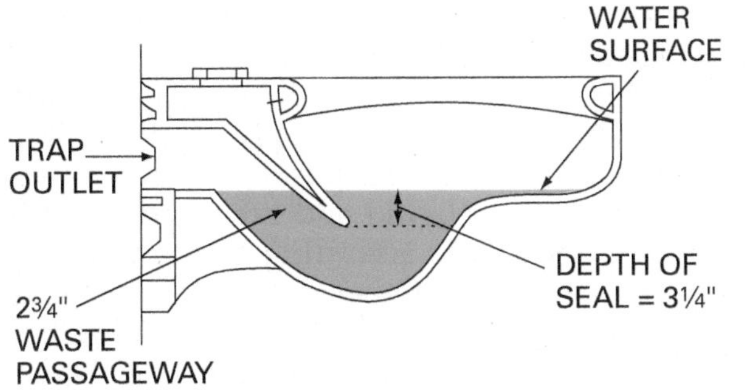

Figure 23 Blowout water closet.

Figure 24 Washout urinal.

Wall-hung washout urinals have built-in, 2" P-traps. The restrictive trap openings are cast into the china at the bottom of the urinal. Wall-hung washout urinals require a 2" waste pipe and a small-diameter vent pipe.

Some wall-hung washout urinals have a bottom outlet. They have a beehive strainer over a drain opening that is attached to a separate, exposed 1½" trap. The bottom-outlet, wall-hung washout urinal requires a small-diameter vent pipe. Some codes prohibit its use because it has an invisible trap seal. Always check applicable codes before installing a urinal.

1.3.5 Siphon-Jet Urinals

Siphon-jet urinals use a flushing action similar to the action used in siphon-jet water closets. This type of urinal has a large opening over the trap inlet located in the bottom of the urinal. To comply with the *Energy Policy Act of 1992*, siphon-jet urinals are designed to flush one gallon or less of water per use. The most common type of siphon-jet urinal is wall hung (*Figure 25*). Wall-hung, siphon-jet urinals usually have a 2" built-in trap. They require a small-diameter vent pipe.

Figure 25 Wall-hung siphon-jet urinal with electric eye.

1.3.6 Flushing Devices

A standard urinal must have an adequate flushing device. The flushing device must be able to remove the urine completely from the fixture after it is used, to prevent the spread of disease, fouling of the urinal, and offensive urine odor. Because the turnover in users is so rapid, and because the bulk of a water tank is not practical in these installations, urinals are usually fitted with a flushometer valve (*Figure 26*).

A flushometer valve, or flushometer, is a flushing device that is connected directly to the water supply, discharges a predetermined amount of water, and requires no storage tank. This device automatically closes by direct water pressure or other mechanical means. Alternatively, some urinals are equipped with a hand valve, flush tank, electric eye (which uses an infrared sensor; refer to *Figure 25*), or an automatic flush device. Infrared sensors on urinals detect when a person is standing in front of the urinal and when the person has stepped away, which automatically triggers the flushing action. Infrared urinals are most common in commercial settings, such as airports and movie theaters, where the number of users is very high.

1.3.7 Waterless Urinals

Waterless urinals use a cartridge filter containing a liquid seal rather than water to remove waste and odor (*Figure 27*). The top of the cartridge is shaped like a funnel to allow wastewater to flow into it, where it passes through the liquid seal. Because the liquid is lighter than water, it floats above the wastewater that is entering the cartridge and thus traps odors. As the wastewater flows through the cartridge, a filter traps sediment suspended in the wastewater. The wastewater then flows through the fixture's trap and into the waste line. A sealing ring on the outside of the cartridge provides an airtight barrier between the cartridge and fixture.

Waterless urinals are installed in locations where access to a water supply is difficult or when required for the purposes of water conservation. The sealant liquid and cartridges must be replaced periodically according to the manufacturer's specifications. Other designs for waterless urinals rely on mechanical traps to eliminate odor. However, plumbing codes in the United States prohibit the use of mechanical traps.

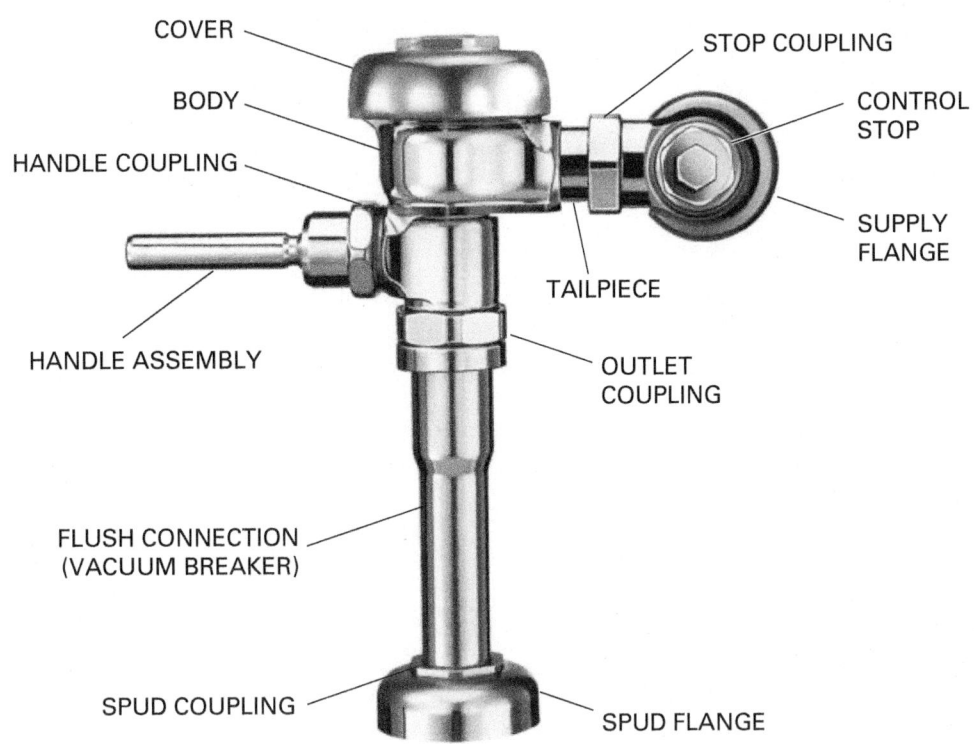

COVER

BODY

HANDLE COUPLING

HANDLE ASSEMBLY

FLUSH CONNECTION
(VACUUM BREAKER)

SPUD COUPLING

STOP COUPLING

CONTROL
STOP

SUPPLY
FLANGE

TAILPIECE

OUTLET
COUPLING

SPUD FLANGE

Figure 26 Flushometer.

1.3.8 Bidets

Bidets (pronounced bih-DAYs) (*Figure 28*) are companion fixtures to water closets. They are used for personal hygiene in bathing the external genitals and posterior parts of the body. Because of their low height and overall convenience, they are particularly useful for older persons or persons recovering from an illness who have difficulty using a shower or tub.

The bidet has long been a popular plumbing fixture in European and South American bathrooms, and it is becoming increasingly common in modern bathrooms in the United States. Bidets are fitted with cold and hot water faucets so that the user can regulate water temperature and rate of flow. The user sits astride the bidet, facing the faucets. The bidet is also fitted with a pop-up stopper that allows the user to let the water accumulate in the bowl. For rinsing the bowl, the bidet has a rim-flushing action similar to that used in a water closet and has a 1½" trap.

1.3.9 Water Conservation

The most practical way to prevent water shortages is to conserve water by using less. A variety of devices have been developed to conserve both water and energy in the face of increasing demand worldwide. Water use also requires energy for pumping, collection, treatment, and distribution. For example, letting a faucet run for five minutes uses as much energy as leaving a 60-watt light bulb on for 14 hours. According to the Environmental Protection Agency (EPA), if just one out of every 100 American homes was retrofitted with water-efficient fixtures, about 100 million kWh of electricity per year would be saved—avoiding 80,000 tons of greenhouse gas emissions. That is equivalent to removing nearly 15,000 automobiles from the road for one year. If 1 percent of American homes had older, inefficient toilets replaced with WaterSense-labeled models, the country would save more than 38 million kWh of electricity—enough to supply more than 43,000 households with electricity for one month.

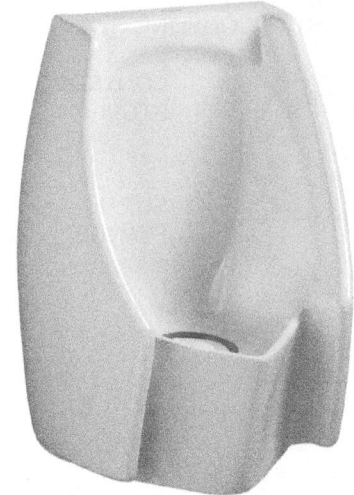

WATERLESS URINAL

CARTRIDGE INSERT, TOOL,
AND ODOR BARRIER LIQUID

Figure 27 Waterless urinal fixture with cartridge insert and odor-barrier liquid.

Figure 28 Bidet.

Many types of fixtures are available to reduce water use in showers and sinks. These include low-flow or aerated fixtures and alternate types of controls. Aeration is the introduction of air into the water flow. It is a common technique to reduce the actual flow of water in faucets and showerheads while maintaining the perception of high-volume flow. It works well for hand washing and showers, but aeration can be frustrating when filling a pot with water for cooking or cleaning. Some aerated kitchen faucets allow the user to control the degree of aeration and to restore full flow when needed.

Other types of alternative controls include foot-operated sinks. These allow precise control of flow timing while leaving the user's hands free. More precise automated flow controls are also becoming available for sinks and toilets. The EPA's WaterSense program provides a labeling system to help identify low-flow fixtures and controls.

Toilet technology has improved considerably. There are now multiple options available to improve the efficiency of water use. In addition to conventional low-flow toilets, several manufacturers now offer dual-flush toilets. These allow a full-volume flush for solid waste and a half-volume flush for liquids. Automated dual-flush control units are also available. These units dispense either a full-volume or low-volume flush depending on how long the user is within range of the flush sensor. Manual controls are also included to allow users to control the flush level if desired.

When selecting water sources for landscaping, consider drip irrigation or irrigation controlled by moisture sensors and timers. Drip irrigation provides a slow release of water directly to the root zone of the plant. This results in less evaporation than a typical sprinkler system. Moisture sensors and timers control the operating periods of irrigation systems. This ensures that water is only dispensed when needed. It also allows watering during the early morning when water has time to reach plant roots before evaporating. With either system, be sure to properly locate and maintain all components of the system to avoid waste.

Water-efficient appliances, such as high-efficiency dishwashers and washing machines, reduce water use by up to 50 percent over traditional units. Since these units use less hot water, they also save energy and are reviewed under the EPA's Energy Star program. See the Energy Star website for a list of washing machines and dishwashers that meet Energy Star water-saving criteria.

The maximum water usages for fixtures manufactured after January 1, 1994 are as follows:

- Gravity tank-type, flushometer tank, flushometer valve, and electromechanical hydraulic toilets: 1.6 gallons per flush (gpf)
- Gravity tank-type white 2-piece toilets for commercial use only (if manufactured between January 1, 1994 and December 31, 1996): 3.5 gpf
- Urinals: 1.0 gpf

The *Energy Policy Act of 1992* defines the steps that state and local governments and manufacturers must take to conserve water. Some states, for example Arizona and California, mandate the use of 1.1 gallons per flush toilets. Always refer to your local applicable code for the water conservation requirements that apply to your area.

1.4.0 Common Kitchen Plumbing Fixtures

While the sink is probably the most important plumbing fixture in kitchens, there are several other common fixtures related to the kitchen plumbing system. These include the food waste disposal units and dishwashing machines.

1.4.1 *Food-Waste Disposers*

A food-waste disposer (*Figure 29*) is an electric device used to grind food waste into a pulp and discharge it into the drainage system. Food-waste disposers should always be used with running water supplied by the kitchen sink faucet. They are mounted under the kitchen sink (or under one compartment of a two-compartment sink) in place of the basket strainer assembly that would otherwise lead to the drainage fixture. Disposers may be connected so that the waste is discharged into a separate P-trap or to a continuous waste fitting. The fitting allows the wastewater to flow

from the dishwasher through the food-waste disposer. Most plumbing codes require this process for dishwashers. Always check applicable codes in your area for specific requirements.

When installing a food-waste disposer in a double sink, you must use a directional tee with an internal baffle that joins the disposal waste and the waste from the other sink compartment (*Figure 30*). The tee and baffle prevent the disposal waste from backing up into the other sink compartment.

1: SINK BAFFLE

2: VIBRATION ABSORBING MOUNT

3: VIBRATION ABSORBING TAILPIPE MOUNT

4: NOISE INSULATION

5: GRINDING RING

6: MOUNTING

7: STAINLESS STEEL GRINDING ELEMENTS

8: 3/4 HP MOTOR

Figure 29 Food-waste disposer.

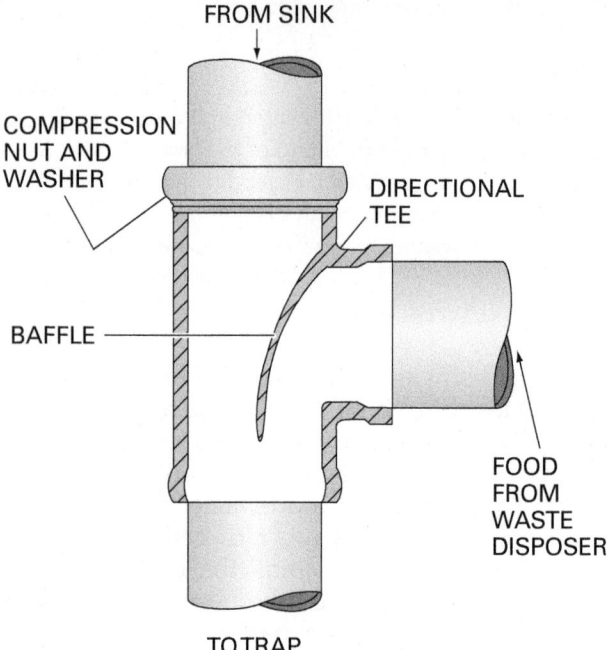

Figure 30 Directional tee and baffle connecting a food-waste disposer to a sink drain.

1.4.2 Domestic Dishwashers

A domestic dishwasher (*Figure 31*) is an electric appliance used in the home for washing dishes. The most common sizes of dishwashers are 18" and 24" wide units. Most dishwashers installed in homes use a pump-out waste disposal system. This system is installed with a rubber hose or copper tubing to connect the dishwasher waste unit to a dishwasher tailpiece under the kitchen sink basket strainer. If the sink is equipped with a food-waste disposer unit, you can connect the dishwasher into the dishwasher drain connection on the food-waste disposer.

In either case, some codes require a vacuum breaker or air gap assembly (*Figure 32*) in the dishwasher discharge piping so that waste does not back up into the dishwasher. Other codes require that you loop the drain hose above the fixture's flood-level rim to reduce the chance of a backup in the dishwasher if the waste line clogs (*Figure 33*).

There is a third option for venting a dishwasher. You may connect a discharge hose to a kitchen sink drain through a single 1½" trap. Use a discharge line that is at least ¾" in diameter. Loop the discharge line and fasten it underneath the counter. You can also connect the discharge line to a deck-mounted dishwasher air gap fitting. To do so, use a wye fitting to connect the drain hose between the sink waste outlet and the trap. Always follow all applicable codes for your installation.

Figure 31 Domestic dishwasher.

1.5.0 Other Plumbing Fixtures

Residential and commercial plumbing may include some less-common fixtures, such as laundry trays, service and mop sinks, floor drains, and water fountains. This section will examine each of these plumbing fixtures.

1.5.1 Laundry Trays

A laundry tray (*Figure 34*) is a tub fixture outfitted with cold water, hot water, and drain connections. Much like a sink but with one or two extra-deep basins, a laundry tray is used for washing clothes and other household items, and for receiving waste from automatic clothes washers. Laundry trays are usually installed in the laundry room or utility area of homes.

Did You Know?

When installing a food-waste disposer in a residence, it is important to remind the homeowners that they should run a large volume of cold water through the food-waste disposer when it is in use. The water helps wash the food waste pulp out of the drainage pipe and keeps the drainpipe from clogging.

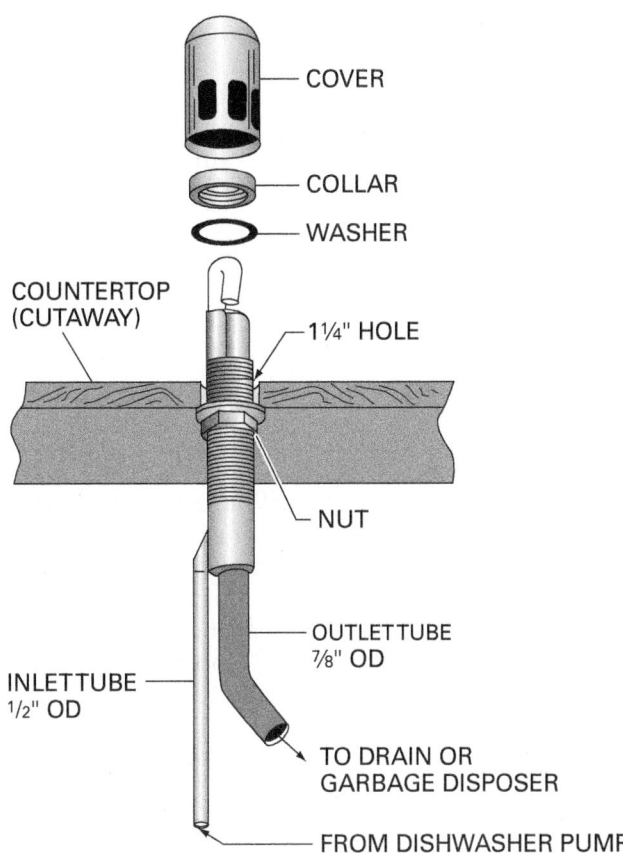

COVER

COLLAR

WASHER

COUNTERTOP (CUTAWAY)

1¼" HOLE

NUT

OUTLET TUBE ⅞" OD

INLET TUBE ½" OD

TO DRAIN OR GARBAGE DISPOSER

FROM DISHWASHER PUMP

Figure 32 Air gap assembly for a domestic dishwasher.

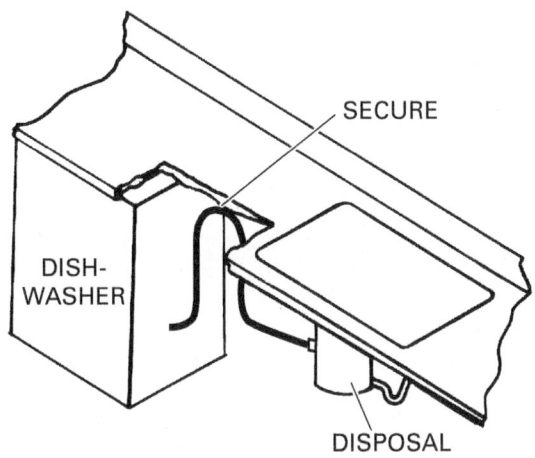

SECURE

DISH-WASHER

DISPOSAL

Figure 33 Looping the dishwasher drain hose.

Laundry trays are available in either single-compartment or double-compartment tubs. They are available as floor models set on steel legs or as wall-hung models mounted on a bracket that attaches to the wall. Manufacturers supply both models with the proper mounting hardware.

Figure 34 Floor-type laundry tray.

1.5.2 *Service Sinks and Mop Basins*

Service sinks (*Figure 35*) and mop basins (*Figure 36*) are plumbing fixtures that are installed in commercial buildings for building maintenance personnel to use. They are commonly installed in janitors' closets and building maintenance areas.

A service sink, also called a slop sink, has a deep basin that is large enough to hold a scrub pail. These sinks are used to fill and empty scrub pails, rinse mops, and dispose of cleaning water. The sinks are usually made of enameled cast iron, and they come supplied with a standard P-trap.

Mop basins, also called mop receptors, are service sinks that are set on the floor of a janitors' closet. Their low position can make them more convenient to use than service sinks because users need not lift the mop pail or other items in and out of the fixture as far compared to a sink.

PHOTOGRAPH COURTESY OF KOHLER

Figure 35 Service sink.

Figure 36 Mop basin.

Mop basins can be made of enameled cast iron, terrazzo, fiberglass, or ceramic tile.

It is also fairly common for plumbers to install refrigerator icemaker connections. Plumbers may also help install private swimming pool water supply and disposal systems.

1.5.3 Floor Drains and Floor Sinks

A floor drain (*Figure 37*) is a fixture installed in the floor that drains water and liquid waste into a plumbing system. A floor sink (*Figure 38*) is a type of floor drain that is coated with vitreous china or another easy-to-clean finish. Often outfitted with a deep seal trap, a floor sink reduces the amount of contaminant that can collect in the grate and body of the drain. Plumbing codes generally require floor drains in certain facilities that require a high level of sanitation. For example, codes may require floor drains to be installed in commercial kitchens, common laundry rooms in commercial buildings, and bathrooms in commercial buildings containing two or more water closets or urinals. Install trap primers on traps that are used infrequently, to ensure that the trap does not evaporate.

1.5.4 Drinking Fountains and Water Coolers

Drinking fountains (*Figure 39*) are plumbing fixtures that deliver a stream or jet of water through a nozzle or bubbler without cooling it. They provide a convenient source of sanitary drinking water. A drinking fountain that includes a water-cooling unit is called a water cooler (*Figure 40*). These units require an electrical hookup to power the water-cooling system. The drinking water is cooled to the desired temperature, usually around 50°F, before it is delivered to the nozzle. Both drinking fountains and water coolers are available in a variety of styles, but the wall-hung fixture is the most common. Both types of fixtures can be manufactured from vitreous china, enameled cast iron or steel, or stainless steel.

More Than Meets the Eye

Plumbers are also called upon to install other fixtures and appliances. Typically, a plumber is required to install the washer box for hooking up the washing machine, as shown in this figure. Plumbers rough-in the hot- and cold-water connections and a drain. A laundry tray provides an alternative method for draining a washing machine. Your local building code will specify the minimum allowable drain height, which is typically 30" above the floor. When installing electrical outlets for washers and dryers, refer to your local electrical code to ensure that the outlets are placed at the appropriate distance from the washer box. Typically, the minimum distance between a washer box and an electrical outlet is 3'.

Figure 37 Floor drain.

Figure 38 Floor sink.

To be ADA-compliant, drinking fountains and water coolers must adhere to many requirements. For example, they must have spouts no higher than 36" from the ground to the spout outlet. In addition, the spouts must be located at the front of the fixture, and there must be clear knee space between the bottom of the fixture's apron and the floor or ground that is at least 27" high, 20" wide, and 17" to 19" deep.

Plumbing codes specify the minimum number of drinking water facilities required for various types of buildings. For example, according to the National Standard Plumbing Code, theaters, public auditoriums, restaurants, and office buildings must have one drinking water facility per 1,000 people. Schools must have one per 100 people. As always, check the applicable codes for your particular installation.

Figure 39 ADA-compliant drinking fountain (lower right).

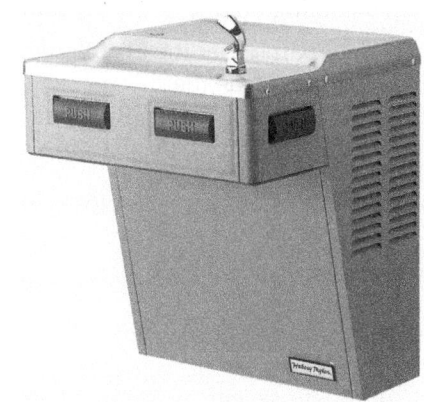

Figure 40 ADA-compliant water cooler.

1.5.5 *Local Plumbing Codes*

Local plumbing codes govern construction and installation of drinking fountains and water coolers. Typical plumbing codes require the following sanitary features:

- The bowl of the fountain must be constructed of a nonabsorbent material.
- A mouth protector must be provided over the mouthpiece.
- The drinking water must be delivered at an angle so that no water can fall back onto the nozzle.
- The nozzle must be located above the flood-level rim of the fountain.
- The water supply to the fountain must be entirely separated from the waste.

Additional Resources

"2010 ADA Standards for Accessible Design, Chapter 6: Plumbing Elements and Facilities." *www.ada.gov.*

EPA WaterSense Website. *www.epa.gov/watersense.*

ICC/ANSI Standard A117.1,Accessible and Usable Buildings and Facilities, Current edition. Washington DC: American National Standards Institute (ANSI).

1.0.0 Section Review

1. Another name for porcelain, which is often used in more expensive fixtures, is _____.

   a. porcelain enamel
   b. cast iron
   c. vitrified china
   d. sheet steel

2. When installing built-in showers, plumbing codes typically require the installation of _____.

   a. a safety pan under the installation
   b. glazed tile inside the shower
   c. mounting frames under the floor
   d. enameled steel in the walls

3. Companion fixtures to water closets that are used for personal hygiene are called _____.

   a. urinals
   b. bidets
   c. lavatories
   d. bath-shower modules

4. A device that prevents waste from backing up into the dishwasher is called a(n) _____.

   a. internal baffle
   b. air gap assembly
   c. directional tee
   d. discharge pipe

5. A plumbing fixture that is fitted with connections for cold and hot water and a drain connection, and which has one or two extra-deep basins is called a(n) _____.

   a. floor sink
   b. mop basin
   c. bidet
   d. laundry tray

2.0.0 FAUCETS

Objective

Describe the different types of faucets used in plumbing systems.

   a. Describe compression and non-compression faucets.

   b. Describe kitchen and bathroom fixture faucets.

   c. Describe utility faucets.

Performance Task

   1. Identify the most commonly installed fixtures and appliances.

Trade Terms

Anti-scald valve: An automatic valve connected to both cold and hot water supplies that uses a sensor to adjust the temperature of hot water to ensure that it does not exceed a specified temperature at the fixture outlet.

Bibb: A faucet with an outlet at a downward angle, generally used for utility or hose connections on the outside of a building.

Diverter: A device that redirects the flow of water, as in a combination shower and bath fitting. When the diverter is engaged, water flows through the showerhead instead of through the spigot.

Seat washer: A flexible, compressible disc that fastened to the valve disc at the end of the valve stem; used to stop and regulate the flow of water through a faucet.

Valve disc: In most compression valve fixtures, the enlarged end of the valve stem that carries and supports the seat washer to create a leak-tight seal when the valve is shut.

Valve seat: The circular or conical surface milled into the internals of a valve against which the valve disc or washer bears to shut off water flow through the valve.

Valve stem: The part of a faucet or valve that connects the valve handle or handwheel to the valve disc inside the valve body. It permits controlling the flow of water through the valve when the handle is turned.

Because faucets are used in all residential, commercial, and industrial buildings, it is important to understand what kinds of faucets are available, how they work, and their applications. Familiarity with the different sizes, shapes, and styles of faucets is essential so that you can choose the right kind for each installation.

Faucets fall into the two basic classifications of compression and noncompression. Both types can be further divided into bathroom, kitchen, and utility faucets.

2.1.0 Compression and Noncompression Faucets

Compression faucets control the flow of water by compressing a seat washer between the valve disc and the valve seat (*Figure 41*). The flow path of water through the valve passes through the valve seat. The valve stem connects the handle to the valve disc at the inner end of the stem, which carries and supports the seat washer. In a compression valve, the stem is the part of the faucet that rises and lowers when you turn the handle to control water flow through the valve. The middle section of the valve stem is enlarged and threaded; it threads into the faucet body so that the stem moves up or down as the handle is turned.

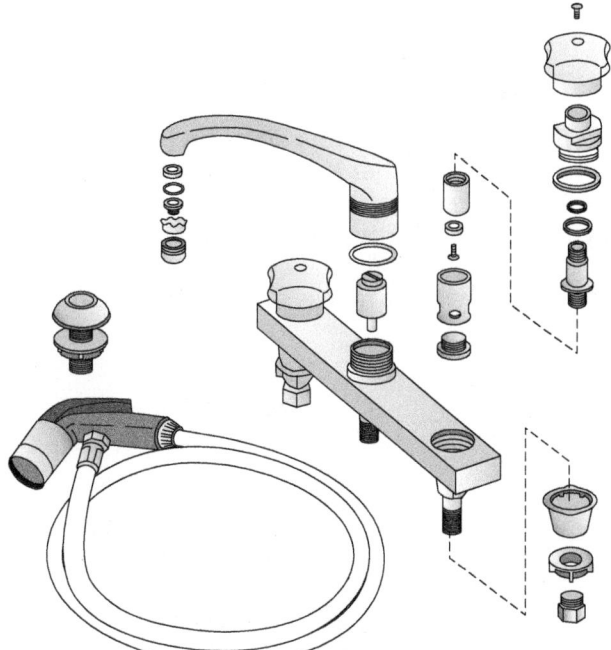

Figure 41 Exploded drawing of a compression faucet with sprayer hose.

Noncompression faucets regulate the flow of water by obstructing the water. When the faucet is open, the openings in two parts of the valve internals are matched, allowing the water to flow through. When the faucet is closed, the holes in the flow path are out of alignment and cut off the flow of water. The flow path is sealed by O-rings between the parts to prevent leakage.

2.1.1 Compression Faucets

Compression faucets for kitchen sinks or bathroom lavatories usually combine two compression valves and a mixer that combines the hot and cold water and delivers it through a mixing spout common to both valves (*Figure 42* and *Figure 43*). A plunger located between supply valves controls the drain valve (*Figure 44*). Some older dwellings may have sinks and lavatories that use individual compression valves and faucets for the hot and cold water. These faucets are installed in pairs (one for hot water and one for cold water).

In some cases, it may be more practical to install a self-closing compression faucet to conserve water (*Figure 45*). The self-closing faucet automatically closes the valve as soon as the handle is released. A self-closing faucet may present some problems over time, though. It has a tendency to receive too little water flow, which can cause lime and other minerals to build up in the valve internals. As a result, the faucet begins to malfunction.

2.1.2 Noncompression Faucets

Noncompression faucets do not rely on the compression of a washer to control the flow of water. Water flow is controlled by rotating balls, rotating cylinders, or rotating discs. Because these faucets do not use the repeated compression of a washer, they usually require less maintenance than compression faucets.

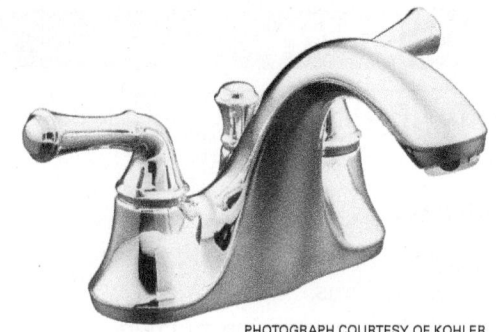

PHOTOGRAPH COURTESY OF KOHLER

Figure 43 Compression faucet for the bathroom.

The most common styles of noncompression faucets are washerless, single control, and push button. Washerless faucets control water flow by matching openings in two separate discs located in the valve. While one disc remains stationary, the other disc rotates with the hand control or lever (*Figure 46*). Although these faucets are called by the term *washerless*, their design usually includes O-rings to prevent water leakage within the valve assembly.

PHOTOGRAPH COURTESY OF KOHLER

Figure 42 Compression faucet for the kitchen with a sprayer hose.

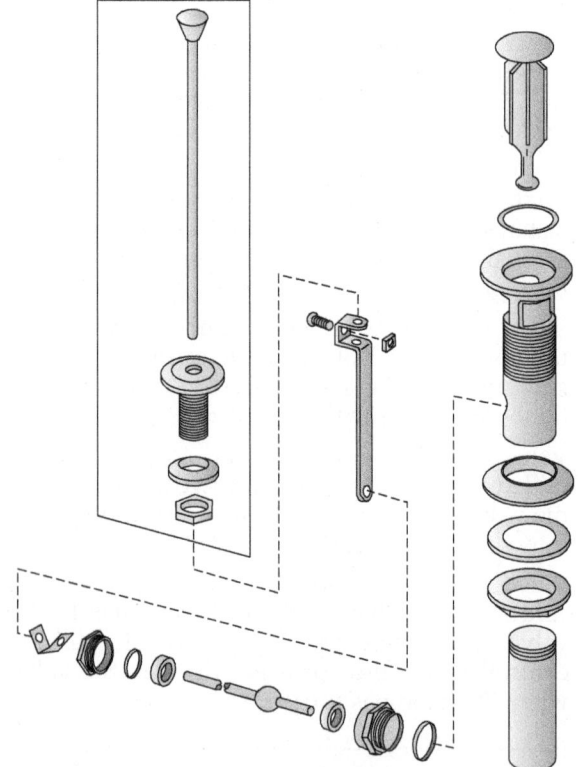

Figure 44 Exploded drawing of a drain valve.

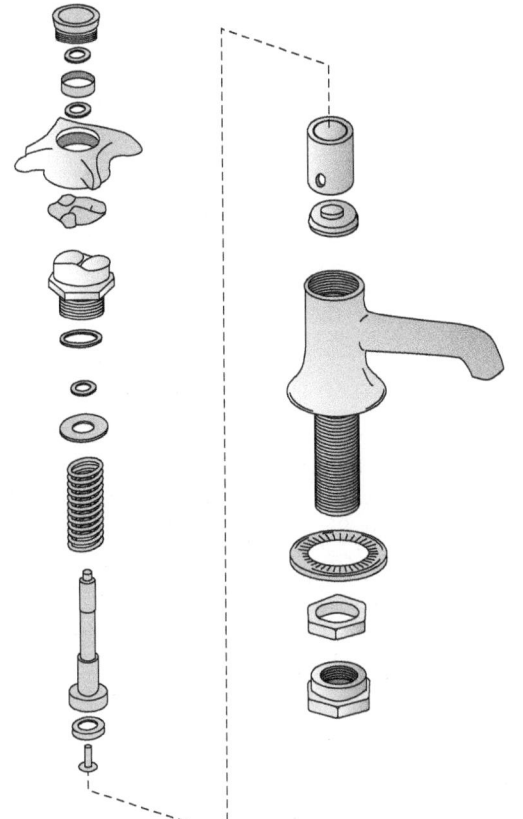

Figure 45 Exploded drawing of a self-closing compression faucet.

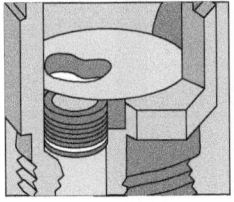

CYLINDER VALVE IS DIRECTLY OVER INLET PORT. WATER FLOWS FREELY.

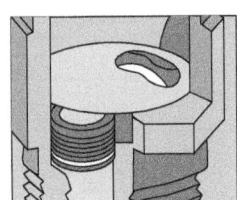

OPENINGS ARE MISALIGNED. WATER IS SHEARED OFF.

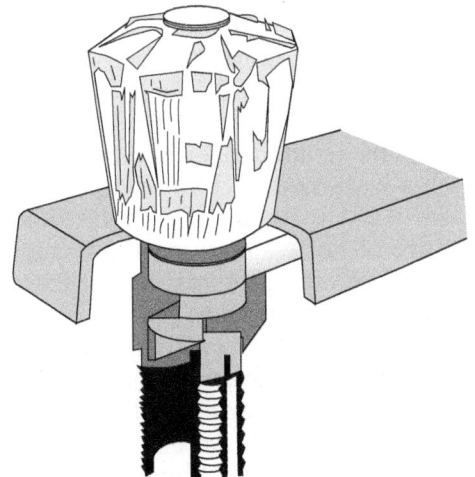

Figure 46 How a washerless faucet controls water flow.

Single-control faucets (*Figure 47*) control both the hot and cold water with one hand control. These faucets have become popular because of their convenience. The most popular designs include the rotating ball faucet and the rotating cylinder faucet. The first type uses a rotating ball in place of a compression washer. The ball is made of metal or plastic and contains several different sizes of holes or orifices. As the user rotates the single-lever control, these openings align with the hot- and cold-water ports. The position of the lever controls water mixture temperature and flow. Moving the lever to the right causes cold water to flow. Moving it to the left causes hot water to flow. Pushing the lever back increases the rate of flow.

The rotating cylinder faucet, also called a cartridge faucet, also controls both the water temperature and the rate of water flow. Some combination cartridge valves, such as those used in showers, include a supply balancing piston. The piston adjusts its position when supply pressure of either hot- or cold-water changes (*Figure 48*) to keep the outlet temperature nearly constant. As the user makes changes in the water temperature and pressure by either turning or pulling the single handle, the balancing piston automatically moves toward the cold or to the hot supply. This ensures that the flow of and temperature of the outlet water remains constant.

Another type of single-control faucet uses a ceramic disc to control the flow of water. The ceramic disc is sealed inside the cartridge. Because ceramic is very durable and unlikely to erode, it provides long-wearing, trouble-free service.

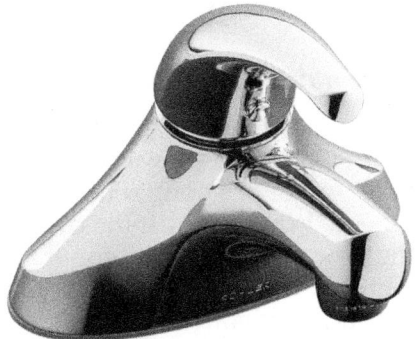

PHOTOGRAPH COURTESY OF KOHLER

Figure 47 Single-control faucet.

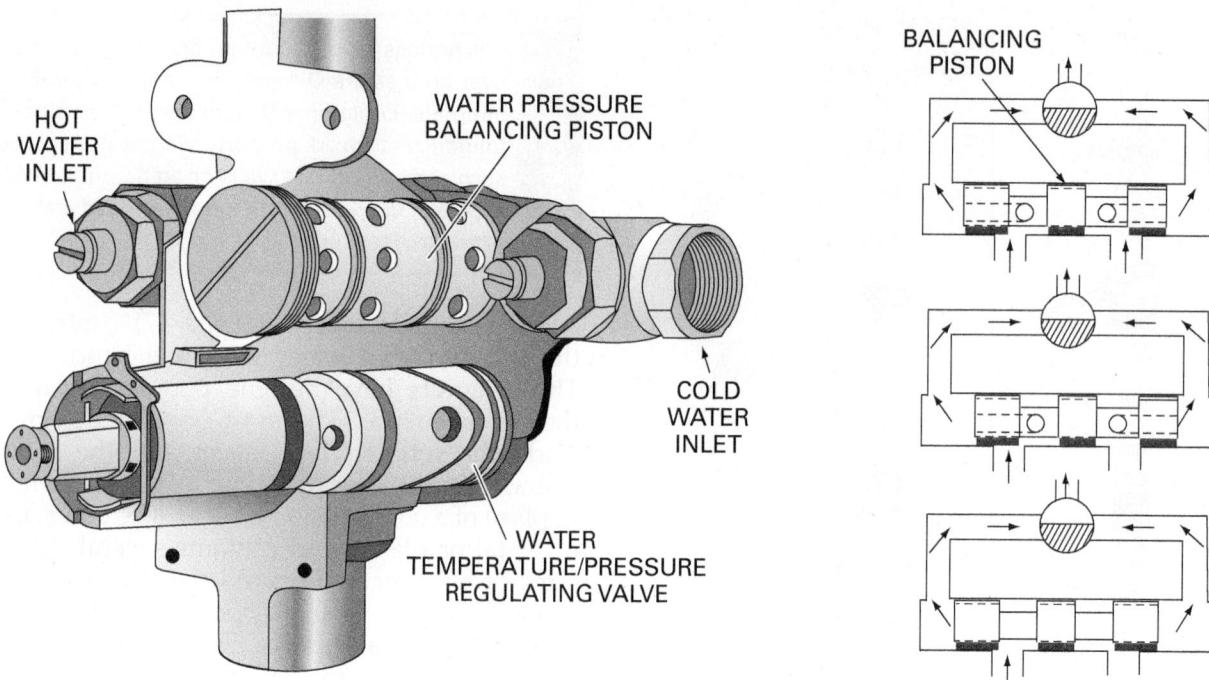

Figure 48 Rotating cylinder faucet.

2.2.0 Kitchen and Bathroom Fittings

Faucets and accessories for kitchen sinks, showers, and bathroom lavatories and fixtures are made of many different materials and come in a variety of decorator shapes and sizes. Materials used for making faucets include brass, white metallic (zinc or aluminum) alloys, and different types of plastics, such as nylons and acrylics.

2.2.1 Sink and Lavatory Faucets

The distance between the hot- and cold-water inlets on faucets varies depending on where the openings are in the sink or lavatory. This distance, usually measured in even numbers of inches, ranges from 4" to 12". The most common distance between cold and hot water supplies for a kitchen sink is 8" (*Figure 49*). The most common centers for bathroom faucets are 4" or 8".

2.2.2 Combination Shower & Bath Fittings

A combination shower and bath fitting (*Figure 50*) controls both a faucet and showerhead. A diverter, located in the spigot or spout, sends the water to the showerhead. When you lift the diverter knob, it blocks the flow of water to the spout and forces the water up into the showerhead. Adjusting water flow rate and temperature may be by means of dual hot and cold valves or a single-control pressure-mixing valve (*Figure 51*).

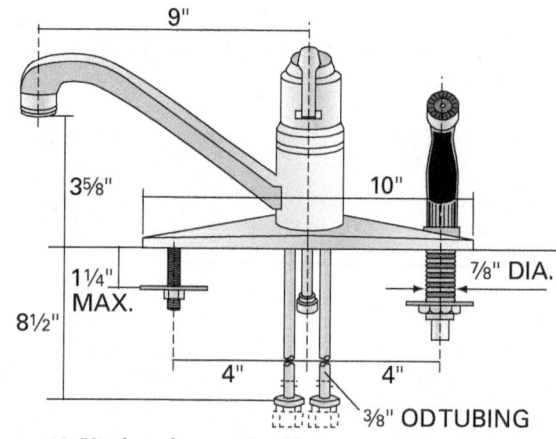

Figure 49 Kitchen faucet installed on a sink with supply holes 8" apart.

Another type of combination shower uses a flexible water line. A flexible water line allows you to raise, lower, or turn the showerhead in any direction. The head can either be handheld or placed on a wall-mounted bracket (*Figure 52*).

An anti-scald valve uses a sensor to adjust the temperature of hot water to a preset value by automatically controlling the mixture of hot and cold water from separate supplies (*Figure 53*). Anti-scald valves ensure that the water temperature does not exceed the specified temperature at the fixture outlet, thus preventing injury or discomfort caused by water that is too hot. Anti-scald valves may be installed at or near the water heater. They are sometimes called mixing valves or tempering valves.

Figure 50 Combination shower and bath fitting.

PHOTOGRAPH COURTESY OF KOHLER

Figure 51 Single-control pressure-mixing valve.

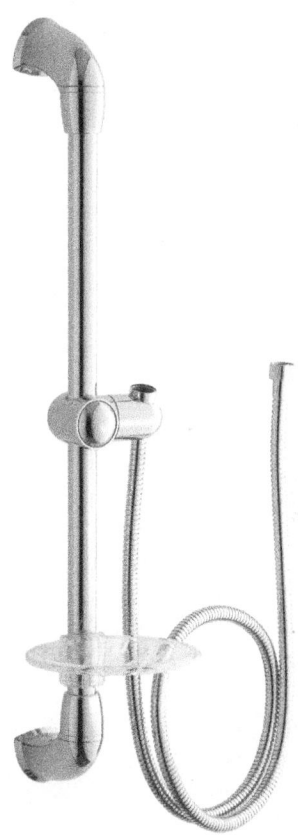

PHOTOGRAPH COURTESY OF KOHLER

Figure 52 Combination fitting with a flexible water line.

Plastic Components in Faucets

Because codes require all faucet materials to be completely lead-free, many manufacturers have switched to using plastic components in faucets. For example, many rotating ball faucets now use balls made of plastic rather than metal.

Figure 53 Anti-scald valve.

2.3.0 Utility Faucets

Utility faucets are found in boiler and laundry rooms, and on a building's outside walls. They are usually manufactured from extra-grade brass, white metallic alloy, or thermoplastic materials. Most utility faucets are not plated because appearance is generally not important for these. Common utility faucets include laundry faucets used for laundry tubs, utility sinks, service sinks, and mop sinks, as well as sediment and boiler drains (*Figure 54*). Unlike other types of faucets that are attached to fixtures, utility faucets often do not have or are not part of fixtures.

The water outlet end for a utility faucet, referred to as a bibb, can be one of two types: plain (threadless) or hose (threaded) (*Figure 55*). The bibb is angled to direct the water downward. Most common utility compression faucets have female iron pipe threads on the inlet end that attach to the water pipe during installation. Bibb faucets generally have a flange designed to fit flush against siding or wall covering by which the fixture is secured to the surface.

Another term for a hose bibb is a sill cock. On new installations, vacuum breakers are required on hose bibbs for backflow prevention.

Figure 54 Boiler drains.

SILL COCK

SILL COCK WITH SEPARATE
VACUUM BREAKER

Figure 55 Threaded hose bibb or sill cock.

In commercial applications, hose bibbs are often specified as frost-resistant wall hydrants (*Figure 56*). These faucets attach to the water pipe with male threads. The faucets run through the outside wall of the structure, leaving only the spigot, or faucet head, exposed to the weather. When you turn off the water, it drains from the spigot through the extension tube to the valve seat, which is inside the building. The valve seat should be well within the protection of the structure and installed according to the manufacturer's specifications.

The weatherproof yard hydrant (*Figure 57*), is used to provide water access beyond the walls of a building. This may be needed in agricultural applications, for example. Similar in function to a frost-resistant wall hydrant, the valve seat is located below the anticipated frost depth in the ground, typically 2' to 6' as needed by the locality. It is important to install the freeze-proof faucets properly so that it does not freeze from exposure to the weather.

COMMERCIAL INSTALLATIONS

RESIDENTIAL INSTALLATIONS

Figure 56 Frost-free wall hydrants.

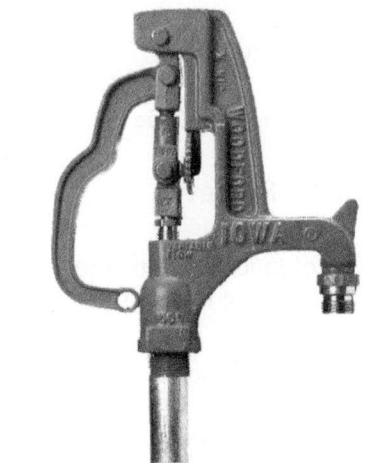

Figure 57 Yard hydrant.

A can-washer sink is a special type of utility sink found in large-scale food preparation cleanup areas. They are used mainly to clean pots and stock cans that are too large for dishwashers. They feature a heavy-duty spray nozzle attached to a flexible stainless-steel hose. The nozzle includes a brush attachment (*Figure 58*). The sink basin is equipped with tubular supports and a second smaller sink next to it, and it has a backsplash. Below the sink, a scrap drawer collects solid waste as it is rinsed off by the nozzle. The tray must be periodically emptied and the solid waste disposed of.

Figure 58 Can washer and sink.

2.0.0 Section Review

1. The type of faucet designed to stop the flow of water by pressing a seat washer between the valve disc and the valve seat is called a _____.

    a. valve faucet
    b. rotating ball faucet
    c. cylinder faucet
    d. compression faucet

2. In a combination shower and bath fitting, the device located in the spigot or spout that sends the water to the showerhead is called a _____.

    a. diverter
    b. divider
    c. fixture riser
    d. water line

3. There are two types of water outlet ends for sill cocks or bibbs: the plain and _____.

    a. hose
    b. flanged
    c. flat
    d. angled

SUMMARY

Fixtures in plumbing installations are devices that receive water from a water supply line. Fixtures are used in every kind of building—residential, commercial, and industrial. In this module, you have learned how these fixtures are manufactured, how they operate, the advantages and disadvantages of each, and how they connect with water supplies and waste outlets.

You also have learned the basics about faucets. It is important to be able to identify various types of faucets, how they operate, and which faucets are used for different plumbing applications.

1. A device that receives water from a water supply line and/or discharges waste water into a connected drainage system is called a(n) _____.

   a. fixture
   b. vent
   c. outlet
   d. inlet

2. Because of its versatility, light weight, and relatively low cost, a material that is often used to make plumbing fixtures is _____.

   a. vitreous china
   b. cast iron
   c. stainless steel
   d. plastic

3. Corner lavatories, raised-back lavatories, and ledge lavatories are examples of _____.

   a. wall-hung lavatories
   b. wheelchair-accessible lavatories
   c. self-rimming lavatories
   d. undercounter lavatories

4. A plumbing fixture designed to hold a semi-reclining bather is the _____.

   a. sink
   b. bathtub
   c. lavatory
   d. combination shower

5. To be ADA compliant, a water closet flushing control must be located above the floor no more than _____.

   a. 17"
   b. 19"
   c. 36"
   d. 44"

6. A urinal with a restricted opening over the trap inlet at the base of the urinal is called a _____.

   a. washout urinal
   b. wall-hung urinal
   c. siphon-jet urinal
   d. water-flushed urinal

7. When installing a food waste disposer in a double sink, to prevent backup in the other sink compartment during use, you must install a(n) _____.

   a. air gap
   b. directional tee
   c. P-trap
   d. flexible coupling

8. To prevent backup of sink waste into an undercounter dishwasher, codes may specify installing a(n) _____.

   a. air gap
   b. directional tee
   c. P-trap
   d. flexible coupling

9. A sink-like plumbing fixture used by building maintenance personnel that is fitted with a drain connection and is installed on the floor is called a(n) _____.

   a. floor sink
   b. mop basin
   c. utility sink
   d. laundry tray

10. A plumbing fixture that is coated with a vitreous china or other easy-to-clean finish and is installed in the floor to drain water and liquid waste into a plumbing system is called a(n) _____.

    a. mop basin
    b. floor sink
    c. laundry tray
    d. utility sink

11. A rotating ball faucet typically uses an O-ring in place of a _____.

    a. valve
    b. compression washer
    c. disc
    d. piston

12. The type of faucet that uses a rotating cylinder and a balancing valve to control the
water temperature and the rate of water flow
is called a _____.

    a. single-control faucet
    b. rotating ball faucet
    c. cartridge faucet
    d. ceramic disc faucet

13. The most common distance between hot-
and cold-water supply connections to a
kitchen sink faucet is _____.

    a. 2½"
    b. 6"
    c. 8"
    d. 12"

14. A diverter is designed to _____.

    a. select water flow between a bathtub spout
    and shower head
    b. prevent discharge of a food waste dis-
    poser from backing up in the other sink
    c. prevent sewer gases from entering a
    plumbing fixture
    d. mix hot and cold water to supply a de-
    sired temperature

15. An important feature of frost-resistant wall
hydrants is that _____.

    a. the valve seat is located in the vacuum
    breaker
    b. the valve seat is located inside the
    building
    c. the faucets attach to the water pipe with
    female threads
    d. their valve seats are located 2' to 6' below
    ground

Trade Terms Quiz

Fill in the blank with the correct term that you learned from your study of this module.

1. Porcelain enamel is a durable material used for fixtures that is created by coating materials such as steel or cast iron with porcelain clay in its liquid state, then firing at high temperature.

2. The Waste pipe, which connects a trap to its vent pipe, conveys only wastewater to the building drain.

3. When reinforced with Fiberglass, plastic materials can be used to create one-piece showers and bathtubs that are lightweight and easy to install.

4. The opening inside a faucet through which water flows and by which flow is controlled is the Valve seat.

5. A(n) Lavatory is a basin designed for the primary purpose of washing hands and face.

6. One of the materials used for the base of shower stalls is Terrazzo, small pieces of marble or other hard stone embedded in a mortar base.

7. Bathtubs are plumbing fixtures designed to hold a person in a semi-reclining position while bathing.

8. Although more expensive than other materials, Vitreous china, also called vitrified porcelain, has many benefits when used in plumbing fixtures, including its glass-like surface, which is strong and easy to clean.

9. The devices that control the flow of water from the water closet flush tanks into the bowl are called the Flush valves.

10. Because the turnover in users of urinals is so rapid, a urinal is usually fitted with a(n) Flushometer valve, which is a flushing device that is connected directly to the water supply and discharges a predetermined amount of water.

11. The Valve stem is the threaded portion of a compression faucet that rises and descends when you turn the handle to open and shut the faucet.

12. Common Plumbing fixtures are receptacles or devices that receive water from a water supply line. They include toilets, bathtubs, and water coolers.

13. Much like sinks but with one or two extra-deep basins, Laundry trays are used for washing clothes and other household items and for receiving waste from automatic clothes washers.

14. The device that redirects the flow of water in a combination faucet, such as a combination bath and shower fitting, is called a(n) Diverter.

15. One method of installing a water closet is by using a floor mount, which involves securing the water closet directly to the floor and mating it to the drainage system pipe at a closet Flange.

16. The water outlet end for a utility faucet, called a(n) Bibb, is angled downward and is available in a threadless or hose style.

17. Fixtures that are used to control the flow of water from pipes are called Faucets.

18. To install a lavatory, you must have a(n) Vent pipe to supply air to its trap, and to provide the circulation of air within the DWV system.

19. The inner end of the stem of a compression valve that supports the seat washer is the Valve disc.

20. When installing lavatories, you should ensure that the lavatory's Flood-level rim (which allows excess water to spill over onto the countertop and floor) is 31 inches above the floor.

21. The Seat washer is the compressible part of a valve stem that creates a water-tight seal when the valve is shut.

22. If a building adheres to ADA requirements regarding the location and spacing of plumbing fixtures, it can be said to be ADA compliant.

23. Guidelines for accessible design are outlined in the standard A117.1, which is published by the American National Standards Institute (ANSI).

24. To avoid injury or discomfort, install a(n) Anti-scald valve to adjust the temperature of hot water to ensure that it does not exceed a specified temperature at the fixture outlet.

25. Sinks are shallow, flat-bottomed basins mainly used for the preparation of food, dishwashing, and for utility purposes.

Trade Terms

ADA compliant	Faucets	Lavatory	Valve seat
American National Standards Institute (ANSI)	Fiberglass	Plumbing fixtures	Valve stem
	Flange	Porcelain enamel	Vent pipe
Anti-scald valve	Flood-level rim	Seat washer	Vitreous china
Bathtubs	Flush valves	Sinks	Waste pipe
Bibb	Flushometer valve	Terrazzo	
Diverter	Laundry trays	Valve disc	

ADA compliant: Follows the accessibility guidelines of the Americans With Disabilities Act of 1990, as described in the Code of Federal Regulations (28 CFR, Part 36, Appendix A—ADA Accessibility Guidelines for Buildings and Facilities)

American National Standards Institute (ANSI): A nonprofit organization that is responsible for creating and implementing standards for many industries, including the construction trades.

Anti-scald valve: An automatic valve connected to both cold and hot water supplies that uses a sensor to adjust the temperature of hot water to ensure that it does not exceed a specified temperature at the fixture outlet.

Bathtubs: Low-profile plumbing fixtures designed to hold semi-reclined bathers. Bathtubs are identified by the location of the drain hole as one faces the fixture. If the drain is on the right end, it is a right-hand bathtub; if on the left end, it is a left-hand bathtub.

Bibb: A faucet with an outlet at a downward angle, generally used for utility or hose connections on the outside of a building.

Diverter: A device that redirects the flow of water, as in a combination shower and bath fitting. When the diverter is engaged, water flows through the showerhead instead of through the spigot.

Faucets: Valve-like fixtures that are used to control the flow of water from pipes into sinks, bathtubs, and similar plumbing fixtures.

Fiberglass: Spun filaments of glass woven into yarn, roving (twisted) strands, and textile materials, such as cloth or mats. Fiberglass cloth saturated with a plastic (vinyl resin or epoxy), called glass-reinforced plastic or GRP, can be pressed into molds to make products such as bathtubs and shower enclosures.

Flange: A rim on one end of a length of pipe that provides a connection point to another length of pipe or to a plumbing fixture, such as a water closet or valve. Bolts or studs with nuts are used to hold two flanges together, with some type of gasket between them.

Flood-level rim: The edge of the basin, bowl, or fixture over which water flows when it is too full.

Flush valves: Devices located at the bottom of tanks for flushing water closets and similar fixtures.

Flushometer valve: A device that discharges a predetermined quantity of water to a fixture for flushing purposes and is automatically closed by direct water pressure or other mechanical means. Also referred to as simply a flushometer.

Laundry trays: Fixed tubs on legs or wall-mounted, usually installed in the laundry room or utility area of homes. They are used for washing clothes and for receiving wastewater from automatic clothes washers.

Lavatory: A basin designed for installation in bathrooms and other locations, primarily for washing the hands and face. Compare with sinks.

Plumbing fixtures: Receptacles or devices which are either permanently or temporarily connected to the water plumbing system of the premises. Also referred to as simply fixtures. They can connect to a supply of water therefrom or discharge used water, liquid-borne waste materials, or sewage either directly or indirectly to the drainage system of the premises. Fixtures may also require both a water supply connection and a discharge to the drainage system.

Porcelain enamel: A coating of vitrified porcelain (or china) that provides an attractive and protective finish to fixtures made of cast iron or steel.

Seat washer: A flexible, compressible disc that fastened to the valve disc at the end of the valve stem; used to stop and regulate the flow of water through a faucet.

Sinks: Shallow, flat-bottomed basins mainly used for the preparation of food, dishwashing, and for utility purposes. Compare with lavatory.

Terrazzo: A type of floor surface (commonly used for showers and other plumbing fixtures) made by pouring in place or precasting a mixture of small pieces of marble, other hard stone, or glass and a mortar base. When hardened, the surface can be ground and polished smooth, although it may be left slightly rough to create a nonslip surface.

Valve disc: In most compression valve fixtures, the enlarged end of the valve stem that carries and supports the seat washer to create a leak-tight seal when the valve is shut.

Valve seat: The circular or conical surface milled into the internals of a valve against which the valve disc or washer bears to shut off water flow through the valve.

Valve stem: The part of a faucet or valve that connects the valve handle or handwheel to the valve disc inside the valve body. It permits controlling the flow of water through the valve when the handle is turned.

Vent pipe: A pipe installed to provide a flow of air to or from a drainage system, or to provide a circulation of air within such a system to protect trap seals from siphonage and back pressure. The combination of all vent pipes in a plumbing system is referred to as the vent system.

Vitreous china: A durable ceramic material usually consisting of a fine clay mineral (kaolinite) and is often a mixture with quartz, feldspar, silica or other materials. The mixture is heated to temperatures between 2,200°F and 2,600°F (1,200°C and 1,400°C) to create a nonporous, glass-like finish. Also referred to as china, porcelain, and vitrified porcelain.

Waste pipe: A pipe in a plumbing system that carries only waste water and sewage.

Additional Resources

This module is intended as a thorough resource for task training. The following reference works are suggested for further study.

"2010 ADA Standards for Accessible Design, Chapter 6: Plumbing Elements and Facilities." **www.ada.gov**.

EPA WaterSense Website. **www.epa.gov/watersense**.

ICC/ANSI Standard A117.1,Accessible and Usable Buildings and Facilities, Current edition. Washington DC: American National Standards Institute (ANSI).

International Plumbing Code (IPC), Current Edition. Washington, DC: International Code Council (ICC).

Figure Credits

Section Review Answer Key

SECTION 1.0.0

Answer	Section Reference	Objective
1. c	1.1.1	1a
2. a	1.2.3	1b
3. b	1.3.8	1c
4. b	1.4.2	1d
5. d	1.5.1	1e

SECTION 2.0.0

Answer	Section Reference	Objective
1. d	2.1.0	2a
2. a	2.2.2	2b
3. a	2.3.0	2c

NCCER CURRICULA — USER UPDATE

NCCER makes every effort to keep its textbooks up-to-date and free of technical errors. We appreciate your help in this process. If you find an error, a typographical mistake, or an inaccuracy in NCCER's curricula, please fill out this form (or a photocopy), or complete the online form at **www.nccer.org/olf**. Be sure to include the exact module ID number, page number, a detailed description, and your recommended correction. Your input will be brought to the attention of the Authoring Team. Thank you for your assistance.

Instructors – If you have an idea for improving this textbook, or have found that additional materials were necessary to teach this module effectively, please let us know so that we may present your suggestions to the Authoring Team.

NCCER Product Development and Revision
13614 Progress Blvd., Alachua, FL 32615

Email: curriculum@nccer.org
Online: www.nccer.org/olf

❏ Trainee Guide   ❏ Lesson Plans   ❏ Exam   ❏ PowerPoints   Other _____

Craft / Level: _____   Copyright Date: _____

Module ID Number / Title: _____

Section Number(s): _____

Description: _____

_____

_____

_____

Recommended Correction: _____

_____

_____

_____

Your Name: _____

Address: _____

_____

Email: _____   Phone: _____

This page is intentionally left blank.

Introduction to Drain, Waste, and Vent (DWV) Systems

OVERVIEW

To design, install, and maintain drain, waste, and vent (DWV) systems, plumbers must be familiar with the factors that affect them. Sanitary drainage systems include the piping system inside the building, the drainpipe buried outside the building, and the public sewer. Knowing how drains, fittings, vents, and pipe move waste out of a building enables plumbers to prevent system malfunctions.

Module 02111

Trainees with successful module completions may be eligible for credentialing through the NCCER Registry. To learn more, go to *www.nccer.org* or contact us at 1.888.622.3720. Our website, *www.nccer.org*, has information on the latest product releases and training.

Your feedback is welcome. You may email your comments to *curriculum@nccer.org*, send general comments and inquiries to *info@nccer.org*, or fill in the User Update form at the back of this module.

This information is general in nature and intended for training purposes only. Actual performance of activities described in this manual requires compliance with all applicable operating, service, maintenance, and safety procedures under the direction of qualified personnel. References in this manual to patented or proprietary devices do not constitute a recommendation of their use.

02111 V4.5

Objectives

Successful completion of this module prepares trainees to:

1. Identify the major components of a DWV system and describe their functions.
   a. Identify and describe the components of DWV systems.
   b. Explain the requirements for sizing of drains and vents.
2. Describe the types, purpose, and construction of traps.
   a. Identify the types of traps.
   b. Identify the parts of traps.
   c. Describe the ways traps can lose their seal.
3. Describe the types of fittings used in DWV systems.
   a. Describe the materials used in making DWV fittings.
   b. Identify the types of DWV fittings and their requirements.
4. Describe the construction of various DWV systems.
   a. Explain the importance of grade.
   b. Describe the construction of sewer and waste treatment facilities.
   c. Identify the health concerns associated with DWV systems.
   d. Explain how plumbing codes affect the construction of DWV systems.

Performance Task

Under the supervision of your instructor, you should be able to do the following:

1. Sketch an isometric drawing of a simple DWV system and label its components.

Trade Terms

Adapters	Elevation	P-trap	Siphonage
Back pressure	Evaporation	Pipe scale	Sludge
Branch interval	Fall	Runs	Spigot
Building sewer	Fixture drains	S-traps	Stack
Capillary action	Grade	Sanitary combination	Test tees
Cleanout	Graywater	Sanitary fittings	Velocity
Double ¼ bends	Heel inlets	Sanitary increasers	Vent branch
Double trapping	Hydraulic gradient	Sanitary upright wyes	Vent ells
Drain, waste, and vent	Interceptors	Sanitary wyes	Vent tees
(DWV) system	Inverted wyes	Short sweep ¼ bends	Weir
Drainage fittings	Long sweep ¼ bends	Side inlets	

Industry Recognized Credentials

If you are training through an NCCER-accredited sponsor, you may be eligible for credentials from NCCER's Registry. The ID number for this module is 02111. Note that this module may have been used in other NCCER curricula and may apply to other level completions. Contact NCCER's Registry at 888.622.3720 or go to *www.nccer.org* for more information.

CODE NOTE

Codes vary among jurisdictions. Because of the variations in code, consult the applicable code whenever regulations are in question. Referring to an incorrect set of codes can cause as much trouble as failing to reference codes altogether. Obtain, review, and familiarize yourself with your local adopted code. Safety codes are developed by the US Occupational Safety and Health Administration (OSHA).

Contents

Figures

1.0.0 DRAIN, WASTE, AND VENT SYSTEM OVERVIEW

Objective

Identify the major components of a DWV system and describe their functions.

a. Identify and describe the components of DWV systems.
b. Explain the requirements for sizing of drains and vents.

Trade Terms

Adapters: Fittings that join pipes of different sizes or materials, such as copper and galvanized pipe or cast-iron and plastic pipe.

Building sewer: The part of the drainage system that extends from the end of the building drain and conveys its discharge to a public sewer, private sewer, individual sewage-disposal system, or other point of disposal.

Drain, waste, and vent (DWV) system: Refers to the combined sanitary drainage and venting systems. This term is technically equivalent to soil-waste-vent (SWV).

Fixture drains: The drains from traps of fixtures to the junction of those drains with any other drain pipe.

Graywater: Wastewater generated from domestic processes other than human-waste disposal. These include laundry, dishwashing, and bathing. Graywater comprises 50 to 80 percent of residential wastewater.

Plumbers need to understand how drainage systems work and know the flow of waste products from a building, to the treatment facilities, and back into the ecosystem (streams, rivers, and lakes). This cycle begins at the fixture drains that connect to the drain, waste, and vent (DWV) system. Waste, either liquid or in solution with solids, enters the DWV system from the fixture drains and flows into the building's sanitary pipe system. The pipe system is designed to remove this waste safely from the building's interior.

Sanitary drainage systems can be divided into three main parts (see *Figure 1*):

- The pipes inside the building, usually referred to as the DWV system
- The drainpipe buried outside the building, which is called the building sewer
- The public sewer, which carries the building wastes to the treatment plant and eventually back to the ecosystem

Certain installations may include a graywater system in addition to the DWV system. This piping collects used water from sinks, showers, and laundry facilities. Because it captures and reuses water that would otherwise be wasted, a graywater system is considered a green application. Graywater is mainly used to flush toilets or provide irrigation for landscapes. Most jurisdictions do not allow graywater irrigation of edible plants and some restrict the use to subsurface irrigation. Graywater systems must be carefully designed to meet local plumbing codes as well as health department requirements.

1.1.0 DWV System Components

Plumbers may design, install, and maintain the DWV systems inside buildings and the building sewers buried outside on the property. Usually, the municipality is responsible for installing and maintaining the public sewers, lift stations, and treatment plants.

1.1.1 Building DWV Systems

The DWV system inside a building is a circuit of piping designed to remove the wastes from plumbing fixtures and drains safely, reliably, and efficiently. There are many names for each type of pipe and fitting in this network. *Figure 2* illustrates the major components of a DWV system, including the following:

- Building drain
- Soil stack
- Stack vent
- Individual vents
- Fixture branches
- Fixture drain or trap arm
- Traps
- Bends or elbows
- Tees
- Wyes
- Couplings, reducers, and adapters

In some applications a graywater system may be added to the DWV system. These systems can redirect graywater for certain uses without treatment if allowed by local codes.

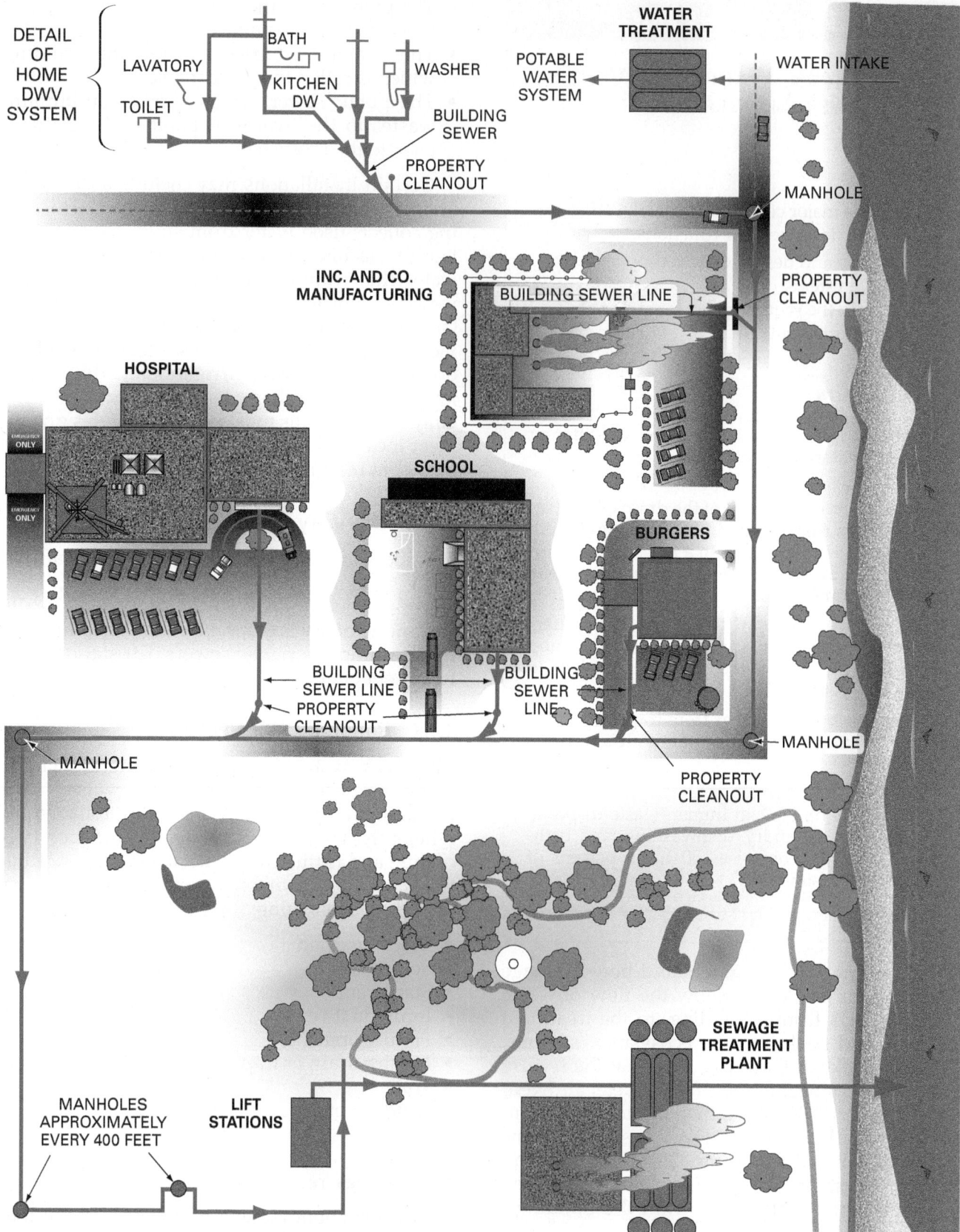

DETAIL OF HOME DWV SYSTEM

LAVATORY
BATH
TOILET
KITCHEN DW
WASHER
BUILDING SEWER
PROPERTY CLEANOUT

WATER TREATMENT
POTABLE WATER SYSTEM
WATER INTAKE

MANHOLE

INC. AND CO. MANUFACTURING
BUILDING SEWER LINE
PROPERTY CLEANOUT

HOSPITAL
EMERGENCY ONLY
EMERGENCY ONLY

SCHOOL
BUILDING SEWER LINE
PROPERTY CLEANOUT
BUILDING SEWER LINE

BURGERS
MANHOLE
PROPERTY CLEANOUT

MANHOLE

MANHOLES APPROXIMATELY EVERY 400 FEET
LIFT STATIONS

SEWAGE TREATMENT PLANT

Figure 1 Overview of a typical community sewer system.

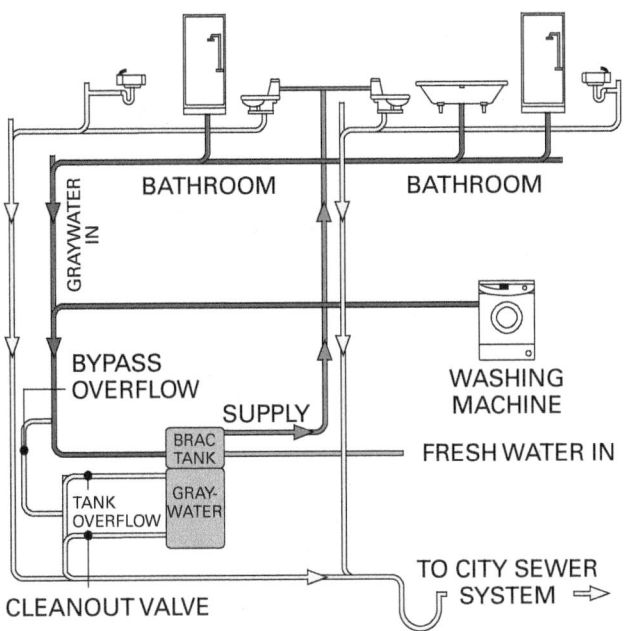

Figure 2 Components of a DWV system and graywater system.

1.1.2 Fixture Drains

Fixture drains connect fixtures to the building's DWV piping system. Many fixtures have drains that strain the wastewater before it enters the drainage piping. Examples of fixture drains include a basket strainer for a kitchen sink, PO (pop-up) plugs for lavatories, and other strainers for bidets and showers (*Figure 3*).

1.2.0 Sizing Drains and Vents

Drainage systems fall into two major categories: storm water drains and building drains. In addition, graywater systems may be used in many areas to address environmental issues. These systems are increasingly being used in commercial or industrial applications to meet LEED (Leadership in Energy and Environmental Design)

Figure 3 Fixture drains.

requirements. Storm water drains collect storm water from roofs and pavement. The water is either held for on-site disposal or is discharged at a certain rate into a storm sewer system. The sizing of storm drainage systems is based on expected rainfall. Building drainage systems must be appropriately sized based on the expected water use in the building. Plumbers must understand the mechanics of fluid flow through pipe to properly size drains and vents.

Graywater Systems

The plumbing industry is increasingly adopting green practices. It recognizes the benefits of recycling and reusing limited resources. However, adopting new practices comes with some challenges. Consider the use of graywater for landscape irrigation. Everyone agrees that it is a good idea, but local jurisdictions tend to differ in their acceptance of these systems. For example, some jurisdictions still define graywater as sewage. As a result, they won't permit using graywater for landscape irrigation. Other jurisdictions may discourage its use with expensive and complex approval requirements. Communities and governments must weigh the benefits of graywater use against potential health risks, additional regulation, and additional costs. As communities develop experience with graywater systems, restrictions and codes will likely change. This situation means plumbers should be paying close attention to code changes that apply in their work area. It also means being aware of other authorities such as the local health department, which also may impose rules for graywater reuse. Complicated? Yes. But many believe that the use of graywater systems will continue to grow in response to environmental concerns.

Vents are used in plumbing systems to balance pressure in the piping network. This balance of pressure is necessary to prevent the fixture traps from losing their seals. For the vents to function properly, plumbers must size the vents correctly according to how many fixtures are connected, the number of drainage fixture units (how much water discharges into the drain per minute), and the length of the vent pipe. If inadequately-sized vents are installed, the plumbing system does not work properly. You will learn how to size drainage and venting systems as you advance through the plumbing curriculum.

Additional Resources

"Design of Sewer System." Civil Engineers PK. Updated 2017. **https://civilengineerspk.com/**

2015 International Plumbing Code Commentary (Includes IPSDC), 2014. International Code Council (ICC).

Plumbing Venting: Decoding Chapter 9 of the IPC, Bob Scott. 2014. Procodeclasses.

1.0.0 Section Review

1. Systems that can capture water that is not heavily contaminated and redirect it for certain uses without treatment if allowed by local codes are called _____.
   a. hard water
   b. ground water
   c. graywater
   d. blackwater

2. What is the main factor when engineers set the sizes of stormwater drain piping in building plans?
   a. The size of the building
   b. The expected rainfall
   c. The frequency of major storms
   d. The capacity of the water treatment plant

2.0.0 DWV TRAPS

Objective

Describe the types, purpose, and construction of traps.

    a. Identify the types of traps.
    b. Identify the parts of traps.
    c. Describe the ways traps can lose their seal.

Trade Terms

Back pressure: A condition which may occur in the DWV system whereby a higher pressure than atmospheric pressure is created in the drain/vent piping, causing a reversal of the normal flow through drain piping and traps. Also referred to as backpressure backflow.

Capillary action: The tendency of water to be drawn into porous or fibrous material against gravity above the level of the water source.

Cleanout: An access point to connected parts of the drainage system for the removal of blockages.

Double trapping: A situation in which one trap is attached to another, creating negative pressure that stops the intended flow of drainage.

Evaporation: The natural change from liquid to vapor of water at a temperature below its boiling point.

Fall: The amount of slope given to horizontal runs of pipe expressed as a height in inches per foot of run.

Hydraulic gradient: The level of the surface of water flowing in a partially-full pipe by gravity alone.

Interceptors: Devices designed and installed so as to separate and retain deleterious, hazardous, or undesirable matter from normal waste, while permitting normal sewage or liquid wastes to discharge into the drainage system by gravity.

P-trap: A trap constructed in the shape of the letter P with the loop facing downward, which provides a water seal in a waste or soil pipe, used mostly at sinks and lavatories.

Siphonage: Loss of water in a trap seal started by unequal pressure inside and outside DWV piping. The water initially flows in the direction of the lower pressure. Sustained siphonage, even in the absence of pressure differences, results from the cohesive property of water.

Stack: A general term for certain vertical DWV pipes, including offsets of soil, waste, vent, or inside conductor piping. This does not include vertical fixture and vent branches that do not extend through the roof or that pass through not more than two stories before being reconnected to the vent stack or stack vent.

S-traps: Traps with long downstream legs, which tend to promote siphonage. S-traps are no longer permitted by code for new installations but are still found in older buildings.

Weir: When referring to plumbing, a ledge or lip in a fixture that controls the level of water inside the fixture. The word comes from Old English, meaning dam.

Traps and vents protect the safety of homes and other buildings. The water seal in a trap protects people from airborne pathogens (germs), foul odors, and potentially explosive sewer gases.

Traps are important components of the DWV piping system and is a fitting or device that provides a liquid seal of 2" to 4". This seal prevents sewer gases from leaking back into the building but should not affect the flow of sewage or wastewater through the drain. As will be explained later in this module, vents protect trap seals.

Modern codes require every plumbing fixture to have a trap that protects the fixture and users from the sanitary drain system. Some fixtures, such as water closets and many urinals, have integral or built-in traps.

A fixture trap is a vital part of any DWV system. To function properly, a fixture trap must flush completely, be self-cleaning, have a smooth interior waterway, and be accessible for cleanout. The depth of the seal and the amount of water normally held in the trap are important factors in trap design. The fixture trap, which creates a water seal, requires a vent system to protect it from siphonage, back pressure, wind, and aspiration. Back pressure is often referred to as backpressure backflow.

2.1.0 Types of Traps

The P-trap is the most commonly used trap. P-traps can be one-piece or two-piece with a union nut. P-traps can also have a cleanout. Several styles of P-traps are shown in *Figure 4*. P-traps designed to be attached directly to fixtures are usually 1¼" to 1½" for sinks and lavatories, 1½" to 2" for tubs and showers, and 2" to 4" for floor drains.

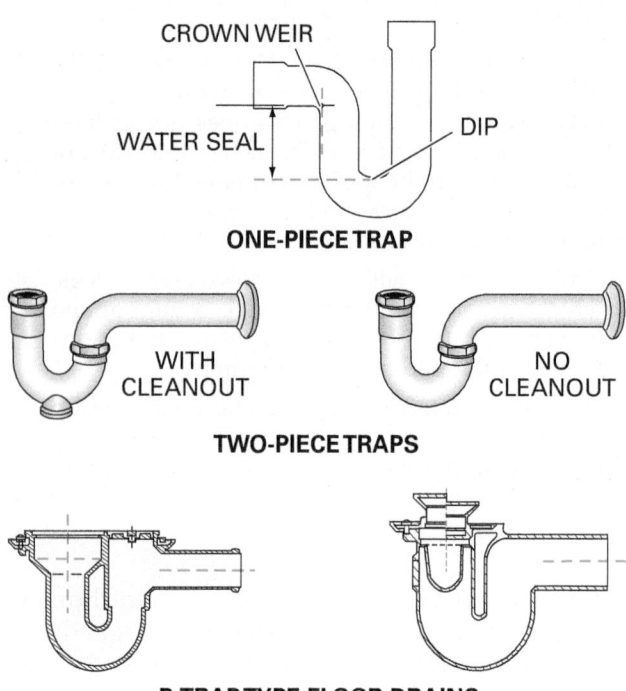

Figure 4 P-traps.

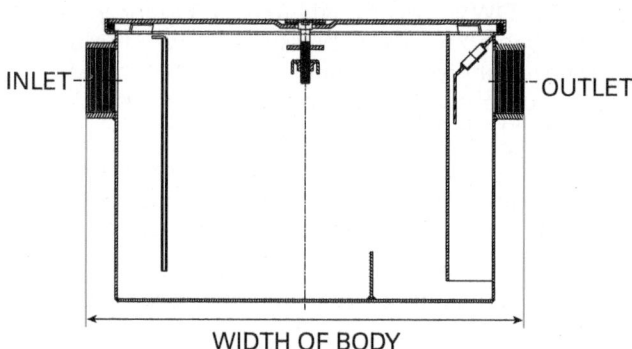

Figure 5 Grease interceptor.

P-traps can be made of brass, brass with chrome plating, plastic, copper, cast iron, malleable iron, or glass. Floor drains may be designed with P-traps.

Interceptors prevent hazardous or undesirable materials from entering building drainage systems, public or private sewers, and sewage treatment plants or processes. Hazardous or undesirable materials include hair, lint, fats, oils, grease, flammable liquids, sand, solids, acid or alkaline waste, and chemicals. Interceptors are available for specific applications. Hair interceptors, for example, may be installed in beauty salons, barbershops, hospitals, or pet grooming shops. Grease interceptors (Figure 5) may be installed in restaurants, commercial kitchens, or auto repair shops. Although they differ in design, all interceptors operate on similar principles. Wastewater flows through a chamber where harmful materials are separated before the wastewater flows out again.

For example, in interceptors designed to capture solid wastes such as grease, wastewater flows into a chamber through screens. Because the solids are heavier than the wastewater, gravity causes them to fall to the bottom of the chamber, where they are retained until the chamber is cleaned out. On the other hand, greases and oils are less dense than water and float. These substances become trapped in a chamber at the top of the fixture.

Current plumbing codes prohibit the use of S-traps (Figure 6). P-traps are more efficient and effective. The long downstream leg of the S-trap tends to promote siphonage. Plumbers could make the trap more effective by increasing the depth of the seal, but this led to other problems. With greater depth, there was a greater chance that solids would stay in the trap. Fungus growth was also a problem in traps that were too deep. Two types of S-traps were used: the full S-trap and the ¾ S-trap. Many older homes still have S-traps. Plumbing supply stores stock these traps for repairs but not for installation in new construction.

> **NOTE**
> Nonsiphon traps are available with deeper trap seals. These are used where the plumbing system is subjected to abnormal changes in pressure or to a lot of evaporation.

Water closets (toilets) have integral traps as part of their design. Water closets (Figure 7) should never be attached to another trap. This is called double trapping and creates a pressure that will stop the flow of drainage.

Figure 6 S-trap.

LONG DOWNSTREAM
LEG TENDS TO PROMOTE
SELF-SIPHONAGE

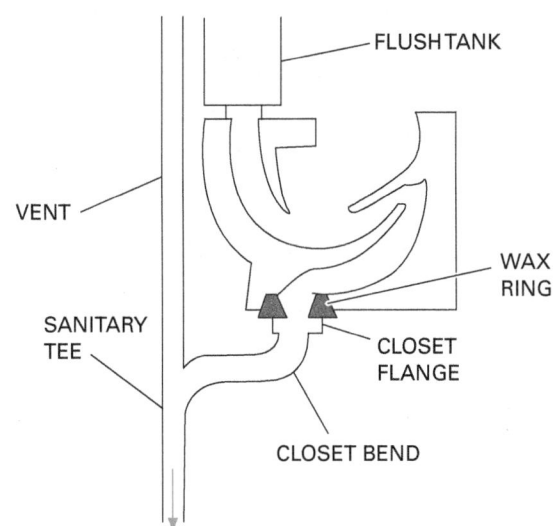

FLUSH TANK

VENT

WAX
RING

SANITARY
TEE

CLOSET
FLANGE

CLOSET BEND

Figure 7 Integral or built-in trap.

2.2.0 Parts of Traps

The following are the basic parts of a trap (*Figure 8*):

- *Inlet*—Where water enters from the fixture
- *Top dip*—The inside curve of the pipe under the inlet
- *Bottom dip*—The bottom of the lowest curve of the pipe beneath the inlet
- *Crown weir* (sometimes called the *trap weir*)—The crown weir is the highest point in the seal of the trap. Crown weirs and trap weirs are often referred to simply as *weirs*.
- *Fixture drain* (also called the *trap arm*)—The point where wastewater leaves the trap and goes into the drainage piping.

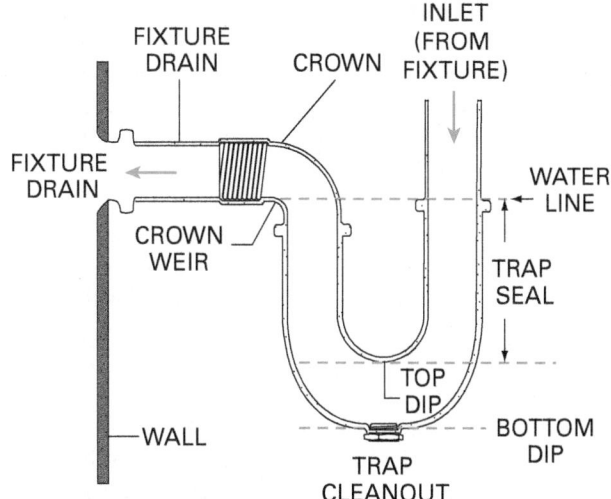

FIXTURE
DRAIN   CROWN   INLET
(FROM
FIXTURE)

FIXTURE
DRAIN   WATER
LINE

CROWN
WEIR   TRAP
SEAL

TOP
DIP

WALL   BOTTOM
DIP

TRAP
CLEANOUT

Figure 8 Parts of a trap.

2.3.0 Trap Installation Considerations

The proper operation of DWV traps can't be over emphasized. Plumbers need to know the principles of trap installation and operation, and the conditions that can reduce or negate their function.

2.3.1 Installation Requirements

Although applicable code governs specific installation requirements, such as dimensions and locations of traps, there are some typical trap installation requirements (*Figure 9*).

Generally, the vertical distance from the fixture outlet to the crown weir may not exceed 24". The second critical dimension is the horizontal distance from the crown weir to the trap vent. This distance varies depending on the diameter of the trap. The third important dimension is the total

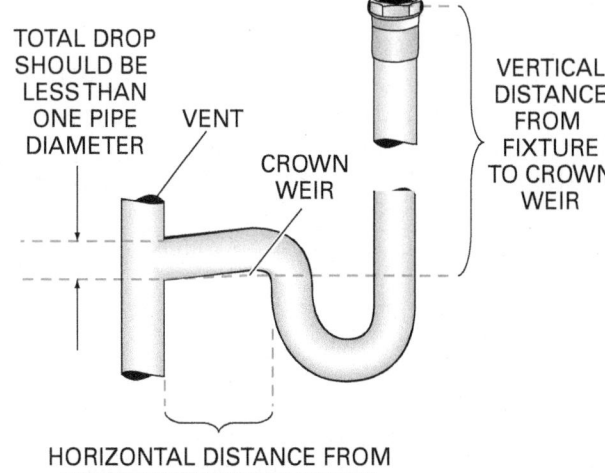

TOTAL DROP
SHOULD BE
LESS THAN
ONE PIPE
DIAMETER   VENT

VERTICAL
DISTANCE
FROM
FIXTURE
TO CROWN
WEIR

CROWN
WEIR

HORIZONTAL DISTANCE FROM
CROWN WEIR TO TRAP VENT

Figure 9 Critical dimensions of a trap vent.

---

drop (or fall) in the fixture drain from the crown weir to the vent. It may not exceed one trap drain pipe diameter. Usually, the fall is ¼″ per foot. If the horizontal leg is installed with greater drop, the trap is likely to siphon. Always refer to your local applicable code for the requirements in your area.

Many P-traps are manufactured in two pieces. The J-bend piece joins with the fixture tailpiece and is secured with a gasket and compression nut called a slip-joint washer and a slip-joint nut. The outlet end, called the trap arm, joins the horizontal drain, which connects to the vent.

Generally, traps should be installed for each plumbing fixture that does not have a built-in trap. Some codes permit one trap to serve more than one fixture. For example, three lavatories that are 30″ or less apart may be connected to one trap. Sinks containing two or three compartments also may be connected to one trap.

2.3.2 Loss of Trap Seal

When a trap functions properly, waste from the fixture flows into and through the trap (*Figure 10*). The trap is refilled with the last of the wastewater to leave the fixture. This water provides the necessary liquid seal. For the trap to function this way, the pressure on both sides of the trap must remain nearly equal. Water tends to flow in a level line, called the hydraulic gradient. The crown weir must always be installed lower than the top of where the fixture drain enters the vent line.

Properly designing and installing the DWV system can prevent siphonage and back pressure. There are multiple ways a trap may lose its seal, including:

- siphonage
- aspiration

- momentum
- oscillation (wind effect)
- back pressure
- evaporation
- capillary action
- cracks

2.3.3 Siphonage

If the trap is not properly vented, it is likely to siphon. Siphonage occurs when there is negative pressure inside the DWV piping. The pressure difference pushes the water that is normally held in the trap into the DWV piping system. Generally, siphonage occurs when the DWV piping is improperly vented or the vent is blocked (*Figure 11*). As the waste leaves the trap, an area of reduced pressure is created in the drainage piping. Because of the difference in pressure, the water is forced from the trap by the higher atmospheric pressure and destroys the trap seal. Siphonage happens when there is too much fall and the crown weir is higher than the top of where the fixture drain enters the vent.

2.3.4 Aspiration

The term aspiration means the drawing in, out, or up of something, usually a fluid. In piping, aspiration takes place when a large volume of water flows past a waste line connecting to a trap. The flow by the connection lowers the pressure in the waste line, drawing the water from the trap, causing the seal to fail. Because of aspiration, failed trap seals can allow the entry of gases and odors from the vent system. When the trap seal has failed or is saturated with the gas or odor, it will emit the same, often unpleasant, odor into the building.

Did You Know?

In the earliest home plumbing fixtures (in the 1850s), the main safeguards against odors and sewer gases were handmade traps that the plumber installed in the drains of individual fixtures. These traps often lost their water seals because of siphonage and back pressure and became ineffective. Efforts to prevent seal loss failed, because the principle of venting fixture drains was not known at the time.

In the early 1900s, the problems with fixture traps led health officials to require a secondary safeguard: the installation of building traps on each sanitary or combined building sewer. Without this additional safeguard, rats were able to travel freely from one building to another. Building traps became the second line of defense against rats in the sewer systems.

This requirement was a big advance at the time. However, since the development of modern collection, drainage, and venting systems, most model codes don't require building traps. In fact, many codes actually prohibit building traps. The only exceptions are in areas where sewer gases are extremely corrosive or where the sewer gases contain high explosive gas content, creating a risk of explosions in the public sewer system that might, for instance, blow off manhole covers and cause considerable damage.

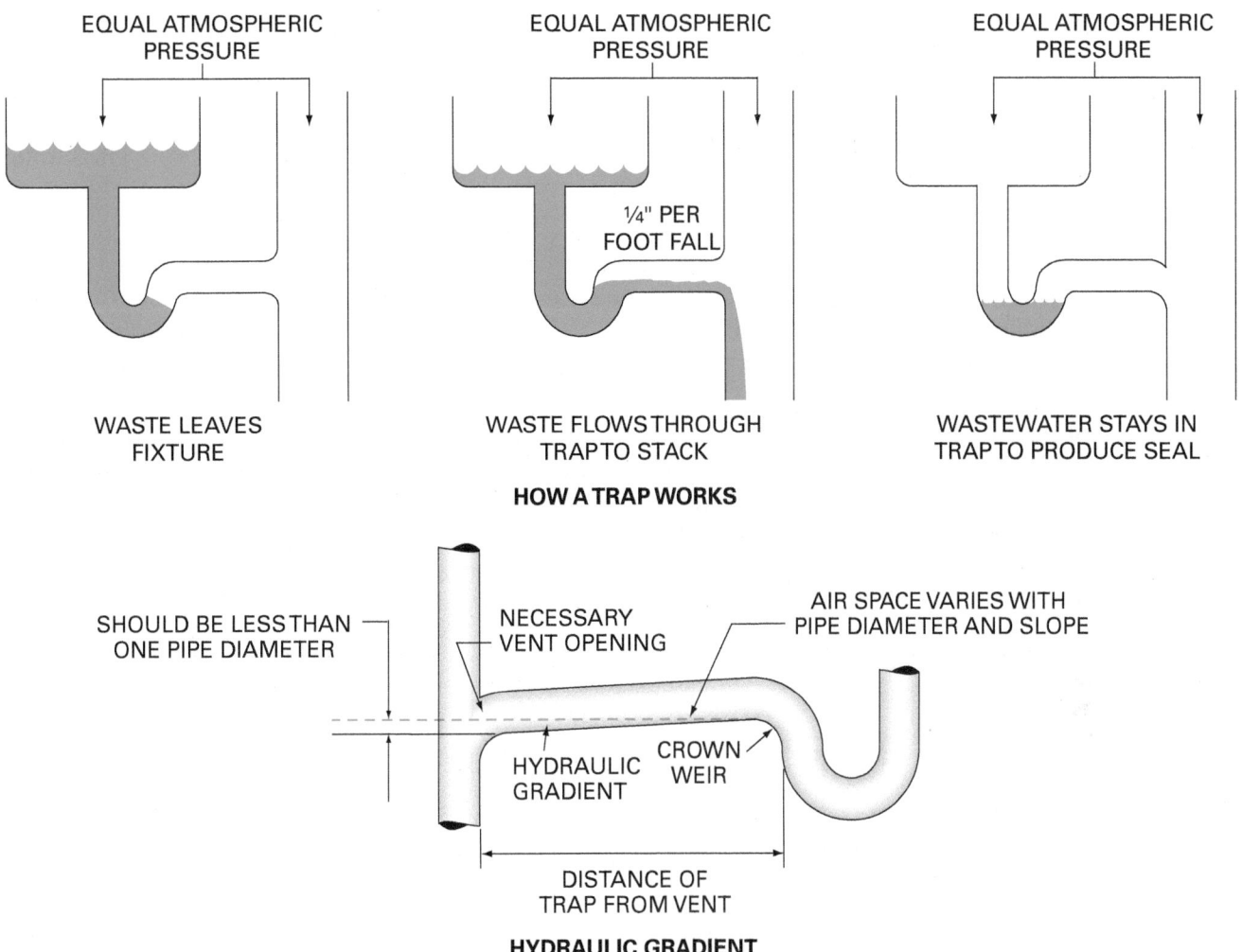

EQUAL ATMOSPHERIC PRESSURE

EQUAL ATMOSPHERIC PRESSURE

EQUAL ATMOSPHERIC PRESSURE

¼" PER FOOT FALL

WASTE LEAVES FIXTURE

WASTE FLOWS THROUGH TRAP TO STACK

WASTEWATER STAYS IN TRAP TO PRODUCE SEAL

HOW A TRAP WORKS

SHOULD BE LESS THAN ONE PIPE DIAMETER

NECESSARY VENT OPENING

AIR SPACE VARIES WITH PIPE DIAMETER AND SLOPE

HYDRAULIC GRADIENT

CROWN WEIR

DISTANCE OF TRAP FROM VENT

HYDRAULIC GRADIENT

Figure 10 How a trap works.

Fixtures with S-traps are the most vulnerable to aspiration. Although modern plumbing codes prohibit S-traps, you may still encounter them in older buildings. Where an S-trap is installed, you can detect a failed seal by the smell of sewer gases or by a gurgling sound in the pipe. Drainage systems must provide adequate circulation of air in the piping to prevent siphonage and aspiration and to protect trap seals.

> **WARNING!**
>
> Many toxic gases do not emit an odor. Always test for gas before working in an area where toxic gases may be present. Use a smoke, water, or peppermint test to locate gas that may be entering the system through a leak in the piping. Use a toxic gas detector to monitor potentially dangerous levels of gas before working in a confined area.

2.3.5 Momentum

The momentum of water—the combination of its speed and mass—rushing through a pipe can force the standing water out of a trap and empty it, thus breaking the seal. Water can gain enough speed to empty a trap when the vertical distance between the fixture outlet and the trap is too long. In most cases, that distance should be limited to around 12", although longer vertical distances may be required for certain types of standpipe.

2.3.6 Oscillation

Oscillation, or wind effect, is one of the least likely ways a trap can lose its seal. Where there are strong upward or downward air currents, the pressure or suction of the moving air may cause the water in the trap to rise or fall in a sloshing or oscillating motion. If it rises enough to spill over into the waste pipe, less water remains in the trap and the seal is weakened. Lower than normal back pressures could break the seal.

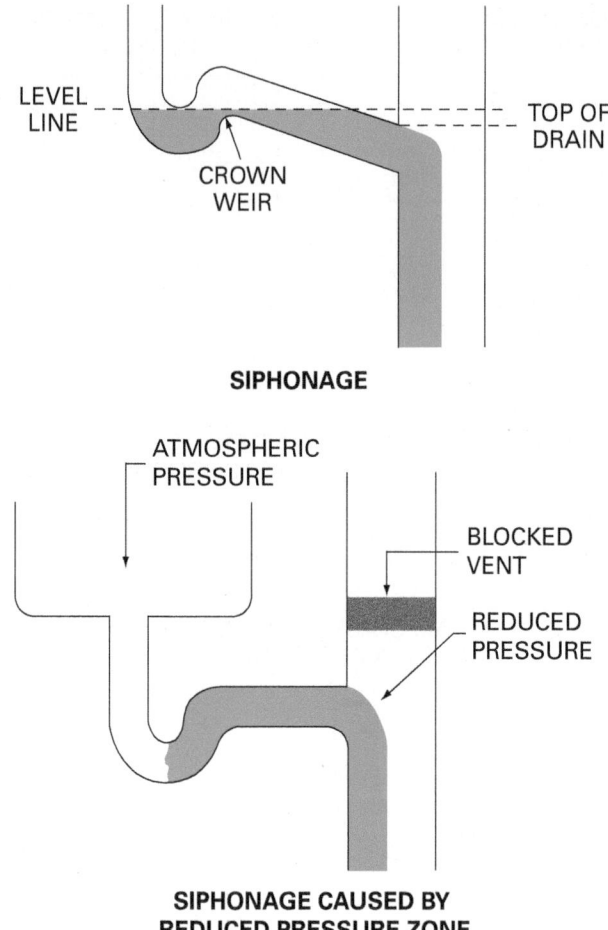

SIPHONAGE

**SIPHONAGE CAUSED BY
REDUCED PRESSURE ZONE**

Figure 11 Siphonage.

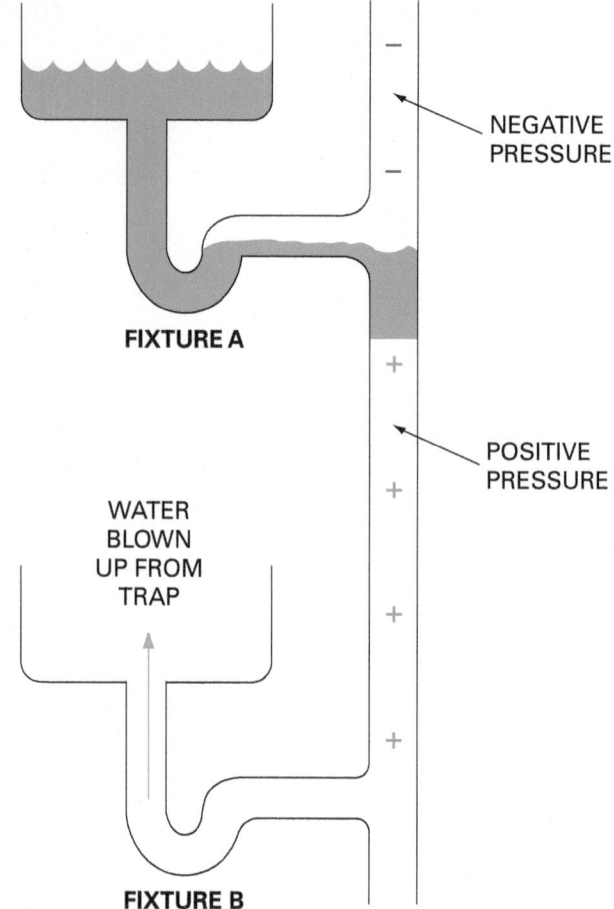

Figure 12 Back pressure.

2.3.7 Back Pressure

Back pressure (*Figure 12*) can cause a trap seal to break. Back pressure is pressure inside the DWV piping that is greater than atmospheric pressure. If enough wastewater from fixture A enters the stack so that a slug of water forms a moving plug, the air in the stack below the plug is compressed. This excess pressure tries to escape through the trap in fixture B illustrated in *Figure 12*. To prevent normal back pressure from destroying the trap seal, the stack must be properly sized, and the trap must be properly sized and protected by a vent. Also, the seal must be deep enough; generally, a trap seal of 2" to 4" is required.

2.3.8 Evaporation

A trap may lose its seal as a result of evaporation. This is most likely to happen in traps that are seldom used. The water evaporates, causing the seal to break. If the DWV piping is properly designed, evaporation only becomes a problem during long periods of nonuse. When sewer gas enters a struc-

ture, unused floor drains are often the cause. If you anticipate long periods of nonuse, you can install extra-deep traps. Many codes require trap primers where evaporation of the trap seal is likely. Trap primers are connected to a regularly used water line. Water flows through a small tube into the trap so that it can keep its seal.

2.3.9 Capillary Action

Capillary action (*Figure 13*) may cause a trap seal to break if a porous material, such as string or paper, is caught in the trap and waste line. The porous material acts as a wick and draws the water out of the trap by capillary action. Cleaning the trap solves this problem.

2.3.10 Cracks

A more common cause of waste and sewer gas leaking into a building is a crack in the trap. Cracks can be caused by worn washers, or by a broken nut, solder joint, or glue joint.

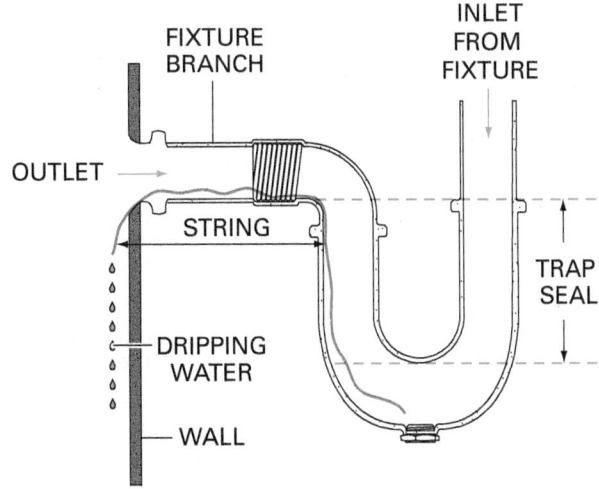

FIXTURE BRANCH

INLET FROM FIXTURE

OUTLET →

STRING

TRAP SEAL

DRIPPING WATER

WALL

Figure 13 Capillary action.

Additional Resources

Plumbing Venting: Decoding Chapter 9 of the IPC, Bob Scott. 2014. Procodeclasses.

2015 International Plumbing Code Commentary (Includes IPSDC), 2014. International Code Council (ICC).

2.0.0 Section Review

1. Which of the following is the most common type of DWV trap?

   a. K-trap
   b. P-trap
   c. R-trap
   d. S-trap

2. The level of water in a trap is determined by the _____.

   a. crown weir
   b. top dip
   c. crown
   d. drain connection

3. Loss of a trap seal due to flow of water past a trap's drain line connection is called _____.

   a. siphonage
   b. oscillation
   c. aspiration
   d. back pressure

3.0.0 DWV FITTINGS

Objective

Describe the types of fittings used in DWV systems.

- a. Describe the materials used in making DWV fittings.
- b. Identify the types of DWV fittings and their requirements.

Performance Task

1. Sketch an isometric drawing of a simple DWV system and label its components.

Trade Terms

Double ¼ bends: Fittings used to collect and combine the flow from two opposite runs into a single run of pipe.

Drainage fittings: Fittings installed in the drainage sections of the DWV system to remove waste from a building.

Heel inlets: Openings in the curved, heel portion of bend fittings in line with the bend, used to connect smaller lines to the main line.

Inverted wyes: Fittings used to join the upper end of vents to the top of soil and waste stacks. The have the appearance of the upside-down letter Y.

Long sweep ¼ bends: Long-radius bends commonly used at the base of DWV stacks and elsewhere when the longer radius is needed to greatly decrease flow resistance and back pressure. Their use is regulated by code.

Pipe scale: A flaky, adherent coating on pipe walls resulting from the corrosion of metals, especially iron or steel. Also, a heavy oxide coating on copper or copper alloys resulting from exposure to high temperatures and oxygen.

Runs: Lengths of pipe that continue in a straight line.

Sanitary combination: A fitting that combines a wye and ⅛ bend. It is used to connect horizontal branch lines that intersect other horizontal branch lines. It offers less resistance to the flow of material than a sanitary tee. Also called a tee-wye.

Sanitary fittings: Fittings used to connect DWV branches to the main DWV system and to serve as cleanouts.

Sanitary increasers: Fittings used to enlarge the diameter of vent stacks. They are usually placed at least 1' below the penetration of the stack through the roof. Their use is regulated by code.

Sanitary upright wyes: Fittings used to connect separate vent stacks to the lower ends of soil and waste stacks.

Sanitary wyes: Drainage fittings, shaped like the letter Y, that join branches to the main run of pipe at an angle.

Short sweep ¼ bends: Bend fitting with a short radius used at the base of a DWV stack and elsewhere. Their use is regulated by code.

Side inlets: Openings in ell or tee fittings at right angles to the line of the run, used to connect smaller lines to the main line.

Spigot: When referring to DWV systems, the pipe end or the male portion of a fitting that inserts into the hub of a downstream fitting.

Test tees: Tees installed as test locations for pressurizing the DWV system to test for leaks before placing it in operation.

Vent branch: A vent connecting one or more individual vent lines with a stack vent.

Vent ells: Plastic fittings with a sharp turn radius, used only in vent piping systems. Their use is regulated by code.

Vent tees: Fittings used in venting systems or as cleanouts. They may not be used in the drainage system because it restricts the flow of material. Their use is regulated by code.

Fittings are devices used to connect pipe. Those used in DWV systems or in specialty drainage systems are called drainage fittings. This section describes the various types of drainage fittings and their uses within the DWV piping system.

3.1.0 DWV Fitting Materials

DWV fittings are made from many different materials, including copper, brass, lead, steel, cast iron, clay, glass, and various types of plastic. Cast iron and plastic are the most commonly used materials (*Figure 14*). Not all fittings are available in all materials.

Although the fitting materials may vary, fittings of the same design have the same names. For example, a plastic sanitary tee and a cast-iron sanitary tee are basically identical, even though they are made from different materials.

CAST-IRON PIPE AND FITTINGS

PLASTIC PIPE AND FITTINGS

Figure 14 DWV fittings.

A number of fittings are available for copper pipe for DWV purposes. Those fittings include 90-degree ells, 45-degree ells, 22½-degree ells, male adapters, tees, cleanout tees, reducing tees, and reducers.

3.2.0 DWV Fittings and Applications

Most codes state that drainage fittings may not slow or block the flow of materials in the pipe. Because of this, sanitary drainage fittings are made with a sweeping design (*Figure 15*) to allow for the smooth flow of material in the system.

Another code requirement is that the direction of hub-type fittings should not go against the flow of the system—that is, the wastes should flow from the bell end to the spigot end of the pipe.

3.2.1 Vents

Every trap requires a vent of some type. Vent pipes are critical for plumbing fixtures to function correctly as part of the sanitary drainage system. Venting prevents back pressure or siphonage from breaking the water trap seals that serve the fixtures. All the vent pipes of a building create the vent system and are connected to the drain pipes. The system may include one or more pipes. Vents are installed to provide a free flow of air and to maintain equalized pressure throughout the drainage system. There are many types of vents, one of which is shown in *Figure 16*.

As *Figure 16* shows, sanitary fittings are available in single and double patterns. The double pattern is used to connect two branch lines entering the system from opposite directions. This allows for central placement of horizontal runs and vertical runs. Sanitary branch fittings consist of tees, wyes, and combinations of the two.

3.2.2 Bends

Sanitary fittings are used to connect DWV branches to the main DWV system. The branch inlets of these fittings may be reducing (going from a larger pipe to a smaller pipe). If so, they can be joined to the system without reducers.

The term *bend* is often used in reference to cast-iron fittings. With other types of fittings, the terms *elbow* or *ell* are more common. Bends are used to change the direction of a run of pipe. A run is one or more lengths of pipe in a straight line. Bends are available in different sizes. Some sizes may be available in only one type of material, such as cast-iron. Pipe and fitting manufacturers' catalogs show all available sizes according to the material of the fitting. You can see examples of common bends in *Figure 17*.

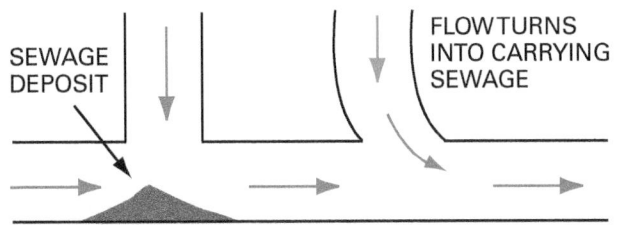

SEWAGE DEPOSIT

FLOW TURNS INTO CARRYING SEWAGE

Figure 15 Sweeping design.

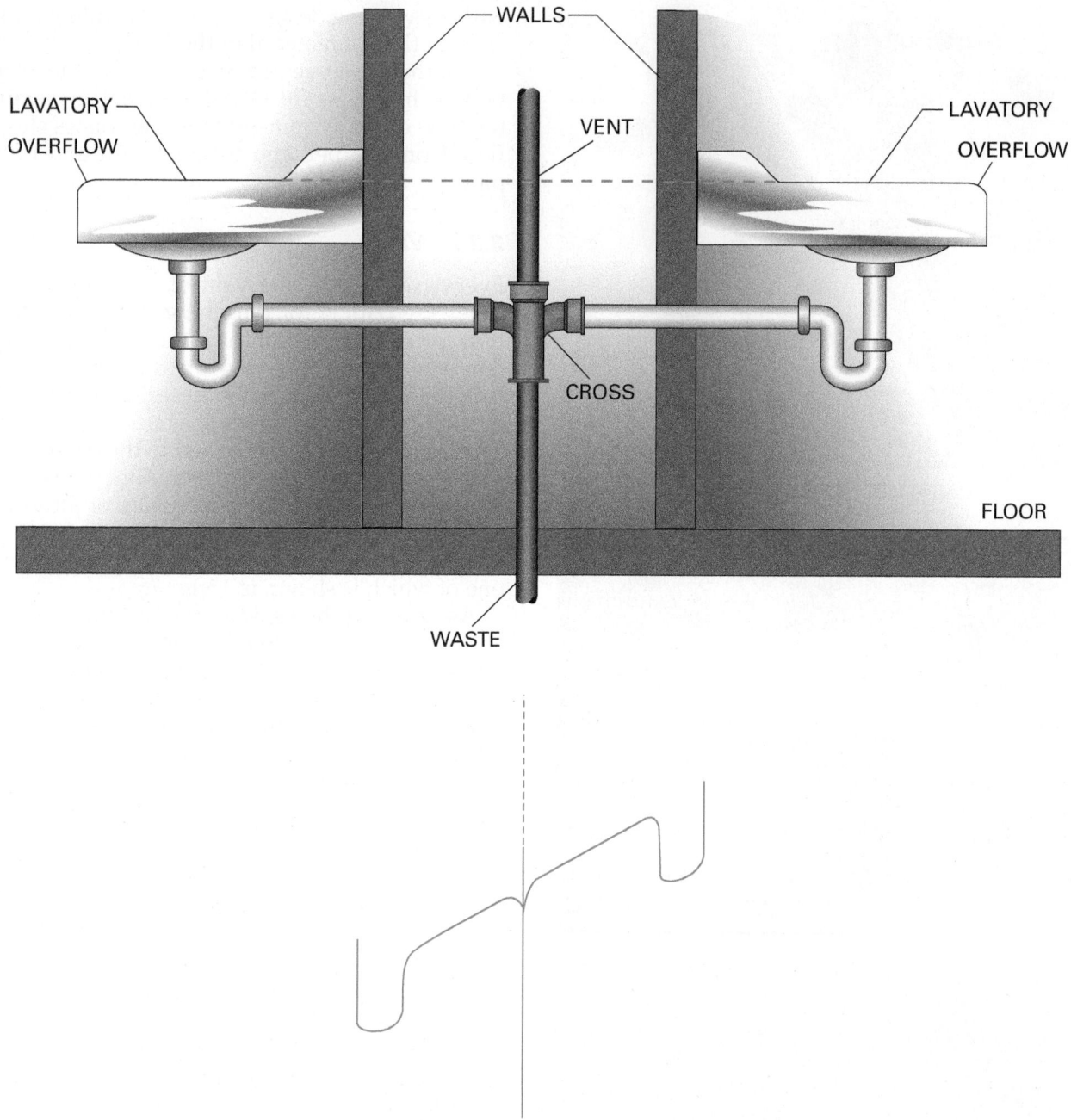

Figure 16 Common vent.

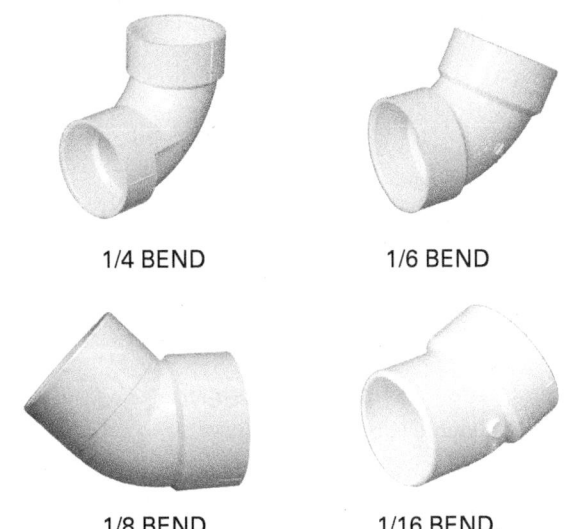

1/4 BEND   1/6 BEND

1/8 BEND   1/16 BEND, STREET

Figure 17 Bends.

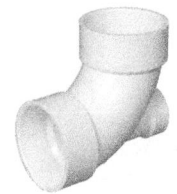
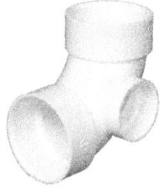

1/4 BEND WITH LOW HEEL INLET (PLASTIC)   1/4 BEND WITH SIDE INLET (PLASTIC)

LONG SWEEP 1/4 BEND WITH HIGH HEEL INLET (PLASTIC)

1/4 BEND WITH SIDE INLET (CAST IRON)   1/4 BEND WITH HEEL INLET (CAST IRON)

Figure 18 Bends with heel and side inlets.

The plumbing design tells you which bends are used and where they are placed in the system. Bends are expressed as fractions of a complete circle. A circle contains 360 degrees. You can easily determine the number of degrees a given bend turns or changes the direction of flow by multiplying the bend fraction by 360 degrees. For example, to determine the bend angle of a ¼-bend, multiply the bend type (¼) by 360 degrees:

$$\tfrac{1}{4} \times 360° = 360° \div 4 = 90°$$

Bends are available with heel inlets and side inlets to allow smaller lines to be connected to the bend (*Figure 18*). It is important to note that a high- or low-pattern side inlet bend cannot be used as a vent if the inlet is horizontal.

Bends with side inlets are available with single and double side inlets. To determine whether the inlets are right or left inlets, place the spigot of the bend down and look through the hub (or bell) end (this is the same direction water would be flowing down the drain) (*Figure 19*). Left inlets are on the left side, and right inlets are on the right side.

Three patterns of ¼ bends are available (*Figure 20*). The basic fitting is simply called a quarter bend.

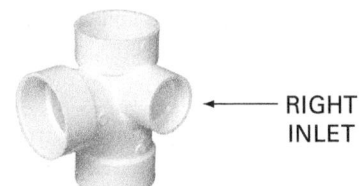

SANITARY TEE WITH RIGHT SIDE INLET

← RIGHT INLET

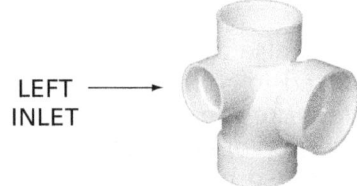

LEFT → INLET

SANITARY TEE WITH LEFT SIDE INLET

Figure 19 Determining left- or right-side inlet.

QUARTER BEND

SHORT SWEEP BEND

LONG SWEEP BEND

Figure 20 Variations of ¼ bends.

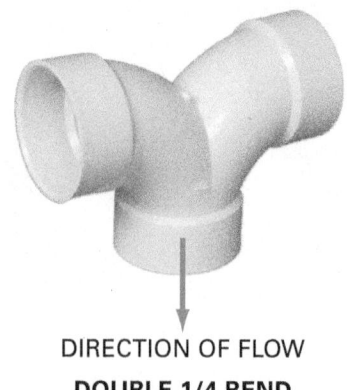

DIRECTION OF FLOW

DOUBLE 1/4 BEND

Figure 21 Double ¼ bend.

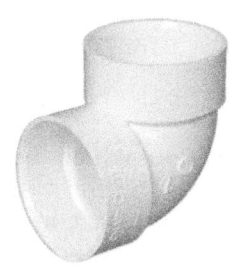

VENT ELL

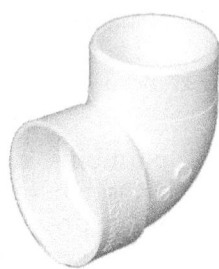

VENT ELL, STREET

Figure 22 Vent ells.

Short sweep ¼ bends and long sweep ¼ bends may be used at the base of DWV stacks. They are required by various codes because the longer radius of their turn greatly decreases flow resistance and back pressure. Always check the applicable codes to determine which bend must be used in your area.

Double ¼ bends (also called twin ells), shown in *Figure 21*, are used to collect and combine the flow from two opposite runs into a single run of pipe. Note the direction of flow in *Figure 21*.

Codes allow the use of vent ells (*Figure 22*) only in vent lines. If they were placed in drainage or waste lines, their sharp turn radius would severely restrict the flow of materials. They are available only in plastic fittings.

3.2.3 Adapters

DWV fittings are most often used with pipe made from the same material. However, adapters (*Figure 23*) can be used to join pipe of different materials. For instance, adapters can be used to join copper tubing to galvanized iron pipe or plastic to cast-iron pipe. Adapters are also useful for joining pipes of different sizes, when permitted by code.

3.2.4 Cleanouts

Fittings are also available for cleanouts. Cleanout fittings have internal threads on the branch fitting to accept a threaded cleanout plug. A cleanout adapter (*Figure 24*) may be installed in one end or branch of a fitting to provide a cleanout access. In both cases, a cleanout plug is provided to permit access to the DWV piping system to remove blockage and to prevent leakage.

3.2.5 Tees

Sanitary tees, shown in *Figure 25*, are used for branches that run from horizontal to vertical. Model codes restrict their use to sanitary drainage systems where the flow of material is from the horizontal to the vertical. Double sanitary tees are also called sanitary crosses.

Sanitary tees are available with side inlets. The side inlet allows smaller drains to be connected from the right or left.

Most codes restrict the use of vent tees (*Figure 26*) and they may not be used to vent lines or as cleanout fittings. They are prohibited for use in drainage systems because their design restricts the flow of material. This design restriction may allow wastes and pipe scale to collect within the fitting.

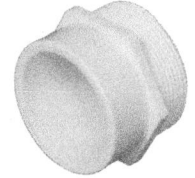

TRAP ADAPTER – MALE

TRAP ADAPTER – MALE
(SPG X SLIP
WITH CHROME NUT)

FEMALE ADAPTER

NO-HUB ADAPTER

TRAP ADAPTER – MALE WITH 1-1/2" PLASTIC NUT
AND WASHERS TO FIT 1-1/2" AND 1-1/4" TRAPS

Figure 23 Adapters.

FITTING CLEANOUT
ADAPTER

FITTING CLEANOUT
ADAPTER WITH
CLEANOUT PLUG

Figure 24 Cleanout adapters.

Test tees (*Figure 27*) are installed in the DWV plumbing system as required by various codes. These tees serve as test locations from which the system is pressurized to test for leaks.

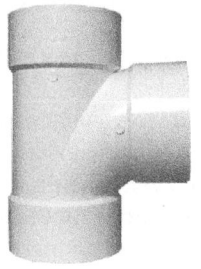

SANITARY TEE

DOUBLE SANITARY TEE

Figure 25 Sanitary tee and cross.

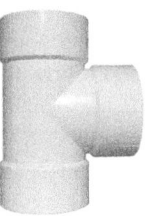

Figure 26 Vent tee.

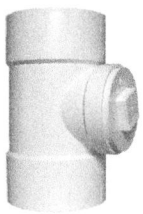

CLEANOUT TEE
WITH CLEANOUT PLUG

Figure 27 Test tee.

3.2.6 Wyes

Sanitary wyes (*Figure 28*) are used to provide a smooth-flowing DWV system, in keeping with code requirements. Codes require vertical branch lines that intersect with horizontal branch lines to connect with long-turn fittings, such as wyes, with 45-degree angles or sweeps. If the correct fittings are not used, soil and wastes may collect on the pipe wall opposite the branch.

Sanitary upright wyes (*Figure 29*) are used to connect the vent stack to the lower end of the soil and waste stacks.

A **vent branch** (or branch vent) (*Figure 30*) is used to join the upper end of the vent to the top of the soil and waste stacks.

Pressure Testing the DWV System

Plumbers pressure test the installed DWV system to ensure that it is free of leaks. Generally, local codes require this test, and the local building inspector may also conduct a follow-up test. To perform a pressure test, turn off the main water supply and block the vent and drain pipes at T-fittings close to the main stack. You must also block all openings such as drains, fixtures, and stubouts using solvent-glued plastic caps or inflatable test balloons. Following the manufacturer's instructions, connect the air compressor and pressure gauge at an access point (usually a cleanout fitting). Run the compressor until the pressure gauge registers the correct pounds per square inch (psi) recommended by the manufacturer or by local code. Check the pressure gauge from time to time for at least 15 minutes. If the pressure remains unchanged at the correct psi, the DWV lines are leak free. Falling pressure indicates a leak.

Even if the pressure does not change, you should also listen for signs of a slow leak along the piping. Slow leaks may not register on the gauge. If you locate a leak, repair any loose or faulty fittings and then repeat the pressure test. You must follow the manufacturer's safety guidelines when working with the compressor.

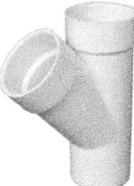   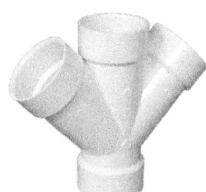

WYE, STREET   DOUBLE WYE

Figure 28 Sanitary wyes.

Figure 29 Sanitary upright wye.

Inverted wyes (*Figure 31*) may be used in place of vent branches. A sanitary combination is a fitting that combines a wye and a $\frac{1}{8}$ bend. Also called a tee-wye, this fitting is available with single or double inlets (*Figure 32*). Sanitary combinations are used to connect horizontal branch lines that intersect at 90 degrees with other horizontal branch lines. They are also used to connect a vertical stack with a horizontal drain. Whenever space permits, they may be used in place of sanitary tees because they offer less resistance to the flow of materials than sanitary tees do. Combination fittings also reduce the number of fittings needed. *Figure 33* shows that a sanitary wye and a $\frac{1}{8}$ bend would be needed to do the same job as a combination fitting. Reducing the number of fittings also reduces the number of joints to be made and the amount of time needed to make them.

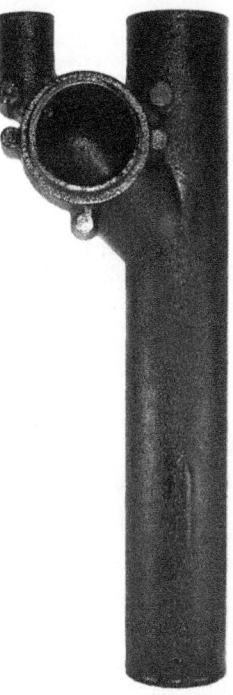

NO-HUB STARTER FITTING
WITH OR WITHOUT 2" NO-HUB INLETS
(Designed for use above the floor with back-outlet water closets)
DOUBLE, LEFT, OR RIGHT

Figure 30 Vent branch.

WYE

Figure 31 Inverted wye for venting.

3.2.7 Miscellaneous Fittings

Sanitary increasers (*Figure 34*) are used to enlarge the diameter of vent stacks and are usually placed at least 1 foot below the stack's intersection with the roof, although this distance may vary by code. Sanitary increasers are necessary in cold climates to keep condensing water vapor from freezing and gradually closing the vent opening. The loss of the vent could cause the loss of the trap seals, allowing sewer gas to enter the building.

COMBINATION WYE AND 1/8 BEND

DOUBLE COMBINATION
WYE AND 1/8 BEND

Figure 32 Sanitary combinations.

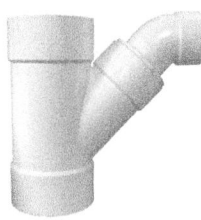

COMBINATION WYE AND
1/8 BEND, REDUCING
(TWO PIECE)

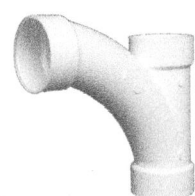

COMBINATION WYE
AND 1/8 BEND
(ONE PIECE)

Figure 33 Combination fittings.

As you have already learned, offsets (*Figure 35*) are used to change the path of the pipe to avoid obstruction. They can offset the run of the pipe from 2" to 12".

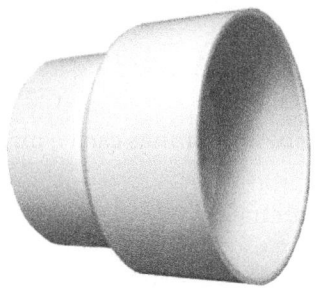

PIPE INCREASER-REDUCER

Figure 34 Sanitary increaser.

OFFSET CLOSET FLANGE, ADJUSTABLE
(WITH METAL RING)
1-1/2" OFFSET

Figure 35 Offset.

Additional Resources

Plumbing Venting: Decoding Chapter 9 of the IPC, Bob Scott. 2014. Procodeclasses.

2015 International Plumbing Code Commentary (Includes IPSDC), 2014. International Code Council (ICC).

3.0.0 Section Review

1. DWV pipe and fittings can be made of all the following materials *except* _____.

    a. cast iron
    b. melamine
    c. clay
    d. glass

2. A ⅛ bend changes the direction of flow _____.

    a. 22½ degrees
    b. 30 degrees
    c. 45 degrees
    d. 90 degrees

4.0.0 DWV CONSTRUCTION

Objective

Describe the construction of various DWV systems.
   a. Explain the importance of grade.
   b. Describe the construction of sewer and waste treatment facilities.
   c. Identify the health concerns associated with DWV systems.
   d. Explain how plumbing codes affect the construction of DWV systems.

Trade Terms

Branch interval: A distance along a soil or waste stack corresponding, in general, to a building story height, but in no case less than 8', within which the horizontal branches from one floor or story of a building are connected to the stack.

Elevation: The height above an established reference point, such as a grade reference point on a construction drawing.

Grade: When referring to DWV and especially sewer systems, the slope of a run of pipe. Also referred to as slope or percent of grade.

Sludge: Semi-liquid matter that settles out in a holding tank during the waste treatment process.

Velocity: Technically, the speed and direction of a moving object. Commonly used in place of speed for describing the motion of fluids in pipes.

Plumbers need to be familiar with the bigger picture of waste water system design and sewage management. Connections to municipal and private sewage systems must be made so that the effectiveness of these systems is not compromised.

4.1.0 Grade

Drainage and waste systems (*Figure 36*) rely on gravity to move solid and liquid wastes, so these piping systems must be installed at a slope toward the point of disposal. In the plumbing industry, this slope is called grade. Grade is also often referred to as percent of grade. Drainage and waste piping systems are designed with the grade engineered into the system. The architects and engineers who design the piping system normally determine the grade. However, in some cases, such as in residential plumbing, the plumber selects the grade according to applicable code.

4.1.1 The importance of Grade

In a system with the proper grade, the liquid wastes flow at the right velocity, or speed, to scour the insides of the pipe, and the solids are carried away. If too much grade is used, the liquid wastes may flow too fast, leaving the solids behind. If too little grade is used, the liquid wastes will not flow fast enough to scour the pipe and remove the solid wastes. If the grade of a pipe does not remain constant, the velocity of the liquid wastes change at the point where the grade changes. In any of these cases, the pipe soon becomes blocked with solid wastes.

Plumbers must determine the grade before they begin their work. This information may be in the local plumbing codes, in the specifications (specs) for the structure, or in the construction drawings. If it is not in any of these places, contact the local plumbing inspector for a decision.

4.2.0 Sewer Construction and Waste Treatment

DWV systems within a building collect wastes to be delivered for treatment. Plumbers need to be familiar with the systems beyond the building walls that collect, treat, and discharge processed wastes to the environment. Familiarity with the health issues and codes relating to sewage and waste water is essential for the professional plumber.

4.2.1 The Building Drain

The building drain is the main horizontal pipe inside a building. It carries all sewage and other liquid wastes to the building sewer. Codes usually define the building sewer as standing 2 to 3 feet outside the building foundation. The building drain is the principal artery to which other drainage branches of the sanitary system may be connected.

Any vertical pipe, including the waste and vent piping of a plumbing system, is considered a stack. A soil stack is a vertical section of pipe that receives the discharge of water closets, with or without the discharge from other fixtures. The soil stack is connected to the building drain.

Figure 36 Grade.

A branch interval is a section of a stack. The branch interval usually corresponds to a story height (the height of one floor in a building), but it can never be less than 8′ long.

A horizontal branch is the part of a drain pipe that extends laterally (sideways) from a soil or waste stack and receives the discharge from one or more fixture drains.

4.2.2 Cleanouts

Cleanouts (*Figure 37*) are fittings with removable plugs. The plugs provide access to the inside of drainage and waste piping systems so that blockages can be removed. As mentioned earlier, cleanout adapters may be used to convert other fittings so they can be used as cleanouts.

PVC DWV TWISTLOK ™ PLUG

Figure 37 Cleanout adapter with recessed hubs.

Torpedo Level

The torpedo level is a tool similar to a general-purpose level, but it is much smaller—usually less than 1-foot long—and more streamlined. Its major advantage is in measuring grade for small runs of pipe—for example, fixture branch lines that run a short distance to the stack. The torpedo level is also light and easy to manipulate in tight places.

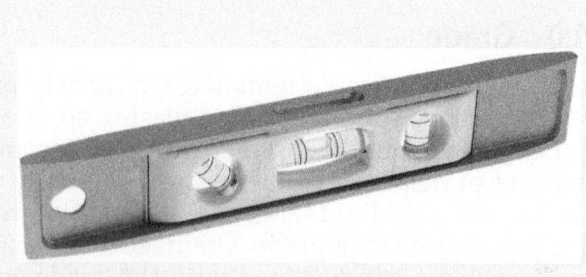

4.2.3 Building Sewer

The building sewer or house sewer is the drainage piping that runs from the building's foundation to the sewer main or septic tank (private waste-disposal system). It normally starts approximately 2' to 3' outside the building.

Building sewers are commonly made using ABS (acrylonitrile-butadiene-styrene) or PVC (polyvinyl chloride) plastic, cast-iron, or vitrified clay (a hard, nonporous clay) pipe. When sewers are laid, the ground must be tamped to keep the pipe from settling and losing its grade. Sometimes the pipe is installed over a bed of gravel that supports it. In some parts of the country, plumbers lay the entire sewer from the foundation wall to the sewer main. In other areas, the municipal sewer crew may be in charge of the installation from the property line to the sewer main. All operations associated with building sewers are regulated by code.

Vitrified Clay

Vitrified clay is typically only used in sewers and falls under the purview of municipality work. Plumbers will rarely work with vitrified clay directly but may encounter it on a jobsite after it's already been placed by a pipelayer. Vitrified clay is used because it is corrosion resistant to common sewer chemicals; it stands up to abrasive cleaning; and has a an extremely long life.

Manholes must be provided for underground piping that is 8" or larger in diameter. They should be located at intervals not more than 400' apart and at every major change in direction, grade, elevation, or pipe size. To meet applicable codes for traffic and loading conditions, the manholes must have metal covers of sufficient weight and strength.

4.2.4 Sewer Main

A public or municipal sewer is installed, maintained, and controlled by the local municipality or town. The sewer main is usually located in a street or alleyway or within an easement on privately owned land. Sewer mains carry waste to the treatment plant.

Municipalities or towns usually install a 6" sewer laterally from the sewer main to the edge of each building lot. This lateral pipe connects the building sewer to the public sewage system.

4.2.5 Municipal Waste Treatment Systems

Many municipalities have sewer systems in which wastes are collected and treated at a sewage plant, then discharged back into the ecosystem. Municipal waste treatment plants (*Figure 38*) are highly sophisticated facilities, as public health and safety depend on their operation.

Backwater Valves

The backwater valve is a specially designed check valve for drainage pipe systems. A check valve prevents fluid from going backwards in the pipe system; the backwater valve prevents the backflow of sewage in the system. When plumbers install backwater valves underground, they usually install them in a valve box or vault so they can be accessed easily. Backwater valves are for special applications only, and they are the only type of valve that some jurisdictions allow.

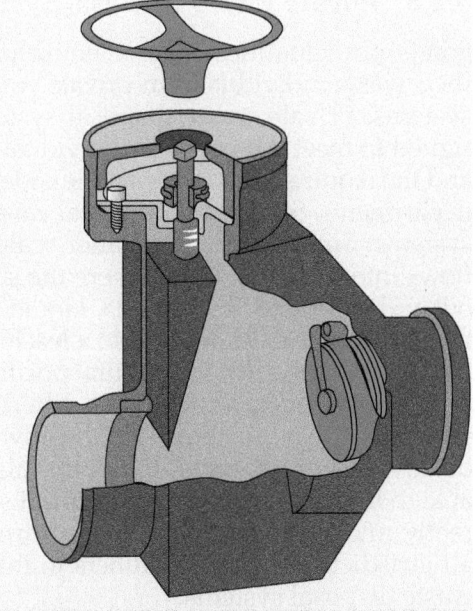

Figure 38 Municipal sewage treatment plant.

These systems are designed to handle thousands of gallons of sewage each day. The treatment facilities receive sewage into huge holding tanks, where heavier substances settle to the bottom and lighter substances float to the top. The heavier layer is called sludge. Both the sludge and the wastewater are then treated.

4.2.6 Private Waste Disposal Systems

Some municipalities require households to treat their waste individually in private waste disposal systems. Private waste disposal systems are designed to meet the needs of individual households and the requirements of applicable codes and health departments. Similar to municipal waste treatment systems, but on a much smaller scale, the waste flows into a holding tank, where the sludge settles out and is digested by bacteria. Liquid waste flows through a distribution box into a leach field, where it seeps into the earth in a natural purification cycle.

Plumbers must know about different types of private disposal systems and the advantages and disadvantages of each. Plumbers also must be able to install private waste disposal systems correctly and according to code requirements. Not all jurisdictions allow plumbers to install private waste disposal systems.

A conventional septic system consists of a septic tank, a distribution box, a leach field, and piping between those parts (*Figure 39*). A septic tank system provides partial treatment of raw wastewater. It protects the soil absorption system from becoming clogged by solids that are suspended in the raw wastewater. Applicable codes strictly regulate the use of these systems.

4.3.0 Health Concerns

Properly designed and installed DWV systems are essential to public safety. Without DWV systems, the public would be at great risk of waste-borne illness and disease. The plumbing profession has improved public health, safety, and comfort over the past 150 years. Many serious health risks have been dramatically reduced as a direct result of good plumbing, especially properly designed and installed DWV systems, and the enforcement of plumbing codes.

The health issues related to the improper design, installation, and maintenance of DWV systems are significant. In the past, diseases such as cholera, typhoid fever, typhus, and dysentery have been traced to failures in DWV systems. These diseases can spread rapidly when bacteria from the sewer system enter buildings through damaged or improperly installed vents and traps.

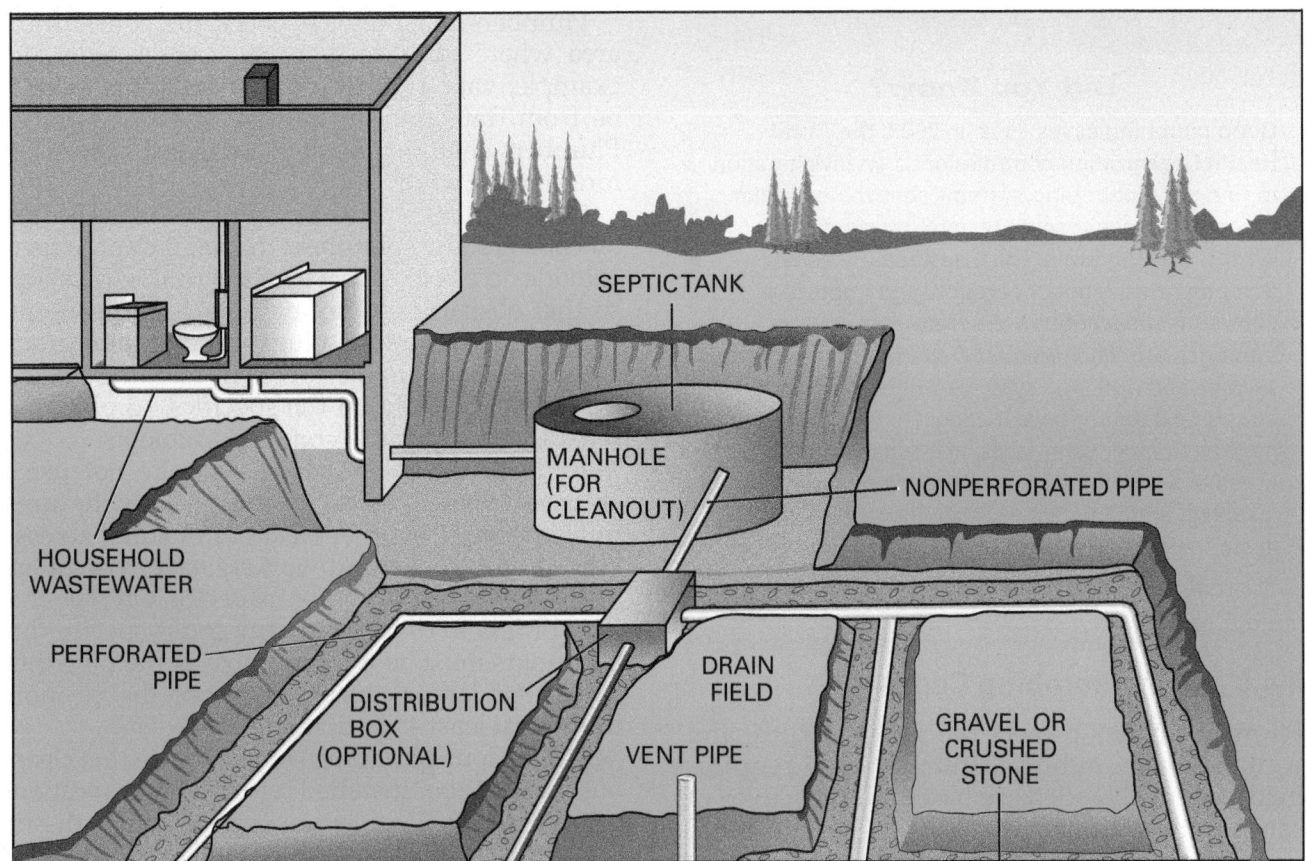

Figure 39 Parts of a septic system.

Another health issue related to DWV systems is the accumulation of toxic sewer gases. Explosions, fires, and suffocation can occur when sewer gases are not properly released through good ventilation. Control of sewage through DWV systems can eliminate many public sanitation and health problems.

Did You Know?

The Babylonians had sewer systems around 5,000 years ago. The Romans later built sewer systems for both storm water and wastewater. However, water treatment methods were unknown then, so wastewater was often returned to the river just below a city or town.

GOING GREEN

Energy Efficient Fixtures

Each year, the average US household may waste up to 450,000 gallons water. Plumbers can help reduce the flow of water "down the drain" by installing modern, energy efficient fixtures. Low-flow showerheads combine air with water pressure to deliver a good shower using less water. Instant hot water dispensers and tankless water heaters can reduce the amount of water consumers run in sinks, tubs, and showers while waiting for the water to warm up. High-efficiency toilets may use as little as 1.28 gallons of water per flush. Learning how to help your customers use sustainable features adds a lot of personal satisfaction to your work while also helping to sustain our natural resources.

4.4.0 DWV Plumbing Codes

Plumbing codes protect the safety, health, and welfare of the public. Although code requirements vary, all codes are based on principles of sanitation and safety.

There is no single national plumbing code with requirements adopted by all states and localities. Model codes are developed and revised on a regular basis. They can recognize international and national codes as well as other standards. States and other jurisdictions can use these model codes as a basis for developing their own plumbing codes. DWV piping requirements are also defined by counties and municipalities and enforced by their inspectors.

Plumbers must always check the codes in the area where you are working. Model codes, for example, vary widely on how far vents have to be from traps and on the size of fixture drains. Plumbers should consult the applicable code before starting a job.

Codes establish minimum standards. Almost all codes require plumbers to install clean-outs to provide access to all parts of the drainage system so that obstructions can be removed. Clean-outs range from removable plugs in horizontal drainage piping to manhole covers in building sewers. Codes vary widely on the specifics, so plumbers must check applicable code requirements.

Under most codes, plumbers may not use a cleanout plug opening to install new fixtures, unless there is another cleanout of equal accessibility and capacity and workers have written approval from the plumbing inspector. On pipe that is 4" or less in diameter, many codes specify that clean-outs must be the same size as the pipe they serve. For larger pipe, the size of the cleanout must be at least 4" in diameter.

The various plumbing codes require that cleanout fittings be installed at specified locations within the DWV piping system and that the fittings be accessible. Most codes have requirements for cleanout locations in horizontal runs of pipe. In horizontal drain lines 4" in diameter or less, cleanouts generally must be installed no more than 50' apart. For larger lines, cleanouts cannot be more than 100' apart. Local codes may differ regarding these distances. Always refer to the applicable code.

Some codes also require cleanouts at or near the foot of each waste or soil stack and near the junction of the building drain and the building sewer.

Additional Resources

"Design of Sewer System." Civil Engineers PK. Updated 2017. **https://civilengineerspk.com/**

2015 International Plumbing Code Commentary (Includes IPSDC), 2014. International Code Council (ICC).

Plumbing Venting: Decoding Chapter 9 of the IPC, Bob Scott. 2014. Procodeclasses.

4.0.0 Section Review

1. The drop of a drain or sewer pipe for a given horizontal distance traveled is called _____.

   a. rise
   b. elevation
   c. altitude
   d. grade

2. Ordinarily, the distance along a DWV stack between the horizontal branch drains on one floor and those on the next is called the _____.

   a. drop
   b. hydraulic gradient
   c. branch interval
   d. elevation

3. What has been reduced as a result of modern DWV systems?

   a. Health risks to the general public
   b. Volume of sewage
   c. The types of infectious bacteria and viruses
   d. The production of explosive gases from sewage

4. Most plumbing codes require cleanouts in piping 4" or less to be spaced apart no more than _____.

   a. 10'
   b. 25'
   c. 50'
   d. 100'

SUMMARY

The DWV system of a building is part of a plumbing system designed to protect the health and safety of the people who use its facilities. The system carries wastewater out of the building, treats it, and returns it to the ecosystem.

Drain and waste piping removes wastewater from a building. The type and size of the piping system selected depends on critical factors such as the amount of fluids expected to flow through the piping, the types of fluids to be carried, and the grade. Drainage and waste systems rely on gravity to move solid and liquid wastes, so the plumber must install piping at the correct grade toward the building sewer where the wastes leave the building and enter the public or private waste disposal system. Grade determines the velocity of the liquid waste flowing through the piping. If the grade is too shallow, the liquid waste moves too slowly and does not scour the pipe and remove solid wastes. If the grade is too steep, the liquid waste flows too fast and leaves solids behind in the piping.

The vent system is an important part of the overall DWV system. Vent piping provides for the free flow of air in the drainage system. This air equalizes the atmospheric pressure inside the pipes and prevents back pressure or siphoning from destroying the water trap seals in the fixtures. The water trap seals keep sewer gas and odors out of the building. Traps provide a way for the wastewater or sewage to flow through the fixture and into the piping system while protecting the occupants of the building from bacteria and potentially explosive gases.

The DWV piping system relies on different kinds of fittings to join the lengths of pipe. The fittings may be made of copper, brass, lead, steel, cast iron, clay, glass, or plastic. The type of installation determines the best piping and fitting material to use. The fittings are designed so they do not block or slow the flow of materials in the pipe. The sweeping design of DWV fittings allows for the smooth flow of material within the system. Different fitting shapes serve specific purposes, depending on whether a pipe runs horizontally or vertically. Correct selection and installation of fittings are vital for the system to function properly.

1. An example of a fixture drain is a(n) _____.
   a. P-trap
   b. pop-up plug
   c. sanitary bend
   d. interceptor

2. LEED standards for commercial and industrial applications encourage the development and use of _____.
   a. graywater systems
   b. sewage treatment plants
   c. septic systems
   d. vent stacks

3. Which DWV component should be considered specifically for a fast-food restaurant?
   a. Fixture drain
   b. P-trap
   c. S-trap
   d. Interceptor

4. The size of the trap seal is the distance between the _____.
   a. top dip and bottom dip
   b. the crown weir and the top dip
   c. crown and the bottom dip
   d. entrance to the fixture drain and the crown weir

5. The total drop from a trap's crown weir to the bottom of the trap's connection to the vent must be _____.
   a. less than one drain pipe diameter
   b. more than one drain pipe diameter
   c. less than one vent pipe diameter
   d. more than one vent pipe diameter

6. Trap siphonage occurs when _____.
   a. gusts of wind affect stack pressure
   b. water flows by the trap's connection to the vent
   c. vent pressure is less than air pressure at the fixture
   d. porous materials get caught in the trap connection to the vent

7. If a fixture outlet drain line is too long, which effect can break a trap seal?
   a. Siphonage
   b. Oscillation
   c. Momentum
   d. Back pressure

8. The most common materials for DWV systems are cast iron and _____.
   a. glass
   b. copper
   c. clay
   d. plastic

9. To maintain the water seal in a trap, every trap needs a(n) _____.
   a. interceptor
   b. vent
   c. side inlet
   d. drain

10. The main difference between a vent ell and a drain bend is their _____.
    a. turn radius
    b. diameter
    c. material
    d. pipe size

11. If a drain's grade does not remain constant, what could be the result?
    a. Drain flow could cause siphonage.
    b. Waste treatment could become inefficient.
    c. Solid wastes could accumulate in the pipe.
    d. Sewer gas could accumulate above the flow.

12. A building's sewer connection begins _____.
    a. at the building foundation
    b. about 2' to 3' from the foundation
    c. at the point where the building's drain line connects to the sewer main or septic tank
    d. at the bottom of the building's DWV stack

13. Conventional private waste disposal systems include a _____.

    a. composting system
    b. graywater system
    c. connection to the municipal waste treatment plant
    d. septic tank and leach field

14. Common entry points in the DWV system for serious diseases are _____.

    a. damaged or faulty vents and traps
    b. septic tanks
    c. manhole covers
    d. leach fields

15. With regards to plumbing codes you will be required follow, they comply with the _____.

    a. international plumbing code
    b. national plumbing code
    c. state plumbing code
    d. codes established and recognized by state, local, and applicable jurisdictions

16. The drainpipe buried outside a building is called a(n) _____.

    a. underground sewer
    b. building sewer
    c. main sewer
    d. out sewer

17. All of the following are major components of a DWV system except for _____.

    a. soil stack
    b. traps
    c. tees
    d. flanges

18. According to code requirements, the direction of hub-type fittings should not _____.

    a. be installed horizontally
    b. be larger than three inches in diameter
    c. go against the flow of the system
    d. be made of the same material as the pipe

19. The most common materials for DWV fittings are _____.

    a. copper and lead
    b. cast-iron and copper
    c. cast-iron and plastic
    d. plastic and lead

20. Bends are available with _____ inlets.

    a. heel
    b. forward
    c. branch
    d. wye

Fill in the blank with the correct term that you learned from your study of this module.

1. DWV branches are connected to the main DWV system using _____.

2. _____ describes the attachment of one trap to another trap, such as with a water closet.

3. The substance that settles on the bottom of holding tanks is _____.

4. The underground drain pipe that carries waste from the building to the public sewer is called a(n) _____.

5. _____ allow pipes made from different materials to be connected.

6. The flow from two opposite runs into a single run of pipe are collected and combined in _____.

7. _____ are bend fittings with a short radius that are used at the base of a DWV stack.

8. In cold climates, _____ can prevent vent openings from closing as a result of frozen condensation in pipes.

9. _____ may not be used in drainage systems because their design can cause waste to collect in the fitting.

10. The sharp turning radiuses of _____ severely restricts the flow of materials, making their use allowable only in vent lines.

11. _____ are designed to keep undesirable or hazardous materials from entering a building drainage system, a public or private sewer, or sewage treatment plant or process.

12. Vertical branch lines that intersect with horizontal branch lines must be joined at an angle with _____.

13. _____ serve as locations to conduct leakage tests.

14. _____ collects on fittings, especially on iron or steel, as a result of metal corrosion.

15. Horizontal branch lines that intersect with other horizontal branch lines at a 90-degree angle are connected using _____ fittings.

16. _____ is one of four factors that must be considered when determining the intervals for manholes.

17. The _____, also called slope, of the piping system works with gravity to move solid and liquid waste through DWV systems.

18. The discharge overflow lip or ledge at the trap outlet is the crown _____.

19. _____ strain the wastewater before it enters the drainage piping.

20. _____ are lengths of pipe that continue in a straight line.

21. When pressure inside the DWV piping is greater than atmospheric pressure, it is called _____.

22. A(n) _____ is the access point to connected parts of the drainage system for the removal of blockages.

23. _____ allow smaller lines to be connected in line to a bend.

24. A(n) _____ is the section of stack between points where branch pipes connect to the main DWV stack.

25. An imbalance in pressure between the inside and outside DWV piping can cause _____.

26. Code generally requires the use of _____ at the base of stacks.

27. _____ is a general term for most vertical line including offsets of soil, waste, vent, or inside conductor piping.

28. A one-piece or two-piece trap with a union nut is called a(n) _____.

29. A(n) _____ consists of a circuit of piping inside a building.

30. Improvements in plumbing have replaced _____ with the P-trap.

31. Porous or fibrous material caught in a trap can cause a seal to be broken by _____.

32. The infrequent use of a trap can cause _____ of the seal.

33. _____ are devices used to connect pipe in DWV systems.

34. The upper ends of vents are joined to the top of soil and waste stacks by a(n) _____.

35. Vent stacks are attached to the lower end of the soil and waste stacks by _____.

36. Connectors that may be used in place of vent branches are called _____.

37. The level at which water tends to flow through the trap is the _____.

38. _____ is the speed at which waste flows through pipes.

39. The _____ in a horizontal pipe from crown weir to vent must not exceed one pipe diameter.

40. _____ is wastewater generated from domestic processes such as laundry and bathing.

41. _____ permit connecting smaller lines to wyes and tees.

42. The downstream end of a pipe or the male end of a fitting that can be inserted into another DWV fitting is called the _____ end.

Trade Terms

Adapters	Evaporation	Pipe scale	Sludge
Back pressure	Fall	Runs	Spigot
Branch interval	Fixture drains	S-traps	Stack
Building sewer	Grade	Sanitary combination	Test tees
Capillary action	Graywater	Sanitary fittings	Velocity
Cleanout	Heel inlets	Sanitary increasers	Vent branch
Double ¼ bends	Hydraulic gradient	Sanitary upright wyes	Vent ells
Double trapping	Interceptors	Sanitary wyes	Vent tees
Drainage fittings	Inverted wyes	Short sweep ¼ bends	Weir
DWV system	Long sweep ¼ bends	Side inlets	
Elevation	P-trap	Siphonage	

Trade Terms Introduced in This Module

Adapters: Fittings that join pipes of different sizes or materials, such as copper and galvanized pipe or cast-iron and plastic pipe.

Back pressure: A condition which may occur in the DWV system whereby a higher pressure than atmospheric pressure is created in the drain/vent piping, causing a reversal of the normal flow through drain piping and traps. Also referred to as backpressure backflow.

Branch interval: A distance along a soil or waste stack corresponding, in general, to a building story height, but in no case less than 8 feet, within which the horizontal branches from one floor or story of a building are connected to the stack.

Building sewer: The part of the drainage system that extends from the end of the building drain and conveys its discharge to a public sewer, private sewer, individual sewage-disposal system, or other point of disposal.

Capillary action: The tendency of water to be drawn into porous or fibrous material against gravity above the level of the water source.

Cleanout: An access point to connected parts of the drainage system for the removal of blockages.

Double ¼ bends: Fittings used to collect and combine the flow from two opposite runs into a single run of pipe.

Double trapping: A situation in which one trap is attached to another, creating negative pressure that stops the intended flow of drainage.

Drainage fittings: Fittings installed in the drainage sections of the DWV system to remove waste from a building.

Drain-waste-vent (DWV) system: Refers to the combined sanitary drainage and venting systems. This term is technically equivalent to soil-waste-vent (SWV).

Elevation: The height above an established reference point, such as a grade reference point on a construction drawing.

Evaporation: The natural change from liquid to vapor of water at a temperature below its boiling point.

Fall: The amount of slope given to horizontal runs of pipe expressed as a height in inches per foot of run.

Fixture drains: The drains from traps of fixtures to the junction of those drains with any other drain pipe.

Grade: When referring to DWV and especially sewer systems, the slope of a run of pipe. Also referred to as slope or percent of grade.

Graywater: Wastewater generated from domestic processes other than human-waste disposal. These include laundry, dishwashing, and bathing. Graywater comprises 50 to 80 percent of residential wastewater.

Heel inlets: Openings in the curved, heel portion of bend fittings in line with the bend, used to connect smaller lines to the main line.

Hydraulic gradient: The level of the surface of water flowing in a partially-full pipe by gravity alone.

Interceptors: Devices designed and installed so as to separate and retain deleterious, hazardous, or undesirable matter from normal waste, while permitting normal sewage or liquid wastes to discharge into the drainage system by gravity.

Inverted wyes: Fittings used to join the upper end of vents to the top of soil and waste stacks. The have the appearance of the upside-down letter Y.

Long sweep ¼ bends: Long-radius bends commonly used at the base of DWV stacks and elsewhere when the longer radius is needed to greatly decrease flow resistance and back pressure. Their use is regulated by code.

Pipe scale: A flaky, adherent coating on pipe walls resulting from the corrosion of metals, especially iron or steel. Also, a heavy oxide coating on copper or copper alloys resulting from exposure to high temperatures and oxygen.

P-trap: A trap constructed in the shape of the letter P with the loop facing downward, which provides a water seal in a waste or soil pipe, used mostly at sinks and lavatories.

Runs: Lengths of pipe that continue in a straight line.

S-traps: Traps with long downstream legs, which tend to promote siphonage. S-traps are no longer permitted by code for new installations but are still found in older buildings.

Sanitary combination: A fitting that combines a wye and 1/8 bend. It is used to connect horizontal branch lines that intersect other horizontal branch lines. It offers less resistance to the flow of material than a sanitary tee. Also called a tee-wye.

Sanitary fittings: Fittings used to connect DWV branches to the main DWV system and to serve as cleanouts.

Sanitary increasers: Fittings used to enlarge the diameter of vent stacks. They are usually placed at least 1' below the penetration of the stack through the roof. Their use is regulated by code.

Sanitary upright wyes: Fittings used to connect separate vent stacks to the lower ends of soil and waste stacks.

Sanitary wyes: Drainage fittings, shaped like the letter Y, that join branches to the main run of pipe at an angle.

Short sweep 1/4 bends: Bend fitting with a short radius used at the base of a DWV stack and elsewhere. Their use is regulated by code

Side inlets: Openings in ell or tee fittings at right angles to the line of the run, used to connect smaller lines to the main line.

Siphonage: Loss of water in a trap seal started by unequal pressure inside and outside DWV piping. The water initially flows in the direction of the lower pressure. Sustained siphonage, even in the absence of pressure differences, results from the cohesive property of water.

Sludge: Semi-liquid matter that settles out in a holding tank during the waste treatment process.

Spigot: When referring to DWV systems, the pipe end or the male portion of a fitting that inserts into the hub of a downstream fitting.

Stack: A general term for certain vertical DWV pipes, including offsets of soil, waste, vent, or inside conductor piping. This does not include vertical fixture and vent branches that do not extend through the roof or that pass through not more than two stories before being reconnected to the vent stack or stack vent.

Test tees: Tees installed as test locations for pressurizing the DWV system to test for leaks before placing it in operation.

Velocity: Technically, the speed and direction of a moving object. Commonly used in place of speed for describing the motion of fluids in pipes.

Vent branch: A vent connecting one or more individual vent lines with a stack vent.

Vent ells: Plastic fittings with a sharp turn radius, used only in vent piping systems. Their use is regulated by code.

Vent tees: Fittings used in venting systems or as cleanouts. They may not be used in the drainage system because it restricts the flow of material. Their use is regulated by code.

Weir: When referring to plumbing, a ledge or lip in a fixture that controls the level of water inside the fixture. The word comes from Old English, meaning dam.

Additional Resources

This module is intended as a thorough resource for task training. The following reference works are suggested for further study.

"Design of Sewer System." Civil Engineers PK. Updated 2017. **https://civilengineerspk.com/**.
2015 International Plumbing Code Commentary (Includes IPSDC), 2014. International Code Council (ICC).
Plumbing Venting: Decoding Chapter 9 of the IPC, Bob Scott. 2014. Procodeclasses.

Figure Credits

SECTION 1.0.0

Answer	Section Reference	Objective
1. c	1.0.0	1a
2. b	1.2.0	1b

SECTION 2.0.0

Answer	Section Reference	Objective
1. b	2.1.0	2a
2. a	2.2.0	2b
3. c	2.3.4	2c

SECTION 3.0.0

Answer	Section Reference	Objective
1. b	3.1.0	3a
2. c	3.2.2	3b

SECTION 4.0.0

Answer	Section Reference	Objective
1. d	4.1.0	4a
2. c	4.2.1	4b
3. a	4.3.0	4c
4. c	4.4.0	4d

NCCER CURRICULA — USER UPDATE

NCCER makes every effort to keep its textbooks up-to-date and free of technical errors. We appreciate your help in this process. If you find an error, a typographical mistake, or an inaccuracy in NCCER's curricula, please fill out this form (or a photocopy), or complete the online form at **www.nccer.org/olf**. Be sure to include the exact module ID number, page number, a detailed description, and your recommended correction. Your input will be brought to the attention of the Authoring Team. Thank you for your assistance.

Instructors – If you have an idea for improving this textbook, or have found that additional materials were necessary to teach this module effectively, please let us know so that we may present your suggestions to the Authoring Team.

NCCER Product Development and Revision

13614 Progress Blvd., Alachua, FL 32615

Email: curriculum@nccer.org
Online: www.nccer.org/olf

❑ Trainee Guide   ❑ Lesson Plans   ❑ Exam   ❑ PowerPoints   Other _____

Craft / Level: _____   Copyright Date: _____

Module ID Number / Title: _____

Section Number(s): _____

Description: _____

_____

_____

_____

_____

Recommended Correction: _____

_____

_____

_____

_____

Your Name: _____

Address: _____

_____

Email: _____   Phone: _____

This page is intentionally left blank.

Introduction to Water Distribution Systems

OVERVIEW

The water distribution system moves water from its source to the building or structure where it is needed. The path the water takes and the types of materials used depend on the building or structure. Plumbers must understand how these water distribution systems work and the different types of materials that are used in these systems.

Module 02112

Trainees with successful module completions may be eligible for credentialing through the NCCER Registry. To learn more, go to *www.nccer.org* or contact us at 1.888.622.3720. Our website, *www.nccer.org*, has information on the latest product releases and training.

Your feedback is welcome. You may email your comments to *curriculum@nccer.org*, send general comments and inquiries to *info@nccer.org*, or fill in the User Update form at the back of this module.

This information is general in nature and intended for training purposes only. Actual performance of activities described in this manual requires compliance with all applicable operating, service, maintenance, and safety procedures under the direction of qualified personnel. References in this manual to patented or proprietary devices do not constitute a recommendation of their use.

02112 V4.5

Objectives

Successful completion of this modules prepares trainees to:

1. Describe the process by which water is distributed in municipal, residential, and private water systems.
   a. Identify and describe water sources.
   b. Explain water treatment processes.
   c. Describe water distribution systems.
2. Identify the major components of a water distribution system and describe the function of each.
   a. Describe the purpose of backflow preventers.
   b. Identify and describe the various types of valves used in water distribution systems.
3. Explain the relationships between components of a water distribution system.
   a. Identify the major components of a building water system and describe how to determine proper placement.
   b. Explain the requirements for sizing of the main supply lines.

Performance Task

Under the supervision of your instructor, you should be able to do the following:

1. Sketch an isometric drawing of a simple water distribution system and label its components.

Trade Terms

Angle valve
Backing board
Branch
Check valve
Chlorination
Coagulation
Corporation stop
Curb box
Curb stop

Fixture risers
Full flow
Galvanic corrosion
Hammer arrestors
Hose bibb
Pasteurization
pH
Precipitates
Pressure regulator valve

Pressure relief valves
Reservoirs
Service lines
Straight-through flow
Supply stop valves
Thermostatic/pressure
  balancing valve
Throttled flow
Turbidity

Ultraviolet (UV) light
Vacuum breaker
Water hammer
Water meter
Water supply fixture units
  (WSFU)
Water table
Well casing

Industry Recognized Credentials

If you are training through an NCCER-accredited sponsor, you may be eligible for credentials from NCCER's Registry. The ID number for this module is 02112. Note that this module may have been used in other NCCER curricula and may apply to other level completions. Contact NCCER's Registry at 888.622.3720 or go to *www.nccer.org* for more information.

CODE NOTE

Codes vary among jurisdictions. Because of the variations in code, consult the applicable code whenever regulations are in question. Referring to an incorrect set of codes can cause as much trouble as failing to reference codes altogether. Obtain, review, and familiarize yourself with your local adopted code. Safety codes are developed by the US Occupational Safety and Health Administration (OSHA).

Contents

Figures

1.0.0 TYPES OF DISTRIBUTION SYSTEMS

Objectives

Describe the process by which water is distributed in municipal, residential, and private water systems.

a. Identify and describe water sources.
b. Explain water treatment processes.
c. Describe water distribution systems.

Trade Terms

Branch: Any part of a piping system other than a riser, main, or stack.

Chlorination: The use of chlorine gas or compounds to disinfect water.

Coagulation: In water treatment processes, a thickening of suspended or dissolved materials into a soft, semi-solid or solid mass.

Corporation stop: A valve that connects the building water service line to the water main.

Curb box: A cylindrical casing placed in the ground over the curb stop, into which a special key-wrench can be inserted to turn off the curb stop. Also given the term buffalo box.

Curb stop: A control valve installed in building water supply lines between the corporation stop and the building.

Galvanic corrosion: Corrosion caused by a weak electrical current that occurs when an electrical path exists between two different metals.

Pasteurization: The practice of heating water and foods to high temperatures to kill harmful bacterial organisms present.

pH: A measure of the acidity or alkalinity of a solution. A pH of 7 is neutral, being neither acidic nor alkaline; higher numbers are more alkaline, lower numbers are more acidic. The symbol comes from the early twentieth-century chemistry term power of hydrogen.

Precipitates: Solid materials resulting from chemical reactions in water solutions that settle out.

Pressure regulator valve: A valve used to reduce water pressure in a building. The valve is activated by changes in pressure within the system.

Reservoirs: Sources of water collected and stored in natural or artificial (man-made) lakes.

Service lines: The main water supply piping, to which branches are connected. Also referred to as feeder lines.

Turbidity: The presence of particles (sand, mud, silt) suspended in water that give the water a cloudy appearance.

Ultraviolet (UV) light: A form of high-energy light with wavelengths shorter than visible light. In water supply systems, it can disinfect water by destroying microorganisms as the water flows through a chamber containing UV lamps.

Water meter: A device for measuring water volume usage by an individual building or customer.

Water table: The level below the ground's surface where soil becomes saturated with water.

Well casing: Outer tube or pipe sunk into the ground after drilling or driving a well to stabilize the hole.

Water supply and distribution play an important role in plumbing systems. Water supply is either private (from a well) or municipal (supplied through a public water distribution system). Components of the water distribution system include the pipes and fittings that carry hot and cold water in a building, the valves used to regulate the flow of water to the fixtures and other outlets, and water heating and treatment equipment.

Any water distribution cycle begins with a water source. Water for a private system comes from a well sunk into an underground water supply that is usually pure enough to drink. Water for municipal systems comes from reservoirs, wells, rivers, lakes, and other sources. It is then treated and distributed to homes and buildings.

In municipal systems, city water undergoes a purification process before it reaches the faucet. Water is pumped to a treatment plant where harmful impurities are removed through processing. Chlorine, aluminum sulphate, and activated charcoal are added to the water during this cycle. This water and chemical mixture flows into a mixing basin where paddles thoroughly mix the chemicals into the water. From here, the water moves to a settling basin where impurities separate from the water. Moving from the basin, sand and gravel filter the water to screen out most of the remaining suspended materials. Treatment plants may add chlorine again to make sure that the water is free of harmful bacteria. Some cities also add fluoride to help prevent tooth decay in the general population. A reservoir holds the filtered water until it is pumped into the main supply pipes that lead to building service lines. Building piping systems deliver water to sinks, bathtubs, showers, dishwashers, icemakers, hoses, and any other water outlets (*Figure 1*).

1.1.0 Sources of Water

As of 2005, about 14 percent of the US population gets its water from private sources. Most of this water comes from wells. Wells are sunk (dug, driven, drilled, or bored) into the earth to extract the water (*Figure 2*). Dug or driven wells are considered shallow wells. This type of well is generally used where the water table is within 20 to 50 feet of Earth's surface. Surface water easily contaminates dug wells. Driven wells are made by forcing a well point into the earth. These wells are practical only where the soil is loose and fairly free of rocks.

Wells that are drilled are considered deep wells and are typically drilled into solid rock. They may extend hundreds of feet down through the earth. They are made using a drilling rig with a rotating diamond-toothed bit that can penetrate the solid subsurface materials. Bored wells are generally shallower and larger in diameter, using an earth auger (drill). These normally do not extend into bedrock. Once the water table is reached and an adequate amount of water volume is found, a well casing is inserted to stabilize the sides of the well and to protect it from contamination. Casings are made of various materials and in different diameters ($1\frac{1}{2}$ to 2 inches) and depths, depending on how far the water is beneath the surface.

After the well is established, water is pumped to the surface to a storage tank inside or near the building, where it can be treated and used by the owners. Generally, the deeper the well, the more gallons per minute can be pumped, depending on the level of the water table. The pump pressurizes the water in the storage tank so that it

Locating and Drilling a Well

The first consideration before drilling or driving a well is state and local regulations. Wells may be approved for household, domestic, or commercial use. A local well driller is the best source of answers to the critical questions of what type of well is needed, what permits are required, and the history of water wells in the area. A variety of different aquifers (water channels) and water tables exist in various formations below the ground surface. A history of wells drilled in a particular area provides important information for the well driller to locate the best spot to drill.

Other considerations are the distance from property lines and the position of septic tanks and septic leach fields in relation to the well site. Most codes require at least 100 feet between the well and the leach field. Because the plumber will make connections to both the well and the wastewater disposal, the plumber must work closely with the owner, the well driller, and code officials.

provides enough water for the installed fixtures to work properly.

Reservoirs are another source of water. Most reservoirs are made by building dams across rivers or streams. Municipalities can then collect and store water for future use. Reservoirs are particularly important in areas that receive little rainfall for part of the year. Pumps move the water from the reservoir to the water treatment plant.

1.2.0 Water Treatment

Much of the water from open reservoirs, lakes, streams, and some wells is not ready for human use until it has been treated. Water must be tested for the presence of chemicals, turbidity (cloudiness resulting from suspended particles), organic

Reclaimed Water

Reclaimed water is distributed through pipes that are separate from the main water distribution pipes. This dual-piping system keeps contaminants in the reclaimed water from entering the potable water system. In the United States, reclaimed water is always distributed in lavender (light purple) pipes to distinguish it from potable water.

The idea of using reclaimed water is not a new one. Los Angeles County, California has provided treated wastewater for irrigation in parks and golf courses since 1929. But this idea may not be in use for all applications in all locations. In general, each community decides how to collect and reuse reclaimed water and each has established codes for this practice. For example, some communities may use reclaimed water to flush fixtures. Others may restrict its use to landscape irrigation for golf courses or sports stadiums.

TYPICAL MUNICIPAL WATER DISTRIBUTION SYSTEM

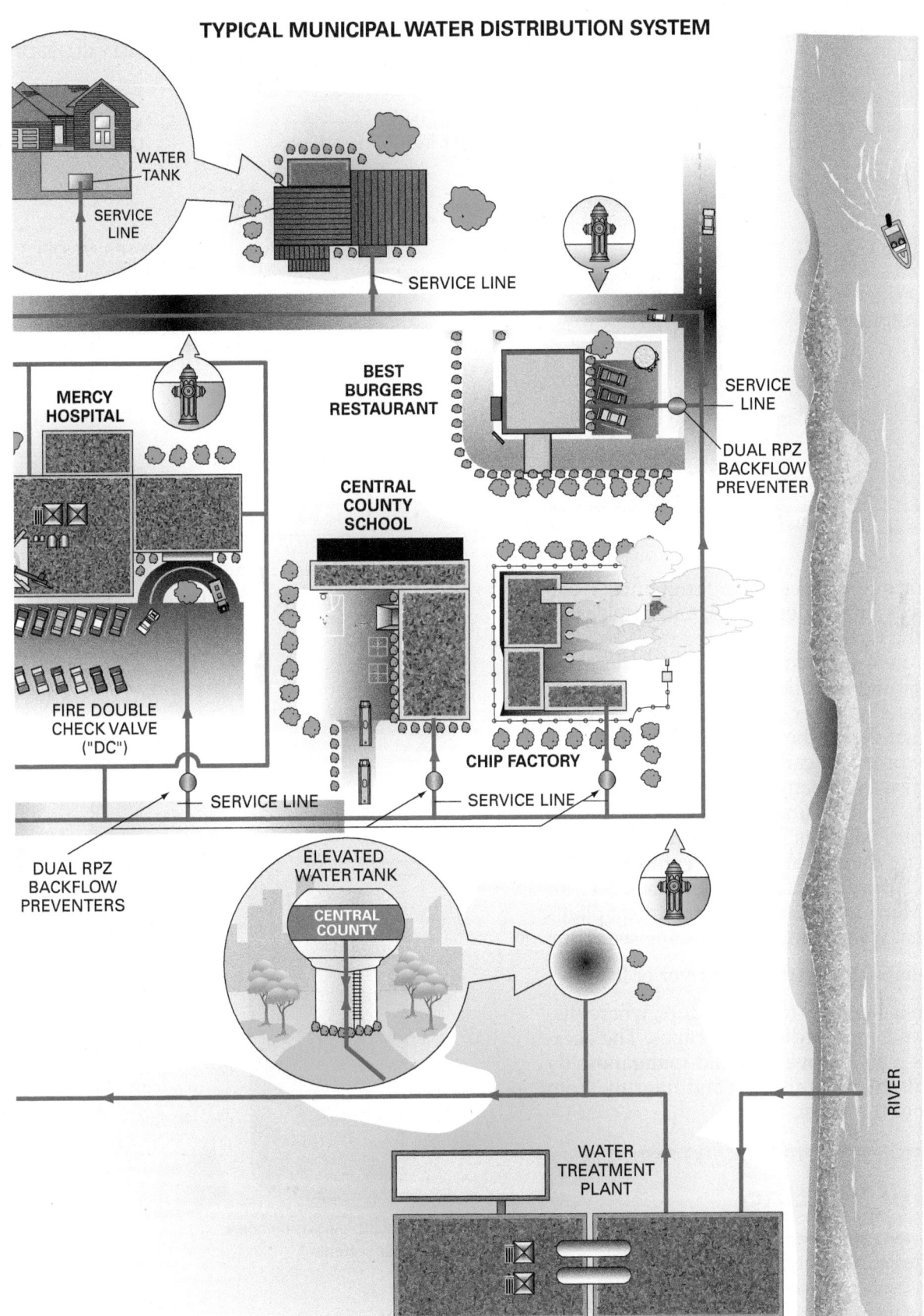

Figure 1 Water distribution.

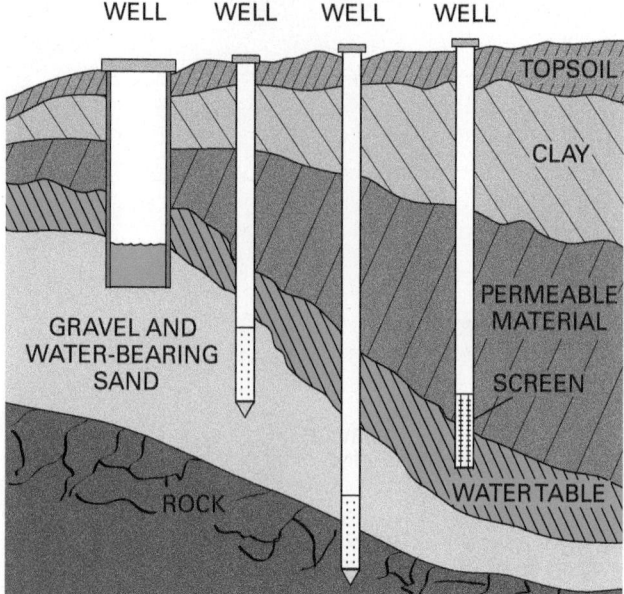

Figure 2 Types of wells.

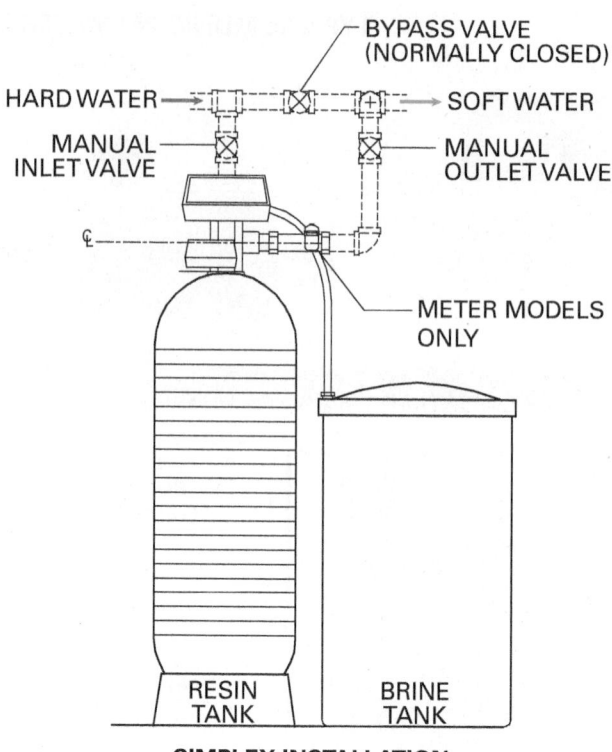

SIMPLEX INSTALLATION

materials, or other types of contaminants. Treatment removes impurities, odors, and unpleasant taste from the water.

Private water treatment for water with inorganic minerals, commonly called hard water, often consists of a water softener (*Figure 3*). Depending on the other qualities of the water, some homes may also have sediment or carbon filters.

Municipal water treatment is more complex because of the high volume of water used by the general public. Millions of gallons of water must be treated in municipal water treatment plants each day to ensure an acceptable level of safety for the public. Many municipal water treatment plants (*Figure 4*) treat water using the following sequence:

Step 1 Water is pumped from a river or lake.

Step 2 The water goes to the aerators, where dissolved carbon dioxide escapes. The aerators also remove iron and manganese by oxidizing the minerals and filtering them out of the water.

Annual Testing

Municipal wells are tested to ensure safety. Wells should be tested annually to check for levels of bacteria, viruses, and microbes. These tests ensure the purity of the water. Usually, the local health department can do this testing. Many tests are performed, but these tests vary, so check with your municipal authority.

COURTESY OF CULLIGAN INTERNATIONAL

Figure 3 Water softener.

Step 3 The aerated water flows to the clarifier, where lime and soda ash are mixed in to cause coagulation, or thickening, which removes precipitates of calcium and magnesium. Precipitates are chemical compounds containing these elements that settle out of solution. This process softens the water.

Step 4 Carbon dioxide is injected to recarbonate the water, and to stabilize and increase the acidity (lower the pH) of the water slightly. The water flows through rapid (or coarse) sand filters to remove any remaining particles.

Step 5 After the filtering process, the water passes to the tanks, where a chlorine compound (e.g., sodium hydrochloride) is added to disinfect it. At this stage, a fluoride compound is usually added to reduce the occurrence of tooth decay. This treated water is then stored in reservoirs until needed.

Step 6 After water leaves the treatment plant, it flows through pipes called water mains. The water mains usually run under the streets and serve many buildings. Permits are required to make connections to the water mains. In some areas, only municipal workers are authorized to make these connections. In other areas, licensed contractors or plumbers install the connections. The connections bring water from the main to the individual buildings through the building water service line.

1.2.1 Disinfecting a Water Supply System

Untreated water contains innumerable microorganisms. Some of these can cause disease and illness if ingested or inhaled. Plumbers must prevent contamination in a treated water supply system by doing their job properly and carefully. Plumbers must ensure that pipes are stored and handled properly and are not stored in dirty or wet locations, which are ideal breeding grounds for harmful organisms.

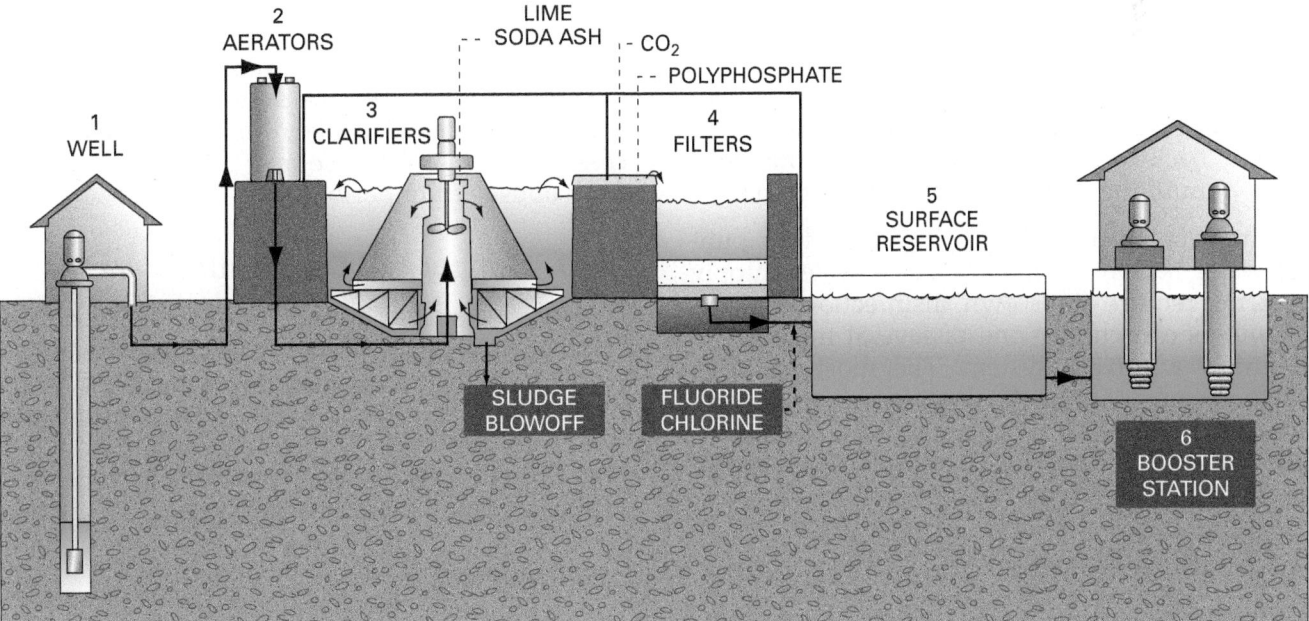

Figure 4 Municipal water treatment plant.

Cap all installed pipe at the end of each workday. When the water supply system has been completely installed, disinfect the system using the following procedure:

Step 1 Flush the completed pipe system with potable water until the water flows clear from all outlets.

Step 2 Fill the entire system with a water/chlorine solution. Valve off the system. Use a solution of at least 50 parts per million (ppm) of chlorine and let the system stand for 24 hours. Alternatively, use a solution of at least 200 ppm of chlorine and let the system stand for 3 hours. Refer to your local code.

Step 3 Flush the system with potable water until the system is completely purged of chlorine. Test the water to verify the system is chlorine-free.

Step 4 Have a testing lab or health department test the system for harmful bacteria. If bacteria remain, repeat the previous steps until the system is free of the bacteria.

Municipal water utilities disinfect their public water supply systems. Private water supply systems require their own disinfection devices. Plumbers can choose one of several methods to disinfect a private system. The most common methods are chlorination, pasteurization, and ultraviolet (UV) light. Each method involves different tools and materials, and local codes govern each method.

1.2.2 Water Pressure

The water pressure in some municipal systems may be either too low or too high for the plumbing system in a building. The plumber has ways to compensate. If the water pressure is too low, a booster system can be installed. One kind of booster has variable capacities to meet increased or decreased demand for water (*Figure 5*). For example, it might have two pumps—one that functions at 33 percent of system capacity, and one that functions at 66 percent. When demand is low, the smaller pump operates. As demand increases, it shuts down and the larger pump starts up. At peak demand periods, both pumps operate.

If the water pressure supplied to the system is too high, the plumber can use a pressure regulator valve (*Figure 6*) to lower it and to eliminate sudden pressure changes.

Pressure-reducing valves are located according to the needs of the system. Some codes may require a valve where pressure buildup at any time of the day exceeds 80 pounds per square inch (psi). A certain appliance (such as a dishwasher) may require a pressure-reducing valve because the operating pressure of the appliance is very different from the system's line pressure. Wherever they are located, the valves must be accessible for maintenance and must be protected from abuse or tampering.

1.3.0 Supply and Distribution

The water distribution system moves water from its source to the building or structure where it is needed. The way the water moves and the type of materials used vary depending on the building or structure. It is important to understand how the water distribution system works for both private wells and municipal water systems.

1.3.1 Materials

Water service lines are available in copper, plastic, and steel. Water mains are available in several materials, including cast iron. The cost of pipe

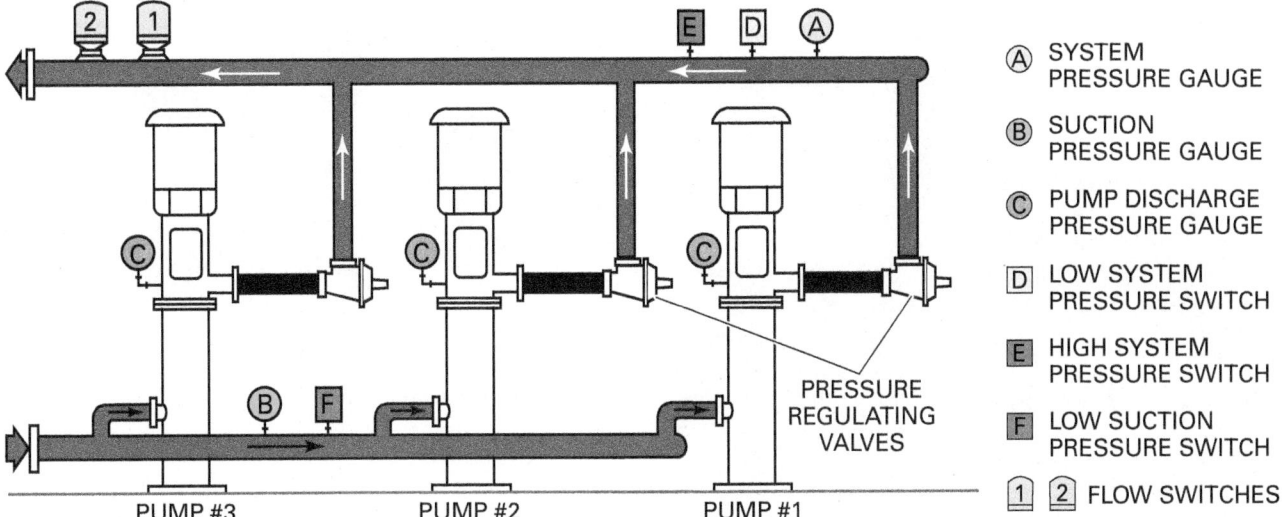

Ⓐ		SYSTEM PRESSURE GAUGE
Ⓑ		SUCTION PRESSURE GAUGE
Ⓒ		PUMP DISCHARGE PRESSURE GAUGE
D		LOW SYSTEM PRESSURE SWITCH
E		HIGH SYSTEM PRESSURE SWITCH
F		LOW SUCTION PRESSURE SWITCH
1	2	FLOW SWITCHES

Figure 5 Variable-capacity system.

Figure 6 Water pressure regulator valve.

and its service lifetime often determine the choice of materials. Other considerations include corrosion resistance, chance of freezing weather, working pressure, local plumbing codes, and ease of installation. In many cases, the life expectancy of galvanized and copper tube depends on soil and water conditions. Galvanized pipe is electroplated with zinc to provide a protective coating that resists corrosion or oxidation (rust). Acidic soils, however, tend to act on galvanized pipe. Soil with higher carbon dioxide levels are also acidic and tend to limit the life of copper tube. Hydrogen

sulfide is harmful to copper pipe as well. The corrosion caused by this gas breaks down the walls of metal pipe and causes the metal to thin or pit.

Another consideration in selecting the water supply pipe is the possibility of galvanic corrosion. This can develop when an electrically-conductive path connects pipes of different metals, such as copper and galvanized steel. It results from the action of a weak electrical current that flows between the different metals. Plumbers must take steps to reduce the likelihood of galvanic corrosion. You will learn more about corrosion prevention as you advance in your training.

1.3.2 Service Line from a Private Supply

In a private water supply, the service line runs from the well point to the house, and the pump can be located at the well or at the house (*Figure 7*).

The installation of a private water supply system usually requires a well, although it may mean getting water from a spring or a cistern. A private water source is required when access to municipal water supplies is not available, such as in rural areas.

The plumber may be responsible for setting (installing) the pump and completing the water service to the structure, or a certified well driller may do it. In some cases, a pump installer is required to set the pump, provide electrical service, and complete the water service to the pressure tank. These responsibilities overlap a great deal. Each state and plumbing code has specific requirements for water service from a well. In most cases, a water sample must be drawn and tested by appropriate authorities after the well is purged and before the final tie-in.

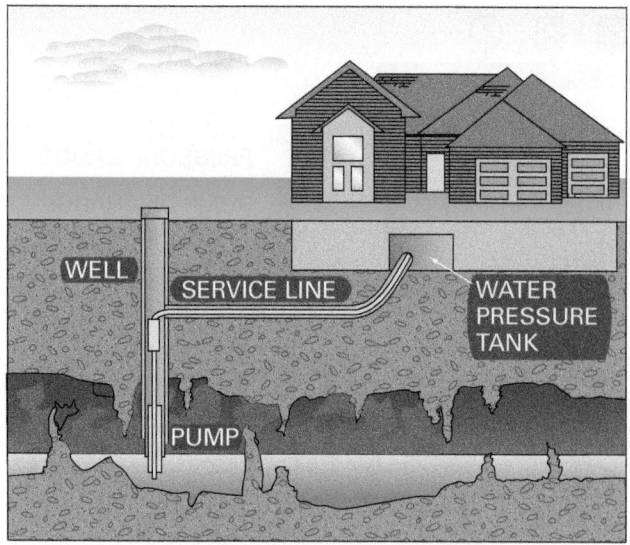

PUMP LOCATED AT THE WELL

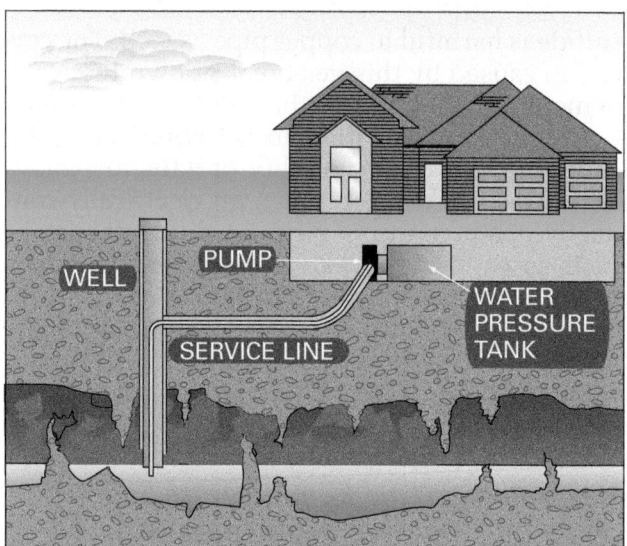

PUMP LOCATED AT THE HOUSE

Figure 7 Well point.

1.3.3 Service Line from a Public Water Main

Water distribution and supply systems in buildings connected to the public water main differ from those used in private water supplies. With private water supplies, pumps are the primary distribution mechanism. For buildings, the building service line is tapped into the water main using a corporation stop (*Figure 8*). This is a valve that is threaded into the main without interrupting service to other locations. A short pipe leads from the corporation stop to the curb stop (*Figure 9*). The curb stop is a control valve installed in the building water supply line between the corporation stop and the building. The curb box is a round casing placed in the ground over the curb stop. The top opening is at ground level. A

Figure 8 Corporation stop.

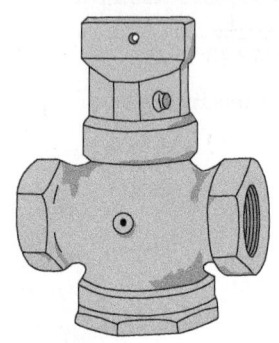

Figure 9 Curb stop.

special key-wrench can be inserted into the curb box to turn off the curb stop in emergencies or for service (*Figure 10*). The curb box, sometimes given the term *buffalo box*, is marked on city plans and drawings. On the property, it can be located by finding the metal marker (*Figure 11*). Because of differences in local procedures and applicable codes, a plumber should contact local municipal, township, or village authorities about connecting to the water main.

The plumber runs the water supply piping from the curb stop to the building's interior and installs a water line between the curb stop and the water meter connection (*Figure 12*). The water meter measures the volume of water used (in gallons or cubic feet). Water meters usually belong to the municipality, and a city or county employee must install them. A meter stop valve (*Figure 13*) is installed on the supply-side of the water meter to allow cutoff of the water service to the entire building. A second stop valve on the building side permits removing the meter for servicing. The building main shutoff valve is installed inside the building after the meter is installed (*Figure 14*). This valve allows the service to be turned off and on during construction. Beyond this point, the building supply system delivers water to the fixtures.

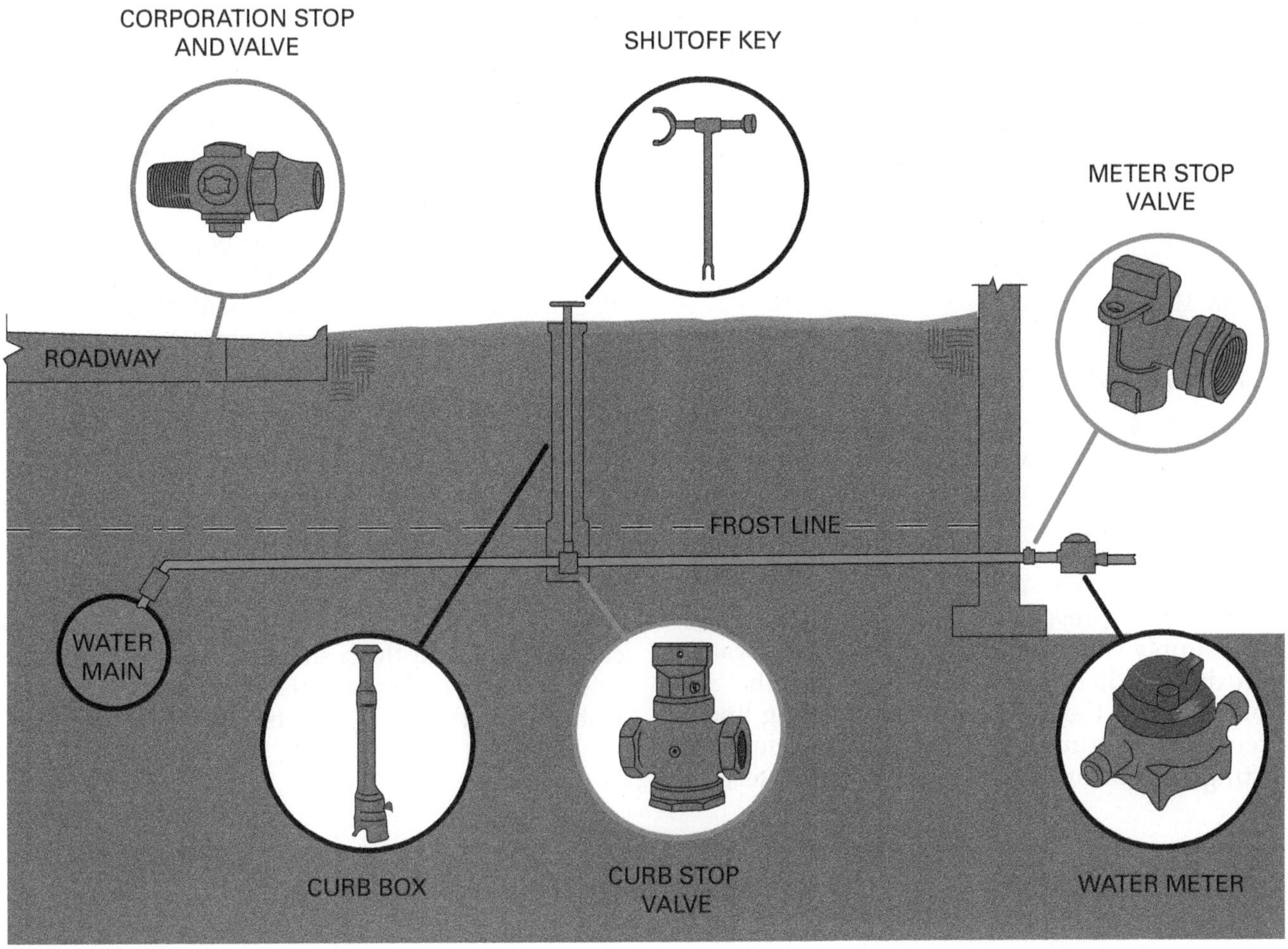

CORPORATION STOP
AND VALVE

SHUTOFF KEY

METER STOP
VALVE

ROADWAY

FROST LINE

WATER
MAIN

CURB BOX

CURB STOP
VALVE

WATER METER

Figure 10 Municipal water supply connection.

USP
WATER

Figure 11 Metal marker for curb box.

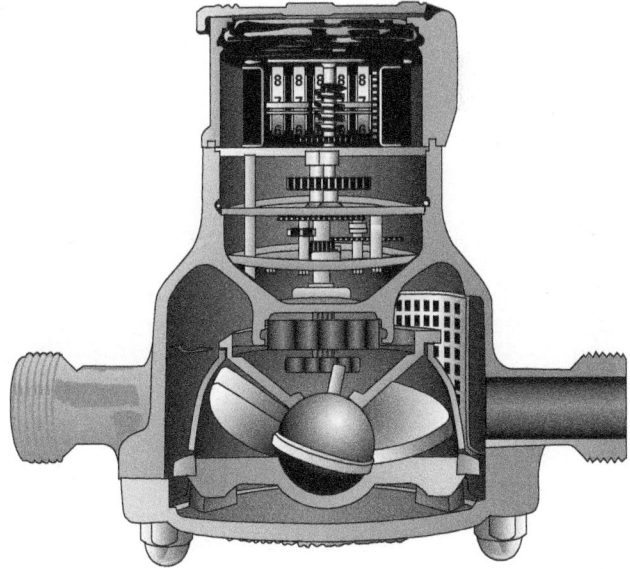

Figure 12 Water meter.

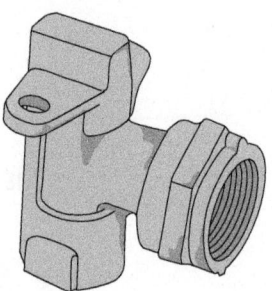

Figure 13 Meter stop valve.

1.3.4 *Water Main Testing*

Municipal governments require several tests on new and established water mains. These tests are done to ensure that the mains are properly pressurized and free of contaminants. The local codes are generally based on standards developed by the American Water Works Association (AWWA).

A backflow-prevention assembly is required on the supply water line when filling a new water main during disinfection and flushing. The assembly isolates the new lines from existing lines. It keeps the new main's water from entering the existing system until the water has been tested.

Once the tests indicate that the water is free of harmful contaminants, the backflow-prevention assembly is removed. It is not used during hydrostatic pressure testing. Hydrostatic pressure testing is done to ensure that sufficient pressure exists in the system to deliver water to taps in homes and businesses.

Established water mains are periodically tested to ensure adequate flow and pressure. They are also flushed to remove sediment from the distribution pipes. In general, these tests are performed under the authority of the local water and sewer department following established guidelines. Fire hydrants and blow-off valves are used as flush-out points. Each community determines how to handle the flushed out water. Local officials must consider public safety and effects on the environment. Some communities, for example, may collect water flushed from hydrants in a truck. The water is de-chlorinated and used to clean streets and flush storm drains and sanitary sewers. Local codes establish methods for reusing water flushed from hydrants.

STREET

MAIN SHUTOFF VALVE

M

CORPORATION STOP

BOX

CURB STOP

MAIN

ALTERNATIVE

METER BOX

METER

CURB STOP

CHECK VALVE

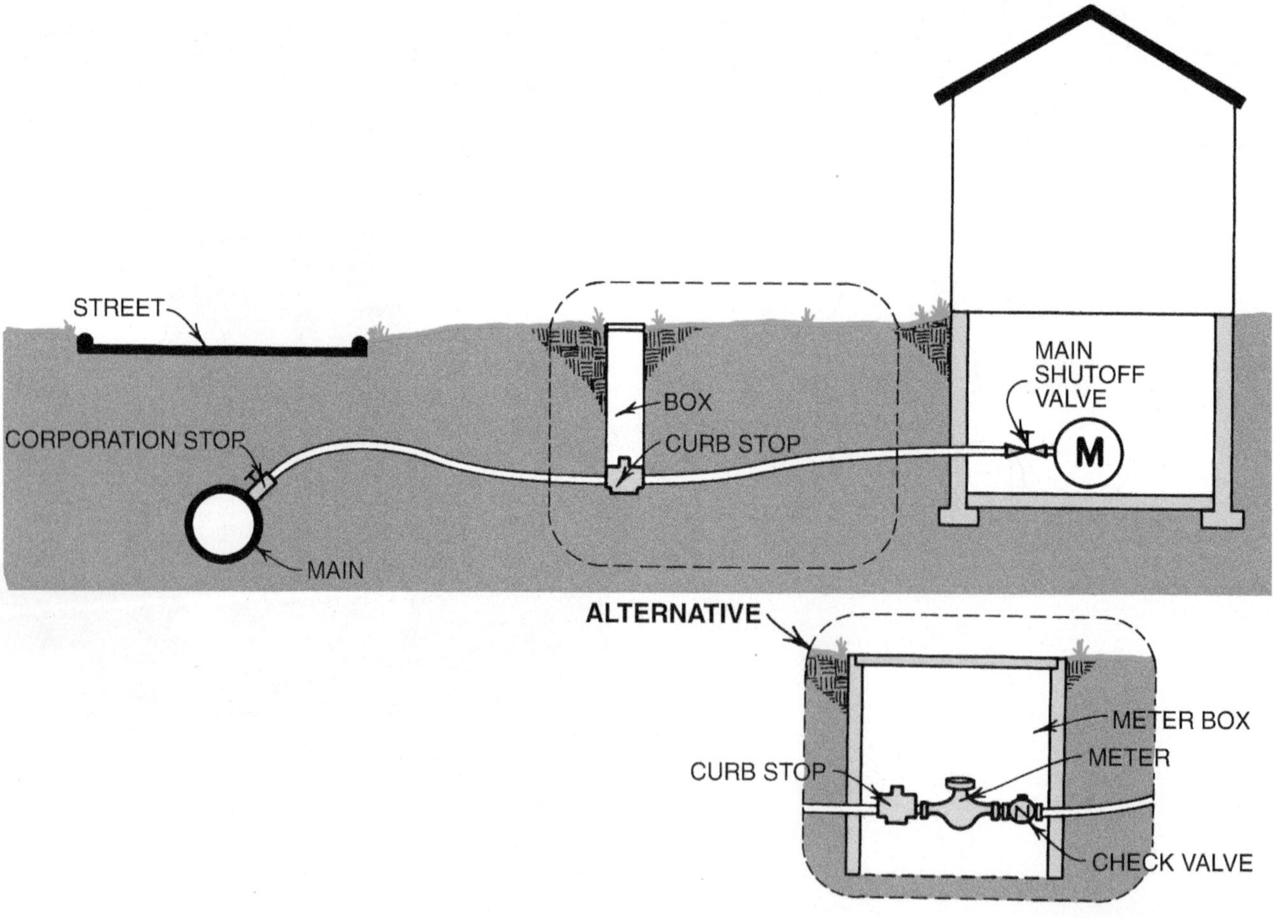

Figure 14 Main shutoff valve.

Frost Protection

Frost protection is an important consideration in many parts of the country. Frost depths, called frost lines, are different for each area. They are determined from historical winter temperature data. The plumber must install the water service line below the frost line to make sure that freezing and thawing temperatures in the earth do not affect the piping. If the water supply line enters the building above grade, it must be insulated to prevent freezing. Check your local code for the frost depth in your area.

Additional Resources

International Plumbing Code, Current Edition. Falls Church, VA: International Code Council.

National Standard Plumbing Code, Current Edition. Falls Church, VA: Plumbing-Heating-Cooling Contractors National Association.

1.0.0 Section Review

1. The deepest types of water wells are generally _____.
   a. dug
   b. drilled
   c. bored
   d. driven

2. To provide water treatment in private systems with water hardness problems, install a _____.
   a. carbon filter
   b. sediment filter
   c. precipitation filter
   d. water softener

3. Using copper tubing for underground applications should be avoided if the _____.
   a. soil is acidic
   b. soil is wet
   c. soil has an absence of hydrogen sulfide
   d. working pressure of the water main is low

2.0.0 WATER DISTRIBUTION SYSTEM COMPONENTS

Objectives

Identify the major components of a water distribution system and describe the function of each.

  a. Describe the purpose of backflow preventers.
  b. Identify and describe the various types of valves used in water distribution systems.

Trade Terms

Angle valve: A valve whose internal design is similar to a globe valve's but its outlet direction is 90 degrees from its inlet.

Check valve: A valve that allows liquid to flow in only one direction. Pressure and flow within the inlet line keeps the valve open. Automatic closure of the valve occurs with the reversal of flow, by the weight of the disc mechanism, or by spring action.

Full flow: Describes a valve that does not restrict the flow of a fluid through it compared to the potential flow through the connected piping.

Hose bibb: A faucet with a threaded outlet, typically set at an angle to the inlet pipe, used to connect a hose. Usually located on the outside of a building. Also referred to as a sill cock or bibcock.

Pressure relief valves: Valves normally used for liquid service to prevent over-pressurization of a system or component. May be slow- or fast-acting depending on their purpose and design.

Straight-through flow: A valve flow design that does not restrict flow through the valve. The element that closes the valve is retracted entirely clear of the passage.

Supply stop valves: Valves that are commonly used to disconnect the hot or cold water supplies to water closets and sinks. They are available in either right-angle or straight design, and with globe- or ball-valve internals.

Thermostatic/pressure balancing valves: Mixing valves which sense outlet temperature and incoming hot and cold water pressure, and compensate for fluctuations in hot and cold water temperatures and/or pressures to stabilize its outlet temperatures. Also referred to as temperature and pressure (T&P) relief valves.

Throttled flow: Reduced flow of water through a valve by positioning of the valve's stopping device; increases both the pressure drop across the valve and flow resistance through the valve.

Vacuum breaker: A type of backflow preventer that inhibits backflow caused by lower pressure in a water supply system by opening a vent path in a cross-connected system to prevent siphonage.

The primary purposes of water distribution systems are to supply clean potable water and to provide the means of controlling the volume, pressure, and rate of water delivery.

2.1.0 Contamination Prevention

Backflow is any unwanted flow of used or nonpotable water back into the potable water distribution system. This reverse flow occurs as a result of cross-connection, which is a direct link between a contaminated liquid and a potable water supply. Backflow can occur as a result of an improper or altered plumbing hookup or a garden hose being left in a pool of contaminated water, for example.

These conditions do not in themselves cause a significant hazard, but they create the potential for serious health threats to the public, especially in terms of waterborne diseases. The danger occurs when a break in the water main or other vacuum potential exists somewhere in a water line. If this happens, the source of contamination could be drawn into the water supply line. Backflow prevention is required in many plumbing installations to keep contaminated water or other liquids from flowing back into the potable water system.

2.1.1 Air Gaps

The most reliable means to protect water supply lines from back-siphonage is by the use of air gaps and restricting overflow levels. Air gaps are measured vertically from the lowest end of the potable water outlet to the flood-level rim of the fixture into which it discharges. In a bathroom sink, for example, the faucet outlet is set above the sink's flood-level rim (the point at which water begins to overflow the top of the sink). This is a designed-in air gap. In addition, the sink's top drain returns water to the bottom drain before the water can rise to the faucet. The minimum required air gap is two times the diameter of the potable water outlet but not less than 1 inch. Depending on the application, the required air gap may be dictated by system flow rate or other factors. You can find the specific rules for air gaps and air-gap fixtures in the *American Society of Mechanical Engineers (ASME) Standard A112.1.3, Air Gap Fittings for Use With Plumbing Fixtures, Appliances, and Appurtenances.*

2.1.2 Vacuum Breaker

Basic backflow prevention devices are designed to safeguard against dangerous cross-connections. The vacuum breaker (*Figure 15*) in a hose bibb (also called a sill cock) connection acts as a backflow preventer by admitting air to the lower-pressure (vacuum), potable-water side while blocking flow from the downstream side of the fitting.

A physical separation must always be maintained between private and municipal water systems. Wells and their associated plumbing systems must not be connected to municipal water supplies. Codes specify the degree of danger and the appropriate device that should be used for various applications.

2.1.3 Other Backflow Preventers

Some backflow preventers protect against both back pressure and back-siphonage, while others can handle only one type of backflow. Backpressure occurs in the water distribution system when a pressure higher than the supply pressure causes a reverse flow into the potable water piping. Back siphonage occurs when contaminated or polluted water flows from a plumbing fixture back into the potable water piping. This can occur when a negative pressure exists in the plumbing fixture. The five most commonly-used mechanical backflow preventers are:

- Atmospheric vacuum breakers
- Pressure-type vacuum breakers
- Dual-check valve backflow preventers
- Double-check valve assemblies
- Reduced-pressure zone (RPZ) principle backflow preventer

Figure 16 shows examples of each of these.

Each type of backflow preventer has specific applications and limitations. These factors must be considered in any type of system design using backflow preventers. Before you select and install any backflow prevention or siphonage prevention device, consult local codes.

Backflow prevention devices must be well-maintained to work effectively. Most devices are field tested before installation and then tested annually. When devices are not working properly, they must be repaired and tested according to the manufacturer's instructions. Testing must be performed by certified personnel.

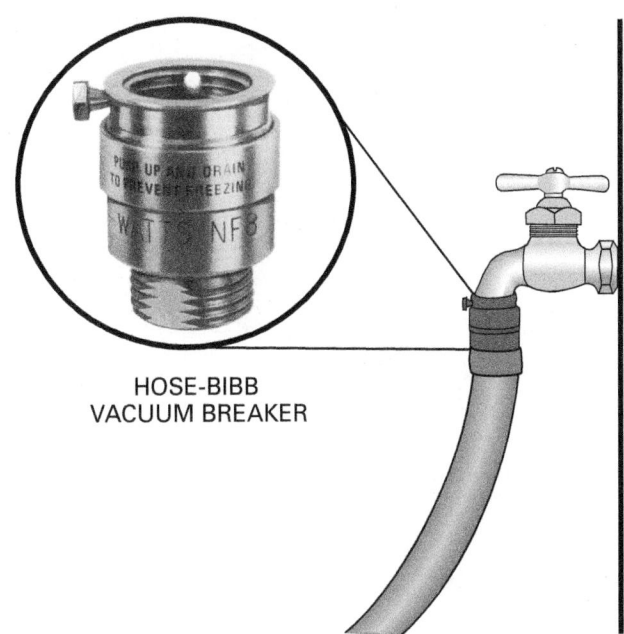

HOSE-BIBB
VACUUM BREAKER

Figure 15 Vacuum breaker.

CAUTION

Some codes require installation of backflow preventers in every new structure, including residences. Always be sure to check architectural plans and all applicable codes before any installation.

2.2.0 Valves

Valves or faucets regulate the flow of water in a water distribution system. They may be used to turn water service on and off, act as throttling devices that control the rate of water flow, regulate the pressure, or prevent a reversal of flow through a line.

RPZ Valves

A reduced-pressure-zone (RPZ) backflow preventer may also be called a reduced pressure zone device (RPZD), RPZ valve, or reduced pressure zone assembly. RPZ valves are designed to prevent backflow and back siphonage. Those qualities make them suitable in applications where backflow into the water supply could cause serious health problems. Local codes outline when, where, and how RPZ valves should be installed. Most codes require these devices in residential applications to prevent back siphonage from outdoor hose bibbs into the household water supply.

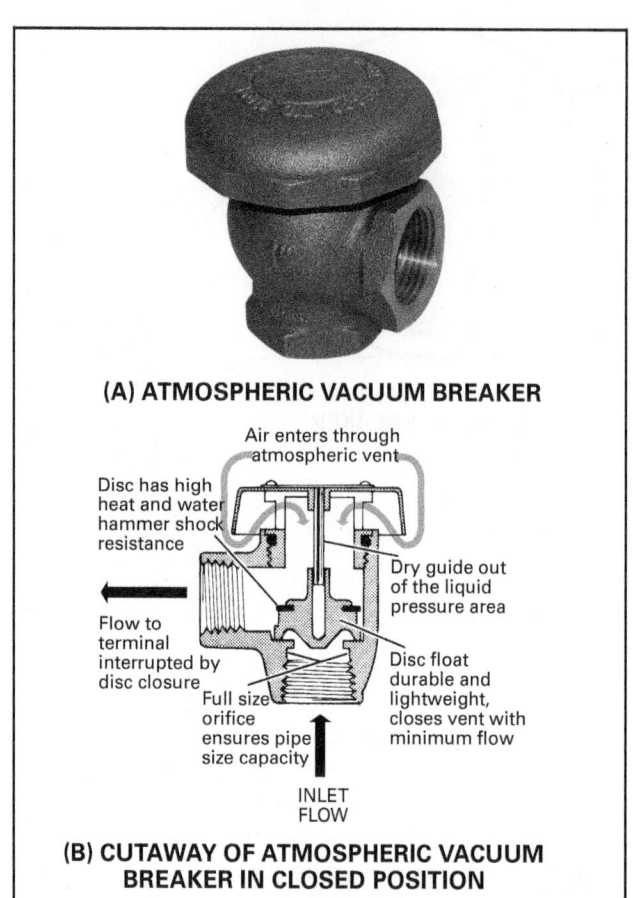

(A) ATMOSPHERIC VACUUM BREAKER

Air enters through atmospheric vent

Disc has high heat and water hammer shock resistance

Dry guide out of the liquid pressure area

Flow to terminal interrupted by disc closure

Disc float durable and lightweight, closes vent with minimum flow

Full size orifice ensures pipe size capacity

INLET FLOW

(B) CUTAWAY OF ATMOSPHERIC VACUUM BREAKER IN CLOSED POSITION

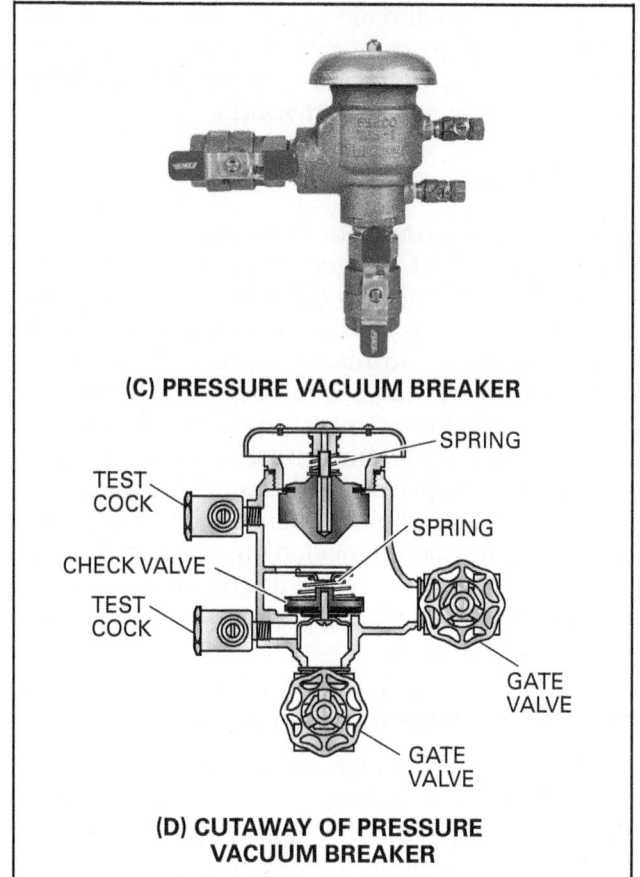

(C) PRESSURE VACUUM BREAKER

TEST COCK

SPRING

CHECK VALVE

SPRING

TEST COCK

GATE VALVE

GATE VALVE

(D) CUTAWAY OF PRESSURE VACUUM BREAKER

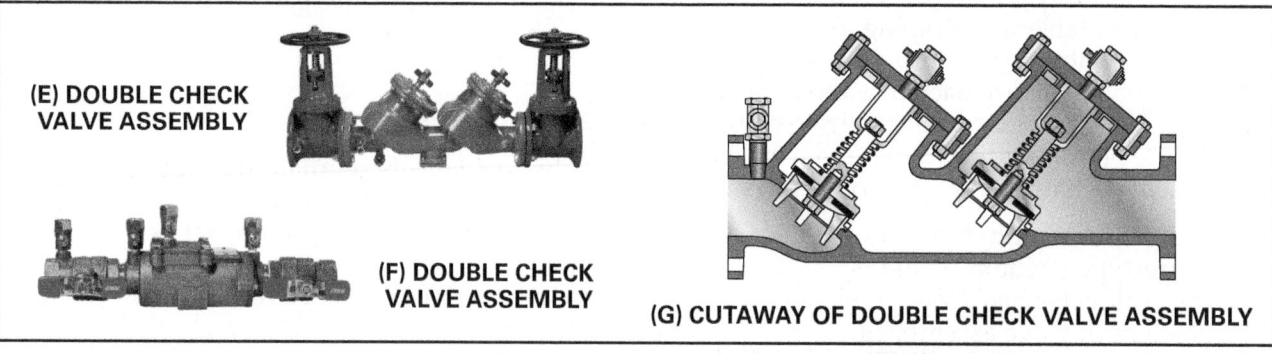

(E) DOUBLE CHECK VALVE ASSEMBLY

(F) DOUBLE CHECK VALVE ASSEMBLY

(G) CUTAWAY OF DOUBLE CHECK VALVE ASSEMBLY

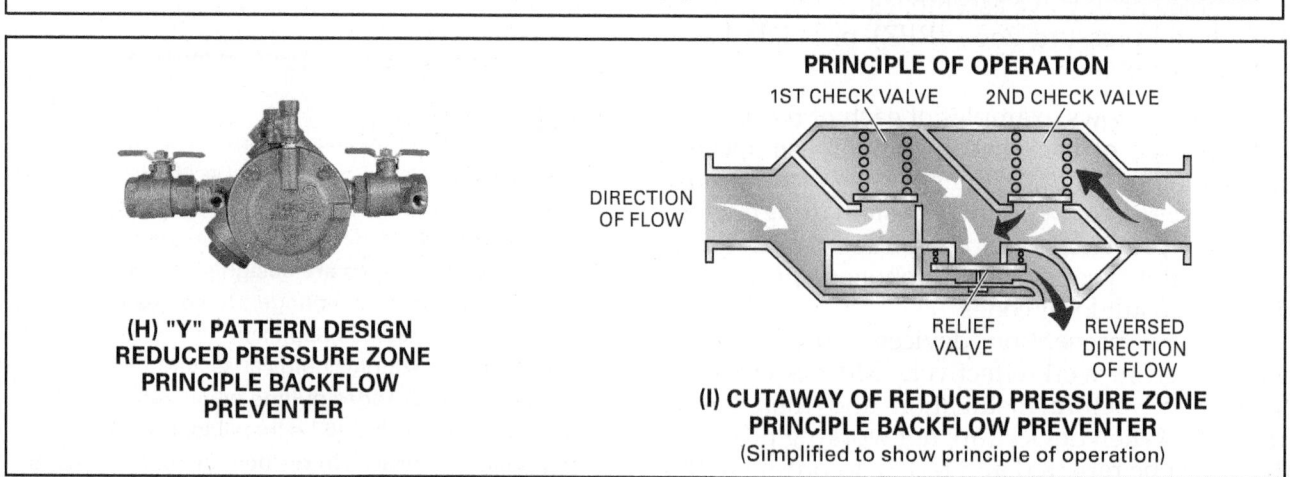

(H) "Y" PATTERN DESIGN REDUCED PRESSURE ZONE PRINCIPLE BACKFLOW PREVENTER

PRINCIPLE OF OPERATION

1ST CHECK VALVE   2ND CHECK VALVE

DIRECTION OF FLOW

RELIEF VALVE

REVERSED DIRECTION OF FLOW

(I) CUTAWAY OF REDUCED PRESSURE ZONE PRINCIPLE BACKFLOW PREVENTER
(Simplified to show principle of operation)

Figure 16 Backflow prevention devices.

Backflow Accidents

Always install proper backflow prevention devices where they are required by your local code. Be sure to install them according to the manufacturer's instructions. Don't let your community become a horror story like the ones discussed in the following case studies.

Backflow Case Study 1

During a routine maintenance check of a Colorado Middle School's hot water-heating boiler, maintenance workers left a valve open between the potable water line and the boiler. This allowed boiler water containing antifreeze to backflow into the school's potable water system. There was no backflow preventer on the feed line to the boiler.

Nine children were sent to the hospital with flu-like symptoms. The hospital treated the students for ethylene glycol (the ingredient in the antifreeze) poisoning. The school was closed while workers flushed the potable water piping and repaired the leak. Workers were instructed to install a backflow preventer in the potable water line to the hot water heating boiler.

Backflow Case Study 2

A municipal office inspected a cross-connection at a block of high-rise apartments. Officials found that a check valve failed to protect three water heaters. The pressure in a high-pressure cold water pump was fluctuating. When the pressure dropped, antifreeze contaminated the potable water system.

Backflow Case Study 3

A Florida home was mistakenly hooked up to treated wastewater lines instead of drinking water lines. As a result, the resident experienced severe stomach cramping and diarrhea. According to a city report, workers accidentally switched the lines when they failed to dig deep enough to expose the identification tape distinguishing the two water lines.

The city was asked to sanitize the home plumbing over a three-month period and monitor the resident's health for 10 years. As part of an agreement with the state health department, the city was also required to spend nearly $75,000 to improve training, installation procedures, and public education efforts.

It is important for plumbers to understand the terms related to valves. The following are commonly used valve terms:

- *Straight-through flow*—Straight-through flow is not restricted as it passes through a valve. The valve shut-off element is positioned entirely clear of the water passage.
- *Full flow*—Full flow describes the flow capacity of a valve as essentially equal to the maximum potential flow through the connected piping.
- *Throttled flow*—Throttled flow identifies reduced water flow controlled by positioning the valve's shut-off element. Not all types of valves are suitable for throttled flow.

2.2.1 Types of Valves

Plumbers must be familiar with a variety of valves. The following are common types of valves:

- Gate valve
- Globe valve
- Angle valve
- Ball valve
- Check valve
- Pressure regulator valve
- Supply stop valve
- Combination thermostatic/pressure balancing valve also called a temperature and pressure (T&P) relief valve
- Tempering valve

These valves are used in different parts of the water distribution system (*Figure 17*).

2.2.2 Gate Valve

A gate valve (*Figure 18*) is a valve in which the flow is controlled by moving a gate or disc that slides in machined grooves at right angles to the flow. The gate is moved by the action of the threaded stem on the control handle.

There are two basic designs of gate valves: rising-stem gate valves (also known as inside screw stem valves) or nonrising-stem valves (also known as outside screw stem valves). A rising-stem gate valve has the stem attached to the gate. The gate and stem rise and descend together as the valve is operated. The stem on a nonrising-stem gate valve is threaded on the lower end into the gate. When the handwheel on the stem is rotated, the gate moves up or down the stem.

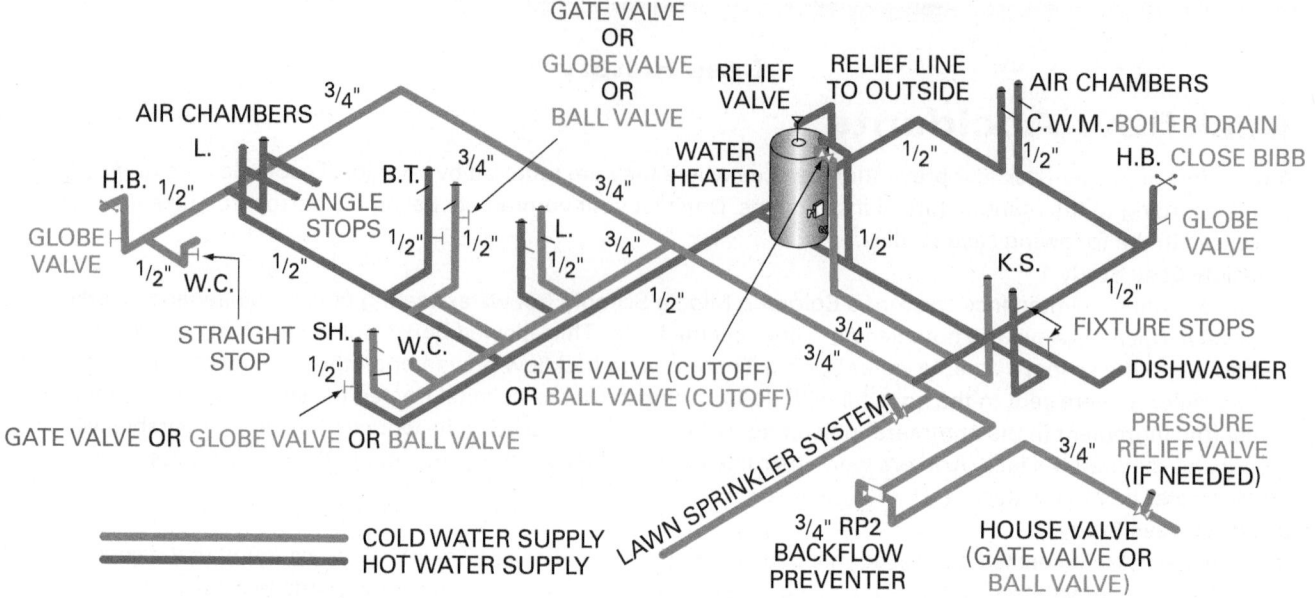

Figure 17 Valves in the distribution system.

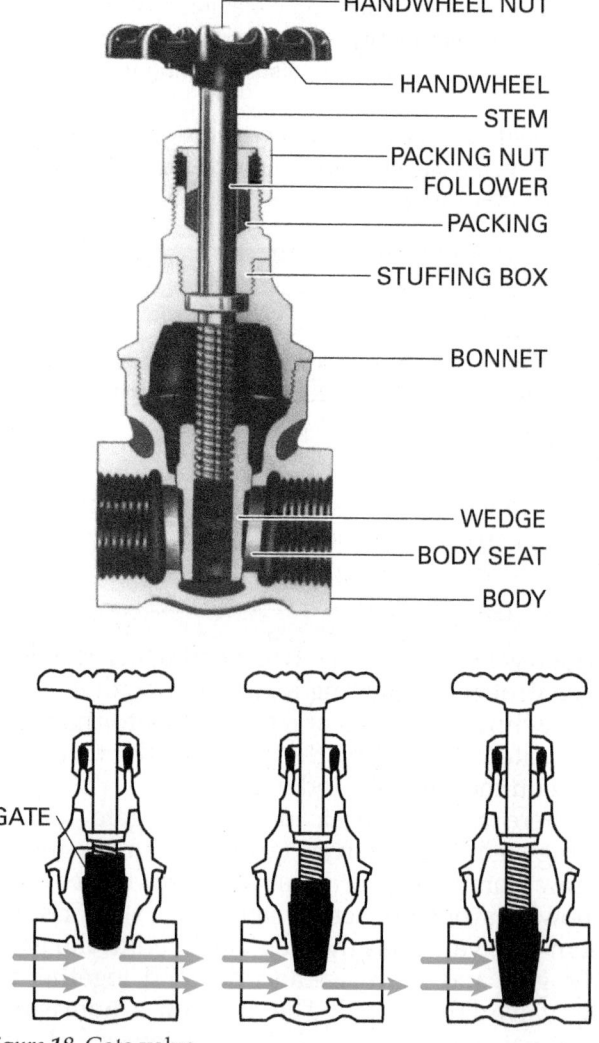

Figure 18 Gate valve.

Gate valves are best suited for main supply lines and pump lines. They provide an unobstructed, full-flow passageway when the gate is fully opened. They can be used on lines containing steam, water, gas, oil, or air. They are unsuitable for precisely throttling fluid flow.

2.2.3 Globe Valve

In a globe valve (*Figure 19*), the flow is controlled by moving a circular or conical disc against a metal seat that surrounds the flow opening. As the handle turns, the stem forces the disc onto the seat or withdraws it by screw action.

Globe valves are useful for general service in systems containing steam, water, gas, or oil, where frequent operation and precise throttled-flow control is required. Inside the globe-shaped body of the valve is a partition. This partition closes off the inlet side of the valve from the outlet side, except for a circular opening called the valve seat. The mating surface of the valve seat is ground smooth so that a proper and complete seal can be made when the valve is in the closed position. Turning the handle clockwise firmly seats the washer or disc between the valve stem and the valve seat, shutting the valve. This stops the flow of gas or liquid. It is a common design practice to install a globe valve so that the high-pressure supply fluid is under the valve seat. This places the stem packing on the downstream, low-pressure side of the valve seat, which lowers the potential for stem-packing leaks.

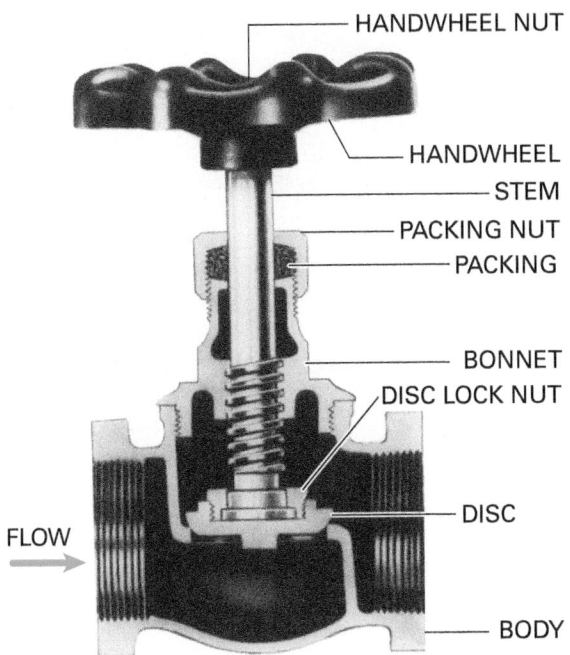

Figure 19 Globe valve.

2.2.4 Angle Valve

The angle valve (*Figure 20*) is similar to the globe valve, but it can serve as both a valve and a 90-degree elbow. Because flow changes direction only once through an angle valve, the angle valve creates less flow resistance and turbulence than the globe valve, in which flow must change direction two times.

There are three types of angle valves: conventional, plug-type, or composition discs. Conventional angle valves have a disc that is in close contact with the body seat. These valves are suitable for less severe services but not for precise throttling. Plug-type angle valves have a long taper and a wide seat bearing surface. These valves are ideal for fine-throttling, severe-service operation and are effective in resisting erosion, especially in steam systems. Composition discs have a tight sealing disc against a raised-crown seat. These valves are well suited for moderate pressure service.

2.2.5 Ball Valve

The ball valve (*Figure 21*) is used to control the flow of gases and liquids. Ball valves are installed in piping systems where quick shutoffs for in-line maintenance may be necessary, or in lines used for mixing various liquids and gases. The ball part of the valve is rotated into the open or shut position by a handle on the outside of the valve body. These valves are straight-through

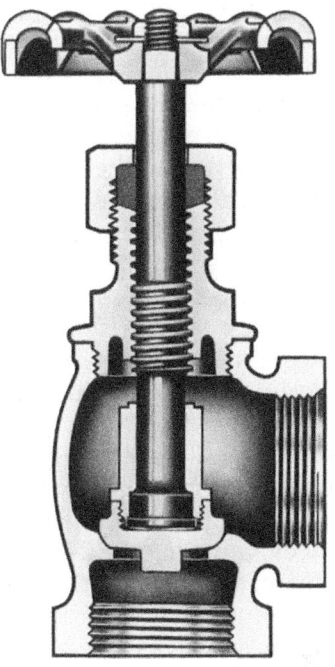

Figure 20 Angle valve.

valves that allow quick action in controlling the flow in piping systems. If the ball bore is the same size as the connected piping, these valves also provide full-flow capacity. Most ball-valve designs do not provide reliable, fine-throttling control.

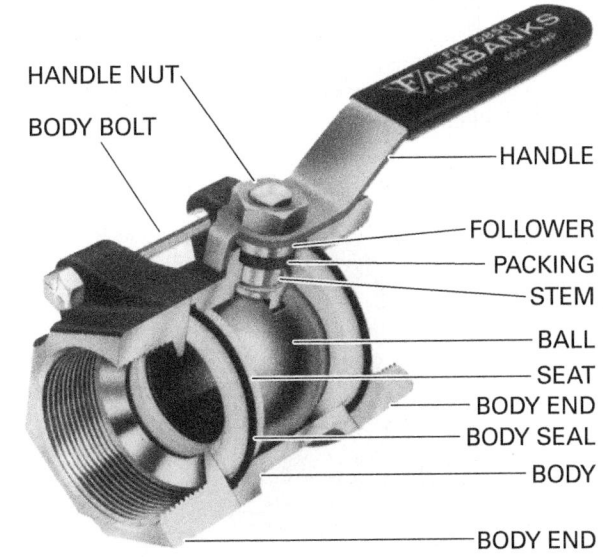

Figure 21 Ball valve.

2.2.6 Check Valve

A check valve prevents reversal of flow in a piping system. Pressure and flow in the upstream line keeps the valve open. The valve automatically closes by the reversal of flow or by the weight of the disc mechanism in no-flow situations. Three types of check valves are available: the ball-check valve, the swing-check valve, and the lift-check valve.

The ball-check valve (*Figure 22*) allows one-way flow in water supply or drainage lines and can be used with extremely low or no backpressure. It seats under the influence of gravity.

The swing-check valve (*Figure 23*) has a low flow resistance that makes it well suited for lines containing liquids or gases with low to moderate pressure. The swing-check valve is available in three different types, depending on the manufacturer: single disc, conventional; dual or split-swing discs; and single disc, angle seating.

The lift-check valve (*Figure 24*) can be used for gas, water, steam, or air. Lift-check valves are recommended for lines that have frequent fluctuations in flow. These valves are available in either horizontal or vertical connection styles, though the check feature in both shown in the figure works by gravity. The horizontal type has an

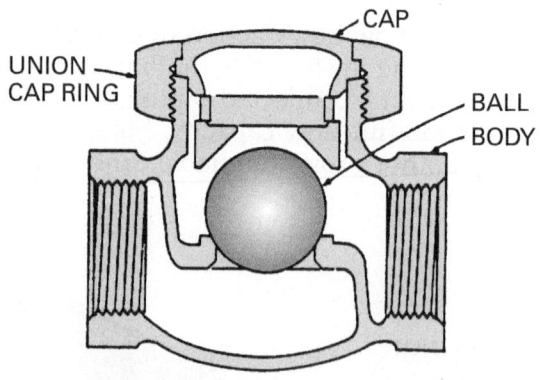

Figure 22 Ball-check valve.

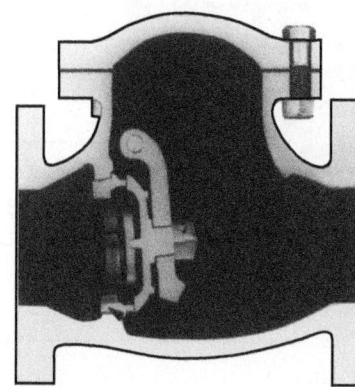

Figure 23 Swing-check valve.

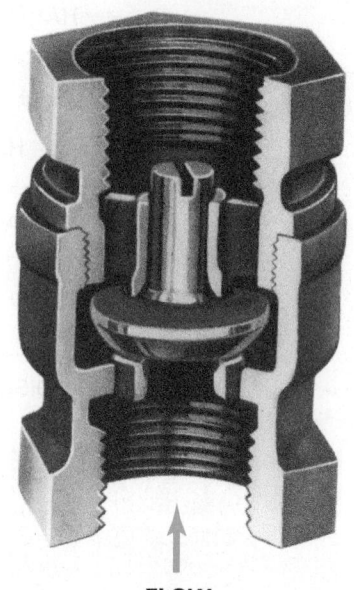

FLOW

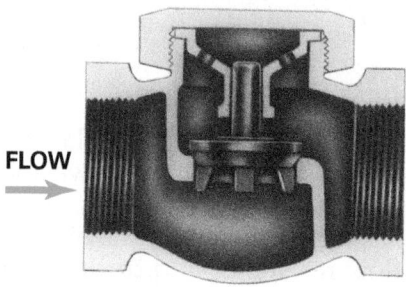

FLOW

Figure 24 Lift-check valves.

internal construction similar to the globe valve, while the vertical type has a straight-through flow design. In higher-pressure systems, the horizontal check valve disc may act against a spring, which can provide an adjustable force to require a minimum flow and pressure to open the valve.

2.2.7 Pressure Regulator Valve

The pressure regulator valve reduces water pressure in a building below that of street main pressure to protect end-use plumbing fixtures from leaks. The valve is activated by comparing delivery and downstream pressures. As the delivery or downstream pressure changes, a spring located in the dome of the valve acts on the diaphragm to move the valve open or shut. In a typical design, the valve opens when the diaphragm presses down away from the valve seat, and it remains open until the pressure in the building reaches a set level. The valve then moves toward the shut position to lower building water pressure. Unless building water demand stops, the valve is always at least slightly open to supply water at the set pressure.

2.2.8 Supply Stop Valve

Supply stop valves, or supply valves, commonly disconnect the hot- or cold-water supply to water closets and sinks. These miniature globe valves make it easy to control the water connection at an individual fixture for repair work. They are available in either right angle or straight design (*Figure 25*). Many manufacturers now offer stop valves with a quarter-turn ball-valve design, which are easier to operate and have longer service lives.

2.2.9 Temperature and Pressure Relief Valve

Temperature and pressure (T&P) relief valves (*Figure 26*) are normally used for liquid service, although safety valves also may be used. T&P relief valves are designed to open on either excessive pressure or temperature, or both. Ordinarily, these pressure relief valves do not have a chamber or a regulator ring for varying or adjusting blowdown, so they operate with a relatively lazy motion. As temperature or pressure increases, they gradually open, releasing up to a cup of water; with a pressure decrease, they gradually close.

2.2.10 Tempering Valve

Hot and cold water can be mixed to reduce the risk of scalding at faucets, such as shower valves, or to reduce sweating at water closet tanks. The best way to do this is to use a tempering valve (*Figure 27*). Tempering valves are available for both residential and commercial/industrial applications. These valves mix the water coming from the water heater to a predetermined temperature. The temperature is set using an adjustable thermostat incorporated into the valve.

Figure 26 Pressure relief valves.

Figure 25 Supply stop valves.

 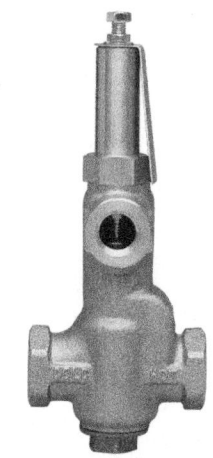

HOT WATER EXTENDER TEMPERING VALVE (RESIDENTIAL)

HOT WATER EXTENDER TEMPERING VALVE WITH HIGH TEMPERATURE RESISTING DISC (COMMERCIAL)

Figure 27 Tempering valves.

Additional Resources

Advanced Home Plumbing, 1997. Black & Decker Home Improvement Library. Minnetonka, MN: Cowles Creative Publishing, Inc.

Air Gap Fittings for Use With Plumbing Fixtures, Appliances, and Appurtenances, Standard A112.1.3, Current Edition. American Society of Mechanical Engineers (ASME)

International Plumbing Code, Current Edition. Falls Church, VA: International Code Council.

National Standard Plumbing Code, Current. Falls Church, VA: Plumbing-Heating-Cooling Contractors National Association.

2.0.0 Section Review

1. For the purposes of preventing cross-connections in drinking water systems, which of the following is *not* a backflow preventer?
   a. Vacuum breaker
   b. Ball check valve
   c. RPZ valve
   d. Air gap

2. Which of the listed valve types is considered a straight-through flow valve?
   a. Angle stop valve
   b. Globe valve
   c. Check valve
   d. Ball valve

3.0.0 BUILDING WATER DISTRIBUTION

Objectives

Explain the relationships between components of a water distribution system.

a. Identify the major components of a building water system and describe how to determine proper placement.

b. Explain the requirements for sizing of the main supply lines.

Performance Task

1. Sketch an isometric drawing of a simple water distribution system and label its components.

Trade Terms

Backing board: A short plank of wood installed between wall framing (or sometimes the unfinished side of the wall interior) used to mount fixture risers and stub-outs behind the fixture.

Fixture risers: Vertical sections of pipe located inside the wall to connect the fixture to the supply pipe beneath the flooring.

Hammer arrestors: Devices installed in a piping system to absorb water hammer by gradually stopping the flow of water against an air or gas cushion.

Water hammer: A loud thumping that results when the piping system deflects against supports as it absorbs the energy in flowing water when it suddenly stops.

Water supply fixture units (WSFU): Design factors to determine the load that different plumbing fixtures produce on the supply side of a plumbing system.

In a typical residential or light commercial building, there are many common fixtures a plumber will frequently install. Each of these has specific purposes and they involve certain considerations when locating and installing them.

3.1.0 Placing Major System Components

Plumbing a small building such as a residence follows the sequence of: main supply installed, locating major components, locating the architectural fixtures, running the DWV piping, running the water service and branch lines, installing risers and stub outs, and finally connecting fixtures. (This list does not include inspections or tests that are required by code.) Once the water supply piping is installed to bring the water from the main into the building, the next step is to locate and install the water heater, hose bibbs, water softener (if needed), and other fixtures. Architects will usually place these components in building plans to efficiently serve their purposes (*Figure 28*). However, the actual location of the major service components will often be left to the plumber depending on as-built circumstances.

3.1.1 Locating the Water Heater

Unless the plans show otherwise, place the water heater in the most efficient location, which is usually as close as possible to the greatest number of hot water outlets. This minimizes the length of hot water piping that runs between the heater and the fixtures. You should also consider the location of the gas supply if installing a gas water heater, or the electrical service entry if installing an electric water heater.

3.1.2 Locating the Water Softener

If the plumbing system includes a water softener, you must first determine the expected use of various fixtures. Not all fixtures or outlets require access to softened water. For example, the hot water supply, including the hot water heater, should always be treated, but the hose bibbs may be served by unsoftened water. Makeup water for a hot-water heating system should also be softened. These decisions affect the placement of pipe runs.

3.1.3 Locating Hose Bibbs

An exterior water outlet, also called a hose bibb or sill cock (*Figure 29*), is frequently required in residential structures. The piping runs serving hose bibbs may be long, and they should bypass the water softener and even the pressure regulating valve, if present. Locating hose bibbs depends on factors such as the size of the structure, accessibility of the basement or crawl space (if any), and intended use.

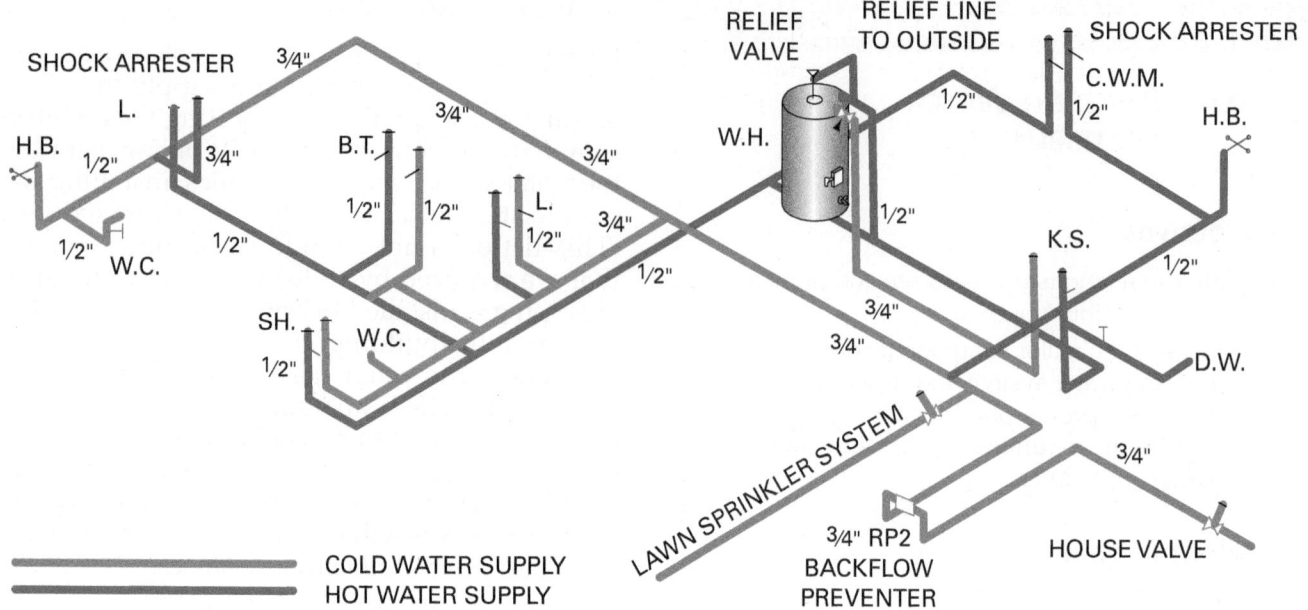

SHOCK ARRESTER   3/4"   RELIEF   RELIEF LINE   SHOCK ARRESTER
VALVE   TO OUTSIDE
L.   3/4"   C.W.M.
1/2"   1/2"   1/2"
H.B.   1/2"   3/4"   B.T.   3/4"   W.H.   H.B.
3/4"
1/2"   1/2"   L.   3/4"   1/2"
1/2"   W.C.   1/2"   1/2"   1/2"   3/4"   K.S.
1/2"   1/2"   1/2"
SH.   W.C.   3/4"   D.W.
1/2"   3/4"   1/2"
LAWN SPRINKLER SYSTEM   3/4"
COLD WATER SUPPLY   3/4" RP2   HOUSE VALVE
HOT WATER SUPPLY   BACKFLOW
PREVENTER

Figure 28 Sample domestic water distribution piping system.

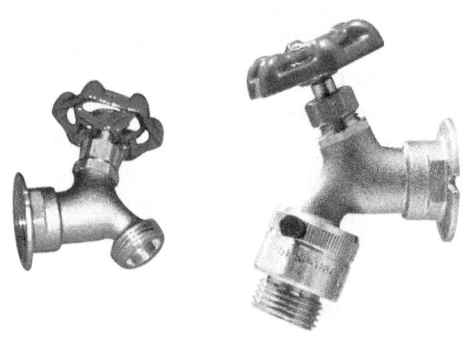

HOSE BIBB WITH
SEPARATE VACUUM BREAKER

Figure 29 Hose bibbs.

3.1.4 Locating Fixtures

After locating the water heater, hose bibbs, and water softener, identify the locations of the other water-using fixtures. For a residence, these will include the water closets, bathtubs, showers, vanity sinks, kitchen sink, and water supplies to ice makers, if required. Refer to the architect's plan, since the locations of these are fixed by the building design. The plumber is responsible for providing the most efficient water distribution system while minimizing cost of materials and time of installation.

When plumbers are installing the plumbing for a structure, they must first install the DWV system. Due to the large size of the pipe and limited flexibility in installation, and because it cannot be moved once it is installed, proper layout of the DWV can eliminate most problems. Water supply piping is located in relation to the already installed DWV piping, not to the walls or corners of the building.

3.2.0 Supply Piping

You must correctly size the supply pipe to the fixtures and appliances and use sizing tables supplied by the applicable plumbing codes. These tables estimate the anticipated demand for water as measured by water supply fixture units (WSFU). The following must be taken into account when sizing supply piping:

- Type of flush devices used on different fixtures
- Water pressure in pounds per square inch (psi) at the source
- Length of pipe in the building
- Types and number of different fixtures installed
- Total number of fixtures in use at any one time

3.2.1 Main Supply Lines

The service lines are the main water supply lines. These lines are also known as main feeder lines. For residential installations, the main feeder line beyond the water heater must be sized to supply the required flow and pressure. Smaller pipe sizes can save both energy and money. Larger pipe sizes carry more water, but water left in the piping system when the fixture is turned off cools down. When the user turns on the faucet again, more cool water must clear the pipe before hot water reaches the fixture. This wastes water and energy. For this reason, hot water service lines are typically $1/2$ inch in domestic service (refer to *Figure 27*).

For hot water distribution systems, one effective method to conserve energy and water is through the use of a hot water recirculation system, which is usually needed when the fixtures are a long distance from the hot water source. At the end of the hot water line, a tee and a pump can be installed to return water into the cold side or into the drain valve opening of the heater. A check valve needs to be installed on the incoming cold water so the return water flows in only one direction.

When you are installing cold water lines, you must consider possible connections to hose bibbs and plan efficiently to save pipe and time. Cold water that is used at the hose bibbs for lawn sprinklers and other outdoor applications should not be softened.

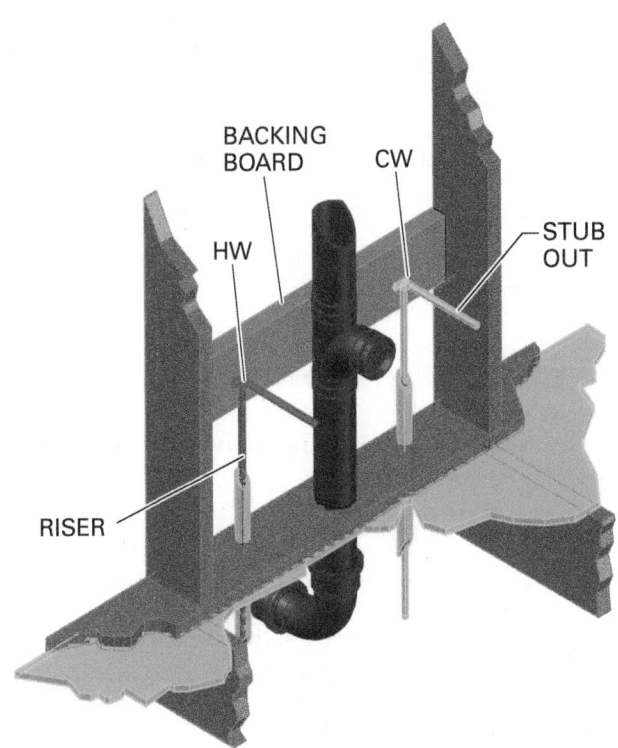

Figure 30 Installation of fixture riser and stubouts.

CAUTION	Keep the water supply piping as clean as possible before installing and using it. Clean, dry storage of pipes helps prevent contamination. Cap all ends of pipe at the end of each workday.

3.2.2 Branch Lines and Risers

Once the runs of the main hot- and cold-water feeder lines are established, determine the branch piping that serves individual fixtures and appliances in the building. Branches are any part of the piping system other than the risers, main, or stack. Then locate the fixture risers and fixture stub-outs. Water supply piping is also located in relation to drains and fixture controls for the fixtures. When plumbers are installing water supply piping, they need to recheck the location of the drains to make sure that they are positioning the piping correctly.

Construct a fixture riser that goes inside the wall as the connection from the fixture to the supply pipe beneath the flooring. Assemble fixture risers and stub-outs, and mount them on a backing board inside the wall behind the fixture (*Figure 30*). When the water supply lines have been located at the fixtures and the stubouts assembled, the assembly is placed through the access hole in the floor. Plumbers should always use approved submittal data to determine measurements when installing fixture risers and stub-outs.

In large water lines or where fixture controls have quick-closing valves, you should install hammer arrestors (*Figure 31*) to absorb the energy when the flow of water suddenly stops, called water hammer. They are usually placed near the

Figure 31 Hammer arrestor.

fixture as an extension of the water pipe riser. If hammer arrestors are not installed, are installed incorrectly, or fail, water hammer may occur when a quick-closing fixture control is used. Hammering could eventually lead to pipe, joint, or valve failure in the water distribution system.

Additional Resources

Advanced Home Plumbing, 1997. Black & Decker Home Improvement Library. Minnetonka, MN: Cowles Creative Publishing, Inc.

International Plumbing Code, Current Edition. Falls Church, VA: International Code Council.

National Standard Plumbing Code, Current Edition. Falls Church, VA: Plumbing-Heating-Cooling Contractors National Association.

3.0.0 Section Review

1. Which of the following should normally be installed first when plumbing a building?
   a. DWV piping
   b. Fixture risers
   c. Stub outs
   d. Plumbing fixtures

2. For planning purposes, the capacity of a water system or its demand is based on _____.
   a. the fixture manufacturer's capacity rating
   b. the size of the fixture water connection
   c. WSFU established by plumbing codes
   d. the plumber's professional experience

SUMMARY

Water comes from a private source, such as a well, or a municipal source, such as a public water distribution system. While well water is usually drinkable, water supplied by municipalities must be treated to remove harmful impurities. Both private and public water-service lines move water from its source to homes and other buildings. These lines, or pipes, may be made of copper, plastic, or steel. Factors such as weather, resistance to corrosion, working pressure, applicable codes, ease of installation, cost, and life expectancy determine the choice of pipe materials. Pumps move water from wells into the water distribution system. Plumbers choose the correct pump based on the depth of the well, the rate at which the water is pumped, and the total distance the water must travel.

The flow of water in a water distribution system is regulated by valves or faucets, which are used to turn water on and off, to control the rate of water flow, to regulate the pressure, or to prevent a reversal of flow through a line. After a water supply to a home or other building is set up, plumbers install the water heater, hose bibbs, water softener (if needed), and fixtures. Plumbers must locate and connect these items efficiently to avoid softening water that does not need to be softened and to save on materials and time.

1. All of the followings are types of wells *except* _____.
   a. drilled
   b. bored
   c. dropped
   d. driven

2. One of the main functions of a municipal water treatment plant clarifier is to _____.
   a. oxidize minerals for filtering
   b. increase the CO_2 levels to acidify the water
   c. add chlorine and fluorine to the water
   d. remove precipitated calcium and magnesium compounds

3. How could a plumber increase the pressure of water inside a building?
   a. Install a pressure regulator valve
   b. Install a booster pump
   c. Increase the size of the supply piping
   d. Reduce the size of the supply piping

4. A building's supply line must connect to a municipal water main by means of a _____.
   a. curb stop valve
   b. corporation stop valve
   c. meter stop valve
   d. supply stop valve

5. The main purpose of a curb stop valve is to _____.
   a. provide a convenient shut off point for water supplying a building
   b. provide a convenient point to connect a building supply to a water main
   c. isolate a water meter for maintenance
   d. isolate a pressure regulating valve or water softener

6. The most reliable method to prevent flow of a contaminated source into a potable water system is by means of a(n) _____.
   a. air gap
   b. vacuum breaker
   c. dual-check valve backflow preventer
   d. RPZ valve

7. Which is a true statement about gate valves?
   a. Gate valves open by rotating the flat disc parallel to the water stream.
   b. Gate valves are excellent for precise flow throttling.
   c. A gate valve is a full-flow, straight-through flow valve.
   d. The flow of water through the valve makes two sharp turns.

8. Which type of valve provides the benefits of both a globe valve and a 90-degree pipe elbow?
   a. Gate valve
   b. Angle valve
   c. Ball valve
   d. Check valve

9. Which of the following statements is true about check valves?
   a. Check valves have a handwheel actuator.
   b. Check valves shut on a reversal of flow.
   c. Check valves are motor-operated.
   d. Only gravity is required to operate all check valves.

10. A standard pressure-regulator valve senses _____.
    a. delivery pressure
    b. downstream pressure
    c. both delivery and downstream pressures
    d. atmospheric pressure

11. Which is the preferred method for reducing the potential for scalding accidents at sinks and in showers?
    a. Lowering hot water heater temperature to 120°F
    b. Installing a tempering valve in the fixture's hot and cold water supplies
    c. Requiring that use of the fixtures involves a responsible individual
    d. Installing a flow restrictor on the hot-water supply line

12. A building's hot water heater will usually be located _____.

    a. as close as possible to the entry point of the building's service water line
    b. in an out-of-the-way place
    c. at the end of the cold-water service line
    d. as close as possible to the largest number of hot-water fixtures

13. The plumbed locations of the standard hot and cold water fixtures normally found in a building are determined by _____.

    a. the plumber
    b. the architect
    c. the location of the DWV piping
    d. national and local codes

14. Which of the following is true when using larger piping when plumbing a building?

    a. Larger piping lowers delivery pressure more than smaller piping.
    b. Larger piping holds heat better than smaller piping.
    c. Larger hot-water piping wastes water.
    d. Larger piping reduces water hammer.

15. Water hammer is mainly caused by _____.

    a. quick-closing valves
    b. loosely-hung piping
    c. long piping runs
    d. small-diameter pipes

Fill in the blank with the correct term that you learned from your study of this module.

1. A(n) _____ is a device that measures the amount of water used.

2. The use of a(n) _____, which acts as a backflow preventer, prevents a vacuum in a water supply from causing backflow.

3. A(n) _____ is any part of the piping system other than the riser, main, or stack.

4. A(n) _____ is a valve on the outside of a building with a threaded outlet, usually set at an angle, that connects to a hose.

5. A(n) _____ is inserted into a drilled or bored well to protect the well from contamination.

6. A buffalo box is properly called a(n) _____.

7. _____, or feeder lines, are the main water supply lines, to which branches are connected.

8. A special key can be inserted into a(n) _____ to stop the flow of water during servicing or in case of an emergency.

9. _____ in waster water treatment is the thickening of liquid mixtures into soft or solid masses.

10. A(n) _____ connects the building water service line to the water main.

11. Fixture risers and stub-outs are mounted on a(n) _____.

12. Deterioration to metal pipe caused by an electric current or a chemical reaction that occurs between different types of metals is _____.

13. _____ are a measure used to properly size pipes in water distribution systems.

14. The acidity of water is identified by its _____ value.

15. When water is allowed to flow without restriction through a valve, the flow is typically referred to as _____.

16. _____ are vertical pipes on the interior portion of the wall that connect the fixture to the supply pipe.

17. The surface of the region of saturated soil underground is the _____.

18. _____ are commonly used to shut off the water supply to water closets and sinks.

19. A(n) _____ relieves water based on the sensed temperature and/or pressure of the connected fixture.

20. A(n) _____ is used to reduce water pressure in a building.

21. A(n) _____ is similar to the globe valve but can serve as both a valve and a 90-degree elbow.

22. A(n) _____ allows liquid to flow in only one direction.

23. _____ is the controlled rate of flow through a valve.

24. _____ are natural or artificial bodies of water used for water storage.

25. _____ are solids that appear and settle out due to chemical changes in water solutions.

26. The flow capacity of valves that is essentially the same as the potential flow through the connected piping is described as _____.

27. _____ is the measure of the cloudy appearance of water due to suspended materials.

28. The banging of pipes when water flow is shut off can be eliminated by using _____.

29. Three methods for disinfecting water are _____, _____, and _____.

30. _____ on hot water heaters gradually open and shut as water pressure changes near the valve setpoint.

31. The loud noises in piping systems that result from the pipes flexing as they absorb the energy of moving water when it is suddenly shut off are called _____.

Trade Terms

Angle valve	Fixture risers	Pressure relief valves	Ultraviolet (UV) light
Backing board	Full flow	Reservoirs	Vacuum breaker
Branch	Galvanic corrosion	Service lines	Water hammer
Check valve	Hammer arrestors	Straight-through flow	Water meter
Chlorination	Hose bibb	Supply stop valves	Water supply fixture units
Coagulation	Pasteurization	Thermostatic/pressure	(WSFU)
Corporation stop	pH	balancing valve	Water table
Curb box	Precipitates	Throttled flow	Well casing
Curb stop	Pressure regulator valve	Turbidity	

Angle valve: A valve whose internal design is similar to a globe valve's but its outlet direction is 90 degrees from its inlet.

Backing board: A short plank of wood installed between wall framing (or sometimes the unfinished side of the wall interior) used to mount fixture risers and stub-outs behind the fixture.

Branch: Any part of a piping system other than a riser, main, or stack.

Check valve: A valve that allows liquid to flow in only one direction. Pressure and flow within the inlet line keeps the valve open. Automatic closure of the valve occurs with the reversal of flow, by the weight of the disc mechanism, or by spring action.

Chlorination: The use of chlorine gas or compounds to disinfect water.

Coagulation: In water treatment processes, a thickening of suspended or dissolved materials into a soft, semi-solid or solid mass.

Corporation stop: A valve that connects the building water service line to the water main.

Curb box: A cylindrical casing placed in the ground over the curb stop, into which a special key-wrench can be inserted to turn off the curb stop. Also given the term buffalo box.

Curb stop: A control valve installed in building water supply lines between the corporation stop and the building.

Fixture risers: Vertical sections of pipe located inside the wall to connect the fixture to the supply pipe beneath the flooring.

Full flow: Describes a valve that does not restrict the flow of a fluid through it compared to the potential flow through the connected piping.

Galvanic corrosion: Corrosion caused by a weak electrical current that occurs when an electrical path exists between two different metals.

Hammer arrestors: Devices installed in a piping system to absorb water hammer by gradually stopping the flow of water against an air or gas cushion.

Hose bibb: A faucet with a threaded outlet, typically set at an angle to the inlet pipe, used to connect a hose. Usually located on the outside of a building. Also referred to as a sill cock or bibcock.

Pasteurization: The practice of heating water and foods to high temperatures to kill harmful bacterial organisms present.

pH: A measure of the acidity or alkalinity of a solution. A pH of 7 is neutral, being neither acidic nor alkaline; higher numbers are more alkaline, lower numbers are more acidic. The symbol comes from the early twentieth-century chemistry term power of hydrogen.

Precipitates: Solid materials resulting from chemical reactions in water solutions that settle out.

Pressure regulator valve: A valve used to reduce water pressure in a building. The valve is activated by changes in pressure within the system.

Pressure relief valves: Valves normally used for liquid service to prevent over-pressurization of a system or component. May be slow- or fast-acting depending on their purpose and design.

Reservoirs: Sources of water collected and stored in natural or artificial (man-made) lakes.

Service lines: The main water supply piping, to which branches are connected. Also referred to as feeder lines.

Straight-through flow: A valve flow design that does not restrict flow through the valve. The element that closes the valve is retracted entirely clear of the passage.

Supply stop valves: Valves that are commonly used to disconnect the hot or cold water supplies to water closets and sinks. They are available in either right-angle or straight design, and with globe- or ball-valve internals.

Thermostatic / pressure balancing valves: Mixing valves which sense outlet temperature and incoming hot and cold water pressure, and compensate for fluctuations in hot and cold water temperatures and/or pressures to stabilize its outlet temperatures. Also referred to as temperature and pressure (T&P) relief valves.

Throttled flow: Reduced flow of water through a valve by positioning of the valve's stopping device; increases both the pressure drop across the valve and flow resistance through the valve.

Turbidity: The presence of particles (sand, mud, silt) suspended in water that give the water a cloudy appearance.

Ultraviolet (UV) light: A form of high-energy light with wavelengths shorter than visible light. In water supply systems, it can disinfect water by destroying microorganisms as the water flows through a chamber containing UV lamps.

Vacuum breaker: A type of backflow preventer that inhibits backflow caused by lower pressure in a water supply system by opening a vent path in a cross-connected system to prevent siphonage.

Water hammer: A loud thumping that results when the piping system deflects against supports as it absorbs the energy in flowing water when it suddenly stops.

Water meter: A device for measuring water volume usage by an individual building or customer.

Water supply fixture units (WSFU): Design factors to determine the load that different plumbing fixtures produce on the supply side of a plumbing system.

Water table: The level below the ground's surface where soil becomes saturated with water.

Well casing: Outer tube or pipe sunk into the ground after drilling or driving a well to stabilize the hole.

Additional Resources

This module is intended as a thorough resource for task training. The following reference works are suggested for further study.

Advanced Home Plumbing, 1997. Black & Decker Home Improvement Library. Minnetonka, MN: Cowles Creative Publishing, Inc.

Air Gap Fittings for Use With Plumbing Fixtures, Appliances, and Appurtenances, Standard A112.1.3, Current Edition. American Society of Mechanical Engineers (ASME)

International Plumbing Code, Current Edition. Falls Church, VA: International Code Council.

National Standard Plumbing Code, Current Edition. Falls Church, VA: Plumbing-Heating-Cooling Contractors National Association.

Figure Credits

Section Review Answer Key

Section 1.0.0

Answer	Section Reference	Objective
1. b	1.1.0	1a
2. d	1.2.0	1b
3. a	1.3.1	1c

Section 2.0.0

Answer	Section Reference	Objective
1. b	2.1.1	2a
2. d	2.2.5	2b

Section 3.0.0

Answer	Section Reference	Objective
1. a	3.1.4	3a
2. c	3.2.0	3b

NCCER CURRICULA — USER UPDATE

NCCER makes every effort to keep its textbooks up-to-date and free of technical errors. We appreciate your help in this process. If you find an error, a typographical mistake, or an inaccuracy in NCCER's curricula, please fill out this form (or a photocopy), or complete the online form at **www.nccer.org/olf**. Be sure to include the exact module ID number, page number, a detailed description, and your recommended correction. Your input will be brought to the attention of the Authoring Team. Thank you for your assistance.

Instructors – If you have an idea for improving this textbook, or have found that additional materials were necessary to teach this module effectively, please let us know so that we may present your suggestions to the Authoring Team.

NCCER Product Development and Revision

13614 Progress Blvd., Alachua, FL 32615

Email: curriculum@nccer.org
Online: www.nccer.org/olf

❏ Trainee Guide   ❏ Lesson Plans   ❏ Exam   ❏ PowerPoints   Other _____

Craft / Level: _____   Copyright Date: _____

Module ID Number / Title: _____

Section Number(s): _____

Description: _____

_____

_____

_____

Recommended Correction: _____

_____

_____

_____

_____

Your Name: _____

Address: _____

_____

Email: _____   Phone: _____

This page is intentionally left blank.

Glossary

Aboveground rough-in: The second phase of a plumbing project. During this phase, holes are cut in walls, ceilings, and floors. Then, supply and waste pipes are attached or hung so they can be connected to fixtures. Also referred to as stack out, top out, or in-wall rough-in.

ABS (acrylonitrile-butadiene-styrene): Plastic pipe and fittings used extensively in drain, waste, and vent (DWV) systems.

ACR tubing: Annealed copper tubing that is manufactured specifically for use in air conditioning and refrigeration work.

ACR: Air conditioning and refrigeration system.

ADA compliant: Follows the accessibility guidelines of the Americans With Disabilities Act of 1990, as described in the Code of Federal Regulations (*28 CFR, Part 36, Appendix A—ADA Accessibility Guidelines for Buildings and Facilities*)

Adapters: Fittings that join pipes of different sizes or materials, such as copper and galvanized pipe or cast-iron and plastic pipe.

American National Standards Institute (ANSI): A nonprofit organization that is responsible for creating and implementing standards for many industries, including the construction trades.

Amperage: A measure of electrical current.

Angle valve: A valve whose internal design is similar to a globe valve's but its outlet direction is 90 degrees from its inlet.

Annealing: Heating a material and slowly cooling it to relieve internal stress. This process reduces brittleness and increases toughness.

Anti-scald valve: An automatic valve connected to both cold and hot water supplies that uses a sensor to adjust the temperature of hot water to ensure that it does not exceed a specified temperature at the fixture outlet.

Apparatus: One tool, which combines a variety of functions, to perform a particular job.

Approved submittal data: The fixtures and fittings in catalog drawings that have been approved by the architect or engineer.

Appurtenances: Accessories or apparatus that require no demand from the water supply side and add no load to the waste side.

Aqueduct: A man-made channel used to carry water.

Aquifer depletion: The use of underground fresh water at a rate faster than it can be replenished.

Architect's scale: A measuring device that uses smaller units, such as ½" or ¼", to represent 1'. Architect's scales are also issued in metric units. The units are used so that all building measurements are in proportion to their actual measurements but able to fit in the drawings.

Asbestos: A fibrous, fire-resistant substance used in pipe insulation, shingles, wallboard, floor coverings, and certain types of insulation. Now banned by government regulation as a health hazard.

Atmospheric hazards: Potential dangers in the air or conditions of poor air quality.

Back pressure: A condition which may occur in the DWV system whereby a higher pressure than atmospheric pressure is created in the drain/vent piping, causing a reversal of the normal flow through drain piping and traps. Also referred to as backpressure backflow.

Back: Part of the fitting that is opposite to the side with an opening or face.

Backflow preventer: A device that prevents nonpotable water from entering a potable supply system.

Backflow: The flow of contaminated water into the freshwater system resulting from a cross-connection between potable and nonpotable water systems.

Backing board: A short plank of wood installed between wall framing (or sometimes the unfinished side of the wall interior) used to mount fixture risers and stub-outs behind the fixture.

Ball valve: A valve used to control the flow of gases and liquids. It is installed in piping systems where quick shutoffs for operation or in-line maintenance may be necessary.

Bar stock: Uncut bar material on hand for use on a project.

Bathtubs: Low-profile plumbing fixtures designed to hold semi-reclined bathers. Bathtubs are identified by the location of the drain hole as one faces the fixture. If the drain is on the right end, it is a right-hand bathtub; if on the left end, it is a left-hand bathtub.

Bell-and-spigot pipe: Pipe that has a bell, or enlargement, also called a hub, at one end of the pipe and a spigot, or smooth end, at the other end. The bell and spigot of two different pipes slide together to form a joint. Also called hub-and-spigot pipe.

Benching: A method of protecting workers from cave-ins by excavating the sides of an excavation to form one or a series of horizontal levels or steps, usually with vertical or near-vertical surfaces between levels.

Bends: Fittings that change the direction of flow in piping.

Bibb: A faucet with an outlet at a downward angle, generally used for utility or hose connections on the outside of a building.

Bioswale: A depression in the ground that filters pollutants from stormwater.

Black-steel pipe: Steel pipe with a black color used for gas and air-pressure pipe. It should not be used where corrosion affects its uncoated surface.

Bladed tools: Tools that use sharp edges to accomplish their tasks. Bladed tools include saws, knives, scissors, tin snips, and wire cutters.

Branch interval: A distance along a soil or waste stack corresponding, in general, to a building story height, but in no case less than 8 feet, within which the horizontal branches from one floor or story of a building are connected to the stack.

Branch: Any part of a piping system other than a riser, main, or stack.

Building sewer: The part of the drainage system that extends from the end of the building drain and conveys its discharge to a public sewer, private sewer, individual sewage-disposal system, or other point of disposal.

Bullhead tee: A tee fitting used on a branch that is larger than the main line, or that has one outlet larger than the run openings.

Burr: Uneven or jagged edge left on metal by certain cutting tools.

Capillary action: The process during soldering in which the molten solder flows into the narrow gap between the pipe and the joint, regardless of whether the solder is flowing up, down, or horizontally.

Capillary action: The tendency of water to be drawn into porous or fibrous material against gravity above the level of the water source.

Catalog drawing: A drawing of a plumbing fixture that is found in manufacturers' catalogs. Also referred to as submittal data.

Cellular-core wall: Plastic pipe wall that is low-density, lightweight plastic containing entrained (trapped) air in solid foam.

Center line: On a drawing, a line that shows the center of an object.

Center point: Point created where the center line of the pipe and the center line of the fitting meet within the fitting. The center point is used to determine the correct length of a pipe.

Center: A point exactly halfway between two other points or surfaces.

Chain vise: A tool that aids in cutting large pipe by supporting the pipe and clamping it in an attached vise with V-jaws and secured by a chain attached to a screw tensioner.

Chain wrenches: Similar in construction and function to pipe tongs, but smaller in size for use on smaller-diameter pipes.

Chamfer: To bevel the edge of construction material to a 45-degree angle.

Channels: Sections of C-, U-, or box-shaped extrusions of steel or aluminum used for supporting pipes and pipe hangers.

Check valve: A valve that allows liquid to flow in only one direction. Pressure and flow within the inlet line keeps the valve open. Automatic closure of the valve occurs with the reversal of flow, by the weight of the disc mechanism, or by spring action.

Chemically inert: Does not react with other chemicals.

Chlorination: The use of chlorine gas or compounds to disinfect water.

Chlorine: A heavy, greenish-yellow gas used as a disinfectant in water treatment. Chlorine should be handled only when wearing appropriate personal protective equipment.

Cleanout: An access point to connected parts of the drainage system for the removal of blockages.

Clevis: An iron, or link in a chain, bent into the form of a horseshoe, stirrup, or letter U, with holes in the ends to receive a bolt or pin.

Closet bends: Elbow-like fittings that connect water closets to the main drainage piping.

Closet flange: A fitting used to mate a water closet to a closet bend or other DWV pipe.

Coagulation: In water treatment processes, a thickening of suspended or dissolved materials into a soft, semi-solid or solid mass.

Code: A requirement published by state and local governments to establish minimum standards for various types of construction. A code carries the force of law.

Combustible: Air or materials that can explode and cause a fire.

Competent person: An individual who is capable of identifying existing and predictable hazards or working conditions that are hazardous, unsanitary, or dangerous to employees, and who has authorization to take prompt, corrective measures to eliminate or control these hazards and conditions.

Compression collars: A piece of hardware that uses compression force to connect sections of polyethylene piping.

Compression joint: A method of connection in which tightening a threaded nut squeezes a compression ring to seal the joint.

Compression joints: Joints formed using a neoprene gasket inserted into the bell end of a hub-and-spigot pipe. The pressure applied between the two joined pipes forces the gasket to compress and fill in any air gaps in the joint.

Computer-aided drafting (CAD): A sophisticated design program used on computers. It allows designers to create drawings in two or three dimensions.

Confined spaces: Spaces that, by design and/or configuration, have limited openings for entry and exit, have unfavorable natural ventilation, may contain or produce hazardous substances, and are not intended for continuous employee occupancy.

Construction drawing: A drawing that shows the design, location, and dimensions of a building and its various components.

Coordination drawing: A dimensioned drawing with elevations and sections that indicate the proposed routing of system components. The dimensioned coordination drawing includes the actual dimensioned equipment as well as the access space this equipment will need.

Corporation stop: A valve that connects the building water service line to the water main.

Corrugated stainless steel tubing (CSST): A gas piping material made from flexible steel tubing, usually but not always supplied with a polyethylene jacket.

Couplings: Fittings used to connect two lengths of no-hub cast-iron pipe.

CPVC (chlorinated polyvinyl chloride): Durable plastic pipe and fittings used extensively in hot and cold water distribution systems.

Cross-connection: An arrangement between a potable water system and a nonpotable water system in which an accidental pressure differential between the two systems causes backflow of contaminated water into the freshwater system.

Crosses: Fittings that connect four lengths of pipe in a cross-like shape in a distribution system.

Curb box: A cylindrical casing placed in the ground over the curb stop, into which a special key-wrench can be inserted to turn off the curb stop. Also given the term buffalo box.

Curb stop: A control valve installed in building water supply lines between the corporation stop and the building.

Cutaway drawing: A section drawing that shows the construction elements of a particular part of a building or fixture.

Decibels (dB): A measure of sound intensity or loudness. The higher the decibel level, the louder and more potentially damaging the sound is.

Details: Sections of construction drawings that are enlarged to make them clearer.

Diameter: The distance across the center of a circle.

Dies: Devices used to cut the threads on steel pipe.

Dimension line: A line on a drawing with a measurement indicating actual length.

Disinfection: The process of destroying harmful organisms in potable water.

Diverter: A device that redirects the flow of water, as in a combination shower and bath fitting. When the diverter is engaged, water flows through the showerhead instead of through the spigot.

Double 1/4 bends: Fittings used to collect and combine the flow from two opposite runs into a single run of pipe.

Double sanitary tee: A fitting with a main pipe and two branch lines that changes the direction of flow 90 degrees from the horizontal to the vertical. Also called a sanitary cross.

Double trapping: A situation in which one trap is attached to another, creating negative pressure that stops the intended flow of drainage.

Double wyes: Fittings with a main pipe and two branches at 45 degrees. These connect horizontal waste pipes and branch drains to the building main.

Double-extra strong: An industry standard measurement of the weight or strength of steel pipe. Also referred to as Schedule 120 and double extra-heavy.

Drain, waste, and vent (DWV): A piping system that combines sanitary drainage with venting.

Drain-waste-vent (DWV) system: Refers to the combined sanitary drainage and venting systems. This term is technically equivalent to soil-waste-vent (SWV).

Drainage fittings: Fittings designed with smooth inside surfaces and shaped for easy flow to form an unbroken, internal contour for use only on drainage systems. Also called recessed fittings.

Drainage fittings: Fittings installed in the drainage sections of the DWV system to remove waste from a building.

Drawn copper: Tubing produced by pulling the tube through dies to reduce its diameter. The drawing process hardens the copper and makes it very rigid.

Drop-forged: A characteristic of a product made when heated metal is pounded or shaped between dies with a drop hammer or press.

Ductile iron: A cast-iron alloy containing magnesium to reduce brittleness.

Easement: A designated right-of-way, such as the access guaranteed to utility companies for repair of utilities that are located on, or cross over, private land, or for vehicles to cross private property.

Elastomeric: Rubberized. Made of an elastic substance such as a polyvinyl elastomer.

Electrical drawings: A drawing that shows the location of outlets, switches, and electrical fixtures. An electrical drawing may be superimposed on the floor plan.

Electrical ground: A conductive connection that provides a path for electrical current to pass from an electrical component into the earth.

Electrically powered tools: Tools that use electrical current to operate.

Elevated pressure systems: CSST systems that operate above the normal gas mains pressure of 7 inches w.c.

Elevation drawings: Drawings of structures showing a side, front, or back view.

Elevation: The height above an established reference point, such as a grade reference point on a construction drawing.

Energy sources: Any sources of electrical, mechanical, hydraulic, pneumatic, chemical, thermal, or other energy.

Energy-isolating device: Any mechanical device that physically prevents the transmission or release of energy. Can include manually operated electrical circuit breakers, disconnect switches, line valves, and blocks.

Ethics: A set of principles and values that guide an individual's conduct.

Evaporation: The natural change from liquid to vapor of water at a temperature below its boiling point.

Exploded drawing: A drawing that shows how to assemble a complex product. It also shows the relationship of the individual parts to the object as a whole. Also referred to as an assembly drawing.

Extension line: A line used on a drawing to locate a dimension away from the actual points of the dimension. This method is used when a drawing would be too crowded or cluttered if the dimension were shown within the two points.

Extra strong: An industry standard measurement of the weight or strength of steel pipe. Also referred to as Schedule 80 and extra heavy

Extra-heavy: A kind of cast-iron pipe. Extra-heavy refers to the pipe's wall thickness, which is thicker than service pipe of the same nominal pipe size.

Face: The open end of a fitting where a pipe is joined to the fitting, such as the opening of the inlet or either end.

Fall: The amount of slope given to horizontal runs of pipe expressed as a height in inches per foot of run.

Faucets: Valve-like fixtures that are used to control the flow of water from pipes into sinks, bathtubs, and similar plumbing fixtures.

Ferrous: Containing iron.

Ferrule: A brass compression ring used for joining copper tube.

Fiberglass: Spun filaments of glass woven into yarn, roving (twisted) strands, and textile materials, such as cloth or mats. Fiberglass cloth saturated with a plastic (vinyl resin or epoxy), called glass-reinforced plastic or GRP, can be pressed into molds to make products such as bathtubs and shower enclosures.

Filtration: The process of cleansing water to remove particles and chemicals.

Finish: The third phase of a plumbing project. During the finish phase, plumbers install fixtures, appliances, water purification systems, water heaters, and controls. Also referred to as trim-out or trim finish.

Fire watch: One or more people who are responsible for preventing and extinguishing fires and notifying the fire department and/or occupants in the event of a fire emergency.

Fitting allowance: The distance from the end of the pipe that goes into a fitting to the center of the fitting.

Fixture drains: The drains from traps of fixtures to the junction of those drains with any other drain pipe.

Fixture drawing: A drawing that shows the components of a fixture in detail.

Fixture risers: Vertical sections of pipe located inside the wall to connect the fixture to the supply pipe beneath the flooring.

Fixtures: Devices that receive water from a water supply line. Common fixtures include sinks, shower stalls, and toilets.

Flange union: A drilled fitting that connects two pipes. A flange is screwed onto the end of each pipe that needs to be joined and then bolted together with nuts and bolts. A gasket in between flanges creates a leakproof seal.

Flange: A rim on one end of a length of pipe that provides a connection point to another length of pipe or to a plumbing fixture, such as a water closet or valve. Bolts or studs with nuts are used to hold two flanges together, with some type of gasket between them.

Flare joint: A fitting in which one end of each tube to be joined is flared outward using a special tool. The flared tube ends mate with the threaded flare fitting and are secured to the fitting with flare nuts.

Flood-level rim: The edge of the basin, bowl, or fixture over which water flows when it is too full.

Floor plan: A construction drawing of a building looking down at the floor from above (bird's-eye view). The plan shows at least the outline of the wall locations and lengths to scale. Normally, this drawing is oriented such that the top of the page represents north.

Flush valves: Devices located at the bottom of tanks for flushing water closets and similar fixtures.

Flushometer valve: A device that discharges a predetermined quantity of water to a fixture for flushing purposes and is automatically closed by direct water pressure or other mechanical means. Also referred to as simply a flushometer.

Flux: A water-soluble substance that facilitates the fusion (joining) of metals and helps prevent surface oxidation (rusting, tarnishing) during welding, brazing, and soldering. Also called soldering paste.

Formability: The ease with which a material can bend.

Foundation plan: A construction drawing showing the placement and dimensions of a building foundation.

Full flow: Describes a valve that does not restrict the flow of a fluid through it compared to the potential flow through the connected piping.

Fusion fittings: A fitting with a butt that has the same outside diameter and inside diameter as the pipe. It is usually joined to a pipe by heat.

Galvanic corrosion: Corrosion caused by a weak electrical current that occurs when an electrical path exists between two different metals.

Galvanized pipe: Pipe electroplated with a zinc alloy to provide a protective coating that resists corrosion or oxidation (rust)

Gas cocks: Valves designed mainly for gas service that provide positive, quick flow control on gas piping systems.

Gassy operations: Working conditions in which one or more of the following conditions exist: higher than minimum levels of methane or explosive gases are present; a gas ignition has previously occurred there; or the area is connected to an underground area designated a gassy operation.

Gate valve: A valve in which the flow is controlled by moving a gate or disc that slides in machined grooves at right angles to the flow. This type of valve is designed to be fully open or fully shut.

Geothermal: Heat that is generated below the earth's surface.

Globe valve: A valve in which the flow is controlled by moving a beveled circular disc or seat washer against a circular or conical metal seat that surrounds the flow opening.

Grade: When referring to DWV and especially sewer systems, the slope of a run of pipe. Also referred to as slope or percent of grade.

Graywater: Wastewater generated from domestic processes other than human-waste disposal. These include laundry, dishwashing, and bathing. Graywater comprises 50 to 80 percent of residential wastewater.

Graywater: Water that comes from baths and washing machines.

Ground-joint union: A fitting that joins two pieces of pipe by screwing the thread and shoulder pieces of the fitting onto the pipe. The collar piece is then tightened to join the sections of pipe into a watertight joint. The fitting permits later disassembly when required.

Guards: Devices that protect tool operators from dangerous moving parts, such as blades, gears, and pulleys.

Guy wires: Ropes, chains, cables, or rods attached to something as a brace or guide.

Hammer arrestors: Devices installed in a piping system to absorb water hammer by gradually stopping the flow of water against an air or gas cushion.

Hangers: For plumbing installations, the pipe attachments, connectors, and structural attachments that support and secure pipes.

Hazard Communication (HazCom) Standard: A federal OSHA regulation requiring employers to educate and inform workers about chemical hazards on the job site (*29 CFR 1910.1200*).

Head: The height of a water column, measured in feet. One foot of head is equal to 0.433 pounds per square inch (psi).

Heel inlets: Bends that have an inlet to connect a smaller pipe to the main line or a branch line. The inlet is located at the base of the outside curve or heel of the bend.

Heel inlets: Openings in the curved, heel portion of bend fittings in line with the bend, used to connect smaller lines to the main line.

Hose bibb: A faucet with a threaded outlet, typically set at an angle to the inlet pipe, used to connect a hose. Usually located on the outside of a building. Also referred to as a sill cock or bibcock.

Hub-and-spigot cast-iron pipe: Pipe that has a bell or enlargement at one end where the spigot (smooth end) of the next pipe slides in to form a joint. Also called bell-and-spigot pipe.

HVAC (heating, ventilating, and air conditioning) drawings: Construction drawings that show the placement of the furnace and air conditioning equipment and the location of ducts and registers or pipes and radiators.

Hydraulic gradient: The level of the surface of water flowing in a partially-full pipe by gravity alone.

Hydronic: A system that heats and cools by circulating water or steam through a closed piping system.

Hydrostatic pressure test: To fill a pipe with water and bleed all air out from the highest and farthest points in the run.

Hypothermia: A life-threatening condition caused by exposure to very cold temperatures.

Impact tools: Tools that must strike or be struck to accomplish their task. They include hammers, chisels, and taps.

Increaser: A fitting used to increase the size of a straight-through line of pipe. It is often used for the vent stack before it goes through the roof to reduce the chance of frost clogging the vent opening in very cold climates.

Insertion length: The length of a pipe that fits into the fitting or other joint when assembling a pipe run. The measurement must be figured into the calculation before cutting a piece of pipe for joining.

Inside diameter (ID): The distance between the inner walls of a pipe; approximates the nominal sizes of piping used in heating and plumbing.

Insulation: A substance that retards the flow of heat.

Interceptors: Devices designed and installed so as to separate and retain deleterious, hazardous, or undesirable matter from normal waste, while permitting normal sewage or liquid wastes to discharge into the drainage system by gravity.

Interference fit: A fit that tightens as the pipe is pushed into the socket.

Inverted wyes: Fittings used to join the upper end of vents to the top of soil and waste stacks. The have the appearance of the upside-down letter Y.

Isometric drawings: Pictorial drawings that create the illusion of a three-dimensional object. All horizontal lines are projected at a 30-degree angle.

Joist: A piece of lumber used horizontally as a support for a ceiling or a floor.

Journey plumber: A plumber who has successfully completed an apprenticeship-training program.

Keel crayon: A waxy crayon used to mark a cutting line on the surface of a tube. Also called soapstone.

Kerf: The cut or groove made by a saw blade, determined by the way the teeth are set on the blade.

Laundry trays: Fixed tubs on legs or wall-mounted, usually installed in the laundry room or utility area of homes. They are used for washing clothes and for receiving wastewater from automatic clothes washers.

Lavatory: A basin designed for installation in bathrooms and other locations, primarily for washing the hands and face. Compare with sinks.

Leadership in Energy and Environmental Design (LEED): A system for certifying that buildings have been designed and constructed to environmental standards.

Level: Straight on a horizontal plane.

Liquid-fuel tools: Tools that use a liquid fuel, such as gasoline or liquid propane, to operate.

Lockout devices: Any devices that use positive means such as a lock to hold an energy-isolating device in a safe position, thereby preventing the energizing of machinery or equipment.

Lockout/tagout procedures: Processes for identifying hazardous equipment, locking it so that no workers can use it until it is certified for safe use, and placing a tag on the equipment that describes the problem and warns against use.

Lockout: The placement of a lockout device on an energy-isolating device, in accordance with an established procedure, ensuring that the energy-isolating device and the equipment being controlled cannot be operated until the lockout device is removed.

Long bends: Bend fittings with one end (typically the spigot end) longer than the other.

Long sweep 1/4 bends: Long-radius bends commonly used at the base of DWV stacks and elsewhere when the longer radius is needed to greatly decrease flow resistance and back pressure. Their use is regulated by code.

Main stack: The principal DWV riser to which branches may be connected.

Manifolds: Devices that allow the supplies to multiple gas devices to branch off of one CSST gas pipe.

Miter: A surface forming the beveled end or edge of a piece where a joint is made by cutting two pieces at an angle and then fitting these pieces together.

Model codes: Construction ordinances that are written by a national construction organization according to suggested national plumbing standards. Model codes that have not been adopted by a jurisdiction do not have the force of law

NFPA warning diamond: A four-color diamond label placed on containers or doors to alert people to specific safety hazards in a product, room, or building.

Nipples: Short lengths of pipe with male threads at both ends, used to make extensions from a fitting or to join two fittings.

No-hub cast-iron pipe: A pipe with no enlargement or bell on either end.

Nominal size: Approximate measurement in inches of the inside diameter (ID) of pipe for most copper tubes. However, nominal size of ACR tubing is based on the outside diameter (OD).

Nonferrous: Not containing iron and therefore not magnetic.

Nonpermit-required confined space: A confined workspace free of any atmospheric, physical, electrical, and mechanical hazards that can cause injury or death.

Oblique drawing: A pictorial drawing that shows the shape of an object. It shows the front of the object with the body of the object at a slight angle.

Occupational Safety and Health Administration (OSHA): The division of the US Department of Labor mandated to ensure a safe and healthy environment in the workplace.

Offset: A combination of elbows or bends that brings one section of the pipe out of line but into a line parallel with the other section.

On-the-job learning (OJL): Field experience used in conjunction with classroom lessons in an apprenticeship program. Office of Apprenticeship requires 144 hours of classroom instruction per year and 2,000 hours of OJL per year.

Orthographic drawing: A construction drawing that shows straight-on views of the different sides of an object. Orthographic drawings are used for elevation drawings.

Outside diameter (OD): The distance between the outer walls of a pipe.

Oxygen-deficient atmosphere: An atmosphere in which there is not enough oxygen to support life. Usually considered less than 19.5 percent oxygen by volume.

Oxygen-enriched atmosphere: An atmosphere in which there is too much oxygen. Usually considered more than 23.5 percent oxygen by volume.

P-trap: A trap constructed in the shape of the letter P with the loop facing downward, which provides a water seal in a waste or soil pipe, used mostly at sinks and lavatories.

Pasteurization: The practice of heating water and foods to high temperatures to kill harmful bacterial organisms present.

PB (polybutylene): Plastic piping that was formerly used for plumbing pipe; it is no longer used but is still found in some residences.

PE (polyethylene): Flexible plastic pipe, tubing, and fittings, usually used for water distribution, that do not deteriorate when exposed to sunlight.

Permit-required confined space: A confined space that has actual or possible hazards. These hazards can be atmospheric, physical, electrical, or mechanical.

PEX (cross-linked polyethylene): Tubing and fittings made with heat and high pressure that resist high temperatures, pressure, and chemicals.

pH: A measure of the acidity or alkalinity of a solution. A pH of 7 is neutral, being neither acidic nor alkaline; higher numbers are more alkaline, lower numbers are more acidic. The symbol comes from the early twentieth-century chemistry term power of hydrogen.

Pictorial drawings: Drawings that show a three-dimensional view of an object.

Pipe riser clamps: A type of clamp-on pipe support that is designed to support the weight of vertical pipe and tubing.

Pipe scale: A flaky, adherent coating on pipe walls resulting from the corrosion of metals, especially iron or steel. Also, a heavy oxide coating on copper or copper alloys resulting from exposure to high temperatures and oxygen.

Pipe tongs: Wrench-like tools that use a length of chain wrapped around a pipe and linked into a cam tensioner to secure the tool to the pipe; used for threading and tightening large pipes.

Pipe-joint compound: A soft, semi-solid sealing material used when joining lengths of pipe. Also called pipe dope.

Plot plan: A drawing of a structure that includes the dimensions of the building site, location of the structure in relation to the property boundaries, elevation of key points, existing and finish contour lines, utility services, and compass directions. Also referred to as a site plan.

Plumb: Straight on a vertical plane.

Plumbarius: The Roman term for someone who works with lead. The root of the modern word plumber.

Plumber: One who installs or repairs plumbing systems and fixtures.

Plumbing drawing: A construction drawing that shows the location of fixtures and pipe runs and gives the size and type of pipe to be installed.

Plumbing fixtures: Receptacles or devices which are either permanently or temporarily connected to the water plumbing system of the premises. Also referred to as simply fixtures. They can connect to a supply of water therefrom or discharge used water, liquid-borne waste materials, or sewage either directly or indirectly to the drainage system of the premises. Fixtures may also require both a water supply connection and a discharge to the drainage system.

Plumbing: According to the National Standard Plumbing Code, plumbing is "the practice, materials, and fixtures within or adjacent to any building structure or conveyance, used in the installation, maintenance, extension, alteration, and removal of all piping, plumbing fixtures, plumbing appliances, and plumbing appurtenances... ."

Plumbum: Latin word for lead.

Polyvinyl chloride (PVC): A thermoplastic material frequently used in tubing for cold water systems and the first type of plastic approved for use in plumbing.

Porcelain enamel: A coating of vitrified porcelain (or china) that provides an attractive and protective finish to fixtures made of cast iron or steel.

Potable: Water that is safe for cooking and drinking.

Pounds per square inch (psi): A common measure of liquid and gas pressure in the United States.

Pounds per square inch gauge (psig): A unit of pressure that measures system pressure above atmospheric (i.e., 0 psig is atmospheric pressure). Used to measure the operating pressures for gas and other plumbing systems.

Power tools: Tools that require a power source, such as electricity, hydraulics, or pneumatics, to operate.

Power-threading machine: An electrically-powered machine useful for cutting threads in large quantities of pipe. The machine rotates, threads, cuts, and reams pipe.

Precipitates: Solid materials resulting from chemical reactions in water solutions that settle out.

Pressure drop: A decrease in pressure from one point to another caused by friction losses in a fluid system, such as a water system.

Pressure fittings: Any type of fitting used for steam, gas, or water supply piping under pressure. Not designed for use in drainage piping, which is typically at atmospheric or a low water-column pressure.

Pressure rating: The maximum pressure at which a component or system may be operated continuously.

Pressure regulator valve: A valve used to reduce water pressure in a building. The valve is activated by changes in pressure within the system.

Pressure relief valves: Valves normally used for liquid service to prevent over-pressurization of a system or component. May be slow- or fast-acting depending on their purpose and design.

Primer: A liquid applied to plastic pipe prior to solvent welding in order to clean and pre-soften the pipe, and ensure a strong solvent weld.

Protective system: A method of protecting employees from cave-ins, from material that could fall or roll from an excavation face or into an excavation, or from the collapse of adjacent structures. Protective systems include support systems, sloping and benching systems, and shielding systems.

PVC (polyvinyl chloride): Plastic pipe and fittings used for cold water distribution and for industrial water and chemicals, as well as for drain, waste, and vent (DWV) systems.

Rainwater harvesting: The collection and storage of rainwater for irrigation.

Reaming: A process that removes burrs from pipe after it has been cut.

Reclaimed water: Wastewater that has had impurities and solids removed from it so that it can be reused for non-potable purposes.

Regulators: Devices used in elevated-pressure systems to reduce and maintain a lower gas pressure at the pipe outlet.

Remainder: The leftover amount in a division problem. For example, in the problem 34 ÷ 8, 8 goes into 34 four times (8 × 4 = 32) and 2 is left as the remainder.

Reservoirs: Sources of water collected and stored in natural or artificial (man-made) lakes.

Ring-tight gasket fittings: Fitting with a rubber O-ring or gasket in the socket.

Riser diagram: A drawing that shows vertical and horizontal piping along with sizes and a riser number that refers back to the full set of plumbing drawings.

Rotary hammer drill: A drill with a pounding action that lets you drill into concrete, brick, or block. It rotates and hammers at the same time and drills much faster than regular drills.

Runs: Lengths of pipe that continue in a straight line.

S-traps: Traps with long downstream legs, which tend to promote siphonage. S-traps are no longer permitted by code for new installations but are still found in older buildings.

Safety data sheet (SDS): A document that must accompany any hazardous material. The SDS identifies the material and gives the exposure limits, the physical and chemical characteristics, the kind of hazard it presents, precautions for safe handling and use, and specific control measures.

Sanitary combination: A fitting that combines a wye and 1/8 bend. It is used to connect horizontal branch lines that intersect other horizontal branch lines. It offers less resistance to the flow of material than a sanitary tee. Also called a tee-wye.

Sanitary fittings: Fittings used to connect DWV branches to the main DWV system and to serve as cleanouts.

Sanitary increasers: Fittings used to enlarge the diameter of vent stacks. They are usually placed at least 1' below the penetration of the stack through the roof. Their use is regulated by code.

Sanitary tee: A tee fitting for DWV systems that consists of a main pipe and a branch has a curved 90-degree side-inlet; this curve helps to channel the flow of wastewater or sewage from a branch line to the main line. This fitting is always used in the vertical position.

Sanitary upright wyes: Fittings used to connect separate vent stacks to the lower ends of soil and waste stacks.

Sanitary wyes: Drainage fittings, shaped like the letter Y, that join branches to the main run of pipe at an angle.

Scale: The relationship of the dimensions on a drawing to the actual dimensions of the structure. For example, in a 1/4 scale, 0.25" represent 1'. Scale is often provided in both English and metric units.

Schedules: Tables relating pipe wall thicknesses to pipe pressure ratings. Higher schedule numbers equate to thicker pipe walls for a given nominal pipe size.

Scribe: A sharply pointed and hardened steel tool used for marking a surface to be cut by etching a line or a point into the surface.

Sealant tape: Tape used to wrap the threads of a pipe before joining to lubricate and ensure a watertight, secure fit. Also called PTFE or Teflon® tape.

Seat washer: A flexible, compressible disc that fastened to the valve disc at the end of the valve stem; used to stop and regulate the flow of water through a faucet.

Sepia: A print or construction drawing with dark reddish-brown lines on a light background.

Service lines: The main water supply piping, to which branches are connected. Also referred to as feeder lines.

Service: A lightweight kind of cast-iron pipe. Service refers to the pipe's wall thickness, which is thinner for a nominal pipe size compared to extra heavy.

Setback: The distance a code requires between a building and a property line, such as the street.

Shoring: A structure such as a metal hydraulic, mechanical, or timber system that supports the sides of an excavation and is designed to prevent cave-ins.

Short sweep 1/4 bends: Bend fitting with a short radius used at the base of a DWV stack and elsewhere. Their use is regulated by code

Side inlets: Openings in ell or tee fittings at right angles to the line of the run, used to connect smaller lines to the main line.

Side yards: The spaces along the sides of a structure that provide access to rear yards, reduce the possibility of fire jumping from one building to the next, and promote ventilation around the structure.

Single-line drawings: Plumbing drawings that use a single line to represent the centerline of a pipe. Single-line drawings can be used to represent pipe of any diameter. Also referred to as schematic drawing.

Sinks: Shallow, flat-bottomed basins mainly used for the preparation of food, dishwashing, and for utility purposes. Compare with lavatory.

Siphonage: Loss of water in a trap seal started by unequal pressure inside and outside DWV piping. The water initially flows in the direction of the lower pressure. Sustained siphonage, even in the absence of pressure differences, results from the cohesive property of water.

Size dimension ratios (SDR): Ratios of the outer diameters of pipes to their pipe wall thicknesses.

Sizing tool: A tool consisting of a plug and a sizing ring that is used to reshape a deformed pipe back to roundness.

Slope: Measurement of the fall of a length of pipe from level. The change in level is expressed in degrees of an angle or distance of fall per foot of run. Also referred to as percent of grade.

Sludge: Semi-liquid matter that settles out in a holding tank during the waste treatment process.

Soapstone: Another term for keel crayon.

Softening: The process of removing magnesium and sodium salts that cause scale on the inside of pipes and fittings.

Soil pipe cutter: A heavy-duty tool used for cutting cast-iron pipe. Soil pipe cutters can be of the snap or ratchet type.

Solar hot water: Water that has been directly or indirectly heated by sunlight.

Solder: An alloy (tin plus antimony, copper, and silver) with a low melting point used to join metals or seal joints.

Soldering: A method of joining metals or sealing joints using solder and heat.

Solid wall: Plastic pipe wall that does not contain trapped air in the form of solid foam.

Solvent weld: A joint created by joining two plastic pipes using solvent cement that softens the material's surface and creates a solid bond of plastic when the solvent evaporates.

Specifications: Written requirements included with the drawings or blueprints of a construction project. They provide more details or descriptions of the technical standards that must be met during construction. Specifications usually override drawings but are overridden by the contract. Also referred to as specs.

Spigot: When referring to DWV systems, the pipe end or the male portion of a fitting that inserts into the hub of a downstream fitting.

Spirit level: A level in which the adjustment to the horizon is shown by the position of a bubble in liquid contained in a nearly horizontal glass tube or a circular box with a glass cover.

Stack: A general term for certain vertical DWV pipes, including offsets of soil, waste, vent, or inside conductor piping. This does not include vertical fixture and vent branches that do not extend through the roof or that pass through not more than two stories before being reconnected to the vent stack or stack vent.

Standard weight: An industry standard measurement of the weight or strength of steel pipe. Also referred to as Schedule 40.

Stock: The component of a hand or power threader that receives and securely holds the pipe-thread die cutter.

Stop-and-waste valve: Valve similar to a globe valve that is opened or closed by raising or lowering a disc oriented perpendicular to the flow through the valve seat using a threaded stem. An elastomeric, or rubberized, washer attached to the valve disc seals the valve seat, closing off water flow. This valve is most commonly used for water faucets.

Straightedge: A length of wood or metal that does not bow or twist along its length.

Straight-through flow: A valve flow design that does not restrict flow through the valve. The element that closes the valve is retracted entirely clear of the passage.

Strap wrench: A wrench consisting of a heavy strap wrapped around a pipe and run through a buckle at the top end of the wrench handle, which creates a cam-lock tensioner. Used to thread and tighten chrome-plated or other types of finished pipe so that there are no jaw marks or scratches left on the pipe.

Striker plates: A heavy manufactured metal plate of a required thickness that is installed behind a wall to protect CSST from penetration by nails, screws, and drill bits.

Subsidence: A depression in the earth that is caused by unbalanced stresses in the soil surrounding an excavation.

Supply stop valves: Valves that are commonly used to disconnect the hot or cold water supplies to water closets and sinks. They are available in either right-angle or straight design, and with globe- or ball-valve internals.

Sweat joint: A pipe joint made by applying solder to the joint and heating it until it flows into the joint.

Sweeps: Fittings having a radius of the curve greater than those for standard 90-degree bends. Used for a smooth change of direction.

Symbols: Marks or drawings used to indicate a specific object, material, class, or entity. A legend shows the symbols used on a drawing and their meanings.

Tagout devices: Any prominent warning devices, such as a tag and a means of attachment that can be fastened securely to an energy-isolating device in accordance with an established procedure. The tag indicates that the machine or equipment to which it is attached is not to be operated until the tagout device is removed in accordance with the energy-control procedure.

Takeoffs: Detailed lists that are compiled, based on drawings and specifications, of all the material and equipment necessary to construct a project. Such a list is also called a material takeoff.

Temper: The strength and resilience of a metal.

Terrazzo: A type of floor surface (commonly used for showers and other plumbing fixtures) made by pouring in place or precasting a mixture of small pieces of marble, other hard stone, or glass and a mortar base. When hardened, the surface can be ground and polished smooth, although it may be left slightly rough to create a nonslip surface.

Test tees: Tees installed as test locations for pressurizing the DWV system to test for leaks before placing it in operation.

Thermoplastic pipe: Pipe that can be repeatedly softened by heating and hardened by cooling. When softened, thermoplastic materials can be molded into desired shapes during manufacturing.

Thermoplastic: A plastic material used in plumbing and sanitary systems that is soft and pliable when heated and hard and rigid when cooled.

Thermoset pipe: Pipe that changes chemically when heated, so that once hardened by heat or chemicals, it is hardened permanently.

Thermoset: A plastic material used in plumbing and sanitary systems that becomes substantially infusible and insoluble when treated by heat or chemicals.

Thermostatic/pressure balancing valves: Mixing valves which sense outlet temperature and incoming hot and cold water pressure, and compensate for fluctuations in hot and cold water temperatures and/or pressures to stabilize its outlet temperatures. Also referred to as temperature and pressure (T&P) relief valves.

Thread engagement: The length of pipe that gets threaded into the fitting when joining threaded pipe.

Thread makeup: The distance that a pipe screws into a fitting. Also called thread engagement or thread-in.

Throat: The part of the fitting where you thread in another pipe or fitting.

Throttled flow: Reduced flow of water through a valve by positioning of the valve's stopping device; increases both the pressure drop across the valve and flow resistance through the valve.

Tolerance: Allowable variation in a given measurement or quantity.

Torque wrench: A wrench with a gauge or other means to indicate the amount of rotating force applied to a fastener, such as a nut, as it is turned.

Torque: Twisting or turning force applied in a rotating motion. Measurements are given in either inch-pounds or foot-pounds.

Transition fitting: A special fitting used to connect plastic pipe to pipe of a dissimilar material, as specified by applicable code.

Traps: DWV fittings or devices that provide a liquid seal to prevent the emission or escape of sewer gases from the system without materially affecting the flow of sewage or wastewater through the fitting.

Trench shields: Structures that are able to withstand the forces imposed on them by a cave-in and can thereby protect employees within the excavation. Shields can be permanent structures or portable and moved along as work progresses. Shields can be either pre-manufactured or job-built in accordance with *29 CFR 1926.652 (c)(3)* or *(c)(4)*.

Tube stock: Uncut tubing material on hand for use on a project.

Turbidity: The presence of particles (sand, mud, silt) suspended in water that give the water a cloudy appearance.

Ultraviolet (UV) light: A form of high-energy light with wavelengths shorter than visible light. In water supply systems, it can disinfect water by destroying microorganisms as the water flows through a chamber containing UV lamps.

Underground rough-in: The phase of a plumbing project during which the plumber locates all supply and waste connections from the building systems to public utilities, and establishes where these systems will enter or leave the building.

United States Green Building Council (USGBC): The non-profit construction trade organization responsible for the development of LEED.

Vacuum breaker: A type of backflow preventer that inhibits backflow caused by lower pressure in a water supply system by opening a vent path in a cross-connected system to prevent siphonage.

Valve disc: In most compression valve fixtures, the enlarged end of the valve stem that carries and supports the seat washer to create a leak-tight seal when the valve is shut.

Valve seat: The circular or conical surface milled into the internals of a valve against which the valve disc or washer bears to shut off water flow through the valve.

Valve stem: The part of a faucet or valve that connects the valve handle or handwheel to the valve disc inside the valve body. It permits controlling the flow of water through the valve when the handle is turned.

Velocity: Technically, the speed and direction of a moving object. Commonly used in place of speed for describing the motion of fluids in pipes.

Vent branch: A vent connecting one or more individual vent lines with a stack vent.

Vent ells: Plastic fittings with a sharp turn radius, used only in vent piping systems. Their use is regulated by code.

Vent pipe: A pipe installed to provide a flow of air to or from a drainage system, or to provide a circulation of air within such a system to protect trap seals from siphonage and back pressure. The combination of all vent pipes in a plumbing system is referred to as the vent system.

Vent tees: Fittings used in venting systems or as cleanouts. They may not be used in the drainage system because it restricts the flow of material. Their use is regulated by code.

Vitreous china: A durable ceramic material usually consisting of a fine clay mineral (kaolinite) and is often a mixture with quartz, feldspar, silica or other materials. The mixture is heated to temperatures between 2,200°F and 2,600°F (1,200°C and 1,400°C) to create a nonporous, glass-like finish. Also referred to as china, porcelain, and vitrified porcelain.

Voltage drop: The tendency of electricity traveling through the length of an extension cord to lose voltage.

Waste pipe: A pipe in a plumbing system that carries only waste water and sewage.

Water column (w.c.): A unit of pressure based on the pressure exerted by the height of water in a tubular column or manometer (e.g., 12 inches w.c. = 0.43 psig). Used for measuring low pressures, such as those in a standard gas supply.

Water efficiency: The managed use of drinkable water to reduce waste.

Water hammer: A loud thumping that results when the piping system deflects against supports as it absorbs the energy in flowing water when it suddenly stops.

Water hammer: An extreme change in water pressure within a pipe that occurs when suddenly stopping a moving mass of water, as when quickly shutting a valve. It can cause a loud, banging sound as pipes flex and even damage the system.

Water meter: A device for measuring water volume usage by an individual building or customer.

Water supply fixture units (WSFU): Design factors to determine the load that different plumbing fixtures produce on the supply side of a plumbing system.

Water table: The level below the ground's surface where soil becomes saturated with water.

Weir: When referring to plumbing, a ledge or lip in a fixture that controls the level of water inside the fixture. The word comes from Old English, meaning dam.

Well casing: Outer tube or pipe sunk into the ground after drilling or driving a well to stabilize the hole.

Wyes: Fittings consisting of a main pipe and a 45-degree branch for connecting a waste branch to a horizontal building main drain.

Yoke vise: A pipe vise with a lower toothed V-jaw topped by a hinged attachment consisting of a jaw positioned by a threaded compression screw. When secured, the assembly creates a yoke that firmly holds a piece of pipe and prevents it from turning.

This page is intentionally left blank.

This page is intentionally left blank.

This page is intentionally left blank.

This page is intentionally left blank.

This page is intentionally left blank.

This page is intentionally left blank.

This page is intentionally left blank.